Separations Using Aqueous Phase Systems

Applications in Cell Biology and Biotechnology

Separations Using Aqueous Phase Systems

Applications in Cell Biology and Biotechnology

Edited by

Derek Fisher

Royal Free Hospital School of Medicine
London, United Kingdom

and

Ian A. Sutherland

National Institute for Medical Research
London, United Kingdom

Plenum Press • New York and London

Library of Congress Cataloging in Publication Data

International Conference on Phase Partitioning on Advances in Separations Using Aqueous Phase Systems in Cell Biology and Biotechnology (5th: 1987: Keble College)

Separations using aqueous phase systems: applications in cell biology and biotechnology.

"Based on the proceedings of the Fifth International Conference on Phase Partitioning on Advances in Separations Using Aqueous Phase Systems in Cell Biology and Biotechnology, held August 23–28, 1987, at Keble College, Oxford, United Kingdom"—T.p. verso.

Includes bibliographies and index.

1. Phase partition—Congresses. 2. Biotechnology—Technique—Congresses. 3. Cytology—Technique—Congresses. I. Fisher, Derek. II. Sutherland, Ian A. III. Title. [DNLM: 1. Biotechnology—congresses. 2. Cell Fractionation—methods—congresses. 3. Cell Separation—methods—congresses. QH585.5.C44 I61s 1987]

TP248.25.P53I57 1987 660′.6 89-8560

ISBN 978-1-4684-5669-1

ISBN-13: 978-1-4684-5669-1 e-ISBN-13: 978-1-4684-5667-7

DOI: 10.1007/978-1-4684-5667-7

Based on the proceedings of the Fifth International Conference on Phase Partitioning on Advances in Separations Using Aqueous Phase Systems in Cell Biology and Biotechnology, held August 23–28, 1987, at Keble College, Oxford, United Kingdom

Softcover reprint of the hardcover 1st edition 1989

A Division of Plenum Publishing Corporation
233 Spring Street, New York, N.Y. 10013

SCIENTIFIC COMMITTEE

Per-Åke Albertsson (Sweden)

Donald E. Brooks (Canada)

Derek Fisher (UK)

Göte Johansson (Sweden)

Werner Müller (Germany)

Ian A. Sutherland (UK)

Harry Walter (USA)

PREFACE

The use of aqueous two-phase systems for the partitioning of macromolecules, organelles and cells was originally developed by Per-Åke Albertsson in Sweden in the mid-fifties [1-3]. These systems were initially applied to separations of plant organelles and viruses but their use has now extended into most areas of cell biology and biochemistry [4,5].

Since 1979 biennial International Conferences on Partitioning in Aqueous Two-Phase Systems have been held in Los Angeles (1979), Sheffield (1981), Vancouver (1983) and Lund (1985). The 5th conference was held in Oxford from 23-28 August 1987 and was entitled "Advances in Separations Using Aqueous Phase Systems in Cell Biology and Biotechnology". It is the formal presentations from this meeting which comprise this volume.

In contrast to earlier books on phase partitioning [4,5] this volume contains, for the first time, worldwide contributions from over sixty partitioners from a variety of scientific disciplines, thereby providing a detailed overview of the widespread application and potential of bioseparations using phase partitioning. Disciplines include Biophysics, Biochemistry, Cell Biology, Microbiology, Biotechnology and Process Engineering, in both academic and commercial establishments. These biennial conferences allow advances in these diverse partitioning fields to be reviewed and compared; they also provide an opportunity for those considering using phase partitioning to obtain information, advice and contacts. Attendance has grown steadily over the years and 140 scientists came to Oxford.

The conference consisted of ten symposia on areas of application of partitioning which have been organised as specific chapters in this volume. Each session comprised one, and sometimes two, keynote lectures, which presented overviews and introductions to each topic, several full papers and a number of short "state-of-the-art" poster presentations. We have kept this structure in organising the individual chapters.

Chapter 1 deals with the partitioning of macromolecules, principally enzymes. This is a logical and practical starting point since macromolecule partitioning occurs between the two bulk phases and is more straightforward to describe than the partitioning of particulates (organelles and cells) which also involves the interface between the phases.

Separations of subcellular organelles and membranes by phase partitioning are described in Chapters 2 and 3. We have kept separate the applications to plant (Chapter 2) and animal (Chapter 3) organelle systems.

Chapter 4 describes the separation of cells and subfractionation of cell populations. Specific bioseparations of cells are obtained by immunoaffinity cell partitioning (Chapter 5). We have allocated a separate chapter to emphasise the potential importance of this developing technique.

Theoretical aspects of partitioning are presented in Chapter 6. Considerable attention is given to interfacial phenomena in aqueous two-phase systems and partitioning, emphasising the importance of surface science in our understanding of phase partitioning.

Important applications of partitioning are being made in biotechnology. Batch processes for the large scale recovery of fermentation products (downstream processing) are described in Chapter 7. Extractive bioconversion (Chapter 8) is a related application. In this process a protein product is extracted from the phase in which it is produced by enzymes, organelles or cells, immobilised in that phase by virtue of their partitioning, into the other phase, thereby providing a means of continuous extraction.

The last two chapters deal with emerging technology. Chapter 9 describes new methodology involving liquid-liquid chromatography and countercurrent distribution techniques. The emphasis of this chapter lies in new apparatus for separations using the well characterised poly(ethylene glycol)-dextran two-phase systems. Chapter 10 concentrates on the emerging technology of new polymers and phase systems. If the full commercial potential of phase partitioning is to be exploited, new cheaper phase systems with improved characteristics will need to be developed. These last chapters focus on the future of phase partitioning.

Derek Fisher and Ian A. Sutherland

London 1988

REFERENCES

1 P.-A. Albertsson Chromatography and partition of cells and cell fragments. Nature 177:771 (1956)

2 P.-A. Albertsson Particle fractionation in liquid two-phase systems. The composition of some phase systems and the behaviour of some model particles in them. Application to the isolation of cell walls from microorganisms. Biochim. Biophys. Acta 27: 378 (1958)

3 P.-A. Albertsson Partition of proteins in liquid polymer-polymer two-phase systems. Nature 182: 709 (1958)

4 P.-A. Albertsson Partition of Cell Particles and Macromolecules (3rd edition) John Wiley and Sons, New York (1986)

5 H. Walter, D.E. Brooks and D.Fisher (eds) Partitioning in Aqueous Two-Phase Systems. Theory, Methods, Uses, and Applications to Biotechnology. Academic Press, Orlando (1985).

ACKNOWLEDGEMENTS

We gratefully acknowledge the facilities and support provided by the Royal Free Hospital School of Medicine, the National Institute for Medical Research and the Medical Research Council's Collaborative Centre. We wish to thank the following organisations for financial support which assisted in providing essential funds for the conference and for providing travel grants for young research workers:

US Navy Research (US and UK Offices), Arlington, USA;

Pfeifer and Langen, Dormagen, West Germany;

E. Merck, Darmstadt, West Germany;

E.I. du Pont de Nemours Co Inc (Medical Products Department), Glasgow, USA;

The British Council, London, UK;

Ciba-Geigy AG, Basel, Switzerland;

The Biochemical Society (Membrane Group, Techniques Group, and Industrial Biochemistry and Biotechnology Groups), London, UK;

Beckman UK Limited (Bioanalytical Systems Group), High Wycombe, UK;

Glaxo Group Research Ltd, Greenford, UK;

Perstorp Biochem, Lund, Sweden;

Lilly Research Laboratories, Indianapolis, USA

Hoechst UK Limited, Milton Keynes, UK.

Domtar Inc (Research Centre), Sennerville (Quebec) Canada

We also thank our colleagues who helped in the build up to the meeting and assisted so wholeheartedly during the conference in Oxford:- Kate Bunker, Dave Stokes, Tom Dempsey, Paul Eggleton, Sophie Bingsmark, Somaiya Sidiqi, and Rob Anderson.

We are particularly indebted to Kate Bunker who, in addition to her role as conference secretary, undertook the retyping of the proceedings to a common format. While this has inevitably introduced a time penalty, the result will we hope justify the pains taken in its production.

Derek Fisher and Ian A Sutherland

London 1988

CONTENTS

Chapter 6: THEORETICAL ASPECTS OF PHASE PARTITIONING

Chapter 7: BIOTECHNOLOGY: DOWNSTREAM PROCESSING

Chapter 8: BIOTECHNOLOGY: EXTRACTIVE BIOCONVERSIONS

Chapter 1

Partitioning of Macromolecules

GENERAL ASPECTS OF AQUEOUS POLYMER TWO-PHASE PARTITION

Per-Åke Albertsson

Department of Biochemistry
University of Lund
Box 124, S-221 00 Lund, Sweden

Aqueous, liquid-liquid, polymer two-phase systems are obtained with mixtures of aqueous solutions of two different polymers. These phase systems are very mild towards biological material and can be used for the partition of biomolecules, membrane vesicles, cell organelles and whole cells. The partition depends mainly on the surface properties of the particles and molecules. These include size, charge, hydrophobicity and the biospecific structure of the surface particles.

The most characteristic feature of these phase systems is that both phases are aqueous. The water content is between 85-99 per cent and if complemented with suitable salts and buffers these systems provide a very gentle milieu for biological material.

Several pairs of polymers can be used forming aqueous two-phase systems:

- o Dextran - polyethylene glycol
- o Dextran - Ficoll
- o Dextran - polyvinylalcohol
- o Ficoll - polyethylene glycol
- o Polyvinylalcohol - polyethylene glycol
- o Polyvinylalcohol - polyethylene glycol
- o Dextran - hydroxypropylstarch

Other combinations will become available as new polymers are introduced.

Properties of aqueous polymer two-phase systems include the following:

- both phases are aqueous (85-99 per cent H_2O)
- density difference is small
- refractive index difference is small
- interfacial tension is small (0,1-0,001 mN/m)
- both phases are viscous
- time of phase separation at unit gravity is long (5 mins - 1 h), but can be reduced to less than a minute by low speed centrifugation

Partition between two phases depends on many factors. This is to be expected since interaction between the partitioned substance and the components of each phase is a complex phenomenon involving hydrogen bonds, charge interaction, van der Waals forces, hydrophobic interactions, and steric effects. The partition depends on the molecular weight and chemical properties of the phase-forming polymers and the size and chemical properties of the partitioned molecules or particles. For particles, it is mainly the exposed groups of the surface which come in contact with the phase components, and partition of particles is, therefore, a surface-dependent phenomenon. Since the ions of a salt have different affinity for the two phases, an electrical potential difference is established between the phases, and this has a strong influence on the partition of charged molecules or particles.

Since partition depends on so many factors it seems to be complicated, making it difficult to plan experiments rationally for separation or to interpret results. However, it also means that partition can be used for a multitude of separations. The different factors which determine partition can be explored separately or in combination to achieve an effective separation. We can also enlarge some of the factors so that they will dominate the partition behaviour. With respect to the partitioned substance the following types of partition can be distinguished:

- Size-dependent partition: molecular size or the surface area of the particles is the dominating factor.
- Electrochemical partition: electrical potential between the phases is used to separate molecules or particles according to their charge.
- Hydrophobic affinity partition: hydrophobic properties of a phase system are used for separation according to the hydrophobicity of molecules or particles.
- Biospecific affinity partition: the affinity between sites on the molecules or particles and ligands attached to the phase polymers is used for separation.

- o Conformation-dependent partition: conformation of the molecules and particles is the determining factor.
- o Chiral partition: enantiomeric forms are separated.

Formally, we can split the logarithm of the partition coefficient into several terms:

$$\ln K = \ln K^{\circ} + \ln K_{el} + \ln K_{hfob} + \ln K_{biosp} + \ln K_{size} + \ln K_{conf}$$

where el, hfob, biosp, size, and conf stand for electrochemical, hydrophobic, biospecific, size, and conformational contributions to the partition coefficient and ln K° includes other factors.

The different factors are more or less independent of each other though none is probably completely independent of the others. For example, when we introduce a hydrophobic group on one of the phase-forming polymers we may slightly influence both the distribution of ions and the electrical potential. When the molecular weight of macromolecules is increased its net charge may also increase. When the conformation of a macromolecule is changed new groups of atoms are exposed to the surrounding phase.

Partition with aqueous polymer phase systems can be applied both on small laboratory scale, down to microliter scale, as well as on a large industrial scale involving several hundred litres of phase volumes. In many cases an efficient separation can be obtained in one or a few extraction steps. For a complex mixture containing several components one has to use a multistage procedure. The following procedures can be used:

- o Single-step partition
- o Repeated batch extractions - gradient extraction
- o Countercurrent distribution
- o Liquid-liquid columns
- o Partition chromatography

REFERENCES

The following references [1,2] will cover different aspects of phase partition aqueous polymer phase systems.

1 P.-Å. Albertsson (1986) Partition of Cell Particles and Macromolecules, Wiley, New York

2 Partitioning in Aqueous Two-Phase Systems, H. Walter, D.E. Brooks and D. Fisher, eds., Academic Press, Orlando, Florida, USA (1985)

AFFINITY PARTITIONING OF ENZYMES

Göte Johansson

Department of Biochemistry
Chemical Center, Box 124
University of Lund
S-221 00 Lund, Sweden

Affinity partitioning is used as a collective name for extraction processes where one (or several) ligands selectively influence the partition of solutes (e.g. proteins) within aqueous two-phase systems. The restriction of the affinity ligand to one phase is generally obtained by covalent attachment of ligand to the dominating polymer of this phase.

INTRODUCTION

Three papers demonstrating affinity partitioning appeared during 1974-1975. These works included: 1) extraction of trypsin with diaminodiphenyl-carbamoyl-PEG [1]; 2) extraction of S-23 myeloma protein by using PEG-bound dinitrophenol [2], also including a model for affinity partitioning (see below); and 3) extraction of albumin from blood serum using PEG-bound palmitate groups [3]. Since then, a great number of publications have been dealing with affinity partitioning of proteins. Table 1 shows a collection of works in this field.

THEORY

The general theory for affinity partitioning, presented by Flanagan and Barondes [2], is based on thermodynamic reasoning and it shows a number of important properties. These are: 1) increasing concentration of ligand-PEG (in a PEG-dextran system) gives rise to a "saturation curve". This is the case if $\Delta\log K$ (= $\log K - \log K_{without\ ligand}$) is chosen as function of the (total) ligand concentration. A reciprocal plot of the two results in a straight line; 2) under saturating conditions (excess of ligand-PEG) the following equation is obtained:

$$\frac{K}{K_{without\ ligand}} = \left[\frac{\mathfrak{N}_B \cdot K_{L-PEG}}{\mathfrak{N}_T}\right]^n$$

or $$(\Delta\log K)_{max} = n.\log K_{L-PEG} + n.\log \mathfrak{N}_B/\mathfrak{N}_T$$

Table 1 *Affinity partitioning of enzymes*

Enzyme	Ligand	Reference
Alcohol dehydrogenase	triazine dyes	4
Colipase	lecithin	5
Formaldehyde dehydrogenase	NADH, dyes	6
Formate dehydrogenase	NADH, dyes	6,7
Glucose 6-phosphate dehydrogenase	triazine dyes	4,8-11
Glutamate dehydrogenase	"	12
Glutamate oxaloacetate dehydrogenase	"	12
Glutamate pyruvate transaminase	"	12
Glyceraldehyde phosphate dehydrogenase	"	13
Glycerol kinase	"	12
Lactate dehydrogenase	"	14-16
Malate dehydrogenase	"	12,16
Nitrate reductase	triazine dyes	17
3-Oxosteroid isomerase	estradiol	18
Phosphofructokinase	Cibacron blue F3G-A	19,20
3-Phosphoglycerate kinase	triazine dyes	4,13
Phosphoglycerate mutase	"	4,13
Pyruvate kinase	"	12,16
Trypsin	p-aminobenzamidine	2

where $\mathfrak{K}_T$ and $\mathfrak{K}_B$ are the association constants in top and bottom phase and n is the number of binding sites on the protein molecule. It is here assumed that the association constant is the same for all binding sites, which in most cases is not so. If we further simplify the model by assuming $\mathfrak{K}_T = \mathfrak{K}_B$ (which is not necessarily true) we obtain:

$$(\Delta \log K)_{max} = n.\log K_{L-PEG}$$

This expression shows that large $\Delta\log K$ values can be expected even with few binding sites. Since K_{L-PEG} can take values of 100 or more the simplified model implies that the K value of the target protein should increase with around 100 times (with one binding site) or 10,000 times (with two binding sites). Experiments have, in several cases, been in accordance with the model [19,20]. In other cases, additional equilibrium processes (e.g. ligand-ligand interaction) may considerably reduce the $\Delta\log K$ value.

Influence of system parameters

A number of parameters have a pronounced effect on affinity partitioning and they should therefore be adjusted (in each individual case) to give optimal result for the extraction. The parameters mentioned below have, in general been studied with triazine dyes as ligands. These dyes are known to bind to a number of enzymes, especially among the groups kinase and dehydrogenase. The parameters are:

Ligand concentration: $\Delta\log K$ approaches a saturation value with increasing concentration of polymer-bound ligand.

Polymer concentration: Increasing polymer concentration gives rise to more extreme $K_{L-Polymer}$ values and therefore larger (absolute) $\Delta\log K$ for the target protein.

Salt and salt concentration: The effects of salts on partitioning of proteins are complex when affinity ligands are used. The salt may influence the base partitioning (without ligand) of protein [21], as well as the partitioning of ligand-polymer and the ligand-protein interaction. Furthermore, protein-protein and ligand-ligand interactions may be affected. It is not only a question of concentration of salt but also, to a high degree, the type of salt. If the salt concentration is chosen to be low (<10 mM) the ligand (if charged) may act as a (non-specific) liquid ion exchanger. High salt concentration (>50 mM) may considerably reduce the affinity partitioning effect in case of some salts (e.g. NaCl, $NaClO_4$) while other show negligible negative influence (e.g. phosphate and acetate) [19].

Temperature: Change in temperature influences the composition of the phases. In general, the $\Delta\log K$ value decreases with increasing temperature [19,20].

Molecular weight of polymers: In some cases the molecular weights of the polymers have dramatic effects on $\Delta\log K$, as has been shown for phosphofructokinase from bakers' yeast [19]. In other cases only low influence has been found.

pH value: The ligand-protein interaction may be influenced if weakly basic or acid groups are involved (either on the ligand or on the protein). In the case of triazine dyes, the lower pH value enhances the binding to most proteins but also reduces the specificity.

Free ligands: Ligands added to the system reduce the affinity partitioning effect. It is therefore of great importance to use preparations of ligand-polymers which are well purified in respect to free ligand. Also crude extracts from tissue or micro-organisms may contain considerable amounts of natural ligands which may be helpful to remove before extraction.

Number of ligands per polymer molecule: When several ligand molecules are attached per polymer molecule the ligand-polymer may partition differently from the original polymer. If the ligand carries charged

Table 2 *Multiple extraction of glucose-6-phosphate dehydrogenase from baker's yeast*[a] *(from reference 11)*

Extraction No	Polymer-bound ligand[b]		Phase transferred to next step	Enzyme recovery (% of original)	Purification factor
	Upper phase	Lower phase			
1	PO	-	Upper	92.1	4.4
2	PO	-	Upper	78.6	6.6
3	PO	-	Upper	73.6	7.5
4	PO	PY	Lower	60.0	20.1
5	PO	PY	Lower	55.7	32.0

[a] Phase system: 9% dextran 70, 5.5% PEG 40000 and 45 mM sodium phosphate buffer, pH 7.0. Temperature: 0°C.
[b] PO = Procion olive MX-3G (bound to PEG) and PY = Procion yellow HE-3G (bound to dextran)

Table 3 *Sequential extraction of enzymes from swine muscle using three affinity ligands in series[a] (adapted from reference 16)*

Extraction No	PEG-bound ligand[b]	Recovery in upper phase (% of original)[c]				
		Protein	LDH	MK	MDH	PK
1+2	PB	23	6	69	16	14
3+4	PY	18	72	28	76	4
5	PN	19	6	4	7	69

[a] System: 6% dextran 500, 9% PEG 8000 and 47 mM sodium phosphate buffer, pH 7.9. Temperature: 0°C
[b] PB = Procion blue MX-2G, PY = Procion yellow HE-3G and PN = Procion navy H-4R
[c] LDH = Lactate dehydrogenase, MK = Myokinase, MDH = Malate dehydrogenase, and PK = Pyruvate kinase

groups salt may considerably affect the partition of highly substituted ligand-polymers [22]. Phenomena like affinity precipitation may be foreseen when the degree of substitution is >2.

Bulk proteins: The presence of high concentration of bulk proteins may affect the partition because of protein-protein interaction. The

Table 4 *Purification of glucose-6-phosphate dehydrogenase from one kg of baker's yeast (from reference 10)*

Purification step	Volume (ml)	Enzyme activity (μkat)	Specific activity (mkat/kg)	Purification factor
Homogenate	3540	123.5	2.78	1
Precipitation with PEG	201	93.5	5.77	2.1
Affinity partitioning[a]				
1st: upper phase	645	87.0	18.5	6.6
2nd: upper phase	645	91.0	24.5	8.8
3rd: upper phase	615	91.0	34.1	12.3
4th: lower phase[b]	510	91.0	121	43.6
DEAE-cellulose treatment	500	59.0	915	329

[a] System: 10% dextran 500, 6.3% PEG 8000, 0.7% Procion yellow HE-3G PEG 8000, and 8 mM sodium phosphate buffer, pH 7.0. Temperature: 0°C
[b] Addition of NADP and Na_2SO_3 to the system

dextran-PEG system can accept protein concentrations to, at least, 150 g per litre. Pre-concentration of proteins by fractional precipitation with PEG is a useful way to "fill up" the systems with material (if desired).

Solvents: Some organic solvents, like ethylene glycol and dimethylsulfoxide, may be included in the systems without changing the volume ratio to any notable extent [23]. Effects on protein partition can in this case not easily be foreseen.

MULTISTEP EXTRACTION

The purification of a protein by affinity partitioning can in most cases be enhanced by using several combined extraction steps. One useful way is to "wash" the first obtained ligand (and protein) containing phase with a series of pure opposite phases. It can also be useful to use washing phases containing a second ligand, either in order to counter-extract the target protein (leaving contaminating protein, with higher affinity for the first ligand) or to more effectively remove the non-wanted proteins, all depending on the relative affinity of proteins for the two ligands. An example of multiple extraction is given in Table 2.

By keeping the bulk protein (e.g. of a tissue extract) in the lower phase of a Dextran-PEG system (by using high polymer concentration) several proteins may be consecutively extracted by introducing a different ligand-PEG in each new upper phase. Extraction of enzymes in this way has been carried out - Table 3.

PREPARATIVE EXTRACTIONS

Affinity partitioning is usually to be applied as one of several steps in a series of purification operations. It might well be used as an early step as has been shown for purification of glucose-6-phosphate dehydrogenase - Table 4.

The first step, fractional precipitation with PEG, both help to concentrate the proteins, to increase the specific activity, and to remove particulate material and part of the nucleic acids (precipitating at low PEG concentration). After affinity partitioning and washing with lower phases the target enzyme is selectively dissociated from the PEG-bound ligand (Procion yellow HE-3G) by addition of its natural ligand, NADP, and thereby recovered in the dextran phase. The final purification step consists of adsorption to DEAE cellulose at 60 mM KCl and desorption at 150 mM KCl. Traces of ligand-PEG present in the final lower phase were effectively removed by the ion exchanger. The SDS gel electrophoresis pattern after each purification step is shown in Fig. 1.
In some cases affinity partitioning can be used as the sole method for purification. This is the case for lactate dehydrogenase extracted from muscle (Table 5). The effectiveness of the extraction is shown by the electrophoresis pattern in Fig. 2.

The economy of large scale extraction of this enzyme has been recently treated [14]. The possibility of recycling PEG and ligand-PEG (with 96% efficiency) makes the process economically attractive.

CONCLUSIONS

Affinity partitioning can be a useful way for protein purification. After initial work with binding of suitable ligand to the polymer and adjustment of the parameters of the system, the main extraction and its scaling-up is straightforward.

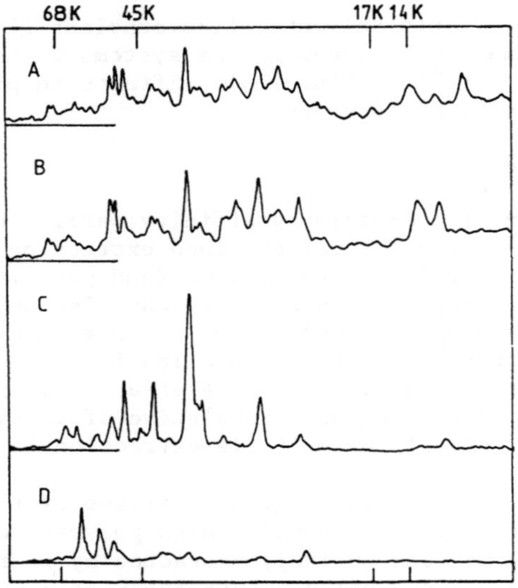

Fig. 1 *SDS gel electrophoresis of the fractions in Table 4 containing glucose-6-phosphate dehydrogenase. A, Yeast extract; B, Pellet after precipitation with PEG; C, Bottom phase after affinity partitioning (4th step); and D, after treatment with DEAE cellulose. Molecular weights are indicated (from reference 10)*

Table 5 *Extraction of lactate dehydrogenase (LDH) from swine muscle using affinity partitioning[a] (from reference 16)*

Extraction No	Recovered Phase	LDH activity			Purification
		Total (kat)	Specific (kat/kg)	Recovery (%)	
1	Lower[b]	9.4	0.32	95	1.2
2	Upper[c]	8.3	1.2	83	3.4
3	Upper[c]	7.5	2.3	75	7.2
4	Upper[c]	6.7	2.9	67	9.1

[a] with Procion yellow HE-3G PEG
[b] without affinity ligand
[c] with affinity ligand

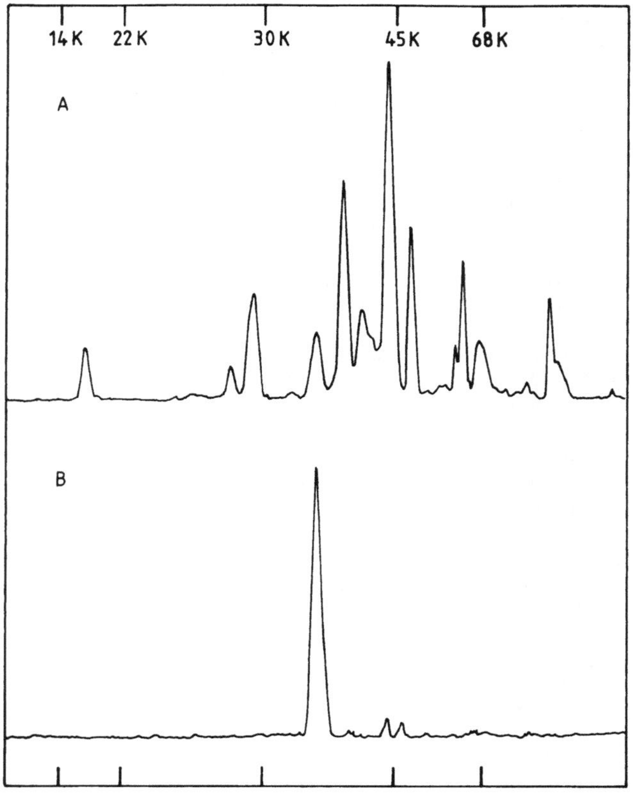

Fig. 2 *SDS gel electrophoresis of lactate dehydrogenase extracted from swine muscle by only using affinity partitioning. A, original protein extract; B, final product (from reference 15)*

REFERENCES

1 G. Takerkart, E. Segard and M. Monsigny, Partition of trypsin in two-phase systems containing a diamino-α-ω-diphenylcarbamyl poly(entylene glycol) as competitive inhibitor of trypsin, FEBS Lett. 42:218 (1974)

2 S.D. Flanagan and S.H. Barondes, Affinity partitioning - a method for purification of proteins using specific polymer-ligands in aqueous polymer two-phase systems, J. Biol. Chem. 250:1484 (1975)

3 V.P. Shanbhag and G. Johansson, Specific extraction of human serum albumin by partition in aqueous biphasic systems containing poly(ethylene glycol) bound ligand, Biochem. Biophys. Res. Commun. 61:1141 (1974)

4 G. Johansson, M. Andersson and H.-E. Åkerlund, Countercurrent distribution of yeast enzymes with polymer-bound triazine dye affinity ligands, J. Chromatogr. 298:483 (1984)

5 C. Erlanson-Albertsson, The importance of tyrosine residues in pancreatic colipase for its activity, FEBS Lett. 117:295 (1980)
6 M.R. Kula, G. Johansson and A.F. Bückmann, Large-scale isolation of enzymes, Biochem. Soc. Transact. 7:1 (1979)
7 A. Cordes and M-R Kula, Process design for large-scale purification of formate dehydrogenase from Candida boidinii by affinity partition, J. Chromatogr. 376:375 (1986)
8 H.K. Kroner, A. Cordes, A. Schelper, M. Morr, A.F. Bückmann and M.R. Kula, Affinity partition studied with glucose-6-phosphate dehydrogenase in aqueous two-phase systems in response to triazine dyes, in: "T.C.J. Gribnau, J. Visser and R.J.F. Nivard, eds., Affinity chromatography and related techniques, Elsevier, Amsterdam (1982)
9 G. Kopperschläger and G. Lorenz, Interaction of yeast glucose 6-phosphate dehydrogenase with diverse triazine dyes: A study by means of affinity partitioning, Biomed. Biochim. Acta 44:517 (1985)
10 G. Johansson and M. Joelsson, Partial purification of glucose 6-phosphate dehydrogenase from baker's yeast by affinity partitioning using polymer-bound triazine dyes, Enzyme Microb. Technol. 7:629 (1985)
11 G. Johansson and M. Andersson, Parameters determining affinity partitioning of yeast enzymes using polymer-bound triazine dye ligands, J. Chromatogr. 303:39 (1984)
12 G. Kopperschläger, G. Lorenz and E. Usbeck, Affinity partitioning in an aqueous two-phase system to the investigation of triazine dye - enzyme interactions, J. Chromatogr. 259:97 (1983)
13 G. Johansson and M. Andersson, Liquid-liquid extraction of glycolytic enzymes from baker's yeast using triazine dye ligands, J. Chromatogr. 291:175 (1984)
14 F. Tjerneld, G. Johansson and M. Joelsson, Affinity liquid-liquid extraction of lactate dehydrogenase in large scale, Biotechnol. Bioeng., 30:809 (1987)
15 G. Johansson and M. Joelsson, Liquid-liquid extraction of lactate dehydrogenase from muscle using polymer-bound triazine dyes, Applied Biochem. Biotechnol. 13:15 (1986)
16 M. Joelsson and G. Johansson, Sequential liquid-liquid extraction of some enzymes from porcine muscle using polymer-bound triazine dyes, Enzyme Microb. Technol. 9:233 (1987)
17 J. Schiemann and G. Kopperschläger, Binding of higher plant NADH-dependent nitrate reductase to different triazine dyes, Plant Sci. Lett. 36:205 (1985)
18 P. Hubert, E. Dellacherie, J. Neel and E-E Baulieu, Affinity partitioning of steroid-binding proteins. The use of polyethylene oxide-bound estradiol for purifying $\Delta_{5\rightarrow6}$ 3-oxosteroid isomerase, FEBS Lett. 65:169 (1976)
19 G. Johansson, G. Kopperschläger and P-Å Albertsson, Affinity partitioning of phosphofructokinase from baker's yeast using polymer-bound Cibacron blue F3G-A, Eur. J. Biochem. 131:589 (1983)
20 G. Kopperschläger and G. Johansson, Affinity partitioning with polymer-bound Cibacron blue F3G-A for rapid large-scale purification of phosphofructokinase from baker's yeast, Anal. Biochem. 124:117 (1982)
21 P.-Å. Albertsson, Partition of Cell Particles and Macromolecules, 3 ed., Wiley, New York (1986)
22 G. Johansson and M. Joelsson, Affinity partitioning of enzymes using dextran-bound Procion yellow HE-3G: influence of dye-ligand density, J. Chromatrogr. 393:195 (1987)
23 G. Johansson and G. Kopperschläger, Effects of organic solvents on the partitioning of enzymes in aqueous two-phase systems J. Chromatogr. 388:295 (1987)

DYE AFFINITY PARTITIONING OF SERUM PROTEINS

Gerd Birkenmeier, Per-Ake Ålbertsson* and Gerhard Kopperschläger

Institute of Biochemistry, Karl-Marx University
7010-Leipzig, Liebigstrasse 16, GDR

* University of Lund, Chemical Center
Biochemistry 1, Box 124, S-122 00 Lund, Sweden

Affinity partitioning in aqueous two-phase systems has been applied to study the interaction of serum proteins with certain dyes. By monitoring the partition coefficient any interaction between proteins and dyes and its alteration under certain conditions can easily be recognized. In detail, the binding of Cibacron Blue F3G-A and Remazol Yellow GGL to albumin and prealbumin was studied. The influence of competing ligands and of dye modification on affinity partitioning of prealbumin was demonstrated. Furthermore, affinity partitioning in three-phase systems composed of PEG, Ficoll and dextran is shown. Attempts were made to fractionate serum proteins by countercurrent distribution using polymer-bound reactive dyes.

INTRODUCTION

Affinity techniques based on the principle of biorecognition have been widely used in separation processes of proteins and particles. Affinity partitioning in aqueous phase systems is a valuable addition to the spectrum of methods employed to recognize and to characterize ligand-macromolecule interactions [1-6]. In principle, one partner of the interacting species, usually the ligand, is immobilized to a polymer and thus restricted to one phase of the two-phase system. Consequently, macromolecules with affinity for the ligand being present in the opposite phase will change their partition due to binding which can easily be monitored by measuring the partition coefficient. The simplicity of this method, together with the high sensitivity in recognizing any interaction, give rise to its application for studying the interaction of proteins with dyes.

Dyes are increasingly being applied in biochemical research and biotechnology due to their ability to complex reversibly with proteins. A number of purification protocols have been developed which include immobilized dyes as affinity chromatography ligands [7-9]. The reasons for this are their low costs, the ease of coupling dyes to soluble and insoluble polymers and their ability to bind to proteins with high specificity. In this paper the interaction of the serum proteins, albumin and prealbumin, with reactive dyes was investigated by means of affinity partitioning. Taking these two proteins as models, the principle of affinity partitioning in three-phase systems is demonstrated. Furthermore, the usefulness of countercurrent distribution (CCD) for fractionation of serum proteins by means of polymer-bound dyes is shown.

EXPERIMENTALS

Dextran T 500 (M_r = 500000, Dextran T 40 (M_r = 40000), Ficoll 400 (M_r = 400000) were supplied by PHARMACIA (Uppsala, Sweden). Poly(ethylene glycol) (PEG) 6000 (M_r = 6000 - 7500) and Cibacron Blue F3G-A were obtained from SERVA (Heidelberg, FRG). Remazol Yellow GGL was from HOECHST AG (Frankfurt, FRG). All other dyes were products of ICI Organics Division (Blackley, Manchester, Great Britain). Dye-PEG derivatives were prepared according to Johansson et al [4]. Dye-Ficoll and dye-dextran derivatives were obtained as described by Johansson and Andersson [10]. The phase systems were prepared from stock solutions of PEG (40% w/w), dextran (20% w/w) and Ficoll (40% w/w), and of stock solutions of salts, buffer and proteins. The partition coefficient, K, is defined as the ratio of the protein concentration of the two phases. The term Δlog K designating the difference between the log K values of a system with and without dye-polymer was taken as a measure of the effectiveness of extraction. Determination of the protein concentration in the phase systems was accomplished by the Laurell technique [11]. The CCD apparatus used was of the standard type with a set of plates with 60 chambers [1]. The chambers were loaded with 0.95 ml of the upper phase and 0.7 ml of the lower phase of a pre-equilibrated PEG-dextran system. Sample system containing 450 μl of serum/g of the phase system was loaded to chamber 0 and 1. The number of transfers was 58, shaking time was 30 s and separation time was 10 min per transfer cycle.

RESULTS AND DISCUSSION

From the present knowledge about the nature of dye-protein interaction it is difficult to predict the binding of a dye to a certain serum protein. A screening of a series of dyes should therefore precede in every case. Fig. 1 demonstrates testing of 10 different dyes for their ability to interact with albumin and prealbumin. As seen, albumin showed a strong affinity to Cibacron Blue F3G-A as well as to several other dyes. Prealbumin, in contrast, displayed binding predominantly to Remazol Yellow GGL but not to the other dyes indicating a high selectivity in the binding behaviour. This property has been exploited for preparation of prealbumin by affinity chromatography using immobilized Remazol Yellow GGL [12].

When the dye-polymer is included in the system at increasing concentration, the partition of the proteins gradually changed towards a limit value. The increase in the logarithm of the partition coefficient, Δlog K, gives rise to a typical saturation curve when plotted versus the concentration of the dye (Fig. 2). From such curves the maximum Δlog K value and the relative affinity (0.5 x $\Delta\log K_{max}$) can be obtained, advantageously, by using double reciprocal plots as shown in the inset of Fig. 2. Thus, in the case of Remazol Yellow GGL a Δlog K of 0.6 and 0.54 and a relative affinity of 110 μM and 38 μM for albumin and prealbumin respectively, were calculated. Albumin displayed rather strong binding to Cibacron Blue F3G-A (Δlog K = 2.6; relative affinity = 35 μM) in contrast to prealbumin which lacked affinity to this dye. Attempts to correlate the maximum Δlog K values to the number of binding sites as proposed by Flanagan and Barondes [3] failed. Usually the total number of binding sites obtained by other methods was found to be higher compared to the values resulting from partition experiments [6].

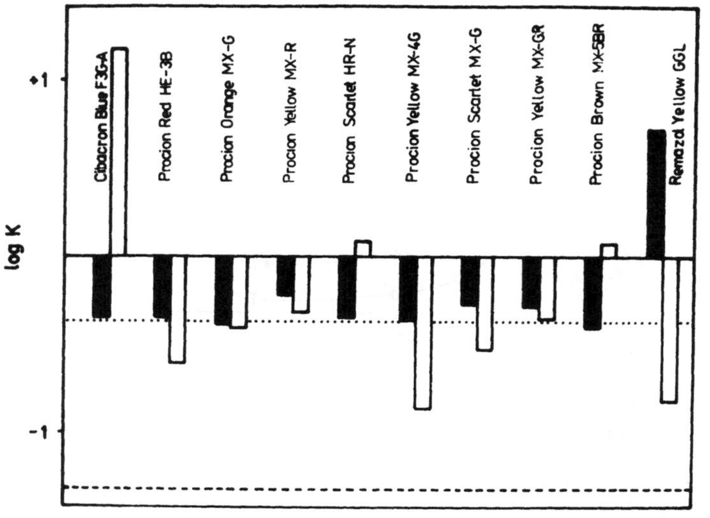

Fig. 1 *Effect of different dye-PEG derivatives on partitioning of prealbumin (black bars) and albumin (white bars). Phase systems were composed of 10% Dextran T500, 7.5% PEG 6000 including 1.6% of dye-PEG, 10 mM sodium phosphate buffer, pH 7.0. Partitioning of prealbumin (.....) and albumin (---) in the absence of the dye-PEG is shown for comparison*

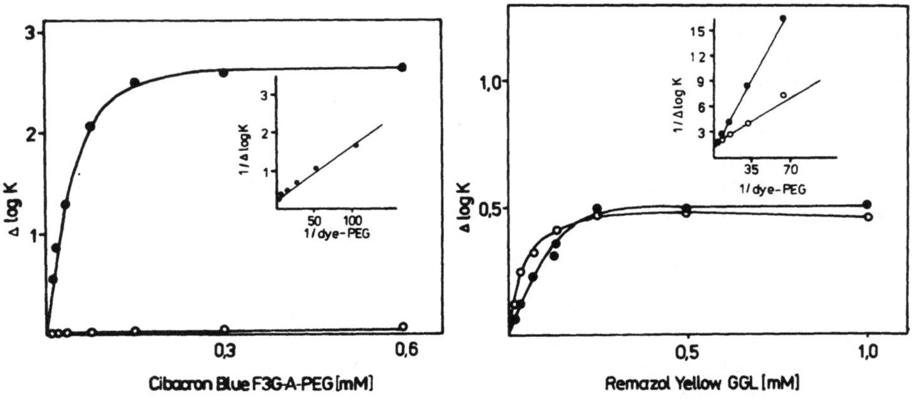

Fig. 2 *Partitioning of albumin (●) and prealbumin (○) is dependent on the concentration of Cibacron Blue F-3G-A-PEG and Remazol Yellow (GGL-PEG). Composition of the phase system was as in Fig. 1. The insets show inverse plots of 1/Δlog K versus 1/dye-PEG concentration using the same symbols*

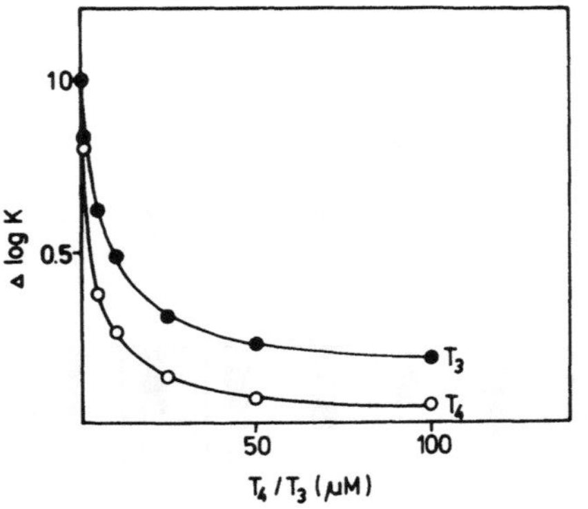

Fig. 3 *Effect of L-thyroxine (T4) and L-triiodothyronine (T3) on binding of prealbumin to Remazol Yellow GGL-PEG. Composition of the phase system was as in Fig. 1 except for pH which was 7.25. The concentration of the bound dye was 0.5 mM. The systems contained 1.6 μM prealbumin and 30 μM albumin*

Although a deeper insight into the nature of the dye-protein interaction has been obtained in the last decade, these findings are still insufficient to establish a generalising concept. In the case of serum proteins only a few constituents interact relatively strongly and/or specifically with certain dyes. Interestingly, these proteins are involved in the binding and transport of specific ligands, e.g. prealbumin, albumin, thyroxine-binding globulin, Gc-proteins etc. The basis of this behaviour might be the common property of specific ligands and dyes to bind to certain pockets on the protein surface by electrostatic, hydrophobic and charge transfer forces [7,8]. The findings that thyroid hormones strongly reduce the binding of prealbumin to Remazol Yellow GGL-PEG (decrease of Δlog K), as shown in Fig. 3, indicate that these naturally bound hormones and the dye may share common binding sites.

Valuable information about the mechanism of dye-protein interaction can also be obtained by chemical modification of the dye. Conversion of dyes is often accompanied by a loss in the affinity, especially when parts or certain groups critical for binding to the protein are removed.

Fig. 4 demonstrates the change in the partition coefficient of prealbumin caused by modification of Remazol Yellow GGL compared to the native dye. Splitting off the substituted terminal ring from Remazol Yellow GGL-PEG drastically diminishes the binding to prealbumin. This indicates that this part of the dye is mainly responsible for the strong and rather specific interaction with binding sites in prealbumin.

As known from the literature most of the partition studies have been performed using aqueous two-phase systems. By using more than two polymers, systems of several phases can be obtained [1]. Three-phase systems composed of dextran, Ficoll and PEG have been applied for partition of enzymes and particles [1,13]. Three phase systems ought to have advantages compared with two-phase systems since mixtures of proteins may be resolved between three or more bulk phases. Furthermore, the possibility of binding affinity ligands to each of the polymers could

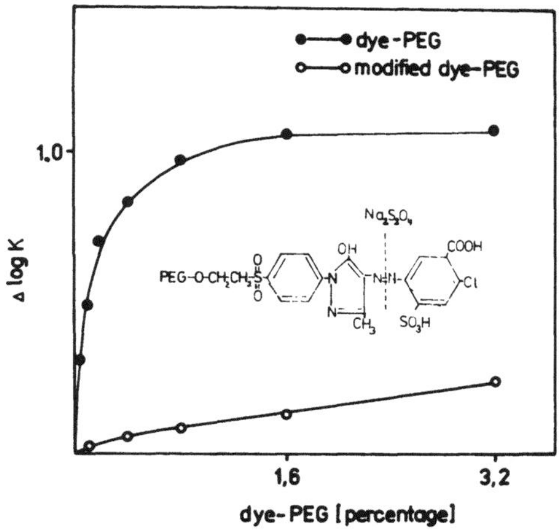

Fig. 4 *Effect of dye modification on affinity partitioning of prealbumin. Composition of the phase system was as in Fig. 1 including 1.6 μM prealbumin and 13.1 μM albumin. The modified dye was prepared by treating Remazol Yellow GGL-PEG with sodium dithionite in aqueous solution followed by extraction into chloroform. The given percentage refers to that portion of total PEG replaced by dye-PEG*

improve the selectivity of partition of proteins. In the following the interacting species albumin-Cibacron Blue F3G-A and prealbumin-Remazol Yellow GGL were used as models to study affinity partitioning in three-phase systems composed of 10% Dextran T 40, 7% Ficoll 400 and of 3.5% PEG 6000. Each of the phases is enriched in one of the polymers, the top phase in PEG, the middle phase in Ficoll and the bottom phase in

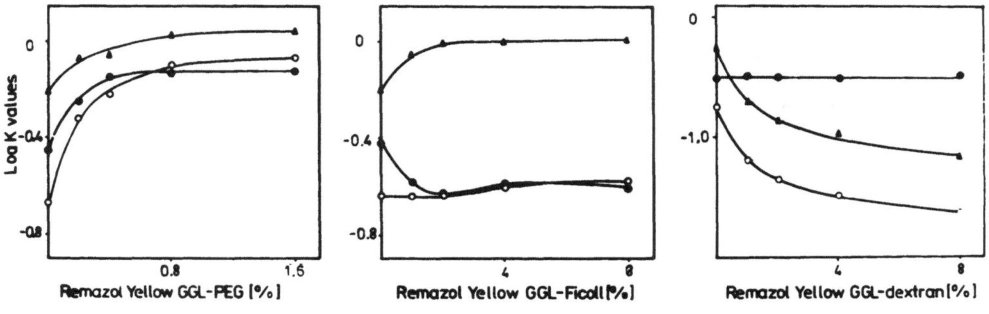

Fig. 5 *Change of the partition coefficient of prealbumin in a three phase system is dependent on increasing concentration of different dye-polymer derivatives. Composition of the phase system was 10% Dextran T 40, 7% Ficoll 400, 3.5% PEG 6000, 20 mM NaCl, 10 mM sodium phosphate buffer, pH 7.4. Symbols: K(t/b) (○); K(t/m) (●); K(m/b) (▲)*

dextran. Consequently, the different dye-polymer derivatives were found to partition also in favour of their mother phases. The partition of a solute between the three phases can be evaluated by the three partition coefficients K(t/b), K(t/m) and K(m/b) where t,m and b are the concentration of the solute in the top, middle and bottom phase. The change in the partition of a protein between three phases in response to increasing concentration of the dye is shown in Fig. 5. As seen, prealbumin is extracted from the bottom through the middle into the dye-containing top phase by the action of Remazol Yellow GGL-PEG. Largest alteration of the partition coefficient was found for K(t/b) followed by K(t/m) and K(m/b). In the presence of the Ficoll-bound dye a more or less hyperbolic extraction curve is obtained with respect to K(m/b) leading to an accumulation of the protein in the middle phase. However, the changes of the other two partition coefficients cannot be described by a saturation curve and seem to be more complex. When dextran is used as target for dye immobilization a constant partition ratio of prealbumin between the top and the middle phase resulted indicating that the protein is extracted from both the top and the middle phase in constant proportions. Similar results were obtained in the case of binding of albumin to different Cibacron Blue F36-A-polymers except that larger changes of the partition coefficients due to stronger interaction were obtained (Fig. 6).

Undoubtedly, one advantage of a three-phase system might be the possibility of including two affinity ligands bound to different polymers in the system. Thus, by the action of dyes at least two proteins could theoretically be concentrated in two phases leaving space for bulk proteins in the third one. This is demonstrated in Table 1. Out of six combinations of dye-polymers a system containing Cibacron Blue

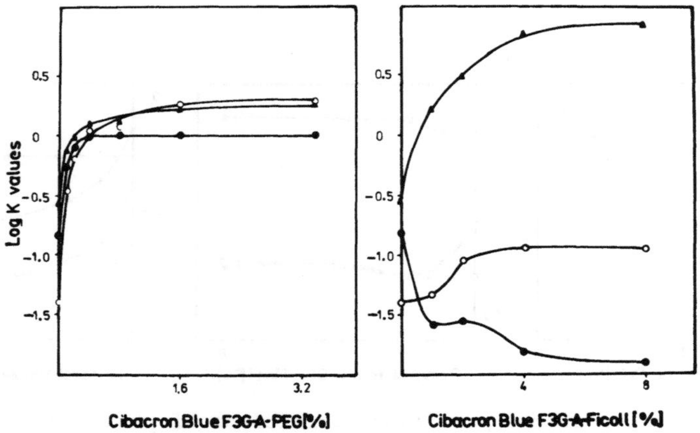

Fig. 6 *Change of the partition coefficient of albumin in a three phase system on increasing concentration of different dye-polymer derivatives. Composition of the system was as in Fig. 5, using the same symbols. The partition coefficient in the dye-dextran system could not be determined with certainty because of the very low concentration of albumin in the top phase*

Table 1 *Partition of albumin (5 μM) and prealbumin (1.1 μM) in a three-phase system in presence of two different immobilised dyes. System composition was as in Fig. 5. The percentage of the dye-liganded polymers in the phase system was as follows: Cibacron Blue F3G-A (Cb-PEG) (1.6%); Cibacron Blue F3G-A-Ficoll (Cb-Fic) (6%); Cibacron Blue F3G-A-Dextran (Cb-dx) (1%); Remazol Yellow GGL-PEG (Ry-PEG) (1.6%); Remazol Yellow GGL-Ficoll (Ry-Fic) (6%) and Remazol Yellow GGL-Dextran (Ry-dx) (6%)*

Dye-polymer combination	Relative amount of protein in the phases (total contents of the system = 100%)					
	Albumin			Prealbumin		
	top	middle	bottom	top	middle	bottom
without dye	0.6	16	83.4	4.7	19	75
Cb-Fic/Ry-dx	2	81	17	1	5	94
Cb-Fic/Ry-PEG	2	76	22	17	26	57
Cb-dx/Ry-Fic	0	2	98	4	27	69
Cb-dx/Ry-PEG	0	2	98	16	24	60
Cb-PEG/Ry-dx	32	34	32	1	7	92
Cb-PEG/Ry-Fic	35	33	32	6	30	64

F3G-A-Ficoll and Remazol Yellow GGL-Dextran was found to be most effective for separation of albumin and prealbumin. Due to their specific binding to the respective dyes more than 90% of the total amount of prealbumin was found in the bottom phase whereas albumin was forced to partition in favour of the Ficoll-rich middle phase (81%). Separation of protein successful only when the ligand added causes a large shift in the partition of the protein [4]. Otherwise several extraction steps, e.g. by counter-current distribution should be performed to get optimal resolution [9]. Fig. 7 demonstrates the effect of PEG-bound dyes on distribution of distinct serum proteins. As seen, due to high affinity of albumin to Cibacron Blue F3G-A-PEG a considerable change in the peak position was found. Gc-proteins, α-2-macroglobulin and haptoglobin exhibited moderate affinity to the dye while transferrin, α-1-proteinase inhibitor and prealbumin lacked binding. Procion Yellow HE-3G-PEG exerted similar effects except for transferrin, haptoglobin and α-2-macroglobulin which were more effectively extracted compared to Cibacron Blue F3G-A-PEG. In the case of α-2-macroglobulin a heterogeneous distribution in the presence of dye-PEG became obvious. By CCD α-2-macroglobulin is separated into two fractions each having different affinity to the dye. It is suggested that the two different forms correspond to the native and complexed (in association with proteases) form of α-2-macroglobulin. Thus, by CCD minute differences in dye binding properties can be recognized.

CONCLUSIONS

The application of aqueous systems in combination with affinity ligands offers a unique arrangement for the study of the interaction of serum proteins with dyes. This will extend our current knowledge about the nature of dye-protein interactions which is still insufficient. There have been several demonstrations on how dyes can be employed to

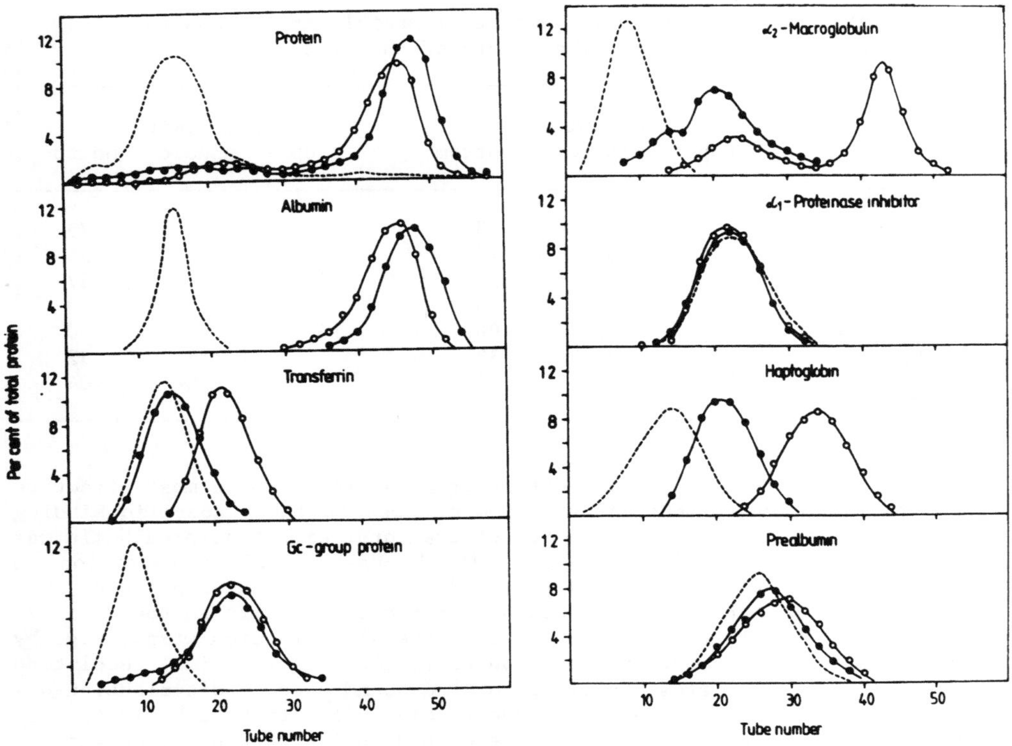

Fig. 7 *Thin-layer countercurrent distribution of human serum proteins in the absence (---) and presence of 5% Cibacron Blue F3G-A-PEG (●) and Procion Yellow HE-3G-PEG (○). The two-phase system was composed of 5% PEG 6000, 7.5% Dextran T 70 and 20 mM sodium phosphate buffer, pH 7.0*

fractionate serum or to solve special separation problems. This seems to be a promising basis to use phase partitioning for preparation of serum proteins on a large scale. The introduction of new, cheaper polymers might be the first step to make it a reality.

REFERENCES

1 P.-A. Ålbertsson, "Partition of cell particles and macromolecules", John Wiley & Sons, New York (1986)

2 H. Walter, D.E. Brooks and D. Fisher, "Partitioning in aqueous two-phase systems. Theory, methods, uses and applications to biotechnology," Academic Press Inc., Orlando (1985)

3 S.D. Flanagan and S.H. Barondes, Affinity partitioning: A method for purification of proteins using specific polymer ligands in aqueous polymer two-phase systems, J. Biol. Chem. 250:1484 (1975)

4 G. Johansson, G. Kopperschläger and P.-A.Ålbertsson, Affinity partitioning of phosphofructokinase from baker's yeast using polymer-bound Cibacron Blue F3G-A, Eur. J. Biochem. 131:589 (1983)

5 M.R. Kula, K.H. Kroner and H. Hustedt, Purification of enzymes by liquid-liquid extraction, Adv. Biochem. Eng. 24:73 (1982)

6 G. Birkenmeier, G. Kopperschläger and G. Johansson, Separation and studies of serum proteins with aid of aqueous two-phase systems containing dyes as affinity ligands, Biomed. Chromatogr. 1:64 (1986)

7 G. Kopperschläger, H.-J. Böhme and E. Hofmann, Cibacron Blue F3G-A and related dyes as ligands in affinity chromatography, in: "Advances in Biochemical Engineering", A. Fichter, ed., Springer, Berlin (1982)

8 S. Subramanian, Dye-affinity chromatography: the interaction of Cibacron Blue F3G-A with proteins and enzymes, CRC Crit. Rev. Biochem. 16:169 (1984)

9 G. Johansson and M. Andersson, Liquid-liquid extraction of glycolytic enzymes from baker's yeast using triazine dye ligands, J. Chromatogr. 291:175 (1984)

10 G. Johansson and M. Andersson, Parameters determining affinity partitioning of yeast enzymes using polymer-bound triazine dye ligands, J. Chromatogr. 303:39 (1984)

11 C.-B. Laurell, Quantitative estimation of proteins in agarose gel containing antibodies, Anal. Biochem. 15:45 (1966)

12 G. Birkenmeier, B. Tschechonien and G. Kopperschläger, Affinity chromatography and affinity partitioning of human serum prealbumin using immobilized Remazol Yellow GGL, FEBS Lett. 174:162 (1984)

13 A. Hartman, G. Johansson and P.-A. Ålbertsson, Partition of proteins in a three-phase system, Eur. J. Biochem. 46:75 (1974)

HYDROPHOBIC AFFINITY PARTITIONING AS A METHOD FOR STUDYING CONFORMATIONAL CHANGES IN PROTEINS

V.P. Shanbhag and L. Backman

Department of Biochemistry
University of Umeå
S-901 87 Umeå, Sweden

INTRODUCTION

Hydrophobic interactions are important for the stability, conformation and function of proteins. The biological function of many proteins, including enzymes and the stability of multimeric proteins and protein complexes is regulated by a change in conformation and a concommitant change in surface hydrophobicity induced by the binding of ions or ligands.

An obvious way of determining surface hydrophobicity of proteins is to study the binding of hydrophobic molecules. One difficulty of doing this by direct binding studies is the very low solubility of hydrophobic molecules in aqueous media. One can circumvent this difficulty by using affinity partitioning in aqueous two-phase systems where the hydrophobic ligand of interest is bound to one of the phase forming polymers. We have earlier shown [1-4] that surface hydrophobicitity of proteins can be measured by partitioning in dextran-polyethylene glycol (PEG) two-phase systems containing PEG bound hydrophobic ligands; usually esters of PEG with fatty acids. In that case we get the affinity for an uncharged non-polar aliphatic group. The affinity for the hydrophobic ligand can be expressed in terms of $\Delta\log K$; the difference in the logarithm of the partition coefficient obtained with and without the hydrophobic ligand in the system. $\Delta\log K$ is a selective measure of the protein surface hydrophobicity and does not reflect some non-specific low affinity type of association common to all proteins as reported for the cooperative binding of amphiphiles to proteins. This conclusion is supported by the following observations:-

1 At constant concentration of PEG-bound acyl group in the system, the $\Delta\log K$ for a protein increases with an increase in the chain length of the acyl group, i.e. the hydrophobicity of the ligand. The increase is different for different proteins and the partition of some proteins is not at all affected ($\Delta\log K = 0$) by the hydrophobic group [1].

2 $\Delta\log K$ for different proteins obtained under a given set of conditions are compatible with measurements done using other methods.

This means that Δlog K provides a useful relative measure of surface hydrophobicity. We have now applied this method for studying conformational changes in two Ca^{2+}-binding proteins, namely calmodulin and α-lactalbumin. The conformation of both these proteins depends on the amount of Ca^{2+} and in both cases the changes in conformation involve a change in surface hydrophobicity and are intimately connected with their biological function [7-10].

MATERIALS AND METHODS

Calmodulin was prepared from bovine brain and labelled with ^{14}C using reductive methylation [11]. Bovine α-lactalbumin was bought from Sigma Chemicals Co., USA, and apo-α-lactalbumin (Ca^{2+}-free form) was prepared according to the method of Murakami et al [12]. Polyethylene glycol 8000, PEG, was from union Carbide, USA, and dextran T 500 from Pharmacia, Sweden. The fatty esters of PEG were prepared and their degree of substitution were determined as described by Johansson [13]. The partition experiments in systems with and without the PEG-bound lignds were carried out as described earlier [1,2]. The concentration of α-lactalbumin was measured photometrically [4] and that of calmodulin by liquid scintillation counting [11] as described earlier. Other experimental details like composition of the phase system are described under each figure.

RESULTS AND DISCUSSIONS

Calmodulin is a small (M = 16 700) ubiquitous protein, with four Ca^{2+} binding sites, that, depending on the number of Ca^{2+}-ions bound activates different proteins [8]. The Ca^{2+}-induced conformational change leads to the exposure of hydrophobic sites on the surface of calmodulin [7,8] and the activation of target proteins by calmodulin apparently involves a hydrophobic reaction. This conclusion is based on the observation that the activation can be inhibited by hydrophobic molecules such as chloropromazine and other antipsychotic drugs [14].

Partition studies show that Ca^{2+}-free calmodulin has no affinity (Δlog K = 0) for esters of PEG with fatty acids, while in the presence of 0.1 mM Ca^{2+} (Fig. 1) the protein exhibits affinity for all esters of PEG used; from PEG-capronate to PEG-stearate, that is for acyl groups ranging in length from 6 to 18 carbons. The affinity (log K) increases with an increase in the concentration of the PEG-bound acyl group and the partition curve has the appearance of a binding curve as in the case of serum albumin and ß-lactalbumin [3]. In analogy with a binding curve, the partition curve is a relative measure of the affinity. To approximate the affinity from the partition data we have plotted log K at two different ligand concentrations as a function of the chain length of the PEG-bound acyl group (Fig. 2).

It is seen that the apparent affinity increases with an increase in chain length up to 16 carbon atoms (palmitate). The slight decrease in affinity noticed for PEG-stearate (C_{18}-group) may reflect a limit in the size of the hydrophobic site or interference from a polar group in the binding site or its vicinity. Alternatively, the effective concentration of the ligand in the system may be less than calculated due to self-association of the ligand [4] as observed for such large hydrophobic groups in an aqueous medium. This could also be the reason for the observed reduction (Fig. 1) in log K at higher concentrations of PEG-myristate (C_{14}) and PEG-palmitate (C_{16}). Self-association could also explain that the maximum of log K decreases in the order from C_{10} through C_{16} while the initial slope, which may be regarded as a relative measure of the binding strength (Fig. 2) increases in the reverse order. At present we lack a model which takes into consideration even ligand-ligand self-association and allows the analysis of the entire

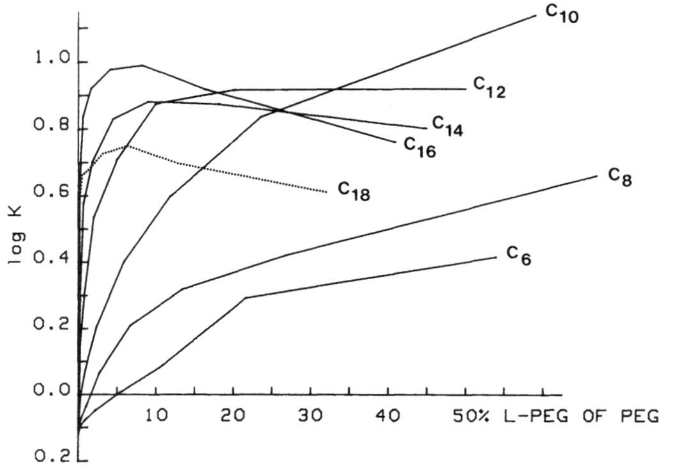

Fig. 1 *Log K of calmodulin as a function of the concentration of PEG-acyl esters (expressed as percentage of the total PEG in the system). The composition of the system was: 4.5% (w/w) Dextran T500, 4.5% PEG 8000 (total), 100 mM KCl, 10 mM PIPES (pH 7.0) and 0.74 μM calmodulin. Temperature 20°C. The different PEG esters were: capronate (C_6), caprylate (C_8), caprinate (C_{10}), laurate (C_{12}), myristate (C_{14}), palmitate (C_{16}) and stearate (C_{18})*

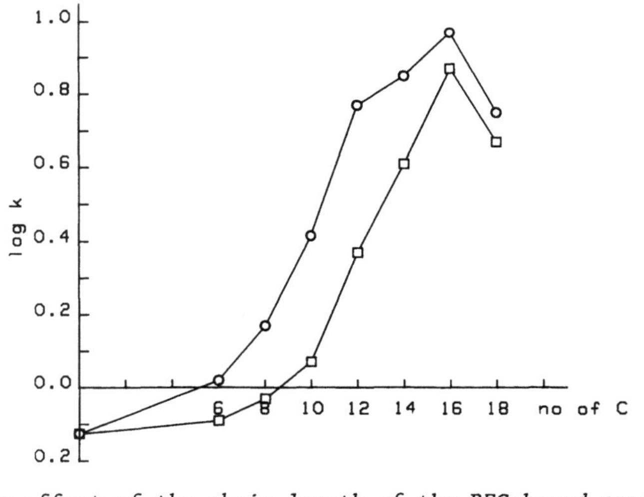

Fig. 2 *The effect of the chain length of the PEG-bound acyl group on log K of calmodulin. Data from Fig. 1 corresponding to the PEG-bound ester concentration of 1% (○) and 6% (□) plotted as function of the number of carbon atoms in the ligand*

partition curve in terms of binding constants and the number of binding sites. Hence, it is best to confine the comparison between ligands to the initial part of the partition curve where the concentration of the ligand is low and the self-association of the ligand is negligible.

The data in Fig. 3 clearly shows that calmodulin exhibits affinity for hydrophobic ligands, PEG-capronate (C_6) and PEG-palmitate (C_{16}), only when the concentration of Ca^{2+} exceeds 1 μM. The affinity then increases with increasing Ca^{2+} up to 6 μM above which it remains constant. This agrees well with the dependence of hydrophobicity on Ca^{2+}-concentration measured by other methods [7,8]. The data also shows that there is no stepwise change in hydrophobicity and that the hydrophobic site is large enough to accommodate PEG-palmitate. The change in conformation underlying the change in hydrophobicity is also reflected in a change in partition in the ligand free system, but the change is considerably smaller than in systems containing acyl-PEG. Thus PEG-capronate causes a 2-fold increase in partition whereas PEG-palmitate increases the partition 15-fold. This shows that hydrophobic affinity partitioning is much superior in sensitivity for detecting and characterising changes in conformation.

α-lactalbumin, a protein from mammalian milk, also undergoes a Ca^{2+} -dependent conformational change which is probably involved in its function as a modulator of the lactose synthase complex [9,10]. In contrast to calmodulin the hydrophobicity of α-lactalbumin decreases with an increase in Ca^{2+} bound to the protein and reaches a minimum at 1 mole Ca^{2+}/mole α-lactalbumin. A similar change in hydrophobicity is observed when pH is changed.

Affinity partitioning in systems containing different fatty esters of PEG confirms that α-lactalbumin has hydrophobic sites on its surface (Fig. 4). The hydrophobicity of α-lactalbumin with 0.3 mole Ca^{2+} bound per mole is, however, low as compared to that of serum albumin or calmodulin. Furthermore, Δlog K is low and almost no change in partition is observed when PEG-bound acyl chains shorter than 12 carbons (laurate) are included in the system. Even the affinity for PEG-palmitate is low when the protein has equimolar Ca^{2+} bound to it (Fig. 5). The complete removal of Ca^{2+} (by the addition of 5 mM EGTA to the system), however, causes a 5-fold increase in the affinity.

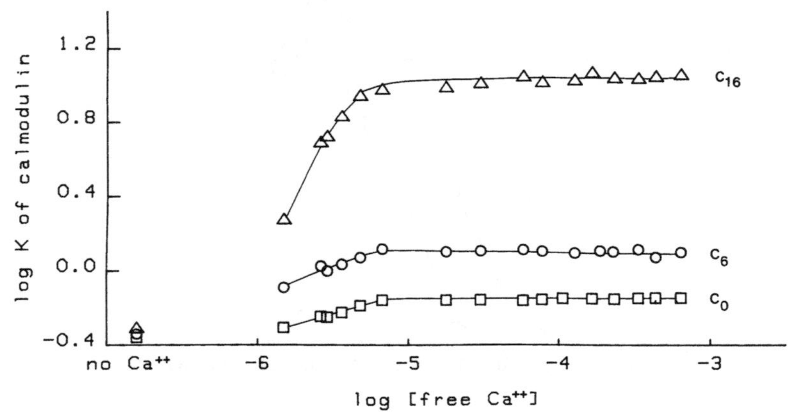

Fig. 3 *Ca^{2+}-dependent change in the hydrophobicity of calmodulin. Log K of calmodulin was determined in systems where 5% of PEG was replaced by PEG-palmitate (Δ) and PEG-capronate (O) and in systems containing no PEG-bound ligand (□). The composition of the system was the same as described in Fig. 1*

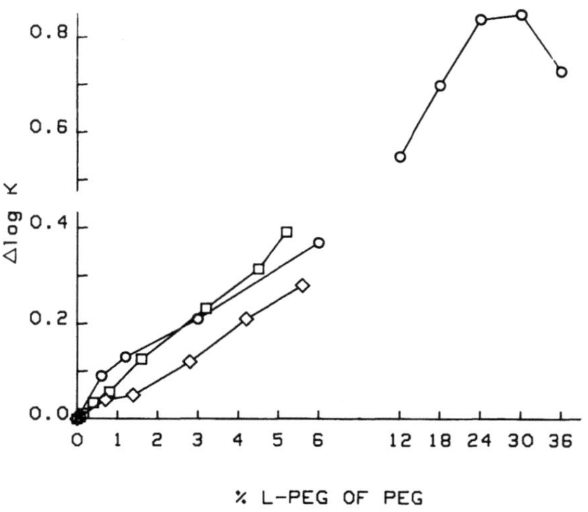

Fig. 4 *Log K of α-lactalbumin as a function of the concentration of fatty esters of PEG (expressed as a percentage of the total PEG). System composition: 8% Dextran T70, 8% PEG 8000, 100 mM K_2SO_4, 2 mM potassium phosphate buffer (pH 7) and 23 μM α-lactalbumin (0.3 mole bound Ca^2/mole protein). Temperature 20°C. The PEG esters were: laurate (◇), myristate (□) and palmitate (○)*

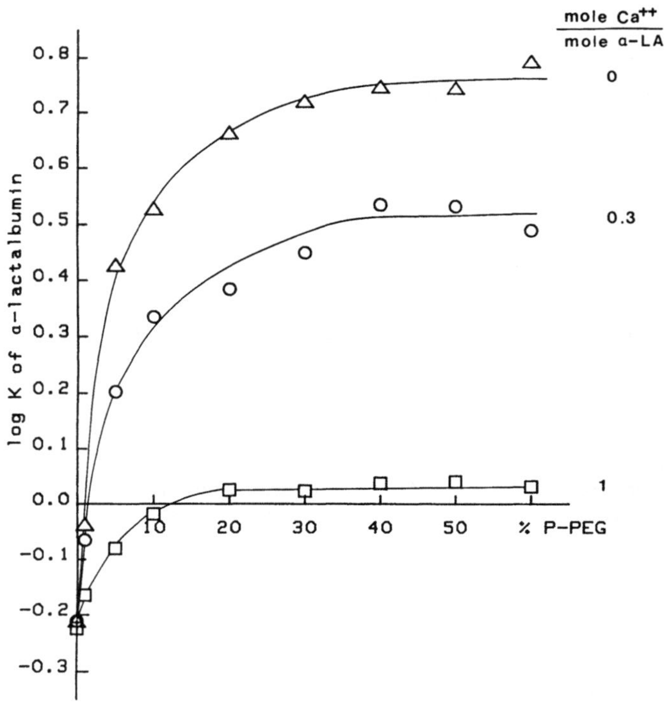

Fig 5 *Log K of α-lactalbumin with 1 (□), 0.3 (○) and 0 (Δ) mole Ca^{2+} bound per mole protein as a function of the concentration of PEG-palmitate in the system. Other conditions were as described in Fig. 4*

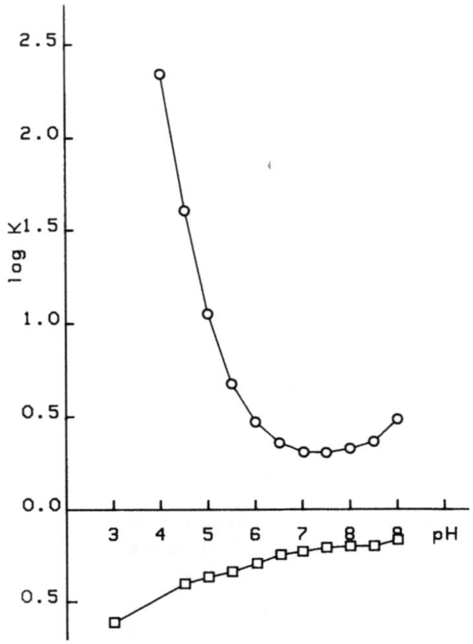

Fig. 6 *The effect of pH on the log K of α-lactalbumin in the presence of 6% (○) and in the absence (□) of PEG-palmitate. Other conditions were as described in Fig. 4*

The affinity of α-lactalbumin for PEG-palmitate and therefore the hydrophobicity also varies with pH in agreement with earlier results [15]; the largest change in hydrophobicity occurs below pH 5 (Fig. 6).

CONCLUSION

The results presented here show that hydrophobic affinity partitioning is a very sensitive method for characterizing conformational changes in proteins when such changes involve a change in surface hydrophobicity. Use of different hydrophobic ligands, such as aliphatic, aromatic hydrocarbons with different types of polar or charged substituents should yield more details about the architecture of the hydrophobic site. Quantitative analysis of the partition data in terms of the number of binding sites and the corresponding binding constants should greatly facilitate a comparison between the different conformations of the same protein or between different proteins. Such an analysis would also be useful in understanding the structure-function relationships in proteins. We hope to contribute to the solution of this problem in the near future.

REFERENCES

1 V.P. Shanbhag and C-G Axelsson, Hydrophobic interaction determined by partition in aqueous two-phase systems. I. Partition of proteins in systems containing fatty acid esters of poly(ethylene glycol), Eur. J. Biochem. 60:17 (1975)

2 C-G Axelsson and V.P. Shanbhag, Partition of histones in aqueous two-phase systems containing polyethylene glycol-bound hydrocarbons, Eur. J. Biochem. 71:419 (1976)
3 C-G Axelsson, Changes in structure and hydrophobic surface properties of ß-lactoglobulin determined by partition in aqueous two-phase polymeric systems, Biochim. Biophys. Acta 5:34 (1978)
4 G. Johansson and V.P. Shanbhag, Affinity partition of proteins in aqueous two-phase systems containing polymer bound fatty acids. Effect of PEG-palmitate on the partition of human serum albumin and α-lactalbumin, J. Chromat. 284:63 (1984)
5 E. Keshavrz and S. Nakai, The relation between hydrophobicity and interfacial tension of proteins, Biochim. Biophys. Acta 576:269 (1979)
6 Å. Kato and S. Nakai, Hydrophobicity determined by fluorescence probe method and its correlation with surface properties of proteins, Biochim. Biophys. Acta 624:13 (1980)
7 T. Tanaka and H. Hidaka, Hydrophobic regions of calcium-binding proteins exposed by calcium, Biochem. International 2:71 (1981)
8 D. Burger, J.A. Cox, M. Comte and E.A. Stein, Sequential conformation changes in calmodulin upon binding calcium, Biochemistry 23:1966 (1984)
9 G. Musci and L.J. Berliner, Probing different conformational states of bovine α-lactalbumin: Fluorescence studies with 4,4'-bis[1-(phenyl-amino)-8-naphthalenesulfonate], Biochemistry 24:3852 (1985)
10 G. Musci and L.J. Berliner, Physiological role of zinc and calcium binding to α-lactalbumin in lactose biosynthesis, Biochemistry 24:6945 (1985)
11 Å. Berglund, L. Backman and V.P. Shanbhag, Calmodulin binding to human spectrin, FEBS Lett. 172:109 (1984)
12 K. Murakami, P.J. Andree and L.J. Berliner, Metal ion binding to α-lactalbumin species, Biochemistry 21:5488 (1982)
13 G. Johansson, The effect of poly(ethylene glycol) esters on the partition of proteins and fragmented membranes in aqeuous biphasic systems, Biochim. Biophys. Acta 451:517 (1976)
14 D.E. Laporte, B.M. Wierman and D.R. Storm, Calcium induced exposure of hydrophobic surface on calmodulin, Biochemistry 19:3814 (1980)
15 P.B. Sommers and M.J. Kronman, Comparative fluorescence properties of bovine, goat, human and guinea pig α-lactalbumin Biophys. Chem. 1:217 (1980)

AFFINITY PARTITION OF ENZYMES WITH DEXTRAN-BOUND PROCION YELLOW HE-3G: EFFECT OF LIGAND DENSITY

Monica Joelsson and Göte Johansson

Department of Biochemistry, Chemical Center
University of Lund, P. O. Box 124
S-221 00 Lund, Sweden

Affinity partition of enzymes has in most cases been carried out in dextran-PEG systems by attaching an affinity ligand to PEG. The ligand is enriched in the upper phase of the two-phase system. The ligand can be bound to the PEG molecule in only two positions, i.e. via the two terminal hydroxyl groups. Many of the works on affinity partition have been done with (mainly) monosubstituted PEG. It is of interest for the practical use of affinity partition for enzyme extraction to know the effects of high densities of ligand. We have therefore studied the effect of the number of ligand molecules per "carrier" polymer molecule (ligand molar ratio) on the affinity partition effect. Dextran has been chosen as ligand carrier because of its high degree of attaching points (hydroxyl groups). As ligand has been used a triazine dye with high affinity for enzymes such as lactate dehydrogenase and 3-phosphoglycerate kinase.

The triazine dye Procion yellow HE-3G has been attached to Dextran 70 in the molar ratios (dye:Dextran) 1.3, 2.3, 5.3 and 8.3. The partition of dye-Dextran was found to depend on the salt included in a Dextran-PEG system. The salt-steering effect increased linearly with the degree of substitution reflecting changes in charge and hydrophobicity of the substituted polymer, Fig. 1. This means that high substituted dye-Dextran can be directed <u>either</u> to the lower <u>or</u> the upper phase by the choice of salt.

The effects of dye-Dextran on the partition of 3-phosphoglycerate kinase, glucose 6-phosphate dehydrogenase and lactate dehydrogenase (with 1, 2 and 4 binding sites respectively) were studied. All three enzymes were extracted into the dye-containing phase. The extraction curves of lactate dehydrogenase as function of the concentration of Dextran-bound ligand are shown in Fig. 2.

The change in the logarithmic partition coefficient of the enzymes (with 1 or 2 binding sites), $\Delta \log K_{enzyme}$, was close to the logarithmic partition coefficient of the polymer-bound dye, $\log K_{ligand}$. For lactate dehydrogenase $\Delta \log K_{enzyme}$ was 1.5 times $\log K_{ligand}$. The results indicate, based on the theory for partition presented by Flanagan and Barondes [2], that only one (or 1-2 in the case of lactate dehydrogenase) Dextran molecules were bound per enzyme molecule at ligand concentrations yielding maximal effects.

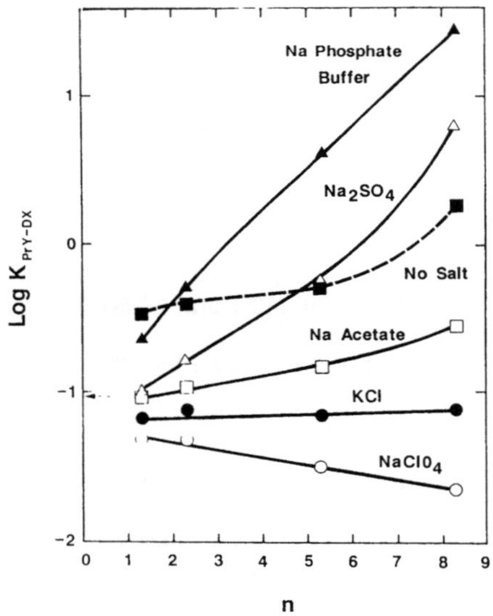

Fig. 1 *Partition of Procion yellow HE-3G Dextran (PrY-Dx) as function of the ligand:Dextran molar ratio (n) in the presence of various salts. System: 8% Dextran 70, 4.5% PEG 8000, 100 mM salt (50 mM in the case of phosphate) and 50 μM Dextran-bound dye, pH 7.9. Temperature 22°C. Arrow = log K for Dextran. From reference 1*

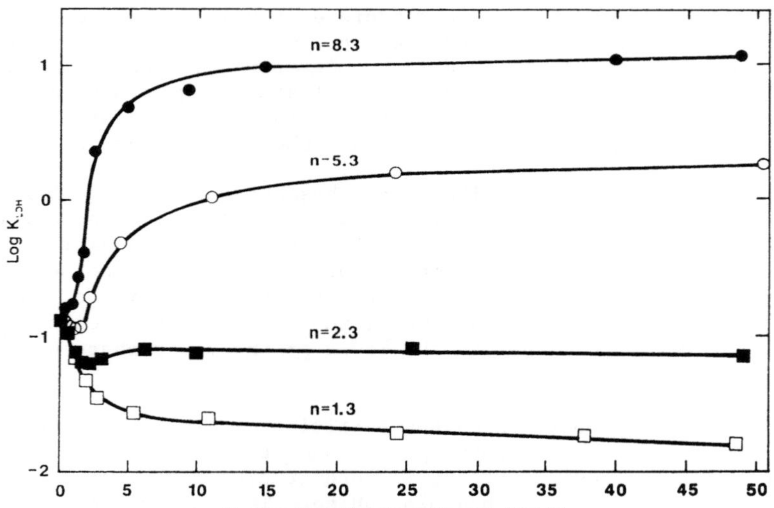

Fig. 2 *Partition of lactic dehydrogenase in systems containing increasing concentrations of Dextran-bound Procion yellow HE-3G. System as in Fig. 1 but with 50 mM sodium phosphate buffer pH 7.9. From reference 1*

REFERENCES

1 G. Johansson and M. Joelsson, J. Chromatogr. 393:195 (1987)
2 S.D. Flanagan and S.H. Barondes, J. Biol. Chem. 250:1484 (1975)

RAPID TWO-PHASE EXTRACTION OF THE CALVIN CYCLE

ENZYMES ALDOLASE AND PHOSPHOGLYCERATE KINASE

Lars-Olof Persson and Björn Olde

Department of Biochemistry
University of Lund, P. O. Box 124
S-221 00 Lund, Sweden

Fructose-1,6-bisphosphate aldolase (EC 4.1.2.13) was extracted from isolated spinach chloroplasts in a procedure consisting of two steps; precipitation of bulk protein with polyethylene-glycol (PEG) with M_r = 6500-8000 and partitioning of remaining soluble protein in aqueous two-phase systems. All manipulations, including chloroplast isolation, activity measurements and protein determination, can be carried out within 6 h and the final preparation is 94% pure, on the basis of electrophoretic criteria, with a yield of 19%.

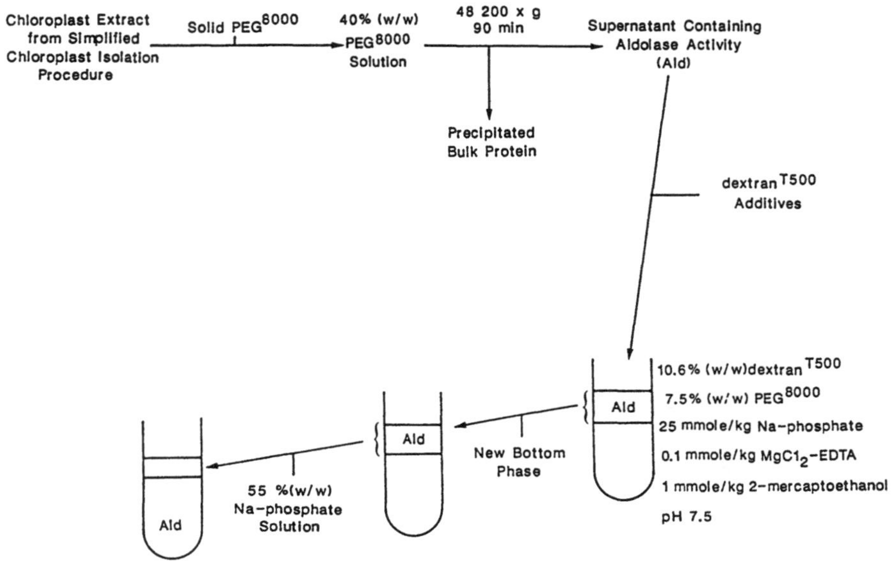

Fig. 1 *Batch procedure for the purification of fructose-1, 6-bisphosphate aldolase from spinach chloroplasts*

A main time-consuming step in this purification procedure, the preparation of chloroplasts, can be avoided if whole leaf extract is used as starting material instead of chloroplast extract. The partition coefficient of the chloroplast aldolase activity varied between 8 and 18 depending on protein concentration. The partition coefficient was considerably reduced (2-3.5) when leaf extract was used while the total protein partitioned as in the case of chloroplast extract (0.1-2). Interaction of aldolase with non-chloroplastic cell constituents is a possible explanation. It is unlikely a result of different partition behaviour of non-chloroplastic isoenzymes since the chloroplast enzyme is reported to constitute as much as 85% of the total leaf activity [3,5].

Affinity partitionings with ATP and the dye Procion Turquoise MX-G as ligands (attached to Dextran and PEG) were performed in order to extract the enzyme 3-phosphoglycerate kinase (EC 2.7.2.3) from spinach leaf extract. By comparing the partition of the target enzyme with the partition of total protein, the usefulness of the different ligand-polymer derivatives for extraction was studied.

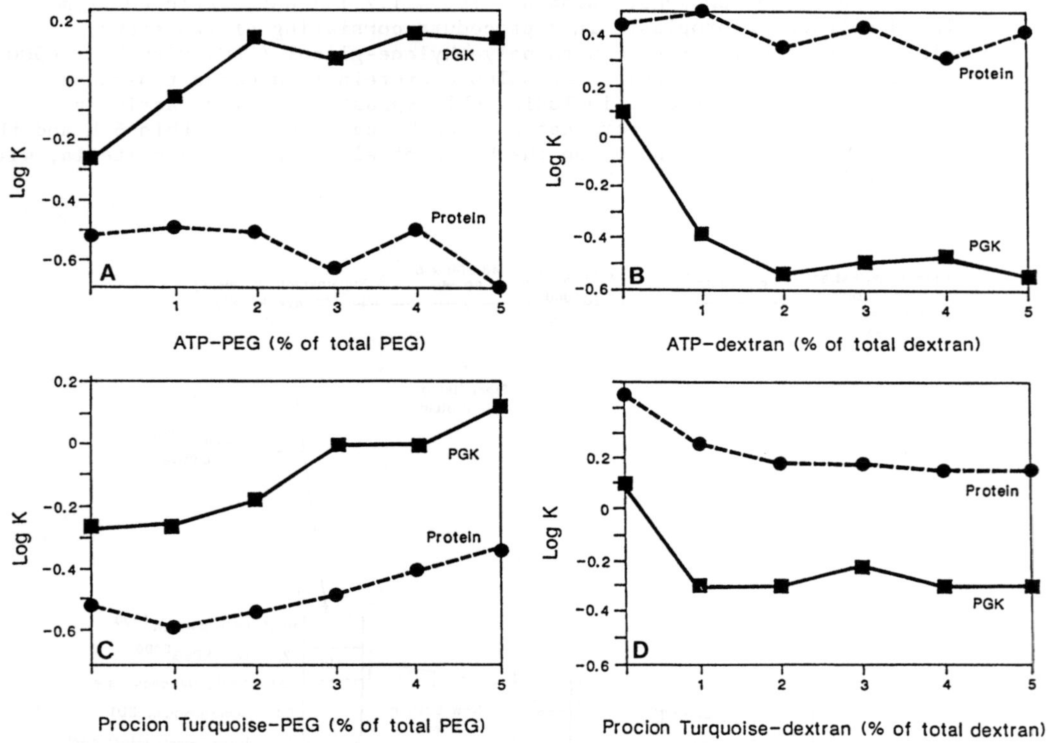

Fig. 2 *Dependence of the partition coefficient (K) of 3-phosphoglycerate kinase (PGK, ■) and total protein (●) from crude spinach leaf extract on the concentration of ATP-PEG (A), Procion Turquoise-PEG (B), ATP-Dextran (C) and Procion Turquoise-Dextran (D)*

Procion Turquoise MX-G was attached to PEG (M_r = 6500-8000) as described by Johansson [1] and to Dextran T500 as described by Johansson and Andersson [2]. The ATP-polymer derivatives was synthesised in the following manner: first adipoyldihydrazide was coupled to tresyl chloride activated Dextran T500 and PEG (M_r = 6500-8000) respectively, then periodate-oxidised ATP was coupled to the formed adipoyldihydrazo-dextran and adipoyldihydrazo-PEG, respectively, as described by Kuntz and Krietsch [4].

Both ligands, whether attached to dextran or PEG, affected the partition of the target enzyme but only ATP showed specificity, that is, did not change the partition of total protein. Optimal separation between the target enzyme and total protein was achieved using ATP-Dextran, which enriches the target enzyme in the lower phase (log K = -0.5), in a two-phase system containing tetrabutylammonium phosphate which increases the partition coefficient of total protein (log K = 0.5-0.6).

REFERENCES

1 G. Johansson, Methods in Enzymology 104:356 (1984)

2 G. Johansson, and M. Andersson, J. Chrom. 303:39 (1984)

3 I. Krüger, and C. Schnarrenberger, Eur. J. Biochem. 136:101 (1983)

4 G.W.K. Kuntz and W.K.G. Krietsch, Methods in Enzymology 90:103 (1982)

5 H.G. Lebherz, M.M. Leadbetter and R.A. Bradshaw, J. Biol. Chem. 259:1011 (1984)

AFFINITY PARTITIONING OF PHOSPHOFRUCTOKINASE FROM RAT ERYTHROCYTES IN POLY(ETHYLENE GLYCOL)-CIBACRON BLUE/DEXTRAN TWO-PHASE SYSTEMS

M.C. Tejedor, M. Grupelli, A. Rubio
C. Delgado and J. Luque

Departamento de Bioquimica y Biologia Molecular
Universidad de Alcalá de Henares
Alcalá de Henares (Madrid), Spain

Partition of enzymes between two liquid aqueous phases can be highly influenced by specific ligands bound to a water-soluble polymer included in the two-phase system. The fact that phosphofructokinase (PFK) binds strongly Cibacron blue (Cb) allow the use of such a ligand to study Cibacron blue-enzyme interaction [1-4] as well as to obtain PFK-enriched fractions from crude extracts [1].

We report the partitioning behaviour of erythrocytic PFK in poly(ethylene glycol) (PEG)/Dextran (D) two-phase systems, containing or not Cb bound to PEG. Different experimental conditions of polymer concentration, pH, ionic strength and PFK effectors were used.

1-g systems of various concentrations of PEG-6000 and D T-500 were prepared by replacing or not PEG by Cibacron blue F3G-A bound to PEG (Cb-PEG). A constant ratio of 1:1.5 PEG:D was maintained throughout. The amount of replaced Cb-PEG was up to 0.5% of total system. Sodium phosphate buffers were used to obtain a range of pH (from 6 to 8) and ionic strength (from 0.01 to 0.30 mM). 0.25 mM EDTA and 2.5 mM 2-mercaptoethanol were kept constant in the systems. Enzyme effectors (F6P, ATP, ATP-Mg and AMP) were present at 1 mM concentrations. A PFK-enriched fraction, previously obtained from rat erythrocyte haemolysates by precipitation with PEG [5], was used as the enzyme source. After shaking for 15 s and letting the system reach the equilibrium PFK concentration was determined in both upper and lower phases. Affinity partitioning is expressed as the increment in logarithmic partition coefficient Δlog K) of the enzyme in the Cb-PEG system relative to the value of log K in the Cb-PEG-free system.

RESULTS AND DISCUSSION

Dependence of the partition coefficient of PFK on the polymer concentration of the phases

Two-phase systems with increasing concentrations of both PEG and D (from 4%/6% to 6%/9%) were used. The higher the polymer concentration in the Cb-PEG-free systems, the lower the partition coefficient. Log K values lower than -1 for a 6%/9% system were obtained.

By replacing PEG by Cb-PEG (0.3% of total system), PFK favours the PEG-rich phase leading to an increase of the partition coefficient between 10-100 times. A maximal log K increment of about 2 units was found for the 5% PEG/7.5% D system containing Cb-PEG. Therefore, this system is suggested as an appropriate one for the extraction of erythrocytic PFK. An increment of about 2 units on the log K has also been reported as useful for the selective extraction of enzymes from Baker's yeast [6].

In order to establish if the purification of PFK could be further optimised other parameters were then changed.

Partition of phosphofructokinase as a function of Cb-PEG concentration

The increase of Cb bound to PEG in the upper phase (up to 0.5% of total PEG) makes the partition coefficient increase to progressively increase in all conditions assayed.

pH Dependence - The affinity partitioning behaviour of erythrocytic PFK depends strongly on pH. By increasing pH values 6 to 8 an increase of the K was observed for all Cb-PEG concentrations used. It means that the most favourable experimental condition for PFK purification was found at pH 8. A maximal Δlog K is obtained (2.83) for the system containing more than 0.2% Cb-PEG. Under these conditions, both a high enzyme purification factor (around 12.8) and a high yield (95%) are being obtained.

Ionic strength dependence - Ionic strength was varied from 0.01 to 0.16 M. No effect of such ionic strength range is observed on the affinity partitioning of PFK. However, for a higher μ value (0.30 M), a clear decrease in Δlog K is observed. The influence of high salt concentrations on the affinity partitioning can be due to the shielding effects on the electrostatic part of the ligand-enzyme interaction and means that the binding strength is reduced by using high salt concentrations. A similar statement has been reported by Johansson et al [2].

Influence of PFK ligands - The presence of tested PFK effectors (F6P, ATP, ATP-Mg and AMP) significantly reduce Δlog K values. This reduction effect is related to the type of effector in the following order: ATP-Mg > ATP > F6P > AMP. ATP acts as the highest competitive effector for the binding of Cb to erythrocytic PFK. In any case, these effectors cannot be suggested as appropriate factors for PFK enrichment in Cb-PEG containing phases. On the contrary, ATP could be selectively used to eliminate enzyme-Cb interactions and so to "strip off" the PFK from the Cb-containing phase.

In summary, the application of affinity partitioning in PEG(Cb-PEG)/D two-phase systems is a rapid and effective method for a partial, but significant, purification of erythrocytic PFK.

REFERENCES

1 G. Kopperschläger and G. Johansson, Anal. Biochem. 124:117 (1982)
2 G. Johansson, G. Kopperschläger and P.-Å. Albertsson, Eur. J. Biochem. 131:589 (1983)
3 J. Frenzel, K. Nissler, G. Schellenberger, G. Kopperschläger and E. Hofmann, Biomed. Biochim. Acta 43:413 (1984)
4 G. Kopperschläger and G. Johansson, Biomed. Biochim. Acta 44:1047 (1985)
5 M.C. Tejedor, C. Delgado, A. Rubio and J. Luque, Bioscience Reports 6:395 (1986)
6 G. Johansson, Methods in Enzymology 104:356 (1984)

STUDIES ON TRIAZINE DYE-ENZYME

INTERACTION BY MEANS OF AFFINITY PARTITIONING

Gerhard Kopperschläger, Jürgen Kirchberger and Thomas Kriegel

Institute of Biochemistry, Karl-Marx-University
7010 Leipzig, Liebigstrasse 16
German Democratic Republic

Reactive dyes have found widespread interest as biomimetic ligands in numerous subjects of enzymology and protein chemistry, taking advantage of the principle of biorecognition. The use of Cibacron Blue F3G-A and other triazine dyes as ligands in affinity chromatography and in related techniques has been successfully demonstrated in diverse isolation procedures of enzymes and proteins. In the last few years affinity partitioning became attractive for studying dye-protein interaction because of its high sensitivity in the recognition of ligand-protein binding and of its methodical simplicity.

In this contribution triazine dye-enzyme interaction was investigated by affinity partitioning of alkaline phosphatase (AP) from calf intestine and of phosphofructokinase (PFK) from baker's yeast as model enzymes. Due to their nucleotide binding both enzymes are potentially able to interact with triazine dyes. In addition, PFK is known to exhibit two distinct sites where ATP is bound as a substrate and as regulatory ligand, respectively, the latter being responsible for the ATP-inhibition of the enzyme. Hence, it was of particular interest to investigate the affinity partition of PFK as a function of its allosteric properties.

Affinity partitioning of aklakine phosphatase - At increasing amounts of dye-PEG in the system, the partition coefficient of AP increased and attained a saturation value. The enhancement of the partition coefficient, $\Delta \log K_{max}$ was determined for 42 individual dye-PEG derivatives. Three of them exhibit a $\Delta \log K_{max}$ higher than 1.4. In order to characterise the binding sites of the enzyme to Procion Red HE-3B the influence of selected effectors of AP on the dye-enzyme interaction was analysed in more detail. Inorganic phosphate and NADPH were found to compete most effectively with the dye. At 2mM, respectively, the affinity partitioning effect was completely abolished. NADH, 5'-AMP and 5'-ADP were also capable of weakening the dye-enzyme interaction but to a lesser extent. The reaction products of the nucleotides adenosine and nicotinamide, however, did not disturb the dye-enzyme complex formation. The same result was obtained when L-phenylalanine was added to the system.

Additionally, the influence of chromophores related to the structure of Procion Red HE-3B but being less complex were studied. Sulfanilic acid (SA) and 8-anilinonaphthalene-1-sulfonic acid (ANS) were found ineffective. In contrast, the chromophore 8-aminonaphthol-3,6-disulfonic acid (ANDS) competes with the dye.

From these results the following conclusion can be drawn: The active centre of AP is involved in the binding of Procion Red HE-3B since substrates and the competitive inhibitor phosphate are able to interfere with the dye-enzyme interaction. The differences in affinity partition of AP are due to different affinities to the dyes. A defined steric arrangement of the anionic groups of a dye seems to be a prerequisite for an optimal docking. In addition, the binding of Procion dyes to the enzyme is predictable if the molecule contains an aminonaphthodisulfonic acid part flanked symmetrically by aromatic ring systems carrying additional negatively charged residues.

Affinity partitioning of phosphofructokinase - Applying affinity partitioning to PFK from yeast a 1000-fold increase in the partition coefficient of the enzyme was observed if dye-PEG of high affinity was introduced into the system. Due to the allosteric properties of the enzyme fructose 6-phosphate is capable of shifting the R - T equilibrium in favour of the R-state thus relieving the inhibitory effect of ATP. If certain triazine dyes mimic the natural nucleotide and bind closely to the ATP-binding site(s) of PFK, any alteration of the allosteric properties of the enzyme accompanied by a change of the ATP inhibition should be paralleled by a change of the dye-enzyme interaction which is reflected in the actual affinity partition behaviour. This was demonstrated by monitoring the partition coefficient of the enzyme as a function of the concentration of fructose 6-phosphate. Thus, a drastic change in the enzyme partition was found in the range of 20 to 40 μM of the substrate. This concentration range coincides with the dissociation constant of the enzyme-fructose 6-phosphate complex. There are two possibilities to interpret this behaviour, either the binding of fructose 6-phosphate causes a steric hindrance of the complex formation with the dye, or, the substrate lowers the affinity to the dye by shifting the allosteric equilibrium towards the R-state.

PFK which is sensitive against inhibition by ATP (PFK_s) is reversibly convertible into an ATP-desensitive enzyme form (PFK_d). Therefore it was of interest to study this desensitisation in respect of the affinity partitioning behaviour of the enzyme. In the course of desensitisation the affinity of the enzyme to the dye decreased. The reversibility of enzyme conversion was indicated by a restitution of the original partition behaviour of PFK.

CONCLUSIONS

Affinity partitioning by applying triazine dyes is a powerful method for the study of PFK-ligand interaction. This is demonstrated by the detection of conformational changes induced by the desensitisation of the enzyme followed by its reconversion to the ATP-sensitive form. Binding of fructose 6-phosphate and of fructose, 2,6-bisphosphate favours the formation of an enzyme form of lower affinity to the dye as reflected by a decrease of the partition coefficient. Both results suggest the existence of two forms of PFK having different sensitivities to the inhibitory effect of ATP.

Chapter 2

Partitioning of Plant Membranes and Organelles

SEPARATION OF ORGANELLES AND MEMBRANES

BY PHASE PARTITIONING - A SHORT OVERVIEW

C. Larsson

Department of Plant Physiology
University of Lund
P. O. Box 7007, S-220 07 Lund, Sweden

INTRODUCTION

Disintegration of animal or plant tissue results in a complex mixture of intact organelles and membrane vesicles. To obtain a homogeneous preparation of a certain organelle or membrane is, therefore, usually a very demanding task, particularly since the organelles we want should not only be morphologically intact but also fully active, and the membrane vesicles should not only be homogeneous with respect to origin (including domains) but also with respect to orientation (sidedness), and, in addition, retain their activities. To fulfil these demands we can usually not rely on one method only but have to combine different methods separating according to different properties of the particles. This makes two-phase partitioning an extremely attractive method since it separates particles according to differences in surface properties [1], and therefore complements differential and gradient centrifugation which separate according to size and density.

Two-phase partitioning, in combination with centrifugation, has already been used for the preparation of a large number of animal and plant organelles and membranes [1-6]. Below I will try to draw some general conclusions from the available data and give some hints on how to use this very powerful method.

A general order in partition behaviour

Summarising the available data in the literature indicates a general pattern in partition behaviour of plant and animal membranes in dextran-PEG biphasic systems. Thus the affinity for the upper phase (the partition coefficient) increases in the following order:

inside-out thylakoids < intact chloroplasts < rough endoplasmic reticulum < mitochondria ~ peroxisomes ~ tonoplast < smooth endoplasmic reticulum <thylakoids ~ Golgi ~ lysosomes ~ inside-out plasma membranes <<< right-side-out plasma membranes

This suggests that the surface properties are similar for corresponding plant and animal membranes. The reason for these similarities are unknown, but may reside in similarities in the lipid head-group composition of corresponding membranes of even vastly different organisms [5]. Whatever the

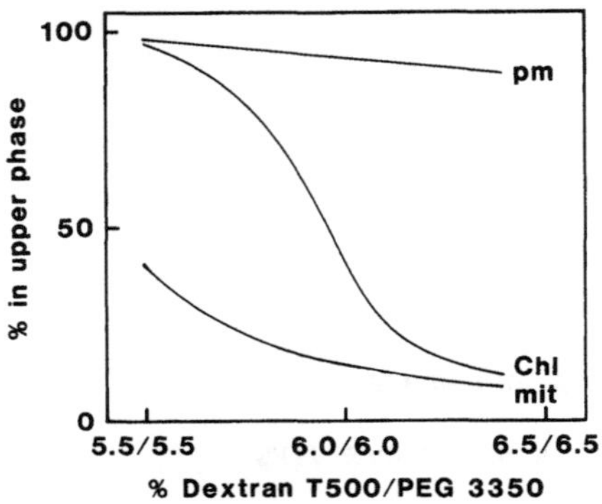

Fig. 1 *Effect of increasing polymer concentration on the partition of plasma membranes (pm), chloroplast thylakoids (Chl) and mitochondria (mit). In addition to polymers the phase systems contained 0.33 M sucrose, 2 mM KCl, and 5 mM potassium phosphate, pH 7.8. Keeping the polymer concentration constant at 5.8/5.8 and varying the chloride concentration from 0 to 5 mM results in similar curves (compare [7])*

explanation is, the knowledge alone of this general order in partition behaviour can serve as a guide in the practical work.

Hence, when the polymer (or Cl^-) concentration is increased the membrane vesicles collect at the interface in the order given above. By changing the polymer and/or chloride concentration, optimal conditions for separation may be found. This is done by partitioning the material in a series of phase systems with systematically altered polymer (or Cl^-) concentrations. This is illustrated in Fig. 1.

For isolation of leaf mitochondria, a polymer concentration of 5.8% was chosen [8]. At this polymer concentration, mitochondria collect at the interface, whereas the main contaminant, chloroplast thylakoids, partition in the upper phase together with plasma membranes. On the other hand, for the isolation of plasma membranes, a polymer concentration of 6.2 - 6.5% is more suitable [7]. At this higher polymer concentration, the plasma membranes are still recovered in the upper phase, whereas chloroplast thylakoids and mitochondria, together with all other intracellular membranes, partition to the interface or in the lower phase. This is the kind of exploratory experiment needed to optimise separation and should precede every attempt to use the method.

Membrane orientation and domains

For the separation of membrane vesicles differing in their orientation only, two-phase partitioning should be the method of choice, since such vesicles should not differ in density but definitely in surface properties, due to the transverse asymmetric arrangement of proteins and lipids. This difference in surface properties has been exploited for the separation of

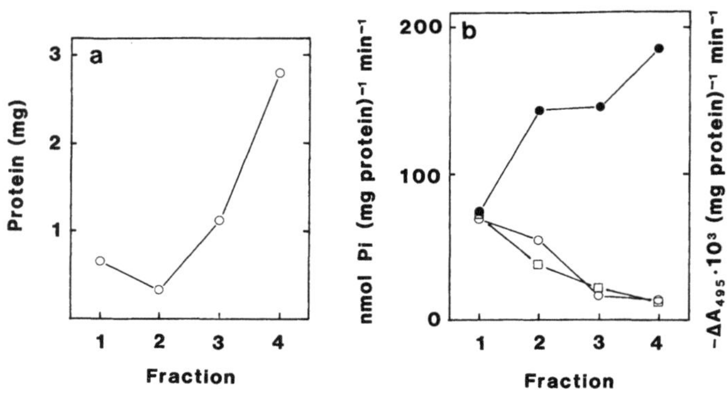

Fig. 2 a) *Distribution of protein, and b) distribution of ATPase activity (○,●) and ATP-dependent H^+-pumping (□) after countercurrent distribution of pure sugar beet plasma membranes in a phase system composed of 6.2% (w/w) Dextran T 500, 6.2% (w/w) polyethylene glycol 3350, 0.33 M sucrose, 5 mM KCl, 1 mM DTT, 5 mM potassium phosphate, pH 7.8. Three transfers of the upper phase were made by hand, resulting in four final fractions. The ATPase was assayed both in the presence (●) and absence (○) of Triton X-100 to permit calculation of ATPase latency (data from [12])*

inside-out and right-side-out membranes of erythrocytes [9], chloroplast thylakoids [10], mitochondria [11], and, just recently, plant plasma membranes [12].

Thus, highly purified plasma membranes obtained from leaves of sugar beet by two-phase partitioning were separated into two fractions by further phase partition steps [Fig. 2a]. The vesicles partitioning to the interface (fraction 1) showed an ATP-dependent H^+-uptake and a negligible latency of the H^+-translocating ATPase, while the vesicles partitioning in the upper phase (fractions 3 and 4) showed only slow H^+-uptake and a high ATPase latency (the active site is on the cytoplasmic surface of the membrane) on addition of detergent (Fig. 2b). Based on these results the material at the interface was estimated to contain ca 90% sealed, inside-out vesicles, and the material in the upper phase ca 90% sealed, right-side-out vesicles [12]. Using a centrifugal countercurrent distribution apparatus [13, see also 14] and 30 transfers (i.e. 10 times more than in Fig. 2 above), a complete separation of inside-out and right-side-out plasma membrane vesicles was obtained, with the peak of inside-out vesicles in fraction 2 and the peak of right-side-out vesicles in fraction 28, and with no material at all in fractions 6 to 24 (C. Larsson, S. Widell and P-Å Albertsson, unpublished). Such an experiment is completed within 1 h and demonstrates how powerful two-phase partitioning may be in resolving vesicle populations differing in sidedness only. A partial separation of inside-out and right-side-out plant plasma membrane vesicles was first achieved by free-flow electrophoresis [15], which showed that the preparations obtained by two-phase partitioning contained a minor proportion of inside-out vesicles. However, judging from the partitioning of the inside-out vesicles (to the interface), they are most likely formed from right-side-out vesicles in the further handling of the plasma membrane preparation (pipetting / pelleting / resuspension / pottering / freezing / thawing) and not present in freshly prepared (and carefully handled) preparations. According to theory, separation by free-flow electrophoresis is linearly related to the difference in surface net charge density, whereas separation by phase partitioning is rather exponentially related to the same difference [1]. This may explain the observed difference in resolving capacity. Alternatively, the explanation may be that the separation of inside-out and right-side-out plasma membrane vesicles is not based on differences in surface charge but rather on differences in hydrophobic/hydrophilic properties. Interestingly, with free-flow electrophoresis inside-out plasma membrane vesicles are well resolved from the bulk of intracellular membranes, in contrast to right-side-out vesicles [15], whereas with two-phase partitioning right-side-out plasma membrane vesicles are much better resolved than inside-out ones. This suggests that the basis for separation in this case differs between free-flow electrophoresis and two-phase partitioning, and that also factors other than charge determine the partitioning of these membrane vesicles.

The results shown in Fig. 2 stress the need to measure activities also in the presence of a detergent. This is not only important in subfractionation experiments using "pure" fractions but as important in separation experiments with crude membrane fractions (e.g. microsomal fractions) since a membrane population may be completely overlooked if membrane orientation is not taken into account.

An additional factor which may give rise to a heterogeneous behaviour of markers and should be taken into account is the lateral asymmetry of membranes, i.e. the existence of membrane domains. This is demonstrated by the recent separation of rat liver plasma membrane domains [16,14]. Furthermore, lateral asymmetry often affects membrane orientation such that certain domains have a high probability to form right-side-out vesicles while other domains preferably form inside-out vesicles. This is the case with the chloroplast thylakoid membrane, where the appressed, photosystem II regions form inside-out vesicles, and the non-appressed, photosystem I regions form right-side-out vesicles upon fragmentation [10].

REFERENCES

1 P.-Å. Albertsson, Partition of Cell Particles and Macromolecules, 3rd edition, John Wiley & Sons, New York (1986)

2 S.D. Flanagan, Partitioning of animal membranes and organelles, in: "Partitioning in Aqueous Two-Phase Systems: Theory, Methods, Uses and Applications to Biotechnology," H. Walter, D.E. Brooks and D. Fisher, eds., Academic Press, Orlando, FA (1985)

3 A. Mircheff, Empirical strategy for analytical fractionation of epithelial cells, Am. J. Phys. 244:G347 (1983)

4 P.-Å. Albertsson, B. Andersson, C. Larsson and H.-E. Åkerlund, Phase partition - a method for purification and analysis of cell organelles and membrane vesicles, in: "Methods of Biochemical Analysis," D. Glick, ed., John Wiley & Sons, New York (1982)

5 C. Larsson, Partition in aqueous polymer two-phase systems: A rapid method for separation of membrane particles according to their surface properties, in: "Isolation of Membranes and Organelles from Plant Cells," Academic Press, London (1983)

6 C. Larsson, B. Andersson and H.-E. Åkerlund, Partitioning of plant cells, cell walls, membranes and organelles, in: "Partitioning in Aqueous Two-Phase Systems: Theory, Methods, Uses and Applications to Biotechnology," H. Walter, D.E. Brooks and D. Fisher, eds., Academic Press, Orland, FA (1985)
7 C. Larsson, S. Widell and P. Kjellbom, Preparation of high-purity plasma membranes, Methods Enzymol. 148:558 (1987)
8 P. Gardeström, I. Ericson and C. Larsson, Preparation of mitochondria from green leaves of spinach by differential centrifugation and phase partition, Plant Sci. Lett. 13:231 (1978)
9 T.L. Steck, Preparation of impermeable inside-out and right-side-out vesicles from erythrocyte membranes, in: "Methods in Membrane Biology," Plenum Press, New York (1974)
10 J.M. Anderson and B. Andersson, The architecture of photosynthetic membranes: Lateral and transverse organisation, Trends Biochem. Sci. 7:288 (1982)
11 I.M. Moller, A. Bergman, P. Gardeström, I. Ericson and J.M. Palmer, Characterisation and purification of inside-out submitochondrial particles obtained from plant mitochondria, FEBS Lett. 126:13 (1981)
12 C. Larsson, S. Widell and M. Sommarin, Inside-out plant plasma membrane vesicles of high purity obtained by aqueous two-phase partitioning, FEBS Lett. 229:289 (1988)
13 H.-E. Åkerlund, An apparatus for countercurrent distribution in a centrifugal acceleration field, J. Biochem. Biophys. Methods 9:133 (1984)
14 P. Gierow, B. Jergil and P.-Å. Albertsson, Two-phase partitioning of rat liver plasma membranes, in: "Separations Using Aqueous Phase Systems: Applications in Cell Biology and Biotechnology," D. Fisher and I.A. Sutherland, eds., Plenum, New York (1989)
15 H. Canut, A. Brightman, A. Boudet and D.J. Morré, Free-flow electrophoresis resolves plasma membrane vesicles isolated by aqueous two-phase partition into two populations of different sidedness, in: "Separations Using Aqueous Phase Systems: Applications in Cell Biology and Biotechnology," D. Fisher and I.A. Sutherland, eds., Plenum, New York (1989)
16 P. Gierow, B. Jergil and P.-Å. Albertsson, Lateral heterogeneity of rat liver plasma membranes analysed by countercurrent distribution, Biochem. J. 249:369 (1988)

PLASMA MEMBRANE ISOLATION BY PHASE PARTITION USING A SINGLE PARTITION STEP: A TOOL FOR THE *IN VIVO* STUDY OF LIPID TRANSFER TO THE PLASMA MEMBRANE

Patrick Moreau, Hélène Juguelin, René Lessire and Claude Cassagne

Institut de Biologie Cellulaire et de Neurochimie du CNRS
1 rue Camille Saint Saëns 33077 Bordeaux Cedex, France

The intermembrane transfer of very long chain fatty acids (VLCFA) from a light membrane fraction (1.08-1.09 g/cm^3) to heavier membranes (1.16-1.18 g/cm^3) was demonstrated in etiolated leek seedlings [1]. By means of biochemical markers, it was determined that the plasma membrane is confined to the 1.16-1.18 g/cm^3 fraction which is, nevertheless, heterogeneous. The plasma membrane was further purified in an aqueous two-polymer phase system, its characterisation is presented. Following pulse-chase experiments, an increase of labeled fatty acids (C16 to C24) in the plasma membrane occurred during the chase, demonstrating the transfer of these molecules to the plasma membrane.

INTRODUCTION

Saturated very long chain fatty acids (more than 18 carbon atoms, VLCFA) are characteristic components of the plasma membrane of leek epidermis cells [2] and may represent up to 20% of the total fatty acid content of this membrane. It was demonstrated that, *in vitro* these molecules are principally synthesized in the endoplasmic reticulum since the bio-synthesis in the ER is 3 to 4 times higher than in the plasma membrane. These results led to the hypothesis of an intracellular transfer of the VLCFA from the ER to the plasmalemma [3].

A methodology was devised to study the existence of intermembrane transfer events of lipids and of VLCFA by seven-day-old etiolated leek seedlings *in vivo* [1]. By *in vivo* pulse-chase experiments, followed by membrane subfractionation of microsomal pellets on linear sucrose gradients, intermembrane transfer events of VLCFA between a light membrane fraction (1.08-1.09 g/cm^3) and the heaviest one (1.16-1.18 g/cm^3) were demonstrated [1]. However, these results did not establish that VLCFA were transferred to the plasma membrane. Effectively, the 1.16-1.18 g/cm^3 fraction contains almost all the membrane vesicles originating from the plasma membrane, but it is contaminated by membrane vesicles of the ER, the Golgi complex and perhaps also the tonoplast.

In order to study the transfer of VLCFA to the plasmalemma further, the purification of this membrane was undertaken by phase partition [4,5].

This method allows to prepare highly purified plasma membrane fractions [4,5], but two or three cycles are generally required to achieve this purification. However, this time-consuming multistep purification is hardly compatible with the short-term pulse-chase transfer studies because of the eventual label redistribution which could occur during the purification. In addition, because of the membrane losses inherent to the technique, the multistep purification is only suitable for the preparation of plasma membrane from large amounts of starting material, which is not always possible when numerous samples have to be handled at the same time. Consequently, it is necessary to evaluate whether a single phase partition yields a reasonably clean plasma membrane fraction from small quantities of membranes.

We show in this paper that a purified plasma membrane fraction may be obtained from etiolated leek seedlings after a single phase partition step of the cell membranes. In a second part, it is shown that this procedure, after *in vivo* pulse-chase experiments, allows us to study the lipid transfer to the plasmalemma of etiolated leek seedlings.

EXPERIMENTAL

Plant material: Leek (*Allium porrum L.*) seeds, stored overnight at 4°C, were sterilised with sodium hypochlorite in the presence of Triton X-100 for 5 min and then washed with distilled water. They were then grown for 7 days in the dark, at 25°C, on a growth medium containing: 5 g agaragar, 900 ml of distilled water and 100 ml of a nutritive solution containing 7.5 g/l KCl, 6 g/l $NaNO_3$, 4.5 g/l $MgSO_4$, 0.95 g/l $CaCl_2$ and 1.25 g/l NaH_2PO_4.

Isolation of plasmalemma: Seedling homogenisation and membrane subfractionations were performed in 0.08 M Hepes buffer, pH 7.8 as previously described [6]. Membranes of the microsomal pellet, or of the membrane fractions obtained on sucrose gradients, were homogenised in the following mixture: Dextran T500 5.6% (w/w), poly(ethylene) glycol (PEG 4000) 5.6% (w/w) in a 0.5 M sorbitol buffer, pH 7.8, containing KH_2PO_4 (10 mM) and NaCl (40 mM). After centrifuging for 10 min at 1000g, the PEG-enriched upper phases were recovered without disturbing the interfaces, which were recovered with the Dextran-enriched lower phases. Membranes were then recovered after centrifugation at 200 000g for 60 mn and resuspended in the appropriate buffer [6]. Marker enzymes were measured as described [6].

Lipid analysis: Lipid extraction and analysis were performed according to Heape et al [7]. Quantitation of lipids was carried out from calibration curves that were established using standard lipids [8].

Fatty acid analysis: Two types of fatty acid analyses were performed. The first was realised with an Intersmat IGC 120 DFL chromatograph fitted with a 10% CP-SIL 58 column (1/8" x 2 m) on WHP (100-200 mesh size), allowing the separation of mono-and poly-unsaturated fatty acids from saturated fatty acids. The second one was realized with a Packard 428 chromatograph fitted with a capillary column, allowing a better quantitation of saturated fatty acids containing up to 30 carbon atoms. For the two types of analysis, the identification of the various fatty acids was performed by comparison with methyl ester standards.

***In vivo* transfer of fatty acids to the plasmalemma**: *In vivo* (1-^{14}C) acetate (55 Ci/mol) incorporations and pulse-chase experiments were performed as described [1]. Plasma membrane isolation was carried out as indicated above. Fatty acid radioactivities were then determined by radio GLC analysis as described previously [1].

RESULTS AND DISCUSSION

Purification of the Plasma Membrane by Phase Partition

The phase partition was realised using either the heavier 1.16-1.18g/cm^3 fraction, or the microsomal pellet. Phase partition allows the isolation of the plasma membrane by making use of its surface charge properties [9]. The plasmalemma has a greater affinity for the PEG-enriched upper phase, whereas other membranes remain confined to the dextran-enriched lower phase and the interface. This one-step method was found to be relatively quantitative, since 85% of the total labeled lipids, or 81% of the total proteins of the initial membranes were recovered in the upper and lower phases after such a partition (Moreau et al, unpublished results). The recovery of proteins after phase partitioning (Table 1) shows that, despite a 20% loss of membranes during this process, the proportion of membranes recovered in the different phases is reproducible (90.4% ± 2.6 in the lower phase and 9.6% ± 2.6 in the upper phase).

Table 1 *Biochemical data showing the purification of the plasma membrane in the upper phase after phase partition realised from the fraction D, or from the microsomal pellet*

	Lower phase	Upper phase
Glucane synthetase II (nmol/min/mg prot.),N=4	3.1 ± 1.5 (71.5%)	11.6 ± 6 (28.5%)
Cholinephosphotransferase (pmol/min/mg prot.),N=3	12.3 ± 2.9 (98.5%)	1.8 ± 1.1 (1.5%)
Inosine diphosphatase (μmol/hr/mg prot.)	3.4 (96%)	1.3 (4%)
ATPase (nmol/min/mg prot.),N=2		
+ $MgSO_4$(3mM)	11.8	27.5
+ $MgSO_4$(3mM) + KCl(50mM)	15.3	74
+ $MgSO_4$(3mM) + KCl(50mM)		
+ Na_3VO_4(50μM)	7.7	5
+ NO_3(10mM)	8.2	66.7
Proteins (%), N=11	90.4 ± 2.6	9.6 ± 2.6

The data given in brackets show the distribution of each activity between the lower and the upper phases. N values give the number of experiments that were realised. The results obtained for the fraction D and for the microsomal pellet were similar and were therefore considered together.

Glucane synthetase (GS II) specific activity (nmol/min/mg) is 4 times higher in the upper phase than in the lower phase, whereas cholinephosphotransferase and inosine diphosphatase (IDPase) activities are essentially located in the lower phase and at the interface (98.5% and 96% respectively). From the total GS II activities, we can estimate that the distribution of plasmalemma vesicles between the upper and lower phases is about 70% in the lower phase and only 30% in the upper phase. However, only 1.5% and 4% of the cholinephosphotransferase and IDPase activities respectively are found in the upper phase. Cholinephosphotransferase activity can be considered as an ER marker, but has also been found in Golgi membrane [10] and mitochondria [10]. Latent IDPase activity is mainly considered as being a Golgi marker [11]. Consequently, these results show that the upper phase is enriched with plasma membrane vesicles, and weakly contaminated by ER and Golgi membranes.

ATPase activity (Table 1) was assayed according to Sze [12]. Firstly, stimulation of the ATPase activity by K+ is greater in the upper phase (+ 169%) than in the lower phase (+ 30%). This stimulation by K+ has been observed essentially for plasma membranes [13,14] and is considered to be a marker for this membrane. The inhibition of the K+-stimulated ATPase activity by vanadate is 93.3% in the upper phase and 49.5% in the lower phase. The inhibition of the ATPase activity by vanadate is specific to the plasmalemma ATPase [12], so an inhibition of 93.3% in the upper phase demonstrates a great enrichment in plasma membrane vesicles. ATPase activity assayed in the presence of nitrate is only inhibited by 9.9% in the upper phase and by 46.3% in the lower phase. Since only tonoplastic and mitochondrial ATPases [12] are affected by this inhibition, the enriched plasma membrane fraction is only weakly contaminated by tonoplastic and mitochondrial membranes.

Table 2 shows the lipid and fatty acid analyses of the lower phase and of the plasma membrane fraction. The sterol and VLCFA contents of the plasma membrane fraction and the degree of saturation of its fatty acids are higher than those of the membranes of the lower phase. These results are in good agreement with the enrichment of the upper phase in plasma membrane vesicles. Indeed, plasmalemma vesicles isolated from maize coleoptiles contain more sterols [15] and VLCFA [16] than do other

Table 2 *Lipid and fatty acid analyses of the lower and upper phases*

Lipids (μg/mg proteins)	Lower Phase	Upper Phase	**Fatty acids** (%)	Lower Phase	Upper Phase
PC	146	94	$C_{14:0}$	0.4	0.2
PI	19	18	$C_{16:0}$	23	39.6
PE	85	74	$C_{18:0}$	0.7	1.2
DGDG	50	12.5	$C_{18:1}$	6.5	5.3
MGDG	70	2.5	$C_{18:2}$	61.9	46.1
Sterols	30	36.5	$C_{18:3}$	5.4	1.2
TG	15	6	$C_{20:0}$	-	0.3
phospholipids / galactolipids	1.9	11.5	$C_{22:0}$	1.1	2.6
			$C_{24:0}$	1	2.6
Total	415	243.5	$C_{26:0}$	-	0.7

PC: phosphatidylcholine, PI: phosphatidylinositol
PE: phosphatidylethanolamine, DGDG: digalactosyldiacylgycerol,
MGDG: monogalactosyldiacylglycerol, TG: triacylglycerol

membranes. Moreover, the amounts of saturated fatty acids are higher in the plasma membrane than in endoplasmic reticulum, or intracellular organelles [16]. In the same manner, plasma membrane vesicles isolated from leek epidermis cells contain more VLCFA than any other membrane [16].

We also found that PE is more abundant in the plasma membrane fraction (30.4%) than in the membranes of the lower phase (20.5%). The galactolipids, DGDG and MGDG, represent only 5.1% and 1% respectively of the total lipids in the plasma membrane fraction. The presence of DGDG (5.1%) in the plasma membrane fraction could be due to a contamination by plastid membranes, but galactolipids have been observed in the plasmalemma [17,18]. The relatively low amount of lipids found in the plasma membrane is probably due to the method of lipid analysis. Indeed, the revelation of lipids is dependent on their degree of unsaturation [7,8], as this degree is low in the case of the plasma membrane (Table 2), the amount of lipids has probably been underestimated. All the biochemical data (marker enzymes and lipid and fatty acid compositions) clearly show that the upper phase is greatly enriched in plasma membrane vesicles after only one step, and can now be used to study the transfer of lipids and fatty acids to the plasmalemma.

In order to study the eventual effect of the polymers (dextran and PEG) on the membranes, the experiment shown in Fig. 1 was realized. The results are shown in Figs. 2a and 2b. It was only possible to realise these experiments with the membrane fractions B (1.10-1.11 g/cm^3), C (1.13-1.14 g/cm^3) and D (1.16-1.18 g/cm^3). For the fractions B and C (Fig. 2b), all the membranes are recovered in the lower phase and, reloaded onto density sucrose gradients, sediment to the same density. For the fraction D (Fig. 2b), membranes recovered in the lower phase show a density of 1.16-1.18 g/cm^3 on sucrose gradients and the plasma membrane-enriched upper phase loaded onto sucrose gradients shows a density of 1.16-1.17 g/cm^3. After phase partition of the microsomal pellet (Fig. 2a), the plasma membrane-enriched upper phase, loaded onto a sucrose gradient, shows a density of 1.17-1.18 g/cm^3, and we note a small shift between the two plasma membrane fractions. The membranes of the lower phase (Fig. 2a) were not separated on sucrose gradients and a peak of membranes was obtained at a density of 1.14 g/cm^3. Consequently, it seems that the phase partition could modify the density of the membranes. It is not impossible that a partial fusion of the membrane vesicles is facilitated by a dehydration by the polymers, explaining the peak of membranes obtained at a density of 1.14 g/cm^3 (Fig. 2a). Nevertheless, the polymer concentrations do not appear to be sufficient for such phenomena to be important [19]. However, the results obtained for the plasma membrane fractions show that the effect of the polymers is only minor in this case.

***In vivo* transfers of lipids and fatty acids to the plasma membrane**: After pulse-chase experiments carried out as described previously [1], the plasma membrane-enriched fraction was prepared. Lipid and fatty acid analyses were then performed according to the experimental section. Phosphatidylcholine, phosphatidylethanolamine and the neutral lipids (including sterols) were transferred to the plasma membrane *in vivo* [20]. Table 3 shows the fatty acid label of the plasma membrane as a function of the chase time. After a chase of 120 min, the amounts of labelled C16, C18, C22 and C24 fatty acids are increased by 2, 3, 4 and 3 times respectively. Consequently, an enrichment of the plasma membrane in VLCFA is observed during the chase. It has been shown that the plasma membrane is unable to synthesize these molecules [6]. These results clearly demonstrate that VLCFA have been transferred to the plasma membrane *in vivo*.

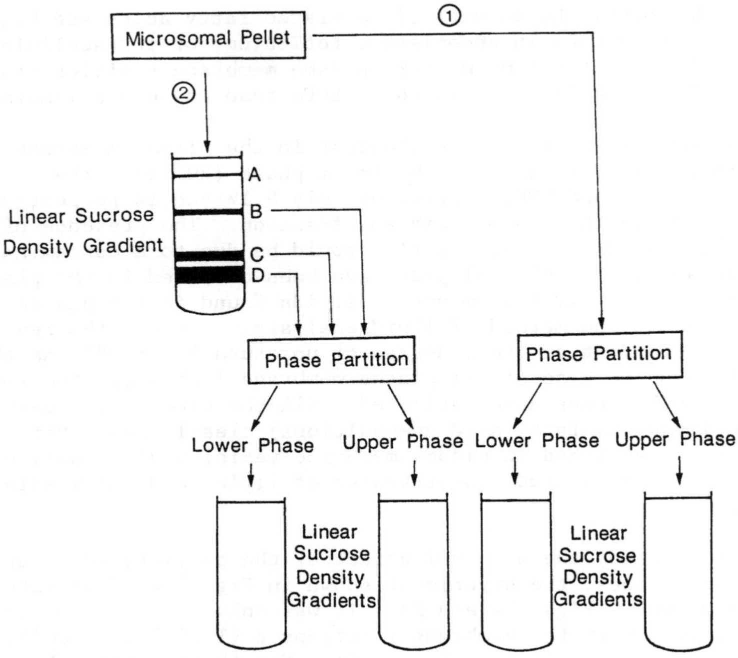

Fig. 1 *Membrane purification steps by phase partition: 1) Phase partition of the microsomal pellet; 2) Phase partition of the membrane fractions isolated on sucrose density gradients*

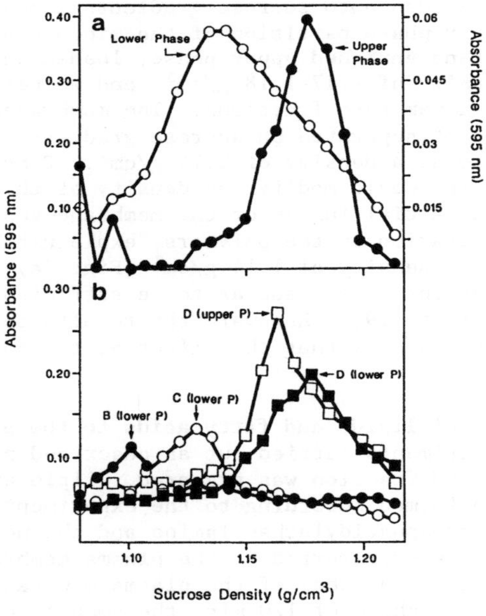

Fig. 2 *Centrifugation on linear sucrose density gradients of the membranes obtained after phase partition of the microsomal pellet (a), or of the membrane fractions B(1.10-1.11 g/cm*^{3}*), C(1.13-1.14 g/cm*^{3} *and D(1.16-1.18 g/cm*^{3}*), (b)*

Table 3 Fatty acid label of the plasma membrane-enriched fraction as a function of the chase time following a two-hour labeling period

Chase time (min)	C_{16}	C_{18}	C_{20}	C_{22}	C_{24}
0	81	38	0	4	5
30	98	36	0	8	3
60	135	46	2	8	9
120	165	110	2	17	14

The results are expressed as cpm x 10^{-3}/mg proteins

CONCLUSION

The phase partition method, with several partition steps, allows the preparation of highly purified plasma membrane fractions for the analyses of the protein composition [9,21], the lipid composition [22,23] or to characterize an enzymatic activity of the plasma membrane [24]. However, the time required for such a procedure is long and unsatisfactory for the study of transfer events of lipids and fatty acids to the plasma membrane. The results presented in this paper demonstrate the possibility of obtaining a plasma membrane fraction which is only slightly contaminated by other membranes by a single phase-partition step. The loss of membranes and their modification by the polymers are reduced. The only requirement is that small amounts of membranes be used. After pulse-chase experiments, the rapid isolation of the plasma membrane allows of study plasma membrane biogenesis and, for example, the transfer of lipids [6,20], and fatty acids to this membrane.

REFERENCES

1 P. Moreau, H. Juguelin, R. Lessire and C. Cassagne, Intermembrane transfer of long chain fatty acid synthesised by etiolated leek seedlings, Phytochem. 25:387 (1986)

2 C. Cassagne and R. Lessire, Biosynthesis of saturated very long chain fatty acids by purified membrane fractions from leek epidermal cells, Arch. Biochem. Biophys. 191:146 (1978)

3 R. Lessire, T. Abdul-Karim and C. Cassagne, Origin of the wax very long chain fatty acids in leek, Allium porrum L., leaves: a plausable model, in: "The Plant Cuticle", D.F. Cuttler, K.L. Alvin and C.E. Price, eds., Academic Press, London, 167 (1982)

4 C. Larsson, P. Kjellbom, S. Widell and T. Lundborg, Sidedness of plant plasma membrane vesicles purified by partitioning in aqueous two phase systems, FEBS Lett. 171:271 (1984)

5 S. Yoshida, M. Uemura, T. Niki, A. Sakai and L.V. Gusta, Partition of membrane particles in aqueous 2 - polymer phase system and its practical use for purification of plasma membranes from plants, Plant Physiol. 72:105 (1983)

6 P. Moreau, Thèse de Doctorat de l'Université de Bordeaux II (1986)

7 A.M. Heape, H. Juguelin, F. Boiron and C. Cassagne, Improved one-dimensional thin-layer chromatographic technique for polar lipids, J. Chromatogr. 322:391 (1985)

8 A.M. Heape,H. Juguelin, M. Fabre, F. Boiron, B. Garbay, M. Fournier, J. Bonnet and C. Cassagne, Correlation between the morphology and the lipid and protein composition in the peripheral nervous system of individual 8-day old normal and trembler mice (BRD 50344), Dev. Brain Res. 25:173 (1986)
9 L.E. Korner, P. Kjellbom, C. Larsson and I.M. Moller, Surface properties of right side-out plasma membrane vesicles isolated from barley roots and leaves, Plant Physiol. 79:72 (1985)
10 A. Sauer and D.G. Robinson, Subcellular localisation of enzymes involved in lecithin biosynthesis in maize roots, J. Exp. Bot. 36:1257 (1985)
11 P.H. Quail, Plant Cell fractionation, Ann. Rev. Plant Physiol. 30:425 (1979)
12 H. Sze, H+ - translocating ATPases: advances using membrane vesicles, Ann. Rev. Plant Physiol. 36:175 (1985)
13 T.K. Hodges, R.T. Leonard, C.E. Bracker and T.W. Keenan, Purification of an ion-stimulated adenosine triphosphatase from plant roots: association with plasma membrane Proc. Natl. Acad. Sci. USA 69:3307 (1972)
14 R.L. Travis and M.L. Booz, Partial characterisation of a potassium-stimulated adenosine triphosphatase from the plasma membrane of meristematic and mature soybean root tissue, Plant Physiol. 63:573 (1979)
15 M.A. Hartmann, G. Normand and P. Benveniste, Sterol composition of plasma membrane enriched fractions from maize coleoptiles, Plant Sci. Lett. 5:287 (1975)
16 R. Lessire, Thèse de Doctorat es Sciences, Bordeaux (1981)
17 P. Mazliak, Glyco- and phospholipids of biomembranes in higher plants, in: "Lipids and Lipids Polymers in Higher Plants", M. Tevini and H.K. Lichtenthaler, eds., Springer-Verlag, Berlin (1977)
18 J.L. Harwood, Plant acyl lipids structure distribution and analysis, in: "The Biochemistry of Plants" (1980) P.K. Stumpf and E.E. Conn, eds., Academic Press, London
19 R.I. MacDonald, Membrane fusion due to dehydration by polyethylene glycol, Dextran, or sucrose, Biochem. 24:4058 (1985)
20 P. Moreau, H. Juguelin, R. Lessire and C. Cassagne, Study of the intracellular transfer of lipids to the plasmalemma, in: "The Metabolism, Structure and Function of Plant Lipids", P.K. Strumpf, J.B. Mudd and W.D. Nes, eds., Plenum Press, New-York (1987)
21 P. Kjellbom and C. Larsson, Preparation and polypeptide composition of chlorophyll-free plasma membranes from leaves of light-grown spinach and barley, Physiol. Plant 62:501 (1984)
22 S. Yoshida and M. Uemura, Lipid composition of plasma membranes and tonoplasts isolated from etiolated seedlings of mung bean (Vigna radiata L), Plant Physiol. 82:807 (1986)
23 C.P. Rochester, P. Kjellbom, B. Andersson and C. Larsson, Lipid composition of plasma membranes isolated from light-grown barley (Hordeum vulgare) leaves: identification of cerebroside as a major component, Arch. Biochem. Biophys. 255:385 (1987)
24 J.D. Clement, J.P. Blein, J. Rigaud and R. Scalla, Characterisation of ATPase from maize shoot plasma membrane prepared by partition in an aqueous polymer two phase system, Physiol. Veg. 24:25 (1986)

FREE-FLOW ELECTROPHORESIS RESOLVES PLASMA MEMBRANE VESICLES ISOLATED BY AQUEOUS TWO PHASE PARTITION INTO TWO POPULATIONS OF DIFFERENT SIDEDNESS

Hervé Canut, Andrew Brightman, Alain Boudet and D. James Morré

Department of Medicinal Chemistry and Pharmacognosy
Purdue University, West Lafayette, IN 47907, USA and
Centre de Physiologie végétale
Université de Toulouse, France

Transport and other studies with isolated plasma membrane vesicles would be facilitated considerably by the availability of purified preparations of known absolute orientations. In this study, we have used ATPase latency [1] and concanavalin A binding [2] to conclude that vesicles obtained in the upper phase by aqueous two-phase partitioning were mixtures of cytoplasmic side out and cytoplasmic side in vesicles. We used the same approach to demonstrate that free-flow electrophoresis fractions nearest the point of sample injection contained purified plasma membrane of cytoplasmic side out orientation, whereas electrophoretic fractions further from the point of sample injection contained plasma membrane vesicles of the opposite, cytoplasmic side in orientation. Used in series, the two techniques of aqueous two-phase partition and free-flow electrophoresis yielded purified homogeneous fractions of plasma membrane vesicles of opposite orientation.

Plasma membrane vesicles were isolated from soybean (*Glycine max L. Merr.*) hypocotyls. Seeds were germinated in darkness and, after 4 to 5 days, the cotyledons and roots were removed and the hypocotyl portions homogenised comparing different procedures (with mortar and pestle, razor blade chopping, Polytron/Waring blender). The plasma membrane vesicles were purified either using a 16 g two-phase system composed of 6.4% (w/w) Dextran T500 (Pharmacia), 6.4% (w/w) poly(ethylene glycol) 3350 (Fisher), 0.25 M sucrose and 5 mM potassium phosphate, pH 6.8 [3] or by the method of preparative free-flow electrophoresis [4]. Determination of vesicle sidedness included both concanavalin A binding [2] as an imposed label and ATPase latency [1].

The upper phase of the aqueous two-phase system contained pure plasma membranes consisting of a mixture of cytoplasmic side out and cytoplasmic side in vesicles. Purity of the fractions was determined by assays of marker enzymes and by a plasma membrane-specific staining procedure utilising phosphotungstic acid at low pH [5]. The starting fractions were more than 90% plasma membrane derived.

When the plasma membrane vesicles, first obtained by aqueous two-phase partition, were subjected next to preparative free-flow electrophoresis, two populations of vesicles were obtained. The least electronegative fraction had low ATPase latency and low concanavalin A binding and corresponded to cytoplasmic side out vesicles. The most electronegative fraction had a high ATPase latency (stimulated 7-fold by addition of detergent) and exhibited a strong reaction with concanavalin A bound to peroxidase when examined in the electron microscope. These vesicles thus corresponded to the cytoplasmic side in orientation.

The relative proportions of the two types of vesicles varied according to the method of tissue homogenisation. Mashing with a mortar gave a higher proportion of cytoplasmic side out vesicles than did the other methods of homogenisation.

Thus, both free-flow electrophoresis and aqueous two-phase partition, while providing separations based on charge density, subfractionate membranes somewhat differently such that perhaps hydrophobic interactions may be more important for two-phase separations. In any event, the aqueous two-phase preparations in our experience are not homogeneous with respect to vesicle orientation when prepared from soybean microsomes. However, when the aqueous two-phase fractions are subjected to free-flow electrophoresis or vice versa, that is when the two methods are used in series in either order, fractions greatly enriched in apparently homogeneously oriented plasma membrane vesicles are obtained [6].

Work supported by the National Institutes of Health CA 36761 and the National Science Foundation/Centre National de la Recherche Scientifique Exchange Program.

REFERENCES

1 C. Larsson, P. Kjellbom, S. Widell and T. Lundborg, Sidedness plant plasma membrane vesicles purified by partitioning in aqueous two-phase systems, FEBS Lett. 171:271 (1984)
2 W.F. Boss and A.W. Ruesink, Isolation and characterisation of Concanavalin A labelled plasma membranes from carrot suspension culture cells, Plant Physiol. 66:835 (1979)
3 P. Kjellbom and C. Larsson, Preparation and polypeptide composition of chlorophyll-free plasma membranes from leaves of light-grown spinach and barley, Physiol. Plant 62:501 (1984)
4 A.S. Sandelius, C. Penel, G. Auderset, A. Brightman, M. Millard and D.J. Morré, Isolation of highly purified fractions of plasma membrane and tonoplast from the same homogenate of soybean hypocotyls by free-flow electrophoresis, Plant Physiol. 81:177 (1986)
5 J.C. Roland, C.A. Lembi and D.J. Morré, Phosphotungstic acid-chromic acid as a selective electron-dense stain for plasma membranes of plant cells, Stain Technol. 47:195 (1971)
6 H. Canut, A. Brightman, A.M. Boudet and D.J. Morré, Plasma membrane vesicles of opposite sidedness from sobean hypocotyls by preparative free-flow electrophoresis, Plant Physiol. 86:631 (1988)

RAISING MONOCLONAL ANTIBODIES AGAINST PLASMA MEMBRANES FROM BARLEY EMBRYOS

Ole Schou

Agricultural Research Department
Riso National Laboratory
P.O. Box 49, DK-4000 Roskilde, Denmark

In vitro propagation and genetic manipulation of plants have vast potentialities for modern breeding. Cultured cells, transgenic or of natural genotype, may be induced to form somatic embryos, leading to a final regeneration into plants with the desired characters. However, regeneration competence is not inherent in every cell, and with the fact that several important crop species are still recalcitrant this calls for studies of embryogenesis and cell differentiation. Barley (*Hordeum vulgare L.*) presents severe obstacles to regeneration, as do most of the cereals. Besides trying out different regeneration techniques for this species [1], molecular markers specific to embryongenic cells are searched for. Some of the markers are likely to be present on the cell surface, and hence the plasma membrane (PM) is to be studied, in our case by means of immunological methods using monoclonal antibodies.

Two weeks old zygotic embryos from *H. vulgare* var. Golden Promise are used as the model material. Due to the lack of detectable marker enzymes in the PM fraction of embryos, the isolation procedure has been set up with green leaves. It is based on partitioning in a 36 g aqueous Dextran-poly(ethylene glycol) two-phase system, essentially as described in [2], though with a ten-fold higher concentration of salts and a batch procedure of three steps. With leaves, a large part of the PM marker, K^+-stimulated vanadate-sensitive ATPase, is recovered in the upper phase (U_3), with an activity of 710 nmol P_i min^{-1}, vs 549 in the lower phase (L_1). Intracellular membranes are virtually absent in U_3, as indicated by the lack of chlorophyll and cytochrome c oxidase, a chloroplast and mitochondria marker, respectively [3]. This is supported by the depletion of prominent chloroplast polypeptides in the U_3 relative to L_1, as seen from the silver stained polypeptide pattern of the fractions in SDS-PAGE gels.

Contrary to the leaves, embryos are homogenised directly in the phase system in order not to waste material. The U_3 is regarded as the PM fraction. It is recovered by centrifugation at 10^5 x g for one hour, and the pellet contains about 0.8 mg protein as the result of a typical preparation starting with 3 g of embryos.

For immunisation the PM pellet is resuspended in phosphate-buffered saline pH 7.4. Female, six weeks old Balb/c mice are immunised with 100 g protein/200 1 buffer emulsified in Freund's complete adjuvant, followed by boosts on days 7, 14, 42 and 70 with the same amount of immunogen, but in buffer only. The formation and culture of antibody secreting hybridomas follow normal procedures [4].

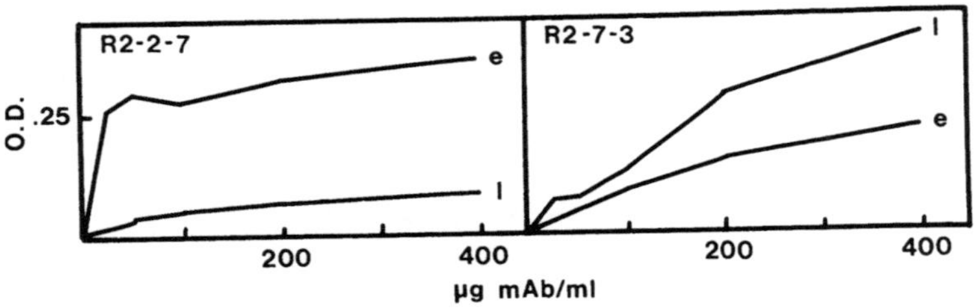

Fig. 1 *Solid-phase ELISA testing of monoclonal antibodies (R2-7-3 and R2-2-7) against embryo (e) and leaf tissue (l) plasma membranes*

A solid-phase ELISA for testing and selection of positive hybridomas has been set up, which is based on the reaction of antibodies with the PM fraction. The assay takes advantage of the strong adherence of PM vesicles to the wells of plastic microtitre plates. The PM fraction, resuspended in immunisation buffer, are added to the wells at a concentration of 1 μg protein/100 μl/well. After incubation overnight, the vesicles that have adhered to the wells are in turn incubated with monoclonal antibodies and goat anti-mouse antibodies conjugated to peroxidase. The reaction is measured photometrically following addition of an appropriate enzyme substrate.

By this method, monoclonal antibodies have been raised which have a much higher affinity to the PM fraction of embryos than to that of leaf tissue, the R2-2-7 antibody being an example. Others, as the R2-7-3, react positively with both. Results from an experiment with serial dilutions of these antibodies are seen in Fig. 1, where e and l denote embryo and leaf tissue PM, respectively.

A considerable similarity between the polypeptide patterns of embryo and leaf PM, as revealed by SDS-PAGE, can explain why most of the antibodies react with both preparations. But a number of distinct peptide bands being embryo specific hold out hopes of more antigenic determinants with the desired properties.

Preparations of plant plasma membranes obtained by phase partitioning are known to consist exclusively of right-side-out vesicles [5]. Consequently, the antibodies found are expected to be directed against epitopes on the outer PM surface, provided that the vesicles remain intact in the ELISA. Conclusive evidence to this could be brought by immunofluorescence labelling of the surface of living protoplasts. However, none of our antibodies so far seem to recognise protoplasts of neither embryo nor leaf origin, a fact that may arise from the epitopes being destroyed by the cell wall degrading enzymes.

Acknowledgements - I am indebted to Drs C. Larsson, S. Widell and P. Kjellbom for their help with the phase partitioning system, and to Dr M. Palmgren for his ATPase assay.

REFERENCES

1 C.J. Jensen and E. Thörn, Hereditas Suppl 3, 144 (1985)
2 P. Kjellbom and C. Larsson, Physiol. Plant 62:501 (1984)
3 J.L. Hall and A.L. Moore (eds) Isolation of membranes and organelles from plant cells, Academic Press, London (1983)
4 J.W. Goding, Monoclonal antibodies: Principles and practice, Academic Press, London (1983)
5 H.D. Grimes et al, Biochim Biophys Acta 862:165 (1986)

SUBFRACTIONATION OF CYANOBACTERIAL THYLAKOIDS

Fredrik Nilsson [1], David J. Simpson [2], Alison C. Stewart [3] and Bertil Andersson [1]

[1] Department of Biochemistry, Arrhenius Laboratory University of Stockholm, S-106 91 Stockholm, Sweden
[2] Department of Physiology, Carlsberg Laboratory Gl Carlsberg vej 10, DK-2500 Copenhagen, Denmark
[3] Department of Biochemistry, University of Cambridge Tennis Court Road, Cambridge CB2 1QW, UK

Cyanobacteria are free-living procaryotes capable of oxygenic photosynthesis, containing both photosystem I and photosystem II [1] including the oxygen evolving complex. Cyanobacteria as a model system for studying the oxygenic photosynthesis has a great advantage compared to plant chloroplasts; being a procaryote organism it permits the use of modern molecular genetic techniques.

Our understanding of the molecular organisation of the cyanobacterial energy transducing system, apart from the phycobilisomes, is mainly based upon assumed analogies with plant thylakoids. The membrane organisation and particularly the transverse assymetry of the cyanobacterial thylakoid are poorly characterised. One reason for this is the lack of efficient subfractionation methods.

In this communication we present a method to generate and isolate inside-out vesicles from the photosynthetic membrane of the cyanobacteria Phormidium laminosum.

According to the mechanism of formation for inside-out vesicles from higher plant thylakoids, the membranes are required to be in a tightly appressed state during the mechanical disintegration [2,3]. Cyanobacterial thylakoids do not have the stacks of appressed membranes characteristic of plant thylakoids. Therefore, in order to generate inside-out vesicles from cyanobacteria, membrane appression has to be induced by manipulations of the membrane and its surrounding milieu.

It was reasoned that two major obstacles might prevent tight membrane pairing of the cyanobacterial thylakoids: the presence of bulky protruding phycobilisomes on the outer thylakoid surface and repulsive intermembrane electrostatic forces. Therefore the phycobilisomes were removed by repeated washing in a high salt buffer. The electrostatic repulsion between the thylakoid membranes was reduced by neutralisation or screening of surface negative charges. Neutralisation was achieved by adjusting the pH to 4.7, which is the isoelectrical point of spinach thylakoid membranes and was assumed to be applicable also in the case of the stripped cyanobacterial thylakoids. Shielding of surface negative charges by addition of divalent ions such as Mg^{2+} can induce membrane pairing at neutral pH [4]. An Mg^{2+} concentration of 50 mM prior to fragmentation was required to achieve membrane pairing. After membrane pairing the thylakoids were fragmented by passage through a Yeda press at a pressure of 15 Mpa.

Phase partition of the appressed and fragmented cyanobacterial thylakoids were performed in a phase system composed of 5.9% (w/w) poly(ethylene glycol) 3350, 5.9% (w/w) Dextran T500, 10 mM Na-phosphate buffer (pH 7.4), 5 mM NaCl, 40 mM sucrose and 10% w/w glycerol. Two membrane populations were resolved, one major partitioning to the upper phase (T1) and one minor partitioning to the lower phase (B1). After the subsequent partition steps the T2 fraction contained 80% of the material on a chlorophyll basis whereas the B3 fraction contained only 2% of the material.

As an analysis of the membrane orientation of the thylakoid vesicles, we used proton translocation measurements with phenyl-p-benzoquinone as electron acceptor. The proton translocation studies on the T2 and B3 fractions revealed that the vesicles of the T2 fraction had a normal light induced reversible proton uptake. In contrast, under the same conditions the B3 fraction showed a light-induced reversible proton extrusion. These results show that the B3 fraction is dominated by vesicles with a reversed proton translocation, suggesting that these vesicles are turned inside-out with respect to the normal orientation of the thylakoid membrane.

An independent way to determine the orientation of membrane vesicles is freeze-fracture electron microscopy. Such studies showed that T2 contained equal proportions of normal and everted vesicles, whereas those in the B3 fraction were enriched (75%) in vesicles of everted sidedness.

The photochemical activity of the cyanobacterial thylakoid membrane was followed throughout the various preparation steps. Both methods of preparation (Mg^{2+} and H^{+} treatment) resulted in some loss of photosystem I and photosystem II activity. Some 60-75% of the control activity (photosystem I or photosystem II) remained after the fragmentation. During phase partition there was further loss of photochemical activity, particularly for the B3 fractions (25-35% of control activity remained). This decrease of photochemical activity in the B3 fractions should be due mainly to the release of extrinsic proteins from the exposed inner thylakoid surface of the everted vesicles. In the case of the photosystem I a release of plastocyanin is a likely reason for loss of activity. This assumption is supported by the fact that addition of spinach plastocyanin partly restores the photosystem II activity. We have previously shown that cyanobacteria contain a 9 kDa photosystem II protein which is easily lost from cyanobacterial thylakoid particles during conditions of lower glycerol concentration [5]. This could account for the relatively low photosystem II activity of the B3 fraction.

Even though the yield of inside out vesicles is rather low and the photochemical activity is reduced, these everted vesicles should be useful tools for studies of the transverse assymetry of the cyanobacterial thylakoid. The low yield might be compensated by scaling up the process, which may easily be done since, in theory, the size of the polymer phase system is the limiting factor.

REFERENCES

1 E. Padan, Annu. Rev. Plant Physiol. 30:27 (1979)
2 B. Andersson, C. Sundby and P.-Å. Albertsson, Biochim. Biophys. Acta 599:391 (1980)
3 B. Andersson, D.J. Simpson and G. Hoyer-Hansen, Carlsberg Res. Commun. 43:77 (1978)
4 J. Barber, Annu. Rev. Plant Physiol. 33:261 (1982)
5 A.C. Stewart, U. Ljungberg, H.-E. Åkerlund and B. Andersson, Biochim. Biophys. Acta 808:353 (1985)

Chapter 3

Partitioning of Animal Membranes and Organelles

USE OF PHASE PARTITIONING IN MULTIDIMENSIONAL SUBCELLULAR FRACTIONATION

Austin K. Mircheff, Michael E. Bradley, Charles B. Hensley, Emile J.J.M. van Corven, Samuel C. Yiu and Ross W. Lambert

Department of Physiology and Biophysics
University of Southern California
School of Medicine, Los Angeles, California 90033, USA

Conventional subcellular fractionation procedures incorporate two separation dimensions, one based on membrane sedimentation coefficient, the other on equilibrium density. Questions about the homogeneity of samples obtained with these procedures have been addressed with additional separations based on manipulation of membrane density or on partitioning in dextran-poly(ethylene glycol) two-phase systems. In some cases it has proven useful to employ a total of 4 separation dimensions, i.e. sedimentation, equilibrium density in 2 differently designed density gradients, and phase partitioning. In other cases, parallel third dimensions analyses in two-phase systems with different pH values have been used to delineate the multiplicity of populations present in conventional subcellular samples.

INTRODUCTION

The basic assumptions underlying the design of subcellular fractionation procedures were articulated by DeDuve in 1964 [1]; they can be summarised as follows: When a cell is disrupted, its constituents generate a mixture of populations of particles. Some populations consist of intact organelles; such organelles as nuclei, mitochondria, lysosomes, and secretory granules typically remain intact during many cell disruption procedures. Other populations are comprised of fragments of organelles. An organelle might, in fact, give rise to more than one population of fragments. For example, the cis-, medial-, and trans-saccules of the Golgi stack might each produce at least one distinct particle population, as might the microvillar and intermicrovillar microdomains of the brush border surface membranes of renal and intestinal epithelial cells. Each particle in the cell lysate is characterised by a set of physical and biochemical properties, and each population of particles might be quite heterodisperse with respect to any given property. Separation procedures which detect membrane physical properties are used to resolve the mixtures of subcellular particle populations. The use of such procedures is complicated by the heterodispersity of the various populations and by the fact that few populations are unique with respect to any single physical property.

The need to combine separation procedures detecting different physical properties was formalised into a concept of two-dimensional subcellular fractionation incorporating equilibrium density gradient centrifugation and differential sedimentation by Anderson and coworkers [2]. It subsequently became evident that fractionation schemes based only on these methods could not adequately resolve all of the particle populations present in a cell lysate. One classical illustration of this point is the isolation of tubulo-vesicular membranes from the gastric parietal cell. These membranes, which are marked by a K+-dependent, H+-translocating ATPase, equilibrate in a density region occupied by at least one other microsomal membrane population, which is a major locus Mg++-dependent ATPase activity [3]. The two populations can be resolved, however, by free-flow electrophoresis, which distinguishes them on the basis of differences in their surface charges.

The practical problems of designing useful subcellular fractionation procedures is further complicated by the fact that few of the membrane constituents which are used as biochemical markers are uniquely localised within the cell. This complexity has several ramifications; it makes it difficult to evaluate the purification of subcellular membrane samples, and, in some cases, it makes it difficult to infer the identities of the membranes comprising such samples.

One approach to this problem was suggested by the early theoretical work of DeDuve [1] and of Anderson [2]. The essence of this approach is the notion that one can combine a battery of separation procedures, each operating on the basis of a different membrane property, to achieve a multidimensional fractionation [4]. When applied systematically, this approach would allow one to delineate and to characterise a large variety of subcellular membrane populations. The work of delineating the membrane populations would be entirely empirical, since it would require no assumptions about the subcellular localisation of particular markers. Once the membrane populations had been resolved, one could draw provisional inferences about their subcellular origins on the basis of their marker contents.

Phase partitioning in aqueous polymer two-phase systems promised to be a particularly useful component of multidimensional separation procedures. Several different epithelial tissues, including exorbital lacrimal gland [5], small intestine [6,7], and parotid gland [8], were subjected to 3-dimensional analytical fractionations based on membrane sedimentation, equilibrium density, and partitioning in dextran-poly(ethylene glycol) two-phase systems. It is becoming apparent, however, that even 3 independent separation dimensions are not sufficient to resolve all of the membrane populations present in a typical tissue homogenate. Strategies for incorporating fourth separation dimensions are now evolving; we will illustrate them with results from a recent study of intestinal epithelial cells and with preliminary results from ongoing studies of the rat exorbital lacrimal gland and renal cortex.

Multidimensional Fractionation of Lacrimal Gland

Although many of the results are still preliminary, recent work with the rat exorbital lacrimal gland illustrates the basic principles outlined above and emphasises the value of introducing a fourth separation dimension. The exorbital gland is a typical tubulo-acinar exocrine gland, in which the acinar cells account for more than 80% of the glandular mass. The acinar cells synthesise a number of macromolecular products and assemble them into secretory granules. They also utilise specific arrays of electrolyte transporters, including Na,K-ATPase, to secrete a primary lacrimal gland fluid which resembles an isotonic ultrafiltrate of plasma. Analytical subcellular fractionation studies were begun with a view toward delineating the mechanisms involved in lacrimal acinar function.

Density Gradient Centrifugation

Recent experiments with this preparation have been performed with gland fragments that are incubated for 85 min in Krebs improved Ringer bicarbonate solution. Both the in vitro incubation and subsequent homogenisation and centrifugation procedures are performed in media which are supplemented with the protease inhibitors phenylmethylsulfonyl fluoride (0.2 mM) and Aprotinin (9 ug/ml). The initial homogenate is subjected to differential sedimentation and equilibrium density centrifugation procedures used previously for fractionation of freshly decapsulated lacrimal tissue [5]. Centrifugation at 1000 x g for 10 min sediments connective tissue fragments, nuclei, and intact secretory granules. The supernatants pooled from 4 cycles of homogenization and low-speed centrifugation are brought to a sorbitol concentration of 55% by addition of a concentrated stock solution, then loaded between hyperbolic gradients of 35% to 55% and 55% to 70% sorbitol in a Beckman Z-60 zonal rotor. The Z-60 rotor can accommodate the relatively large total volume of 330 ml, and it can be run at speeds up to 60,000 rpm, so that relatively short centrifugation times are required for analysis of relatively large samples.

Fig. 1 summarises distributions of a battery of biochemical markers after density gradient centrifugation of the low-speed supernatant from the rat exorbital lacrimal gland fragment preparation. The marker distribution patters are qualitatively similar to those which were obtained when freshly decapsulated lacrimal tissue was homogenized and fractionated in media which lacked protease inhibitors [5]. These distribution patterns illustrate the general principles that were outlined above. First, it is possible to infer that the sample contained a multiplicity of membrane populations. Second, several biochemical markers have multimodal distributions, indicating that each is associated with a multiplicity of different populations. Third, a given region of the density gradient might contain elements of two or more different populations. The clearest example of the third point is found in the region of the density gradient defined as density window IV; this window contains the mode of a well-defined population of intact mitochondria, marked by succinate dehydrogenase, as well as one or more populations of microsomal membranes, marked by NADPH-cytochrome c reductase, Na,K-ATPase, galactosyltransferase, and acid phosphatase. The mitochondrial and microsomal populations can be separated from each other by a single differential rate centrifugation through a column of 17.5% sorbitol [5].

Several features of the distribution patterns in Fig. 1 warrant close examination, because they provide clues about the number and the nature of the membrane populations present in the sample and about the subcellular distributions of the marker enzymes. The Na,K-ATPase distribution could, in principle, be interpreted in terms of 4 partially overlapping populations, one centered in density window I, one centered in window II, one centered in window IV, and one centered in window VI. The galactosyltransferase distribution would suggest the presence of 4 populations, centered in windows II, III, IV and VI, and the acid phosphatase distribution would suggest the presence of at least 3 populations, one centered in window I, one centered in window IV, and one centered in window VI.

Provisional interpretations of the density distributions in Fig. 1 lead immediately to a number of questions. For example, are the Na,K-ATPase and acid phosphatase activities in window I associated with the same membrane population, or does this window contain elements of two distinct populations, one relatively richer in Na,K-ATPase and the other relatively richer in acid phosphatase? Similarly, are the Na,K-ATPase and

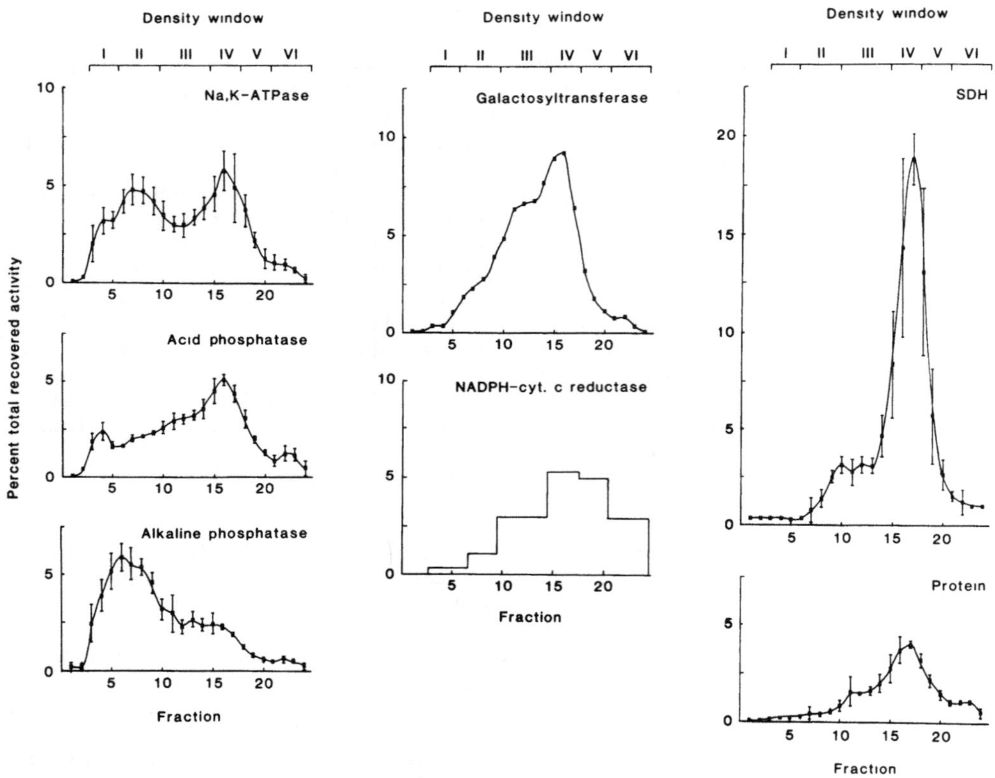

Fig. 1 *Density distribution of markers from lacrimal gland framents. Galactosyltransferase and NADPH-cytochrome C reductase were determined in a single preparation; n = 4 for other markers. Bars indicate S.D.*

galactosyltransferase activities in density window II associated with the same population of membranes, and what is the relationship between the Na,K-ATPase and galactosyltransferase activities in density window IV? In density window III, how much of the Na,K-ATPase activity might be assigned to elements spilling over from the major Na,K-ATPase containing population centered in window II, how much might be assigned to spillover of elements of the population centered in window IV, and how much might be assigned to a distinct population centered in window III?

Third-Dimension Separations by Phase Partitioning

When marker density distribution patterns such as those presented in Fig. 1 were first observed, phase partitioning was seen as a particularly versatile separation procedure for analysing the compositions of the various density windows [4,5]. A two-phase system containing 5% Dextran T-500, 3.5% poly(ethylene glycol) 8000, 5% sorbitol, 10 μM NaEDTA, and 8.33 mM imidazole-HCl, pH 6.55, was selected after a preliminary survey. Countercurrent distribution in this phase system indicated that at least 6 different membrane populations were represented in windows I, II and III. Window I was dominated by a single membrane population, indicating that alkaline phosphatase, acid phosphatase, and Na,K-ATPase were all associated with the same membranes. However, the partitioning properties of this population, the major population from window II, and the major population from window III were all similar. Thus, while it was clear that these populations differed with respect to their modal densities and with respect to their relative contents of Na,K-ATPase, acid phosphatase, alkaline phosphatase, and galactosyltransferase, it was not possible to discern how the marker contents measured in a sample of one population might have been influenced by spillover in the density distributions of neighbouring populations.

Interest in answering these questions more precisely increased when activities of Na/H-antiporters [9], Cl/HCO_3 antiporters [9], and muscarinic acetylcholine receptors [10] were detected in density windows I, II and III. The mode of each distribution was in density window II, but substantial fractions of the total activities were present in density windows I and III. Thus, one would now like to know whether an antiporter or receptor activity in window I or in window III is associated with one or more of the membrane populations which are primarily localised in those windows, or whether it is associated with elements spilling over from a membrane population which is primarily localised in density window II. Current work on this problem focuses on efforts to refine the analyses obtained by partitioning in aqueous polymer phase systems; it is taking advantage of 2 phenomena which, one must admit, remain poorly understood. First, the partitioning properties of some membrane populations have been found to depend on the concentration of sample protein introduced into the phase system. Second, changes in the pH of the phase system have been found to have varying effects on the partitioning properties of different membrane populations.

For the purposes of these studies, care was taken that a constant protein load be applied for all samples. Since density window I contains 9-fold less, and density window II contains 4-fold less protein than window III, a practical decision was made to standardise the sample load at 0.4 mg protein / ml upper phase. This load would allow one 40-step CCD analysis of the density I sample and two separate 40-step CCD analyses for the window II sample from a typical density gradient centrifugation preparation. It represented a 3-fold reduction of the density window III sample protein load from the value used in previous studies [5], but essentially no change in the protein loads for samples of windows I and II.

When phase partitioning was performed at pH 7.3 (Fig. 2), most of the protein from density window III partitioned far to the left of the major membrane population of density window II. Therefore, one can conclude that galactosyltransferase activity is, in fact, associated with the major 2window II membrane population. On the other hand, a close comparison of the Na,K-ATPase, galactosyltransferase, and protein distribution patterns generated by phase partitioning analysis of window III at pH 7.3 suggests that the region between partitioning fractions 1 and 5 might contain a multiplicity of membrane populations, each differing slightly from the others with respect to its partitioning properties. As depicted in the lower half of Fig. 2, this prediction is confirmed by the marker distributions observed after phase partitioning at pH 8.0.

The composite picture that can be constructed on the basis of the phase partitioning analyses of windows I - III summarised in Fig. 2 involves a minimum of 9 distinct membrane populations. The approximate positions of these populations are depicted in Fig. 3, while marker cumulative enrichment factors calculated for samples of these populations are summarised in Table 1. Population a, centered in window I, is the locus of the highest observed cumulative enrichment factors for Na,K-ATPase and acid phosphatase; the sample in window I appears to be largely free of contamination by the major Na,K-ATPase containing population from window II, population c; therefore, the antiporter and receptor binding

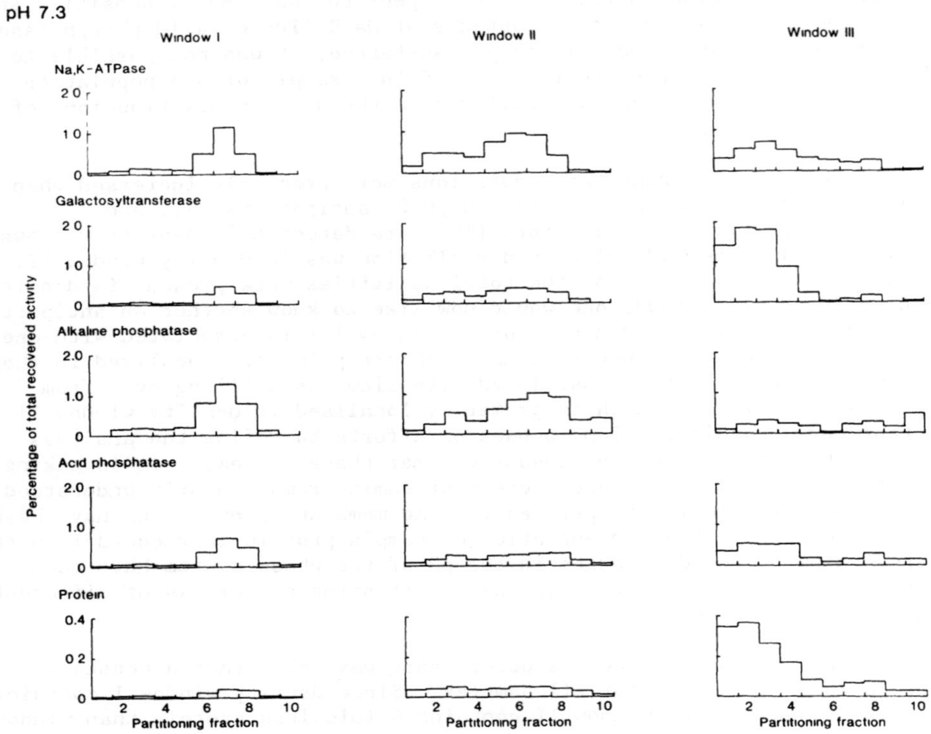

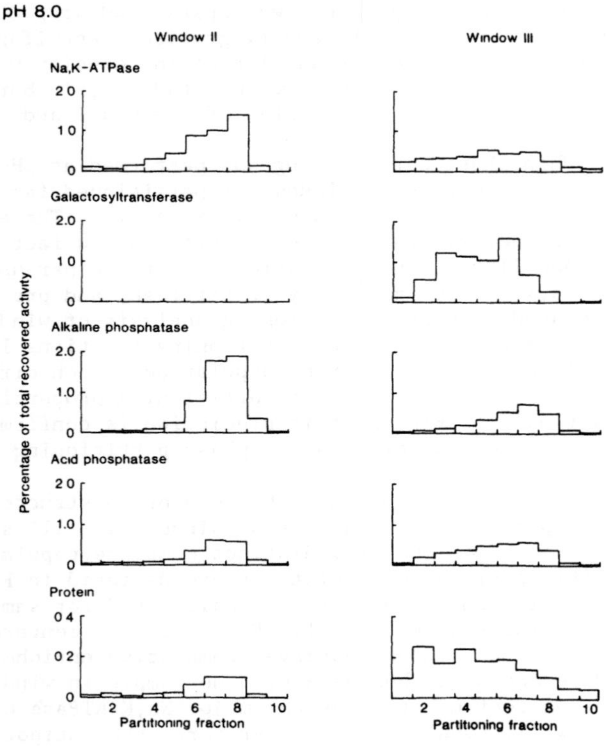

Fig. 2

Phase partitioning analyses of density windows I, II and III

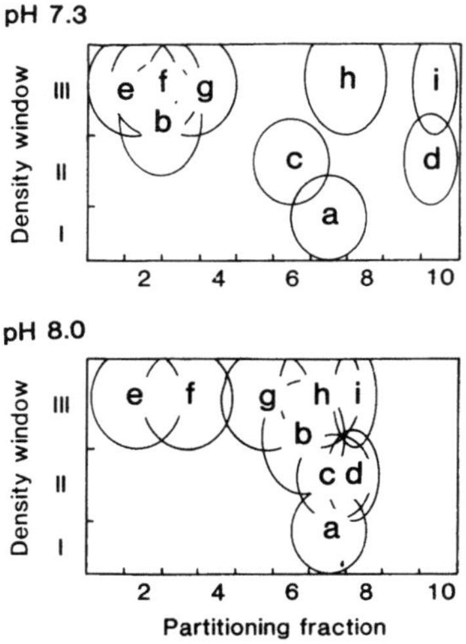

Fig. 3 *Approximate density-partitioning positions of populations delineated from data in Figs. 1 and 2*

activities present in window I can be assigned to population a rather than attributed to spillover of elements of population c. While population c does not spill extensively into window I, population a does appear to spill over into window II. Analysis in the pH 7.3 phase system yields the sample of population c which appears least contaminated by population a.

Populations e, f and g account for most of the protein in density window III. Population e is characterised by relatively low cumulative enrichment factors for all of the markers. It would account for the high ratio of protein to the other markers in partitioning fraction 1 after analysis of window III at pH 8.0. Populations f and g are not resolved from each other when partitioning is performed at pH 7.3, but they can be clearly discerned on the basis of the peaks of galactosyltransferase in fractions 3 and 6 when phase partitioning is performed at pH 8.0. Both populations are characterised by relatively high cumulative enrichment factors for galactosyltransferase and by high ratios of galactosyltransferase to Na,K-ATPase. The true cumulative enrichment factors of Na,K-ATPase in populations f and g remain uncertain, because of the spillover of elements of population b into density window III. This spillover also means that additional data will be required before one can estimate the extent to which the antiporter and receptor activities in window III should be assigned to elements of population b.

Identities of Isolated Membrane Populations

The analytical approach to subcellular fractionation outlined above allows one to visualise the multiplicity of membrane populations present in a subcellular fraction, and it permits at least an approximate characterization of their biochemical properties. This approach can be extended to generate a map of the populations present in a whole cell lysate. The more difficult challenge is to determine the subcellular origins of these populations. This effort has been entirely a matter of deduction, based on traditional concepts of the subcellular distributions of biochemical markers. In the original studies with freshly decapsulated lacrimal tissue [5], a population corresponding to population a in Fig. 3

Table 1 *Maker Enrichment Factors in Lacrimal Gland Membrane Populations Delineated in Density Windows I, II and III*

Population[a]	Na,K-ATPase	Galactosyl-transferase	Alkaline Phosphatase	Acid Phosphatase
a (I, 7, 7.3)	27.3	9.8	30.2	17.0
b (II, 3, 7.3)	9.2	6.0	10.8	5.4
c (II, 6, 7.3)	20.0	7.9	17.5	5.8
d (II, 10, 7.3)	-	-	43.0	5.0
e (III, 1, 8.0)	2.1	1.0	0.1	0.5
f (III, 3, 8.0)	2.0	7.2	0.8	1.8
g (III, 6, 8.0)	2.5	7.0	2.9	2.9
h (III, 8, 7.3)	3.7	2.5	4.1	4.1
i (III, 10, 7.3)	1.5	1.1	14.8	4.8

[a] Coordinates in parenthesis are density windows, partitioning fraction, and pH of the phase system at which the populations were sampled.
[b] Values given are cumulative enrichment factors, calculated as described by Mircheff and Lu [5].

was provisionally identified as a minor plasma membrane population, a population corresponding to population c was identified as the population of acinar cell basal-lateral membrane vesicles, a population corresponding to population e was identified as a subpopulation of endoplasmic reticulum-derived membranes, and a population which now appears to have been a composite of populations b, f, g and h was identified as having been derived from trans-elements of the Golgi complex.

Preliminary density gradient analyses of lysates of isolated lacrimal acini indicate the presence of populations corresponding to a and c. Therefore, it appears that neither population has been derived from a minor cell type in the gland. Given the traditional assumption that the epithelial cell basal-lateral membrane should be the locus at which Na,K-ATPase has its highest specific activity, the data in Table 1 suggest that population a might represent the lacrimal acinar cell basal-lateral plasma membrane, while population c might have been derived from some cytoplasmic structure, perhaps one involved in a late step in the assembly of newly synthesised and recycling basal-lateral membrane constituents.

Cation Antiporter Subcellular Distribution in Renal Cortex

One topic of interest in the physiology of the kidney is the subcellular disposition and regulation of Na/H antiporters. These antiporters are known to be expressed in the apical membrane of the proximal tubular epithelial cell. They serve the dual function of mediating the first step in the transcellular reabsorption of filtered Na^+ ions and of accelerating the reabsorption of HCO_3^- ions in compensation for acidosis. Regulation of Na/H-antiporter activity is believed to be mediated via modulation of the affinity of an H^+ binding regulatory site [11]. However, there has remained a possibility that regulatory mechanisms involving the exocytic insertion of additional antiporters might also be available for increasing the apical membrane antiporter activity.

Several factors have complicated the interpretation of published studies of the Na/H antiporter subcellular distribution in proximal tubule. Equilibrium centrifugation on sucrose density gradients separates apical and basal-lateral membrane populations and indicates that negligible antiporter activities are expressed in the basal-lateral membranes [12]. However, the density distribution of the apical membrane population is overlapped by the distributions of membranes apparently derived from cytoplasmic structures, so that it is not possible to discern how much of the antiporter activity can be assigned to the apical membranes and how much should be assigned to the cytoplasmic membranes. Somewhat more highly purified samples of the proximal tubular apical membrane have been obtained by procedures which exploit the ability of divalent cations to preferentially aggregate membranes derived from structures other than the apical membrane. Phase partitioning analysis demonstrated that the divalent cation-obtained apical membrane preparations are heterogeneous; this analysis also indicated that roughly 25% of the Na/H antiporter activity in such preparations was associated with a population which was distinct from the major population of apical membrane vesicles in terms of its partitioning properties and in terms of its specific contents of a variety of biochemical markers [13]. While this approach suggested that Na/H antiporters had a complex subcellular distribution, a quantitative analysis of that distribution was not possible, since the partially purified apical membrane preparation did not constitute a representative sample of the membrane populations present in the original tissue lysate.

Three-Dimensional Fractionation

In studies aimed at a more comprehensive evaluation of the Na/H antiporter subcellular distribution in proximal tubular epithelium, rat renal cortex was homogenised and subjected to a sequence of differential sedimentation and density gradient centrifugation steps similar to those employed for analytical fractionation of lacrimal gland fragments. A density window was defined which contained a major sample of the apical membrane population, and membranes from this density window were analysed by phase partitioning in dextran-poly(ethylene-glycol) phase systems similar to those described above. As depicted in Fig. 4, these analyses generated complex marker distribution patterns indicating the presence of elements of at least 5 distinct membrane populations. The approximate positions of these populations are indicated in Fig. 5. Na/H antiporter activity paralleled maltase after phase partitioning at pH 6.6 and after partitioning at pH 7.6; both markers are associated with at least 2 different membrane populations, designated b and c in Fig. 5. These populations overlap when phase partitioning is performed at pH 6.6, accounting for the apparent right-ward skewing of the maltase and Na/H antiporter distributions. These are resolved, however, when partitioning is performed at pH 7.6.

Marker cumulative enrichment factors for samples of these populations are collected in Table 2. With the exception of population e, which is well-resolved by phase partitioning at pH 6.6, none of the samples appears to be entirely free of spillover of elements from neighbouring populations. Therefore, the values in Table 2 should be taken as only first approximations of the marker relative specific activities in the various populations. Given these estimates, it is most likely that population b represents the apical membranes; with its apparently lower specific contents of maltase and Na/H antiporter activity, population c might represent a cytoplasmic structure involved in apical membrane assembly rather than the apical membrane itself. In any event, the maximum cytoplasmic pool of Na/H-antiporter activity indicated by these

Table 2 *Maker Enrichment Factors in Samples of Membrane Populations Delineated in Rat Renal Cortex*

Population[a]	Na/H Antiport	Maltase	Alkaline Phosphatase	Na,K-ATPase	Acid Phosphatase
a (20, 7.6)	14.1	11.1	5.0	1.1	3.8 b
(30, 7.6)	21.1	15.8	7.0	1.1	2.7 c
(70, 7.6)	9.4	7.8	9.5	3.9	1.3 d
(70, 6.6)	2.0	2.0	3.9	14.2	1.8 e
(90, 6.6)	-	-	3.5	2.4	1.4

[a] Coordinates in parenthesis are partitioning fraction and pH of the phase system at which the populations were sampled.

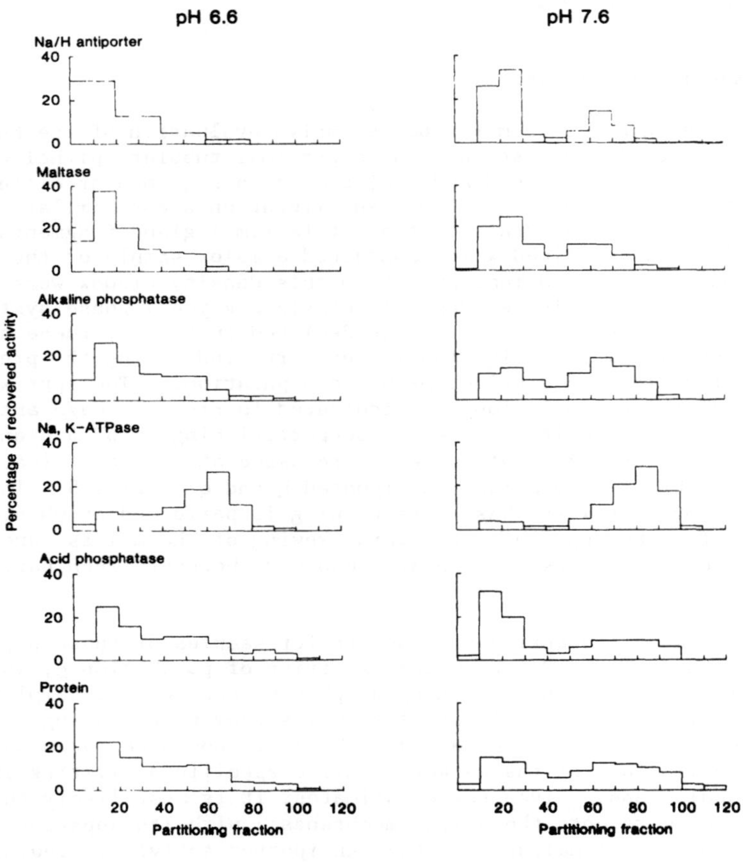

Fig. 4 *Phase partitioning analyses of apical membrane density window from renal cortex homogenates*

results is too small to represent a reserve that could significantly increase the apical membrane antiporter activity.

Multidimensional Fractionation of Intestinal Epithelial Cells

A final illustration of a multidimensional approach to subcellular fractionation is drawn from a recent study of the subcellular distribution of ATP-dependent Ca^{++} transport activity in isolated rat duodenal epithelial cells. The active absorption of Ca^{++} by the intestinal epithelium is a key point in the regulation of Ca^{++} homeostasis. Intestinal epithelial cells utilise Ca^{++} as an intracellular messenger, so they must be able to regulate their cytosolic Ca^{++} activities within narrow limits at the same time that they generate large transcellular fluxes of Ca^{++}. Both processes are believed to involve ATP-dependent Ca^{++} pumps, so it was of interest to survey the distribution of pumps between cell surface and cytoplasmic membranes.

Major populations of endoplasmic reticulum- and Golgi-derived membranes can be partially separated from the basal-lateral membranes by equilibrium density gradient centrifugation of low-speed supernatant fractions from intestinal cell lysates [6,7,14]. The separation achieved is sufficient to demonstrate that ATP-dependent Ca^{++} pumping activity is present in both types of membranes. However, the presence of cytoplasmic membranes in the basal-lateral membrane density window makes it impossible to evaluate the precise contribution of the basal-lateral membrane to the cell's total activity [14].

Analysis of the basal-lateral membrane density window by partitioning in a dextran-poly(ethylene glycol) phase system generated bimodal distributions of Na,K-ATPase and galactosyltransferase. However, close examination of the distribution patterns suggested the presence of at least 3 distinct membrane populations. One population, which accounted for minor fractions of both activities, partitioned between fractions 10 and 40 after a 120-step run. The second population, which was relatively enriched in galactosyltransferase, had its mode between fractions 60 and 70, while the third population, relatively enriched in Na,K-ATPase, had its mode between partitioning fractions 70 and 80. The overlap between the latter two populations made it difficult to conclude more about their relative marker contents. Furthermore, it did not seem that manipulating the composition of the phase system in order to refine this separation would lead to definitive information about the relative distribution of ATP-dependent Ca^{++} transport activity, since less than 10% of the applied activity was recovered after phase partitioning.

Two Separation Dimensions by Density Gradient Centrifugation

A solution to this problem was suggested by the observation that the relative positions of the major basal-lateral membrane population and of the major endoplasmic reticulum and Golgi populations after density gradient centrifugation depended on the density of the medium in which the sample was introduced onto the gradient. Thus, a scheme was devised incorporating two density gradient centrifugation steps [14]. In the first step, the low-speed supernatant fraction was suspended in 5% sorbitol, then introduced onto a hyperbolic gradient of 30% to 80% sorbitol; the major cytoplasmic populations equilibrated at lower densities than the basal-lateral membranes [14]. The membranes harvested from the basal-lateral membrane density window were then brought to a sorbitol concentration of 55%, and loaded beneath a gradient of 30% to 55% sorbitol. The second density gradient centrifugation step yielded a

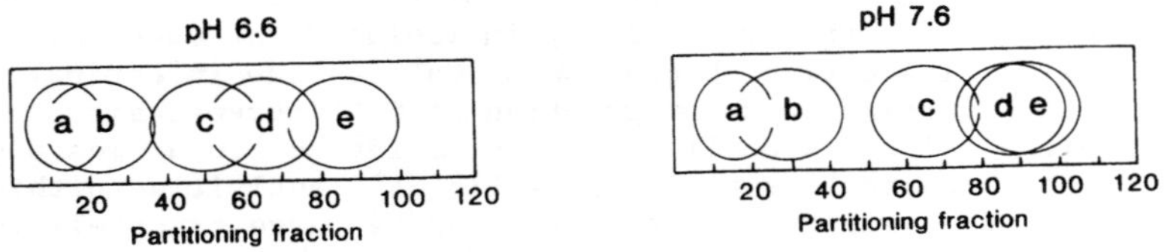

Fig. 5 *Approximate positions of membrane populations delineated from distributions in Fig. 4*

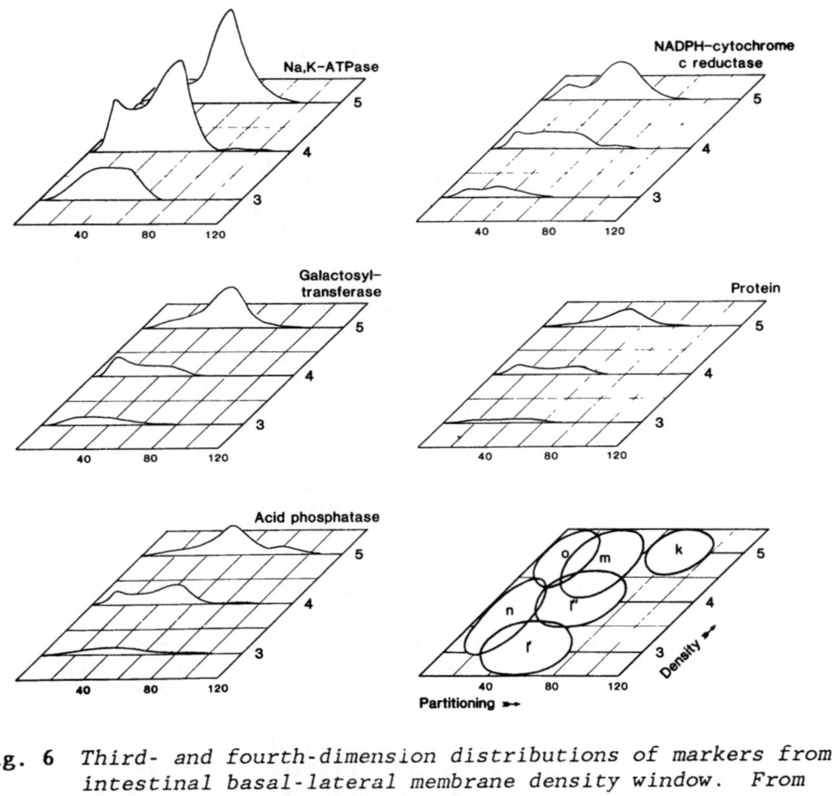

Fig. 6 *Third- and fourth-dimension distributions of markers from intestinal basal-lateral membrane density window. From reference 14, with permission*

partial separation of an Na,K-ATPase-rich population from a galactosyltransferase- and NADPH-cytochrome c reductase-rich population. The resolution between the basal-lateral membranes and the cytoplasmic population was not complete, but it was sufficient to permit the conclusion that a substantial portion of the ATP-dependent Ca++ transport activity equilibrating in the basal-lateral membrane density window from the first gradient was actually associated with cytoplasmic membranes [14].

A Fourth Separation Dimension by Phase Partitioning

Phase partitioning was used to evaluate the homogeneity of several of the membrane samples obtained after the second density gradient centrifugation step. The resulting distributions, summarised in Fig. 6, can be interpreted in terms of a minimum of 6 membrane populations. Populations l' and l'', which are centered in second density gradient fractions 3 and 4, respectively, had Na,K-ATPase cumulative enrichment factors in the range of 10 to 11. These were the highest Na,K-ATPase enrichment factors obtained in the preparation, suggesting that the two populations together represent subpopulations of basal-lateral membranes. The galactosyltransferase enrichment factors in samples of these subpopulations were less than 2.0, indicating that the basal-lateral membranes had been purified substantially with respect to Golgi-derived membranes, represented by populations m, n and o. Of the 6 populations resolved by this 4-dimensional analysis, l' and m are particularly noteworthy. Population l' represents a basal-lateral membrane subpopulation which can be sampled at relatively high purity without phase partitioning. Population m appears to be a cytoplasmic membrane population which could be a major contaminant of basal-lateral membrane samples obtained by conventional 2-dimensional separations.

ACKNOWLEDGEMENT

This work was supported by NIH Grants AM 28408 and EY 05801

REFERENCES

1 C. DeDuve, Principles of tissue fractionation, J. Theor. Biol. 6:33 (1964)
2 N.G. Anderson, W.W. Harris, A.A. Barber, C.T. Rankin and E.L. Chandler, "Separation of Subcellular Components by Combined Rate and Isopycnic Zonal Centrifugation," National Cancer Institute Monograph 26, U.S. Government Printing Office, Washington, D.C. (1966)
3 G. Saccomani, H.B. Stewart, D. Shaw, M. Lewin and G. Sachs, Characterisation of gastric mucosal membranes. IX. Fractionation by zonal centrifugation and free flow electrophoresis technique, Biochim. Biophys. Acta 456:311 (1977)
4 A.K. Mircheff, Empirical strategy for analytical fractionation of epithelial cells, Am. J. Physiol. 244:G347 (1983)
5 A.K. Mircheff and C.C. Lu, A map of membrane populations isolated from rat exorbital gland, Am. J. Physiol. 247:G651 (1984)
6 A.K. Mircheff, D.J. Ahnen, A. Islam, N.A. Santiago and G.M. Gray, Complex subcellular distributions of enzymatic markers in intestinal epithelial cells, J. Membrane Biol. 83:95 (1985)
7 D.J. Ahnen, A.K. Mircheff, N.A. Santiago, C. Yoshioka and G.M. Gray, Intestinal surface aminooligopeptidase. Distinct molecular forms during assembly on intracellular membranes in vivo, J. Biol. Chem. 258:5960 (1983)

8 C.N. Conteas, A.A. McDonough, T.R. Kozlowski, C.B. Hensley, R.L. Wood and A.K. Mircheff, Mapping subcellular distribution of Na^+-K^+-ATPase in rat parotid gland, Am. J. Physiol. 250:C430 (1986)
9 R.W. Lambert, M.E. Bradley and A.K. Mircheff, Sodium/proton and chloride/bicarbonate antiporters in rat lacrimal gland basal-lateral membranes, Invest. Ophthalmol. Vis. Sci. 28s:156 (1987)
10 C.L. Peters, M.E. Bradley, S.C. Yiu and A.K. Mircheff, Muscarinic receptors in cell surface and intracellular membranes of rat exorbital lacrimal gland, Invest. Ophthalmol. Vis. Sci. 27s:26 (1986)
11 S. Grinstein and A. Rothstein, Mechanisms of regulation of the Na^+/H^+ exchanger, J. Membrane Biol. 90:1 (1986)
12 H.E. Ives, V.J. Yee and D.J. Warnock, Asymmetric distribution of the Na^+/H^+ in renal proximal tubule epithelial cells, J. Biol. Chem. 258:13513 (1983)
13 A.K. Mircheff, H.E. Ives, V.J. Yee and D.G. Warnock, Na^+/H^+ antiporter in membrane populations resolved from a renal brush border vesicle preparation, Am. J. Physiol. 246:F843 (1984)
14 E.J.J.M. van Corven, C.H. van Os and A.K. Mircheff, Subcellular distribution of ATP-dependent calcium transport in rat duodenal epithelium, Biochim. Biophys. Acta 861:267 (1986)

COMPARISON OF PHASE PARTITION WITH ALTERNATIVE METHODS FOR ISOLATION OF SUBCELLULAR STRUCTURES

D. James Morré

Department of Medicinal Chemistry
Purdue University
West Lafayette, IN 47907

Preparative free-flow electrophoresis alone or in combination with phase partition provides opportunities for decisive separations of membranous cell components and organelles based on surface charge not easily duplicated by classical differential or density centrifugation techniques. Applications include subfractionation of Golgi apparatus into component cisternae, isolation of plasma membrane and tonoplast vesicles from plant homogenates without significant cross contamination one by the other, separation of cytoplasmic side out and cytoplasmic side in plasma membrane (and tonoplast) vesicles and in the enrichment of minor cell components such as lysosomes and endoplasmic reticulum-derived transition vesicles. The procedures are facile, fast and efficient. Free-flow electrophoresis provides both analytical capabilities and the opportunity for continuous operation on a preparative scale. The procedure is non-destructive and fractions obtained retain full in vitro functional characteristics.

INTRODUCTION

Development of cell-free systems involving cellular membranes has been instrumental in the detailed elucidation of significant biological and biomedical processes. An essential feature of such cell-free systems is the need to prepare requisite membrane fractions in high yield and purity and with full retention of in vitro biological activity. Differences in size and density among different membranous compartments found in cell homogenates have afforded considerable opportunities for decisive separations based on size and density [1]. However, some applications such as subfractionation of Golgi apparatus along its functional axis, endosome isolation, and fractionation of tumours, tissue culture cells and higher plant, algae and fungal homogenates have proven frustrating due to components of different origins having both overlapping sizes and densities such that additional separation parameters have been sought. Imposed labels such as dense ligands to alter centrifugation behaviour have provided one useful approach [2]. Separations, based on differences in surface charge, offer another. It is separations based on surface charge, such as free-flow electrophoresis, and especially in combination with aqueous two-phase partition, that provide the major focus of this contribution.

Surface charges have been used to effect membrane separations in many different ways. Phase partitioning [3] and preparative free-flow electrophoresis [4] however, are among the more useful from a preparative standpoint where rapid, large scale isolation of highly purified membrane fractions in high yield becomes possible. We have utilised both procedures alone and in combination, together with standard centrifugation procedures [5] to provide in many instances, for total fractionation of both plant [6] and animal [7] cells. In the sections below, applications will be summarised where separations based on charge afford unique opportunities for cell fractionation not readily duplicated by traditional methods of differential or sucrose gradient centrifugation. Details and other applications of a more general nature can be found in the references provided.

Principles of Free-Flow Electrophoresis

In the technique of free-flow electrophoresis, the mixture of components to be separated is introduced as a fine jet into a separation buffer moving perpendicular to the field lines of an electric field (Fig. 1). Membranes bearing different electrical charge densities will migrate different distances across the separation chamber and thus may be resolved. The method is preparative and capable of continuous operation. The recoveries of membranes and proteins are high (about 90%). Separations are at 4°C, vesicle integrity is maintained (vesicles remain sealed), and most enzymatic activities are retained at or near initial levels.

For most separations involving animal membranes the electrophoresis medium (chamber buffer) contained 10 mM triethanolamine, 10 mM acetic acid, 5 mM glucose, 0.25 M sucrose and 0.5 mM $MgCl_2$, final pH 6.5 (1 M NaOH), osmolarity 270 mosm, conductivity 5.9 x 10^2 μmoho. The electrode buffer contained 100 mM triethanolamine and 100 mM acetic acid, pH 6.5.

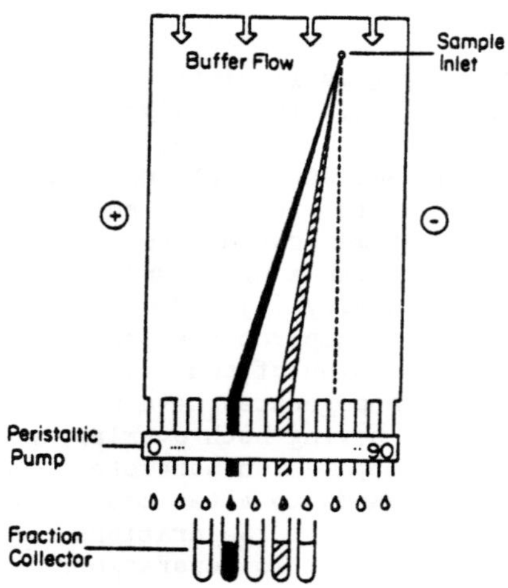

Fig. 1 *Principles of preparative continuous free-flow electrophoresis. Redrawn and reprinted with permission from Bender and Hobein, Munich, FRG*

Electrophoresis conditions were 167 mA (constant current), 131 ± 10% V/cm, buffer flow 2.75 ml/fraction/h, sample injection 3.5 ml/h, temperature 6°C. The electrophoresis equipment used were VAP-5, VAP-21 or VAP-22 continuous free-flow electrophoresis units (Bender & Hobein, Munich, FRG).

For separations involving plant cells, the electrophoresis medium (electrophoresis chamber buffer) contained 0.25 M sucrose, 2 mM KCl, 10 μM $CaCl_2$, 10 mM triethanolamine, and 10 mM acetic acid (pH 7.5) (NaOH). The electrode buffer contained 100 mM triethanolamine and 100 mM acetic acid pH 7.5. Conditions for the electrophoretic separations were constant voltage of 800 V/9.2 cm field, 165 ± 5 mamp, buffer flow 1.7 ml/fraction/h, sample injection 2.7 ml/h, and constant temperature of 6°C.

The distribution of membranes in each separation was monitored from the absorbance at 280 nm. The membranes were collected from individual fractions by centrifugation and fixed directly for electron microscopy or resuspended in appropriate buffer for assay.

Subfractionation of Golgi Apparatus

Golgi apparatus are polarised structures consisting of stacked cisternae. Cisternae at different levels in the stack exhibit different structural and functional characteristics including a gradient of increasing negative surface charge. The membrane constituents responsible for the charge gradient, which should reside on the membranes' cytoplasmic surfaces to influence the electrophoretic separation, are unknown.

Based on the cis to trans gradient of increasing net negative charge, different Golgi apparatus subfractions have been prepared (Fig 2). Cis cisternae and markers occur in the least electronegative fractions and trans cisternae and markers occur in the most electronegative fractions [8-11] NADPase, a marker for the intermediate Golgi apparatus compartments [12], coincides with electrophoretic fractions of intermediate electronegativity.

For the most part, confirmation of a cis to trans separation of Golgi apparatus cisternae from rat liver by free-flow electrophoresis has been provided by morphological and cytochemical criteria. When subjected to osmication (40°, 40 h) according to Friend and Murray [13], membranes and vesicles from the cis face were marked [8,9]. NADPH cytochemistry was used to mark membranes and vesicles from the midregion of the stack [11] while thiamine pyrophosphatase and galactosyltransferase cytochemistry, together with an NADH-ferricyanide reductase resistant to fixation [8,9] were utilised to mark trans face membranes.

In addition to thiamine pyrophosphatase activity (Fig. 3), a cis to trans gradient across the electrophoretic separation was found for the enzymes of terminal glycosylation, galactosyltransferase and sialyltransferase. Lumenal acidification, as evidenced by monensin-induced swelling and, in fibroblasts, by immunocytochemistry [14], is restricted to the trans-most Golgi apparatus cisternae. In the electrophoretic separations of unstacked Golgi apparatus, vesicle acidification, as determined with acridine orange as indicator, and ATPase activity are restricted to the most electronegative electrophoretic fractions.

With rat liver, intact Golgi apparatus stacks are first purified and then unstacked and subfractionated. For plants a similar approach has been followed that provides distinct subfractions enriched in Golgi apparatus markers, latent nucleoside diphosphate phosphatase and fucosyl

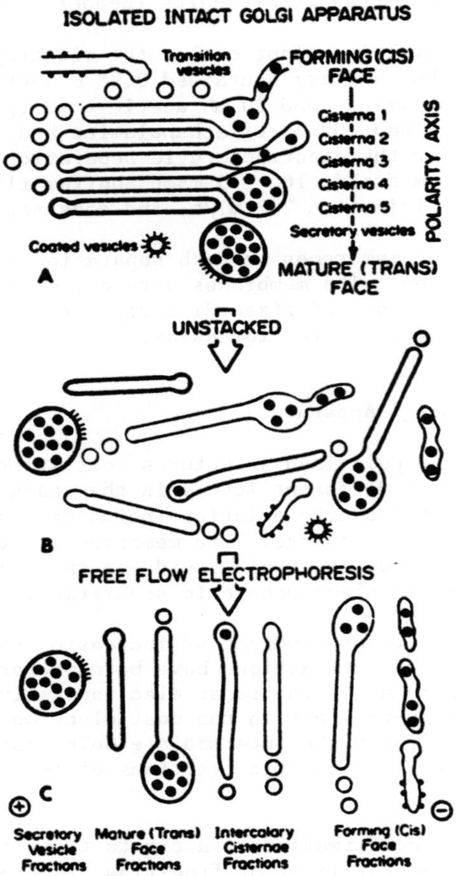

Fig. 2 *Diagrammatic representation of Golgi apparatus subfractionation by free-flow electrophoresis. A. The various components of the Golgi apparatus of the liver cell as displayed in cross-sectional representation from the cis to the trans face across the polarity axis. Differentiation of membranes from ER-like to plasma membrane-like is indicated by an increasing line width of the membranes. B. The same structure after unstacking into individual cisternal components. C. Resorting of the different structures by free-flow electrophoresis along the same polarity axis as seen in the starting material. By providing isolated Golgi apparatus subfractions derived from membranes predominantly from one face of the structure of the other, the technique can help elucidate molecular details of Golgi apparatus function in membrane flux and the processing of secretory products and membrane proteins.*

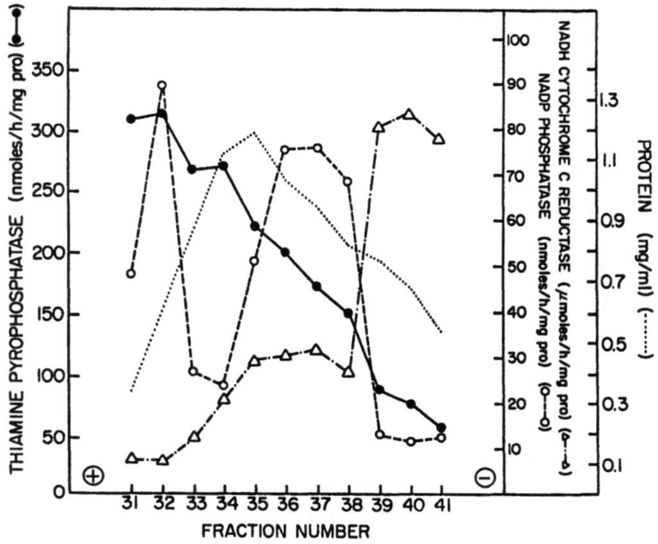

Fig. 3 *Electrophoretic separation of unstacked rat liver Golgi apparatus comparing specific activities of thiamine pyrophosphatase, a trans face marker, with those of NADP phosphatase which marks intermediate cisternae in hepatocytes of our rat strain [11] and NADH-cytochrome c reductase, a putative cis face marker [9]. The distribution of protein is shown by the dotted line which gives the absorbance determined at 280 nm. Unpublished results of A.O. Brightman, P. Navas, N. Minnifield and D.J. Morré, Purdue University.*

transferase. However definitive markers to differentiate cis, trans and intermediate Golgi apparatus compartments in plants remain to be developed.

Preparation of Plant Membranes

Mature plant cells contain two surface membranes, a plasma membrane at the cell surface and a tonoplast membrane surrounding the vacuole. These two membranes, important to transport processes, have been difficult to isolate in quantity, each free of contamination by the other. Using both etiolated stems of soybeans [15] and green leaves of spinach [16] these two membrane types have been completed resolved from a starting function of crude microsomes by preparative free-flow electrophoresis (Fig. 4). Fraction purity is >90% with essentially no cross contamination. The most electronegative fractions (A and B) contain tonoplast whereas the least electronegative fractions (D and E) contain plasma membrane (Fig. 4). Other cell components exhibit intermediate electrophoretic mobilities and are recovered from the central fractions of the electrophoretic separation.

The shoulder toward the anode (Fraction A) has been identified as nearly pure vacuole membrane (tonoplast) on the basis of the following characteristics. The membranes are 9-11 nm thick but do not stain with phosphotungstic acid at low pH (a characteristic of the plasma membrane

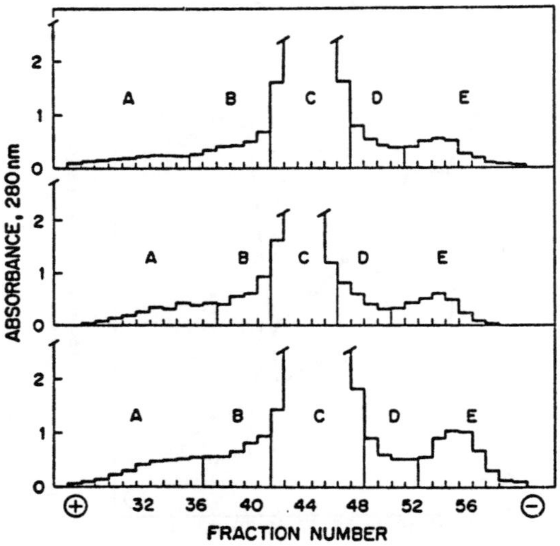

Fig. 4 *Free-flow electrophoretic separation of crude microsomal fractions from hypocotyls of dark-grown seedlings of soybean monitored by the absorbance at 280 nm. The letter designations refer to pooled fractions. From Sandelius et al 1986 [15]*

of plant cells). They contain an ATPase that is 80% inhibited by 50 mM nitrate as is characteristic for the ATPase of the tonoplast membrane [17].

The second shoulder toward the cathode (Fraction E) has membranes also 9-11 nm thick that react with phosphotungstic acid at low pH as is characteristic of plant plasma membranes [18]. The membranes bind N-α-naphthylphthalamic acid, an inhibitor of the polar transport of 3-indoleacetic acid, a growth hormone of the auxin type in plants, and contain an ATPase inhibited about 50% by vanadate but not by nitrate as is characteristic of the ATPase of plant plasma membranes.

While there is no obvious advantage of free-fow electrophoresis over aqueous two-phase partition for the isolation of plasma membranes, free-flow electrophoresis does provide the opportunity to simultaneously isolate both plasma membrane and tonoplast from the same homogenate.

Preparation of Plasma Membrane (and Tonoplast) Vesicles of Opposite Absolute Known Orientations

To isolate plasma membrane vesicles from a wide variety of sources, both plant and animal, aqueous two-phase partition is clearly a method of choice [19]. It is easy, reproducible and rapid. With only minor modifications necessary, the method can be adapted readily to a variety of tissue sources ranging from green leaves to tissue culture cells. For the latter, two-phase partition is especially useful since gradient methods have proven complicated to use due to density alterations with different culture conditions, and the presence of extensive cytoskeletal-membrane interactions that characterise mammalian cells in culture and interfere with the use of centrifugation and separations based on size and density.

We use mixtures of Dextran T-500 (Pharmacia) and poly(ethylene glycol) 3350 (Fisher) in 0.25 M sucrose and 5 mM potassium phosphate with the final pH and polymer concentration depending on the tissue source.

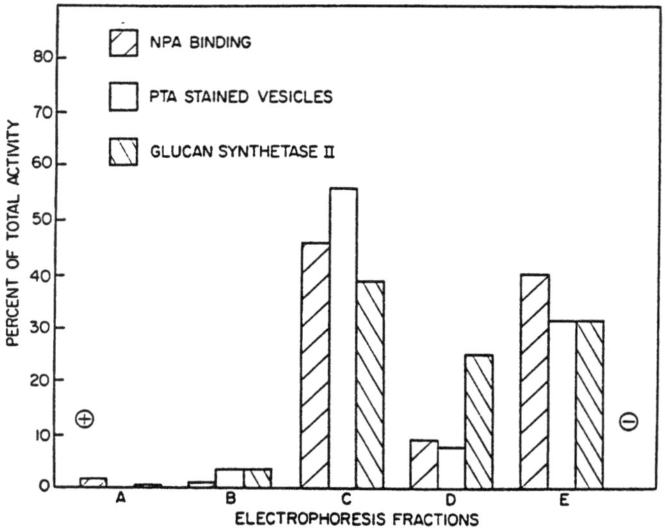

Fig. 5 *Expression of plasma membrane markers as percent of total activity recovered in each of the different pooled free-flow electrophoresis fractions from soybean as illustrated in Fig. 4. Evidence is for two populations of plasma membrane vesicles. One is primarily in fraction D and E. The other appears to overlap fraction C. From [21]*

With plants and all mammalian cell types thus far investigated, the plasma membrane isolates from aqueous two-phase partition are a mixture of right side out and inside out vesicles. The relative proportion may depend on the manner of tissue homogenization. These different populations may be resolved by preparative free-flow electrophoresis as described below.

An application of free-flow electrophoresis to separate plasma membrane vesicles of different orientations was indicated first from work with erythrocyte ghosts [20]. The existence of two plasma membrane populations also was indicated from the work with plants where two populations of plasma membrane (and tonoplast) markers were observed in the free-flow electrophoretic separations (Fig. 5). One population of each was clearly resolved. The other was contaminated by membrane components of the midportion of the separation [21].

To remove contaminating membranes, microsomes were first subjected to aqueous two-phase partition followed by free-flow electrophoresis of the membranes collected from the upper phase. Two populations of vesicles were obtained. The least electronegative showed little or no ATPase latency (the active site was exposed) and weak binding of concanavalin A to the external surface. The vesicles were identified as cytoplasmic side out. A more electronegative fraction showing high ATPase latency (vesicle disruption with detergent required to expose active sites) and strong binding of concanavalin A were identified as cytoplasmic side in (Table 1) [22]. A similar analysis has revealed two populations of tonoplast vesicles (cytoplasmic side out, least electronegative and cytoplasmic side in, more electronegative) also resolved by free-flow electrophoresis [23].

Table 1 *Structure-linked latency of ATPase of plasma membrane vesicles isolated from soybean hypocotyls by different methods*

Fraction	ATPase activity/nanomoles/min/mg protein -Triton X-100	+Triton X-100	Enhancement
Free-flow fraction E	26.7	53.5	2.0
Aqueous two-phase	11.3	38.9	3.4
PM_C from free-flow of aqueous two-phase	7.1	49.6	7.0
PM_E from free-flow of aqueous two-phase	30.0	55.8	1.9

Isolation of Plasma Membranes from Hepatomas and Cultured Cells by Two-Phase Partition and Resolution by Free-Flow Electrophoresis into Subpopulations

Hepatomas were homogenised in a glass/glass tissue homogenizer for 6 min at 600 rpm in a medium (1:2 w/v) containing 37 mM Tris-maleate, pH 6.4, 0.5 M sucrose, 1% dextran, 5 mM $MgCl_2$, 5 mM ß-mercaptoethanol and centrifuged for 15 min at 5,500 rpm (5000 x g). The bottom two-thirds of

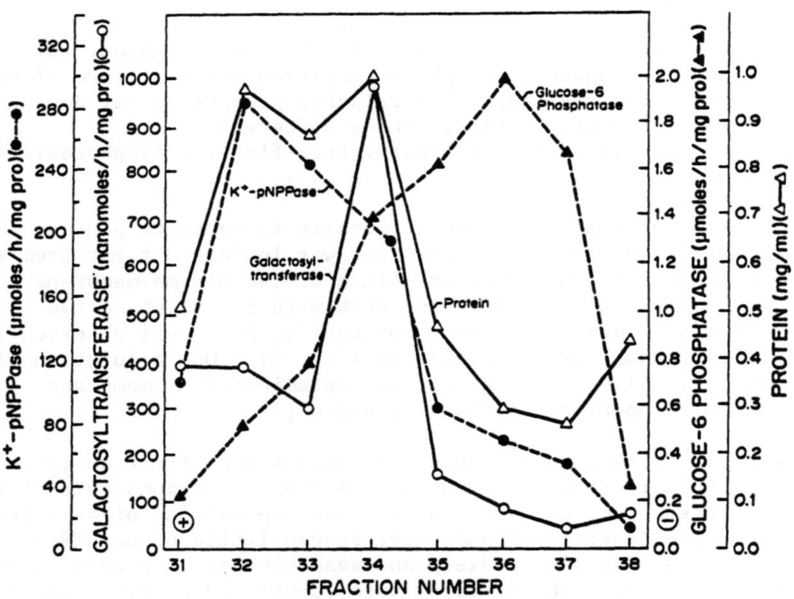

Fig. 6 *Electrophoretic separations of RLT-28 rat hepatoma homogenate prepared with a Teflon/glass homogenizer in hypotonic medium. Glucose-6-phosphatase marks endoplasmic reticulum, galactosyltransferase marks Golgi apparatus and potassium stimulated-, ouabain inhibited-p-nitrophenylphosphatase (K^+-pNPPase) marks plasma membrane. Data of P. Navas*

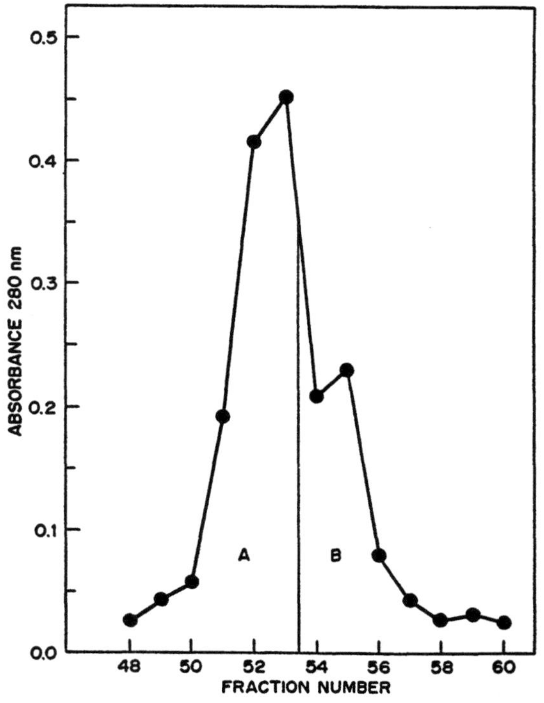

Fig. 7 *Electrophoretic separation of plasma membranes from RLT-28 hepatoma cells grown in culture. The plasma membrane vesicles were first purified by aqueous two-phase partition. The starting plasma membrane fraction and pooled fractions designated A and B were analyzed for vesicle sidedness by determination of concanavalin A binding as illustrated in Fig. 8. The anode (+) is on the left (see Fig. 6) such that the most electronegative membranes would be found in fraction A and the least electronegative membranes in fraction B. The relative amounts of material in the different pooled fractions was estimated from the absorbance determined at 280 nm. Data of D.D. Nowack, Purdue University.*

this pellet was resuspended in 5 ml of 1 mM bicarbonate in a conical glass/glass homogeniser for 2 min at 600 rpm with an additional 5 ml of bicarbonate being added after suspension. This mixture was centrifuged for 15 min at 5,250 rpm. The top half of this pellet was used for the two-phase partition.

For cultured cells, six 150 cm^2 flasks were harvested by rinsing with 10 mls of phosphate-buffered saline, pH 7.2, followed by scraping in 10 mls of PBS. The cells were collected by centrifugation for 10 min at 3,000 rpm, resuspended and swollen in 15 mls of HEPES-KOH, pH 7.2 containing 15 mM KCl and 1.5 mM Mg acetate (HEPES-KOH) for 10 min. The cells were pelleted again for 10 min at 3,000 rpm and resuspended in 2 mls of HEPES-KOH. This suspension was homogenised in a Dounce teflon/glass homogenizer for 2 min at 600 rpm with 20 up/down strokes. Next, 2 mls of nine-times concentrated HEPES-KOH were added and this mixture was centrifuged for 10 min at 1,000 rpm. The resulting supernatant was centrifuged for 30 min at 10,000 rpm and the pellet resuspended in 1 ml of

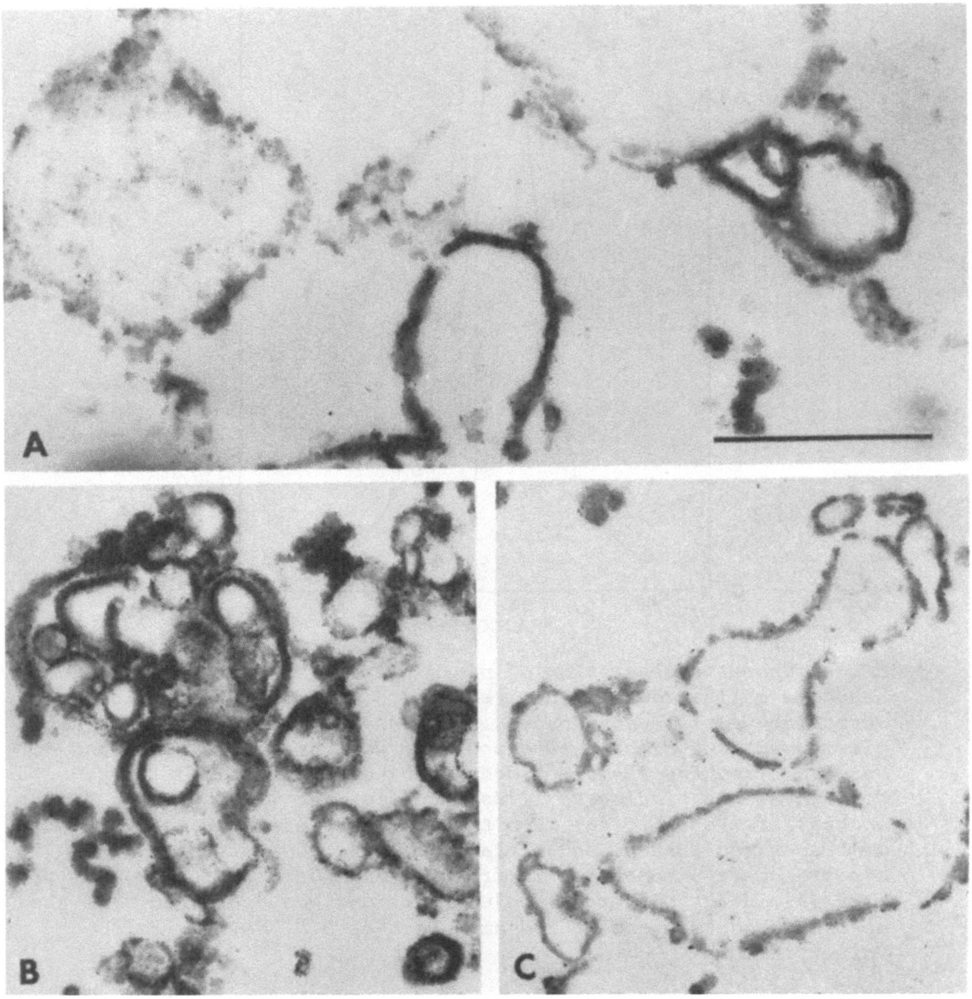

Fig. 8 *Concanavalin A binding of purified plasma membranes of cultured hepatoma (RLT-28) cells isolated by aqeuous two-phase partition with subsequent separation by preparative free-flow electrophoresis. The concanavalin A was linked to peroxidase to generate electron dense deposits via reaction with* H_2O_2, *diaminobenzidine and* OsO_4. *Plasma membrane vesicles oriented cytoplasmic side in bound concanavalin A strongly whereas vesicles oriented cytoplasmic side out bound concanavalin A only weakly. A. Total membranes obtained by aqueous two-phase partition; B. Electrophoresis fractions 47-53 bound concanavalin A and are cytoplasmic side in. C. Electrophoresis fractions 54-58 bound concanavalin A poorly and are presumed to be cytoplasmic side out. Data of D.D. Nowack, Purdue University. Bar = 0.5* μm

phosphate buffer pH 7.2 as the starting material for the two-phase partition.

The two-phase partition system contained 6.6% (w/w) Dextran T500, 6.6% (w/w) poly(ethylene glycol) 3350, 0.25 M sucrose, and 5 mM phosphate buffer, pH 7.2. Two tubes, each containing 7 g of the above mixture, were prepared. Suspended membranes (1 g) were added to one tube and phosphate buffer, pH 7.2 (1 g) was added to a second tube. The tubes were shaken vigorously 40 times in the cold (4°C) and the phases were separated by centrifugation for 5 min at 750 rpm in a swinging bucket rotor. The two upper phases were exchanged, the tubes shaken, and phases separated as before. The upper phase of both tubes were pooled, not taking the membranes at the interphase, and pelleted following dilution with 5 volumes of 1 mM bicarbonate by centrifugation for 30 min at 10,000 rpm.

With mammalian tissue culture cells, a heterogeneity of plasma membrane vesicle was observed as determined by measurement of the plasma membrane marker K^+-stimulated p-nitrophenyl phosphatase (pNPPase) (Fig. 6). Plasma membranes prepared by aqueous two-phase partition were subjected to preparative free-flow electrophoresis (Fig. 7). Vesicle sidedness was deduced from concanavalin A binding activity as determined by electron microscopy using concanavalin A conjugated with peroxidase. The initial plasma membrane isolates from two phase partition were mixtures of strongly and weakly reactive vesicles with concanavalin A peroxidase (Fig. 8A). Following preparative free-flow electrophoresis, the strongly reactive vesicles were concentrated in fraction A of the more electronegative part of the separation (Fig. 8B), whereas the weakly reactive vesicles were concentrated in fraction B in the least electronegative part of the separation (Fig. 8C). These findings suggest the presence of both inside out and right side out plasma membrane vesicles and their resolution by free-flow electrophoresis. The more electronegative population that binds concanavalin A peroxidase most strongly would be cytoplasmic side in whereas the less electronegative population that binds the lectin only weakly would be cytoplasmic side out.

The use of concanavalin A binding to determine vesicle sidedness for isolated plasma membrane preparations is relatively well established since the concanavalin A-binding glycoconjugates of the cell surface should be directed primarily to the cell exterior and thus be exposed only with cytoplasmic side in vesicles. In contrast, plasma membrane vesicles with cytoplasmic side out orientation would be unmarked since the concanavalin A is unable to penetrate the sealed vesicles. The availability of both right side out and inside out vesicles of known absolute orientation and resolved one from the other in separate fractions would facilitate greatly certain types of transport studies, for example, as well as to augment investigations of the topographical organization of constituents within the membrane.

The resolution of two vesicle populations from mammalian tissue culture cells will require independent confirmation from enzymatic studies such as ATPase latency or other criteria. However, the presence of both right side out and inside out vesicles in the preparations would help explain the heterogeneity of the plasma membrane distribution observed with homogenates of culture cells upon free-flow electrophoresis (Fig. 6).

In liver and other tissues characterised by junctional complexes, large sheets of plasma membrane are obtained. The dominant orientation is right side out (cytoplasmic side in). Whether these large sheets, once isolated from the cell, will retain their cytoplasmic side in orientation or also form a population of vesicles with mixed orientation if further reduced after isolation to smaller vesicles by sonication or other means has not been determined.

The lower phase membranes obtained after two-phase partition may provide a plasma membrane-free source of other membranes for use as reference fractions. After several washings to remove the dextran, the membranes have been subjected to preparative free-flow electrophoresis. In preliminary experiments with tissue culture cells (D.D. Nowack, Purdue University), this approach has provided fractions enriched in rough endoplasmic reticulum and fractions enriched in mitochondria and mitochondrial fragments. Even though the Golgi apparatus marker enzyme galactosyltransferase appears most concentrated in the mid-portions of the electrophoretic separations (e.g. Fig. 6), most Golgi apparatus, if present, are unstacked and represented by single cisternae and/or vesicles. This makes morphological identification difficult. Because of heterogeneity of membrane within the stack, electrophoretic methods as a primary method to purify Golgi apparatus membranes once separated from the stack may not be practrical. Here a final sucrose gradient may be advantageous.

Other Applications

In addition to Golgi apparatus and plasma membrane subfractions and isolation of tonoplast vesicles of known absolute orientation, free-flow electrophoresis has found other applications [7]. These include isolation of endosomes [24] and lysosomes [25,26], preparation of mitochondria [27] and plastid [28] subfractions, separation of right-side out and inside out red cell ghosts [20], purification of endoplasmic reticulum [29] and endoplasmic reticulum - derived subfractions such as transition vesicles induced in vitro [30].

Work supported by grants from the NIH CA 18801 and CA 36761.

REFERENCES

1 J.B.C. Findlay and W.H. Evans, eds., "Biological Membranes - A Practical Approach," IRL Press, Oxford-Washington DC (1987)

2 P.J. Courtoy, J. Quintart and P. Bauduin, Shift of equilibrium density induced by 3,3'-diaminobenzidine cytochemistry: a new procedure for the analysis and purification of peroxidase-containing organelles, J. Cell Biol. 98:870 (1984)

3 P.-Å. Albertsson, B. Andersson, C.K. Larsson and H.-E. Åkerlund, Phase partition - a method for purification and analysis of cell organelles and membrane vesicles, Meth. Biochem. Anal. 28:115 (1982)

4 K. Hannig and H.-G. Heidrich, Continuous free-flow electrophoresis and its application to biology, in: "Cell Separation Methods, Part IV. Electrophoretic Methods," H. Bloemendal (ed.) Elsevier-North Holland, Amsterdam (1977)

5 E.M. Croze and D.J. Morré, Isolation of plasma membrane, Golgi apparatus and endoplasmic reticulum fractions from single homogenates of mouse liver, J. Cell Physiol. 119:46 (1984)

6 D.J. Morré, A. Brightman and A.S. Sandelius, Membrane fractions from plant cells, in: "Biological Membranes - A Practical Approach," J.B.C. Findlay and W.H. Evans, eds., IRL Press, Oxford-Washington DC (1987)

7 D.J. Morré, K. Howell, W.H. Evans and G.M.W. Cook, eds. "Cell-Free Analysis of Membrane Traffic," Alan R. Liss, New York (1988)

8 D.J. Morré, D.M. Morré and H.-G. Heidrich, Subfractionation of Golgi apparatus by free-flow electrophoresis, Eur. J. Cell Biol. 31:263 (1983)

9 D.J. Morré, K.E. Creek, G.R. Matyas, N. Minnifield, I. Sun, P. Baudoin, D.M. Morré and F.L. Crane, Free-flow electrophoresis for subfractionation of rat liver Golgi apparatus, BioTechniques 2:224 (1984)
10 N. Minnifield, K.E. Creek, P. Navas and D.J. Morré, Involvement of cis and trans Golgi apparatus elements in the intracellular sorting and targeting of acid hydrolases to lysosomes, Eur. J. Cell Biol. 42:92 (1986)
11 P. Navas, N. Minnifield, I. Sun and D.J. Morré, NADP phosphatase: A marker in free-flow electrophoretic separations for cisternae of the Golgi apparatus midregion, Biochim. Biophys. Acta 881:1 (1986)
12 C.E. Smith, Ultrastructural localisation of nicotinamide adenine dinucleotide phosphatase (NADPase) activity to the intermediate saccules of the Golgi apparatus in rat incisor ameloblasts, J. Histochem. Cytochem. 28:16 (1980)
13 D.S. Friend and R.A. Murray, Osmium impregnation of the Golgi apparatus, Amer. J. Anat. 117:135 (1965)
14 R.G.M. Anderson and R.K. Pathak, Vesicles and cisternae in the trans Golgi apparatus of human fibroblasts are acidic compartments, Cell 40:635 (1985)
15 A.S. Sandelius, C. Penel, G. Auderset, A. Brightman, M. Millard and D.J. Morré, Isolation of highly purified fractions of plasma membrane and tonoplast from the same homogenate of soybean hypocotyls by free-flow electrophoresis, Plant Physiol. 81:177 (1986)
16 G. Auderset, A.S. Sandelius, C. Penel, A. Brightman, H. Greppin and D.J. Morré, Isolation of highly purified fractions of plasma membranes and tonoplast from spinach leaves by preparative free-flow electrophoresis and effect of photoinduction, Physiol. Plant 681 (1986)
17 R.R. Walker and R.A. Leigh, Characterisation of a salt-stimulated ATPase activity associated with vacuoles isolated from storage roots of red beets (Beta vulgaris L), Planta 153:140 (1981)
18 J.C. Roland, C.A. Lembi and D.J. Morré, Phosphotungstic acid-chromic acid as a selective electron-dense stain for plasma membrane of plant cells, Stain Tech. 47:195 (1972)
19 P. Kjellbom and C. Larsson, Preparation and polypeptide composition of chlorophyll-free plasma membranes from leaves of light-growth spinach and barley, Physiol. Plant 62:501 (1984)
20 H.-G. Heidrich and G. Leutner, Two types of vesicles from the erythrocyte-ghost membrane differing in surface charge. Separation and characterisation by preparative free-flow electrophoresis, Eur. J. Biochem. 41:37 (1974)
21 D.J. Morré, A. Brightman, G. Scherer, B. vom Dorp, C. Penel, G. Auderset, A.S. Sandelius and H. Greppin, Highly purified tonoplast fractions by preparative free-flow electrophoresis, in: "Proc. 1st International Workshop on Plant Vacuoles," B. Marin, ed., Plenum, New York (1987)
22 H. Canut, A. Brightman, M. Boudet and D.J. Morré, Plasma membrane vesicles of opposite sidedness from soybean hypocotyls by preparative free-flow electrophoresis, Plant Physiol. 72:631 (1988)
23 H. Canut, A. Brightman, A. Boudet and D.J. Morré, Preparation of highly purified tonoplast vesicles of opposite orientation from soybean hypocotyls, in preparation
24 M. Marsh, S. Schmid, H. Kern, E. Harms, P. Male, I. Mellman and A. Helenius, Rapid analytical and preparative isolation of functional endosomes by free-flow electrophoresis, J. Cell Biol. 104:875 (1987)
25 R. Henning and H.-G. Heidrich, Membrane lipids of rat liver lysosomes prepared by free-flow electrophoresis, Biochem. Biophys. Acta 345-326 (1974)

26 E. Harms, J. Kartenbeck, G. Darai and J. Schneider, Purification and characterisation of human lysosomes from EB-Virus transformed lymphoblasts, Exp. Cell Res. 131:251 (1981)
27 H.-G. Heidrich, The sidedness of mitochondrial membranes and their subparticles as characterised by preparative free-flow electrophoresis, FEBS Lett. 17:253 (1971)
28 J.-P. Dubacq and J.-C. Kader, Free-flow electrophoresis of chloroplasts, Plant Physiol. 61:465 (1978)
29 G.R. Matyas and D.J. Morré, Subcellular distribution and biosynthesis of rat liver gangliosides, Biochim. Biophys. Acta 921:599 (1987)
30 D.D. Nowack, D.M. Morré, M. Paulik, T.W. Keenan and D.J. Morré, Intracellular membrane flow: Reconstitution of transition vesicle formation and function in a cell-free system, Proc. Natl. Acad. Sci., USA 84:6098 (1987)

ANALYSIS OF THE DOMAIN STRUCTURE OF MEMBRANES BY FRAGMENTATION AND SEPARATION

Per-Åke Albertsson

Department of Biochemistry
University of Lund, Box 124
S-221 00 Lund, Sweden

Our aim with separation is often preparative. We want to obtain a purification of a desired substance. Separation can also be used as an analytical tool if we want to learn about the composition of a mixture or the heterogeneity of a cell population. I would like to point out the possibility of using separation in combination with fragmentation for gaining information on the structure of a cell organelle or a biological membrane. Selective fragmentation with specific enzymes followed by separation and puzzle analysis is used for the determination of the sequence of proteins and nucleic acids. A similar approach should be possible in the case of larger structures such as membranes. Since we do not yet have specific enzymes available which can cleave the membranes selectively, we have to resort to random fragmentation. Consider for example an object (Fig. 1) composed of three domains A, B and C. If we break this to pieces by a mechanical random fragmentation we will get fragments varying in size and composition. If we have in our hands two separation methods, one of which can separate according to size and the other according to chemical composition, we can get the diagrams of Fig. 1. The large fragments obtained by centrifugation will not give distinct peaks (Fig. 1a) while the smallest fragments will give three well separated peaks when separated by a method, such as aqueous phase partition, which separates according to surface chemical composition. Between peak A and B we will find fragments composed of A and B, and between peaks B and C we will find fragments composed of B and C. However, since we will not find any fragments composed of only A and C, we can conclude that A and C cannot be neighbours in the original object. Thus, fragmentation and separation has given us information of the structure of the original object.

It can be shown that if a membrane consists of two domains, one enriched in component A and the other enriched in B, then there is a linear relation between the concentration of A against that of B for different fragments. We may compare with a phase diagram of a two phase system. The two domains of the membrane can be considered as two two-dimensional phases in equilibrium and the composition of fragments containing both these domains will have a composition which is represented by points on the tie line connecting the two

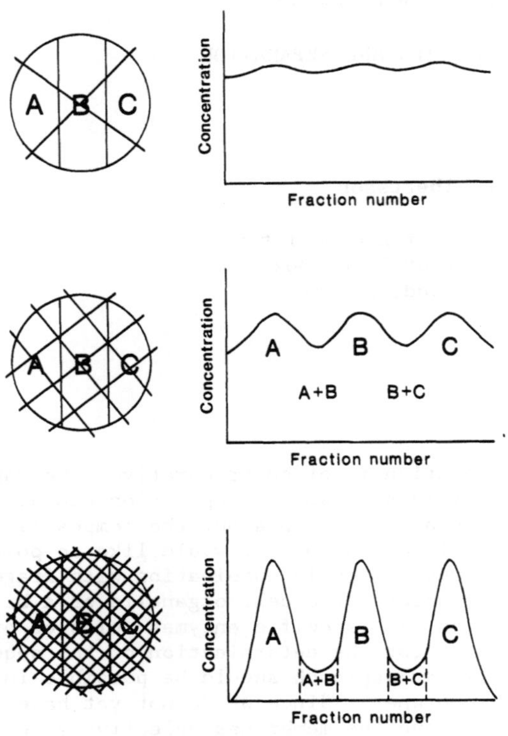

Fig. 1 *An object which consists of three domains A, B and C, is randomly fragmented. Upper: large fragments give rise to a continuous variation of composition; most fragments contain parts from all three domains. No distinct peaks are obtained in the separation diagram to the right. Middle: smaller fragments will give rise to some fragments enriched in the components A, B and C respectively; peaks are obtained in the separation diagram. Lower: even smaller fragments will give rise to three distinct peaks in the separation diagram representing pure A, B and C respectively. Some fragments contain both A and B and some B and C. No fragments contain A and C (from reference 1)*

phases. Fig. 2 shows such a plot for the thlakoid membrane. P 700 represents photosystem I which is enriched in the stroma exposed domain of the thylakoid and is plotted versus chlorophyll b which is enriched in the photosystem II rich partitions. From this plot we can conclude that the thylakoid consists of at least two different domains. Moreover by extrapolation we can predict the composition of the photosystem II domain

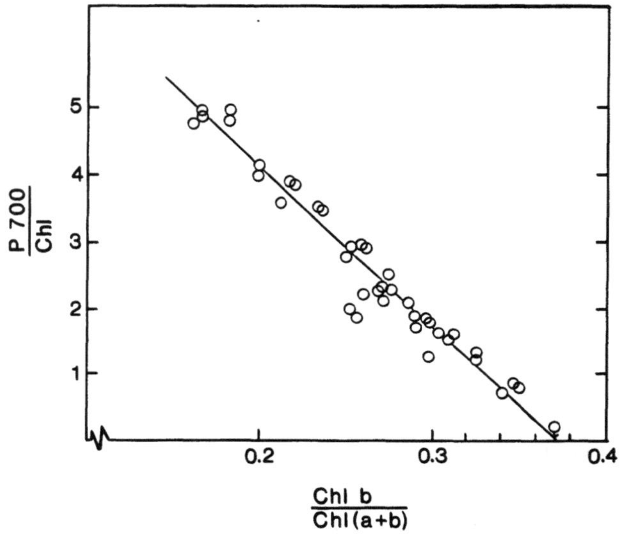

Fig. 2 *Thylakoids were fragmented by sonication and separated by counter-current distribution. The figure shows a plot of P700/chlorophyll (a+b) against chlorophyll b per chlorophyll (a+b) of the different fractions. The linear relation suggests that PSI (P700) and PSII (chlorophyll b) are located in two separate domains (reference 2)*

with respect to the chlorophyll composition if we assume the P 700 concentration to be zero.

If we assume that a membrane consists of three or more domains the situation will be much more complex. Depending on the size of the fragments we will get different diagrams when we plot the concentration of one component against the concentration of one of the others of the different fragments obtained by random fragmentation. Fig. 3 shows such a plot for a model of a membrane containing three different domains. A more detailed discussion has been published elsewhere [1]. As shown in the next paper and in reference [3] an application of this approach has been used on the rat liver plasma membrane.

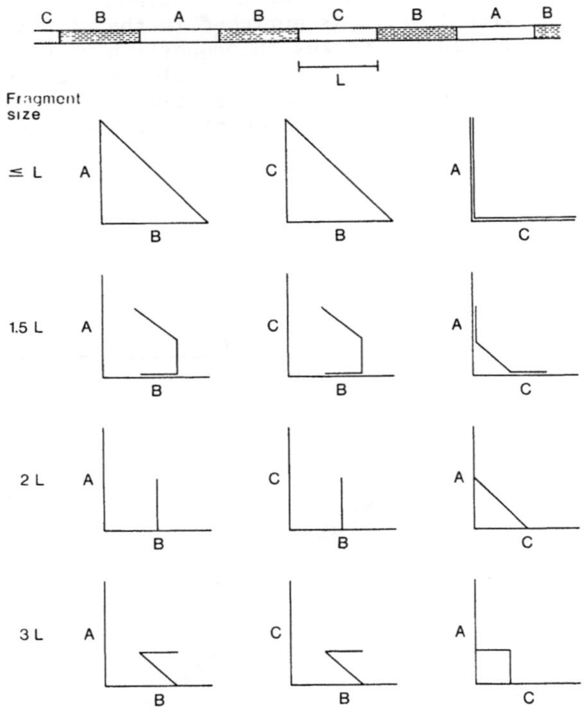

Fig. 3 *On top, cross-section through a model of a membrane with three different domains containing the components A, B and C, respectively. When the membrane is fragmented randomly different fragments with different contents of A, B and C are obtained, depending on the size of the fragments and the location of the cuts. The plots show the concentration of one component versus the concentration of one of the other components for different fragments within four different size classes, i.e. the sizes of the fragments are* <L, *1.5* L, *2* L *and 3* L *respectively (indicated to the left of the figure).* L *is the size of each domain (reference 1)*

REFERENCES

1 P.-Å. Albertsson, Quart. Rev. Biophysics 21:61 (1988)
2 P.-Å. Albertsson, Chemica Scripta 27B:189 (1987)
3 P. Gierow, B. Jergil and P.-Å. Albertsson, Biochem. J. 249:369

TWO-PHASE PARTITIONING OF RAT LIVER PLASMA MEMBRANES

Peter Gierow, Bengt Jergil and Per-Åke Albertsson

Department of Biochemistry
Chemical Centre, University of Lund
POB 124, S-221 00 Lund, Sweden

Rat liver plasma membranes were subjected to countercurrent distribution in a dextran/poly(ethylene glycol) two-phase system. The lateral heterogeneity was examined by plotting the specific activities of marker components, for each distribution fraction, against each other. The patterns obtained were compared with those of different model membranes. By this method 5'-nucleotidase and leucine amino-peptidase were found in separate but adjacent domains, whereas asialo orosomucoid receptors resided in a domain separated from those containing 5'-nucleotidase or leucine aminopeptidase by a domain lacking these markers.

INTRODUCTION

In the plasma membrane of rat hepatocytes at least three different domains can be recognised; the blood-sinusoidal, the bile canalicular and the contiguous domain. By various centrifugation techniques it is possible to obtain fractions enriched in different domains as judged by analysis of marker components [1-6].

We have previously reported on the separation of blood-sinusoidal and bile-canalicular membranes by thin-layer countercurrent distribution [7]. Whatever method is used, a complete separation of marker components is difficult to attain, despite the fact that different membrane properties are used for fractionation. These difficulties could be due not just to cross-contamination, but also to heterogeneity within the membranes.

A method for analysis of lateral heterogeneity has recently been described by Albertsson [8]. The general basis is that random fragmentation of a membrane, containing two or several domains, will result in fragments of different composition depending on domain size and organisation. After fractionation, e.g. by counter-current distribution, the shape of a plot of the concentration of one marker component against that of another would depend on fragment size and the lateral distribution of the markers in the membrane.

This procedure was applied on isolated rat liver plasma membranes, fragmented by sonication for various periods of time and fractionated by centrifugal countercurrent distribution. The resulting plots were compared with those of model membranes.

EXPERIMENTAL

Stock solutions of 20% (w/w) Dextran T500 (Pharmacia Fine Chemicals AB) and 40% (w/w) poly(ethylene glycol) 3350 (Carbowax 3350, Union Carbide) were prepared and checked as described by López-Perez [9]. ^{125}I-iodine and ^{3}H-AMP were purchased from Amersham International plc. All other reagents used were of analytical grade.

A microsomal plasma membrane fraction was prepared from male Sprague-Dawley rats (around 200 g) by the method of Aronson and Touster [2], modified as described earlier [7,10]. The final fraction (the P_2-fraction of Aronson and Touster) was suspended in 15 mM Tris/H_2SO_4, pH 7.8, by 20 up and down strokes in a Dounce-homogeniser (tight-fitting pestle). The suspension was sonicated (0, 1.5 or 6 minutes) in a Branson Sonifer cell disruptor, model B15, equipped with a microtip (intensity output 2, 50% pulsed duty cycle) and used directly.

Two-phase partitioning was performed in a centrifugal countercurrent distribution apparatus [11]. Briefly, the partition block consists of an outer ring for the lower phase and an inner ring for the upper phase, together making up 60 partition cavities. Separate top and bottom phases were prepared in advance as described previously [7].

Plasma membranes (20 mg of protein) were added to a sample-system with a total weight of 10.00 g, containing 6.0% Dextran T500, 6.0% poly(ethylene glycol) 3350, 3 mM KCl and 15 mM Tris/H_2SO_4, pH 7.8. The volume ratio was adjusted to 1:1 by addition of bottom phase. The volume of the lower phase compartment was 1.00 ml, but only 0.90 ml was used in order to keep the interface and the lower phase stationary. Thus, 1.80 ml of well mixed sample-system was added to each of cavities 0-4, whereas the remaining cavities were each loaded with 0.90 ml of bottom phase and an equal amount of top phase. Shaking was set to 30 s, centrifugation to 90 s and the whole cycle was completed within 2 h after 53 transfers. The phase system of each cavity was broken by the addition of 0.90 ml Tris/HCl, ph 7.4, and mixed by shaking for 1 min. The unloaded fractions were analysed immediately.

Protein was determined by the method of Lowry et al [12], modified as described earlier [7]. The plasma membrane markers measured were: 5'-nucleotidase [13] (EC 3.1.3.5.), leucine aminopeptidase [14] (EC 3.4.11.1.) and ^{125}I-asialo orosomucoid binding [7,14].

RESULTS AND DISCUSSION

Countercurrent distribution of unsonicated plasma membranes (Fig. 1) resulted in a two-peak pattern in which more than 85% of the protein and the asialo orosomucoid receptors, the blood-sinusoidal marker, were recovered in the left hand peak. In contrast, the bile-canalicular marker, leucine aminopeptidase, distributed into two distinct peaks. 5'-Nucleotidase, a more general plasma membrane marker, distributed in a way similar to the latter. Both enzymes were more or less equally divided between the two peaks.

Sonicating the membranes for 1.5 min before countercurrent distribution dramatically shifted leucine aminopeptidase and 5'-nucleotidase from the left to the right hand peak. Protein also shifted to the right resulting in an even distribution between the two peaks, whereas asialo orosomucoid binding was less affected.

Sonication for 6 min caused no major changes in the distribution pattern. Increased amounts of protein and, to some extent, asialo-orosomucoid binding were recovered in the right hand peak.

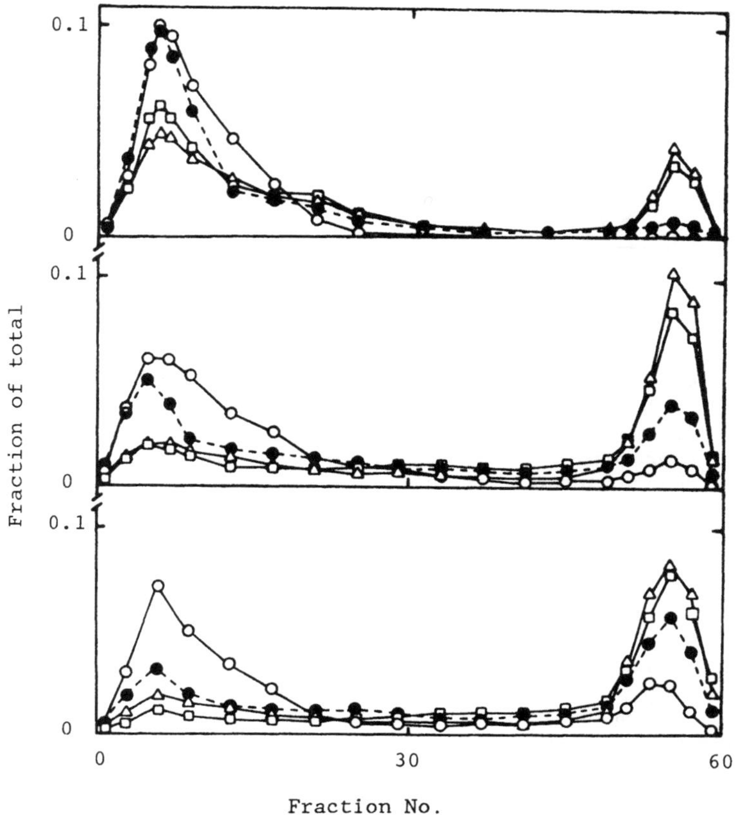

Fig. 1 *Countercurrent distribution of plasma membranes sonicated for 0-6 min. The distributions were performed as described in Experimental. The sonication periods were 0 min (top), 1.5 min (middle) and 6 min (below). Marker enzyme activities and protein were calculated as percentages of total amounts recovered and expressed as fractions of total. Recoveries were above 90%. Protein (●), 5'-nucleotidase (□), leucine aminopeptidase (Δ) and asialo-orosomucoid binding (○)*

Plotting the specific binding of asialo-orosomucoid binding against the specific activity of either 5'-nucleotidase or leucine-amino peptidase resulted, in the case of unsonicated membranes, in a rather complex pattern (Figs. 2 and 3). Similar results were obtined after 1.5 min of sonication. In both cases a vertical line was connected with a horizontal line by a line with negative slope. Prolonged sonication shifted them into right-angled configurations.

These results are in good agreement with a hypothetical model with three domains, each having the length L in cross-section (Fig. 4). Two domains

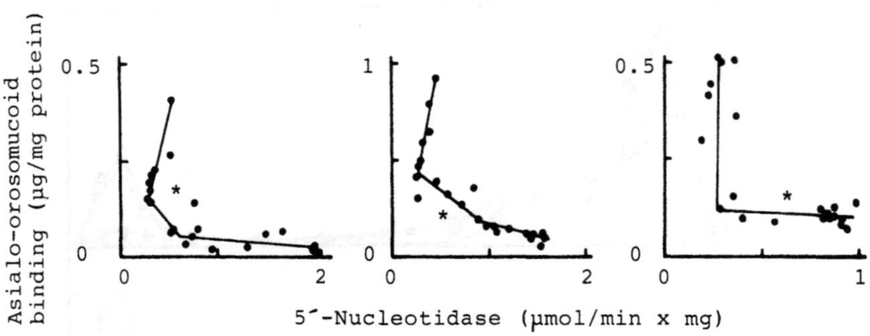

Fig. 2 *Specific binding of asialo-orosomucoid against specific activity of 5′-nucleotidase after different sonication periods. The membranes were sonicated for 0 (left panel), 1.5 (middle) and 6 min (right), respectively. The * represents the specific activity of material prior to countercurrent distribution*

containing either component A or B are separated by the third domain. Decreasing the fragment size changes the plot of the concentration of A vs that of B, from a straight line with a negative slope to a right-angled configuration. The latter will coincide with the abscissa and the ordinate if component A and B are completely restricted to respective domains.

Plotting the specific activities of leucine aminopeptidase and 5′-nucleotidase against each other (Fig. 5) produced a straight line with a positive slope in the case of unsonicated membranes. Decreasing the fragment size by sonication for 1.5 min introduced a break in the line and further fragmentation shifted the plot into a right-angled configuration.

These results could be compared with a model in which the components B and C are located in adjacent domains but well separated from component A (Fig. 6). Large fragments result in a straight line with a positive slope, as with unsonicated membranes above (Fig. 4). Reducing the fragment size results in plots of various complexity, but with the plot of fragment size 2L well corresponding to that of membranes sonicated for 6 minutes. This

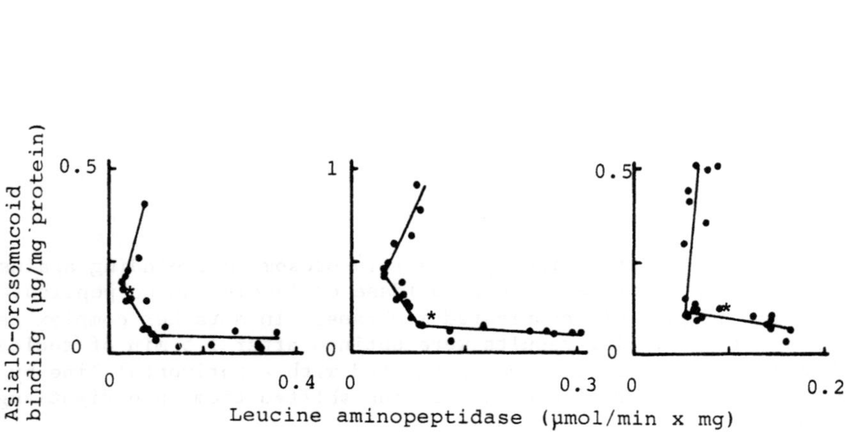

Fig. 3 *Specific binding of asialo-orosomucoid against specific activity of leucine aminopeptidase after different sonication periods. See Fig. 2 for details.*

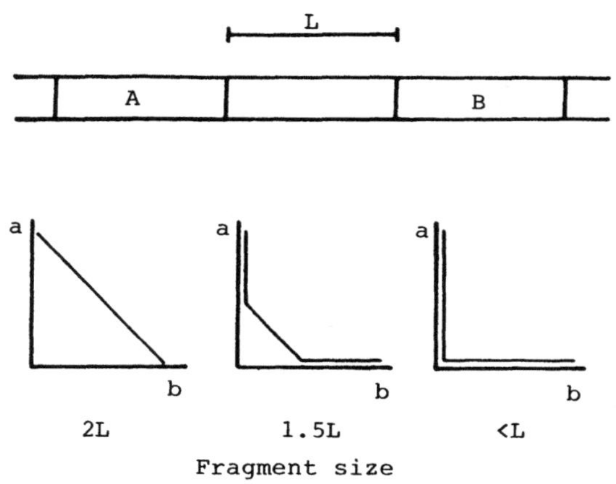

Fig. 4 *Upper: cross-section of a hypothetical membrane consisting of three alternating domains (each of the size L), one containing component A, one containing component B and in between one domain devoid of these components. Lower: Plots of concentrations of A (a) against B (b) for different fragment sizes.*

indicates that 5'-nucleotidase and leucine aminopeptidase are located in close but separate domains.

The results obtained are in good agreement with the current knowledge of the distribution of these markers in rat hepatocytes [3,14-17]. However, the observed differences in the distribution of 5'-nucleotides and leucine aminopeptidase have not been shown before even though some discrepancies in density-gradient profiles have been reported [18].

CONCLUSION

Two-phase countercurrent distribution has previously been shown to be useful for the separation and purification of cell organelles and membrane vesicles. The present work also shows it to be a tool in the analysis of lateral heterogeneity of membranes. The plotting method in conjunction with the countercurrent distribution diagram can be used to outline models of membranes.

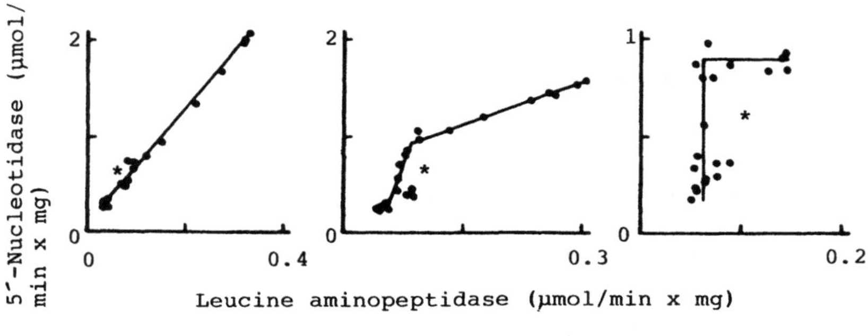

Fig. 5 *Specific activities of 5'-nucleotidase and leucine aminopeptidase against each other after different periods of sonication. For details see Fig. 2*

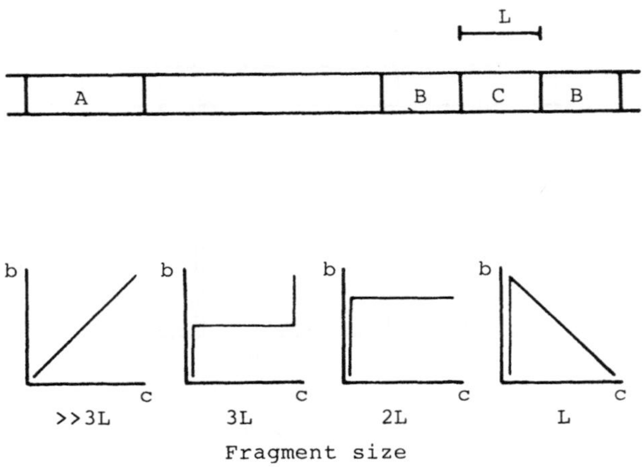

Fig. 6 *Upper: cross-section of a hypothetical membrane consisting of domain C surrounded by domain B (both having the size L). These domains are well separated from a third domain, A, by a fourth domain devoid of any of the corresponding markers. Lower: plots of concentrations of B (b) against C (c) for fragments of different sizes, obtained by random fragmentation*

REFERENCES

1 D.M. Neville, Jr., The isolation of a cell membrane fraction from rat liver, J. Biochem. Biophys. Cytol. 8:413 (1960)

2 N.N. Aronson and O. Touster, Isolation of rat liver plasma membrane fragments in isotonic sucrose, Methods Enzymol. 31:90 (1974)

3 M.H. Wisher and W.H. Evans, Functional polarity of the rat hepatocyte surface membrane. Isolation and characterisation of plasma-membrane subfractions from the blood-sinusoidal, bile canalicular and contiguous surfaces of the hepatocyte, Biochem. J. 146:375 (1975)

4 A.L. Hubbard, D.A. Wall and A. Ma, Isolation of rat hepatocyte plasma membranes. 1. Presence of the three major domains, J. Cell Biol. 96:217 (1983)

5 J.A. Taylor, D. Lawson and J.D Judah, Analysis and purification of the blood sinusoidal domain of rat liver plasma membrane, Biochim. Biophys. Acta 732:154 (1983)

6 P.J. Meier, E.S. Sztul, A. Reuben and J.L. Boyer, Structural and functional polarity of canalicular and basolateral plasma membrane vesicles isolated in high yield from rat liver, J. Cell Biol 98:991 (1984)

7 P. Gierow, M. Sommarin, C. Larsson and B. Jergil, Fractionation of rat liver plasma-membrane regions by two-phase partitioning, Biochem. J. 235:685(1986)

8 P.-Å. Albertsson, Analysis of the domain structure of membranes by fragmentation and separation in aqueous polymer two-phase systems, Quart. Rev. Biophys. 21:61 (1988)

9 M.J. López-Pérez, G. Paris and C. Larsson, Highly purified mitochondria from rat brain prepared by phase partition, Biochim. Biophys. Acta 635:359 (1981)

10 P. Gierow, B. Jergil and P.-Å. Albertsson, Lateral heterogeneity of rat liver plasma membranes analysed by countercurrent distribution, Biochem. J. 249:369 (1988)

11 H.-E. Åkerlund, An apparatus for countercurrent distribution in a centrifugal acceleration field, J. Biochem. Biophys. Methods 9:133 (1984)
12 O.H. Lowry, N.J. Rosebrough, A.L. Farr and R.J. Randall, Protein measurement with the folin phenol reagent, J. Biol. Chem. 193:265 (1951)
13 J. Avruch and D.F.H. Wallach, Preparation and properties of plasma membrane and endoplasmic reticulum fragments from rat liver cells, Biochim. Biophys. Acta 233:334 (1971)
14 S. Matsuura, H. Nakada, T. Sawamura and Y. Tashiro, Distribution of an asialoglycoprotein receptor on rat hepatocyte cell surface, J. Cell Biol. 95:864 (1982)
15 L.M. Roman and A.L. Hubbard, A domain-specific marker for the hepatocyte plasma membrane. Localisation of leucine aminopeptidase to the bile canalicular domain, J. Cell Biol. 96:1548 (1983)
16 S. Matsuura, S. Eto, K. Kato and Y. Tashiro, Ferritin immunoelectron microscopic localisation of 5'-nucloetidase on rat liver cell surface, J. Cell Biol. 99:166 (1984)
17 J.R. Bartles, L.T. Braiterman and A.L. Hubbard, Endogenous and exogenous domain markers of the rat hepatocyte membrane, J. Cell Biol. 100:1126 (1985)
18 C.W. Slife, M.D. Dorsett, G.T. Bouquett, A. Register, E. Taylor and S. Conroy, Subcellular localisation of a membrane-associated transglutaminase activity in rat liver, Arch. Biochem. Biophys. 241:329 (1985)

PLASMA MEMBRANES FROM HEPATOMAS AND CULTURED CELLS BY AQUEOUS TWO-PHASE PARTITION

D. David Nowack, Dorothy M. Morré and D. James Morré

Department of Foods and Nutrition and
Department of Medicinal Chemistry and Pharmacognosy
Purdue University, West Lafayette, IN 47907, USA

Neoplastic cultured cells and tissues often exhibit altered plasma membrane morphologies and biochemical activities when compared to non-transformed counterparts [1]. However, accurate interpretation and reliability of data collected for such comparisons may be influenced greatly by the purity and composition of the plasma membrane fractions utilised. Non-uniform homogenisation, contamination by organelles other than plasma membrane and changes in culture conditions , together with the transformation process itself, all tend to confound the fractionation of plasma membranes from cells and hepatomas.

By using a mixture of dextran (6.6%, w/w), poly(ethylene glycol) (6.6%, w/w), 0.25 M sucrose and 5 mM potassium phosphate as the partitioning system, plasma membrane vesicles in high yield and purity were obtained from total homogenates of hepatomas and from a variety of cultured cell lines both transformed and non-transformed. Potassium-stimulated, ouabain-inhibited p-nitrophenylphosphatase was used as an enzymatic marker of plasma membrane to assess yield and recovery. Starting fractions and ending materials were monitored by electron microscopy, together with morphometric methods to quantitate plasma membrane recoveries from cells and tissues.

Hepatomas were homogenised in a glass/glass tissue homogeniser for 6 min at 600 rpm in a medium (1:2 w/v) containing 37 mM tris-maleate, pH 6.4, 0.5 M sucrose, 1% dextran, 5 mM $MgCl_2$ and 5 mM ß-mercaptoethanol and centrifuged for 15 min at 5,000 X g. The bottom two thirds of this pellet was resuspended in 5 ml of 1 mM sodium bicarbonate in a conical glass/glass homogeniser for 2 min at 600 rpm with an additional 5 ml of bicarbonate being added after suspension. This mixture was centrifuged for 15 min at 5,000 X g. The top half of this pellet was used for the two-phase partition.

For cultured cells, six 150 cm^2 flasks were harvested by rinsing with 10 ml of phosphate-buffered saline, pH 7.2, followed by scraping in 10 ml of the PBS. The cells were collected by centrifugation for 10 min at 3,000 rpm, resuspended and swollen in 15 ml of HEPES-KOH, pH 7.2 containing 15 mM KCl and 1.5 mM Mg acetate (Medium A) for 10 min. The cells were pelleted again for 10 min at 3,000 rpm and resuspended in 2 ml of Medium A. This suspension was homogenised in a Dounce teflon/glass homogeniser for 2 min at 600 rpm with 20 up/down strokes. Next, 2 ml of Medium A were added and this mixture was centrifuged for 10 min at 1,000 rpm. The resulting supernatant was centrifuged for 30 min at 10,000 X g and the pellet resuspended in 1 ml phosphate buffer, pH 7.2 as the starting material for the two-phase partition procedure.

The two-phase partition system contained 6.6% (w/w) Dextran T500, 6.6% (w/w) poly(ethylene glycol) 3350, 0.25 M sucrose and 5 mM phosphate buffer, pH 7.2. Two tubes, each containing 7 g of the above mixture, were prepared. To one tube, one gram of suspended membranes was added. To the second tube, one gram of phosphate buffer, pH 7.2, was added. The tubes were shaken vigorously 40 times in the cold and the phases were separated by centrifugation for 5 min at 750 rpm in a swinging bucket rotor. The two upper phases were exchanged, the tubes shaken, and phases separated as before. The upper phases of both tubes were pooled, not taking the membranes at the interface, and pelleted following dilution with 5 volumes of 1 mM bicarbonate by centrifugation for 30 min at 10,000 X g.

The K^+-stimulated, ouabain-inhibited, p-nitrophenylphosphatase (EC 3.1.3.1) used to assess the yield of plasma membrane was, according to Kashiwampta et al (1979) [2]. Other enzymatic markers were as described by Croze and Morré (1984) [3].

The aqueous two-phase partition procedure provided plasma membranes of about 90% purity as determined from electron microscope morphometry from a variety of solid tumours and cell grown in culture. Purity was verified for one solid hepatoma and a cultured hepatoma line from a detailed analyses of enzymatic markers. Enrichments, based on K^+-stimulated, ouabain inhibited p-nitrophenylphosphatase were 7- to 30-fold depending on the starting material. Recoveries of 25% or greater may be achieved. Most important, however, is that the procedure appears to be completely general in that it is applicable to most, if not all, types of cultured cells and solid tumours.

ACKNOWLEDGEMENTS

Work supported in part by grants from the National Institutes of Health CA 36761 (to DJM) and from the American Diabetes Association (to DMM).

REFERENCES

1 D. Wallach, Membrane Molecular Biology of Neoplastic Cells, Elsevier, New York (1975)

2 S. Kashiwampta, S. Goto, R. Samba and F. Suzuki, Inhibition by bilirubin of (Na^+-K^+)-activated adenosine triphosphatase and K^+-activated p-nitrophenyl phosphatase activities of nal-treated microsomes from young rat cerebrum J. Biol. Chem. 254:4577 (1979)

3 E.M. Croze and D.J. Morré, Isolation of plasma membrane, Golgi apparatus, and endoplasmic reticulum fractions from single homogenates of mouse liver, J. Cell Physiol 119:46 (1984)

AFFINITY PARTITIONING AND CENTRIFUGAL COUNTERCURRENT DISTRIBUTION OF SYNAPTOSOMES FROM CALF BRAIN CORTEX

B. Olde and G. Johansson

Department of Biochemistry
Chemical Center
Box 124, S-221 00 Lund, Sweden

The aim of this work was to test the possibility of using affinity partitioning, in combination with centrifugal-CCD, for fractionation of a crude synaptosomal preparation.

EXPERIMENTAL

The following ligand polymers were synthesised and their effects on the partitioning of synaptosomes were tested.

HE-PEG 6000 Hexaethonium-PEG 6000 was synthesised as described elsewhere [1].

ABT-PEG 6000 and ABT-Dextran T500 - ABT,(3-(2'-aminobenzhydryloxy)-tropane), were prepared as described by Haga and Haga [2]. The ABT was then coupled by reacting it with tresylated PEG 6000 and tresylated Dextran T500.

ACR-PEG 6000 and ACR-Dextran T500 - ACR,(9-aminoacridine) purchased from Janssen was coupled in the same way as ABT.

A new and efficient route was used for tresylation of dextran.

The polymers were tested for their ability to influence the partition of a synaptosome preparation in a phase system containing 6.2% (w/w) Dextran T40, 7.4% (w/w) PEG 4000, 2 mM phosphate-tris pH 7.4 and 0.32 M sorbitol. When testing the PEG-derivatives, the phase-system also contained 2 mM NaCl. ABT-PEG, ACR-PEG and ACR-Dextran had no detectable effect on the partition under the conditions used. The effect of ABT-Dextran was weak and uncertain. The HE-PEG, however, proved to be efficient in displacing material into the upper phase. A centrifugal-CCD was carried out with synaptosomes prepared essentially as described by [3]. The composition of the phase system was 6.2% (w/w) Dextran T40, 7.4% (w/w) PEG 4000, 2 mM phosphate-tris pH 7.4, 2 mM NaCl and 0.32 M sorbitol. The material was included in the first five wells which also contained HE-PEG (0.2% of total PEG).

RESULTS AND DISCUSSION

The material displayed inhomogeneity both in respect of distribution of material and in the distribution of marker enzymes. Glutamine synthetase (a marker of astroglia) displayed its highest specific activities to the right of the CCD-train while specific activities of acetyl transferase (a marker of cholinergic synaptosomes) and of acetylcholine-esterase were more evenly distributed. This suggests that the high affinity binding site for hexaethonium-PEG is only partly the cholinergic receptor or the acetylcholine esterase. This was confirmed when both eserine and quinuclidinebenzilate (QNB) were found able to block only a minor part (10%) of the extracting effect of the HE-PEG. The nature of the remaining high affinity binding sites is not known.

REFERENCES

1 G. Johansson, R. Gysin and S.D. Flanagan, Affinity partitioning of membranes. Evidence for discrete membrane domains containing cholinergic receptors, J. Biol. Chem. 256:9126 (1981)

2 K. Haga and T. Haga, Affinity chromatography of the muscarine acetylcholine receptor, J. Biol. Chem. 258:13575 (1983)

3 M.J. Lopez-Perez, G. Paris and C. Larsson, Highly purified mitochondria from rat brain prepared by phase partition, Biochem. Biophys. Acta 635:359 (1981)

ISOLATION OF PLASMA MEMBRANE VESICLES FROM FAT CELLS OF EPIDIDYMAL FAT PADS OF THE RAT BY AQUEOUS TWO-PHASE PARTITION

Dorothy M. Morré, D. James Morré, Timothy Snyder, William Li, Amy Erb and David W. Sammons

Department of Foods and Nutrition
Department of Medicinal Chemistry
Purdue University, W. Lafayette, Indiana 47907
and Center for Separation Science
The University of Arizona, Tucson, Arizona 85721

The responses of fat cells to insulin involve potentially important membrane alterations. There is now considerable evidence to support the concept that insulin stimulates glucose transport by inducing a rapid and reversible translocation of glucose transporters from an intracellular membrane pool to the plasma membrane [1-4]. However, little is known about the cellular mechanism of insulin and the process or biochemical events whereby glucose transporter levels at the cell surface may be modulated. Initial efforts [5] to better characterise the glucose transport system involved use of a sensitive silver staining method to visualise membrane proteins in two-dimensional gel electrophoretograms in conjunction with computerised scanning and quantification. A 90 kDa protein was lost from the plasma membranes upon incubation of fat cells with insulin and calcium. We are investigating the dynamics and identity of the 90 kDa protein in the fat cell, and especially the mechanism whereby the protein is dissociated from the plasma membrane as well as its fate and function in relation to the oncoming glucose transporter vesicles from the cytoplasm.

Fat cell plasma membranes usually have been purified by differential and density-gradient centrifugation [6-7]. However, such techniques are based on using 16-24 rats for each preparation. An alternative procedure, such as partitioning in aqueous polymer two-phase systems, might be useful for purification of plasma membrane from fat cells isolated from as few as 1 or 2 rats. Use of two-phase partition in dextran/poly(ethylene glycol) is well established in the purification and characterisation of plant membranes [8] and, to a degree, has been used for animal membranes, particularly plasma membranes of liver [9]. In our study, this technique has been utilised to prepare highly purified plasma membranes from isolated fat cells.

Male Wistar rats (fed standard laboratory chow) weighing approximately 200 g were killed by decapitation and the epididymal fat pads removed. Fat cells were isolated by collagenase digestion according to the method of Rodbell [10] in Krebs-Ringer bicarbonate buffer (pH 7.4) containing 0.6% calcium, 3% bovine serum albumin, 0.2% glucose and 7.0 mg collagenase for 30 min at 37°C. The isolated cells were washed in 10 volumes of homogenisation medium containing 10 mM Tris-HCl (pH 7.4 at 4°C), 0.25 M sucrose and 0.1 mM PMSF and centrifuged for 1 min at 400 g. The isolated fat cells were then incubated in buffer in the presence or absence of insulin, resuspended in 3.5 volumes of the homogenisation medium, transferred to a glass homogeniser fitted with a Teflon pestle and homogenised with 10 up-and-down strokes by means of a variable-speed motor driven at 2,600 rpm. The homogenate centrifuged at 35,000 g for 15 min to obtain a "heavy" microsomal pellet was then resuspended in 0.5 ml phosphate buffer, pH 7.4, and used for two-phase separation of plasma membranes. The 4 g two-phase system contained equal portions (5.4 to 6.8%, w/w) of Dextran T500 (Pharmacia, Sweden) and Polyethylene Glycol 3350 (Fisher, USA) plus 0.1 M sucrose, 5 mM potassium phosphate, pH 7.2, and 0.5 g of resuspended microsomes (0.5 mg protein). The mixture was shaken and centrifuged at 1,500 g for 5 min to separate the phases. The upper phase was collected, diluted in 1 mM sodium bicarbonate and pelleted at 20,000 g as the source of plasma membranes.

Both upper and lower phases were analysed for the plasma membrane marker, K^+-stimulated p-nitrophenylphosphatase (EC 3.1.3.1) [11], the mitochondrial marker succinate-INT reductase (EC 2.3.99.1) [12] and protein ([13], with BSA as standard). As the polymer concentration of the two phases was increased, mitochondria and mitochondrial enzyme activities were partitioned to the lower phase. K^+-stimulated p-nitrophenylphosphatase specific activity increased to an optimum at about 6.6% of each polymer. The recovery of plasma membrane in the upper phase obtained by aqueous two-phase partition was about 2% of the total protein. Based on concanavalin A binding, both right side out and inside out plasma membrane vesicles were present in the upper phase. These could be resolved using free-flow electrophoresis with the cytoplasmic side in vesicles exhibiting the greatest electrophoretic mobility.

Sufficient plasma membranes were obtained from fat pads of 1-2 rats for both enzymatic assays and analysis by two-dimensional gel electrophoresis with detection by silver staining. The insulin-responsive 90 kDa protein was detected only in the upper phase membranes of cells not treated with insulin and was not detected in either lower phase untreated membranes or in both upper and lower phase membranes of cells treated with insulin. The findings confirm the previous observations of Schoenle et al [5] and establish the utility of aqueous two-phase partition as a preparative method for plasma membranes of fat cells.

Work supported in part by grants from the American Diabetes Association, Inc. and NASA NAGW-693. Insulin was a gift from Eli Lilly and Company, Indianapolis, IN.

REFERENCES

1 L.J. Wardzala, S.W. Cishman and L.B. Salans, Mechanism of insulin action on glucose transport in the isolated rat adipose cell. Enhancement of the number of functional transport systems, J. Biol. Chem. 253:8002 (1978)

2 S.W. Cushman and L.J. Wardzala, Potential mechanism of insulin action on glucose transport in the isolated rat adinose cell. Apparent translocation of intracellular transport systems to the plasma membrane, J. Biol. Chem. 255:4758 (1980)

3 K. Suzuki and T. Koni, Evidence that insulin causes translocation of glucose transport activity to the plasma membrane from the intracellular storage site, Proc. Natl. Acad. Sci. USA 77:2541 (1980)

4 Y. Oka and M.P. Czech, Photoaffinity labeling of insulin-sensitive hexose transporters in intact rat adipocytes, J. Biol. Chem. 259:8125 (1984)

5 E.J. Schoenle, L.D. Adams and D.W. Sammons, Insulin-induced rapid decrease of a major protein in fat cell plasma membranes, J. Biol. Chem. 259:12112 (1984)

6 D.W. McKeel and L. Jarett, Preparation and characterisation of plasma membrane fraction from isolated fat cells, J. Cell Biol. 44:417 (1970)
7 L. Jarett, Subcellular fractionation of adipocytes, Methods Enzymol. 31:60 (1974)
8 P.-Å. Albertsson, B. Andersson, C. Larsson and H.-E. Åkerlund, Phase partition: A method for purification and analysis of cell organelles and membrane vesicles, Methods Biochem. Anal. 28:115 (1982)
9 P. Gierow, M. Sommarin, C. Larsson and B. Jergil, Fractionation of rat liver plasma membrane regions by two-phase partitioning, Biochem. J. 235:685 (1986)
10 M. Rodbell, Metabolism of isolated fat cells. 1. Effects of hormones on glucose metabolism and lipolysis, J. Biol. Chem. 239:375 (1964)
11 S. Kasiwampta, S. Goto, R. Samba and F. Suzuki, Inhibition by bilirubin of (Na^+-K^+)-activated adenosine triphosphatase and K^+-activated p-nitrophenylphosphatase activities of Na^+-treated microsomes from young rat cerebrum, J. Biol. Chem. 254:4577 (1979)
12 R.J. Pennington, Biochemistry of distrophic muscle. Mitochondrial succinate-tetrazolium reductase and adenosine triphosphatase, Biochem. J. 80:649 (1961)
13 O.H. Lowry, N.J. Rosebrough, A.L. Farr and R.J. Randall, Protein measurement with the Folin phenol reagent, J. Biol. Chem. 293:265 (1951)

Chapter 4

Partitioning of Cells

SEPARATION AND SUBFRACTIONATION OF CELL POPULATIONS BY PHASE PARTITIONING - AN OVERVIEW

Derek Fisher

Department of Biochemistry and Chemistry
Royal Free Hospital School of Medicine
University of London
Rowland Hill Street
London NW3 2PF

The general principles of cell partitioning in aqueous two-phase systems are reviewed. Particular attention is given to the ability of countercurrent distribution (CCD) to subfractionate cell populations which by other means may seem to be homogeneous. General strategies for demonstrating heterogeneity in cell populations are outlined and examples provided from recent studies on K-562 cells by Walter and his colleagues. The influence of the conditions of phase separation upon the efficiency of cell fractionation is discussed.

CELL PARTITIONING: GENERAL PRINCIPLES

The partitoning of cells differs from that of macromolecules in that partitioning to the bulk interface can occur. Macromolecules partition between the bulk phases of aqueous two-phase systems whereas cells generally partition between one of the phases and the interface. As we shall see, this difference has important consequences for the methodology of cell separations by countercurrent distribution (CCD) in aqueous two-phase systems.

To be able to separate cells it is necessary to adjust the phase composition so that appreciable distribution occurs between a bulk phase and the interface.

In a phase system with polymer concentrations some distance from the critical point in the presence of salts that partition evenly between the phases (e.g. NaCl) virtually all the cells will be at the interface. This arises from the fact that the free energy change associated with adsorption of cells at the interface (interfacial tension of the two-phase system x cross- sectional area of the cell) is of the order of the average thermal energy kT [1].

By appropriate choice of the composition of the phase system, cell partitioning can be altered in favour of one of the bulk phases. This partitioning is dependent on cell surface properties and cell separations can be obtained on the basis of small differences in these surface properties. This area of partitioning has been developed extensively by Harry Walter at the VA Medical Center, Long Beach, California [2,3].

There are essentially three ways in which cells can be moved from the interface into one of the bulk phases.

1) If salts are introduced that partition unevenly between the phases an electrostratic potential difference between the phases ($\Delta\psi$) is generated. In the particular case of phosphate the top phase becomes positive relative to the bottom phase; under these conditions cells, which are generally negatively charged, will to some extent partition into the top phase, the rest will stay at the interface. This is called charge-sensitive partitioning.

2) In phase systems with NaCl as the main salt rather than phosphate there is virtually no $\Delta\psi$ and cells partition to the interface. However, if the polymer concentrations are reduced one can again reach a point when cells partition into the top phase. This is called non-charge-sensitive partitioning. The parameters governing the partitioning are quite different; they are not yet clearly defined and continue to be a subject of investigation. In the case of red cells from different species there is a good correlation between their partition ratio in charge-sensitive partitioning and their relative electrophoretic mobilities, whereas in non-charge-sensitive partitioning their partition ratios correlate with the ratio of poly- to mono-unsaturated fatty acids in membrane phospholipids, a lipid parameter.

3) One can use a phase system which has no $\Delta\psi$ and at some distance from the critical point. By incorporating ligands that are attached to either PEG or dextran one can "pull" cells out of the interface into the phase in which the ligand carried by the polymer predominates, on the basis of the interaction of the ligand with a particular cell surface. This is called affinity partitioning. Early examples used fatty acids linked to PEG, in particular PEG-palmitate, which associate hydrophobically with cells (hydrophobic affinity cell partitioning). More recently, immunoglobulins coupled to PEG have provided immunoaffinity cell partitioning. This approach offers the possibility of highly selective bioseparations. This is dealt with in detail elsewhere in this volume [4,5].

SENSITIVITY OF CELL PARTITIONING

Partitioning is a sensitive process because the parameters that influence the partitioning of cells are related in an exponential manner to the partition ratio [1,2,6]. In other separatory methods this relationship is linear and therefore less sensitive (e.g. electrophoretic mobility is proportional to surface charge).

CELL SEPARATION AND SUBFRACTIONATION OF CELL POPULATIONS

CCD of cells can be performed just as one does for soluble materials, provided one procedural aspect is observed; it is important to remember that cells partition between one of the bulk phases and the interface. The implications for CCD can be seen from considering the example of cell partitioning between the top phase and the interface. In performing the transfer step in the CCD the interface must be left behind when the top phase is moved; the interface must not be cut through else the resolution of the CCD will be lessened. This is achieved by appropriate choice of the volume of the bottom phase so that a proportion of the top phase is left behind to ensure that the cells at the interface are not transferred. By contrast in the partitioning of soluble material the whole of the top phase can be transferred and the whole of the bottom phase can be left behind.

Although CCD can be used to separate cell populations, by far the most interesting and useful aspect of CCD and cell partitioning is the ability of this method to subfractionate cell populations which by other methods may seem to be homogeneous.

Simple tests can demonstrate that one is dealing with different cell populations. If the two cell populations have major differences in partition ratios the appearance of two distribution curves will be obvious. However, if a single CCD curve is obtained, measurement of the partition ratios of the fractions comprising the peak is informative. The contents of the CCD cavities are emptied directly into tubes without the addition of isotonic medium to "break" the phase system (render the two-phase system a single phase). The tubes are then mixed and the cell partition determined. An increase in the partition ratio through the profile indicates that the population has been subfractionated. If the ratio is the same through the peak, the population has not been subfractioned.

Another way to demonstrate subfractionations of populations is to pool fractions from the left hand side of the profile and subject the cells to a second CCD. If the peak obtained is to the left of the peak obtained on subjecting pooled fractions from the right hand side to a second CCD it follows that the cells from the left hand side of the profile have a lower partition ratio than the cells from the right side of the profile, and therefore that the peak is heterogeneous. However, if the two profiles overlap then no differences in partition ratio exist and therefore, in this phase system at least, there has been no subfractionation.

Differences in functional properties of the cells through the profile also serve to show that the population is heterogeneous. For example we have found functional heterogeneity in human neutrophils [7,8] by charge-sensitive phase partitioning. Cells in the profile showed increasing electrophoretic mobility with increasing fraction number which was associated with a progressive decrease in their ability to phagocytose bacteria and to undergo chemotaxis.

CCD can also be used to detect small differences in the partitioning behaviour of two closely related cell populations [9,10]. One population of cells is labelled with an appropriate radioactive isotope, mixed with an excess of the second population, and the sample subjected to CCD.
A displacement in the profile of radioactivity from the profile of the main cell population indicates a difference in partition behaviour, and consequently a difference in the surface properties of the two cell populations. Control experiments are required to demonstrate that each cell population, labelled and unlabelled, has CCD curves which are coincident.

EXAMPLES OF CELL SEPARATIONS AND POPULATION SUBFRACTIONATIONS

Recent work by Walter and colleagues on the charge-sensitive partitioning studies on K-562 cells grown in suspension culture serves to illustrate the general stratergy one can apply to cell separations [11,12].

CCD curves of the cells during cell culture show a change from a single peak of viable cells to two peaks (Fig. 1). The second peak had a lower partition ratio and consisted of non-viable cells. Its proportion increased with the culture time. Although the partition ratio of the peak of viable cells did not alter with culture time, because of the increasing contribution of the non-viable cells, the overall partion ratio showed a decrease. Such a decrease in overall partition ratio with culture time has also been reported by others; the CCD studies shows that it arises from the increasing contribution of dead cells to the cell population. Thus, as well as separating dead cell from viable cells, CCD shows its importance in assessing the homogeneity of a cell population.

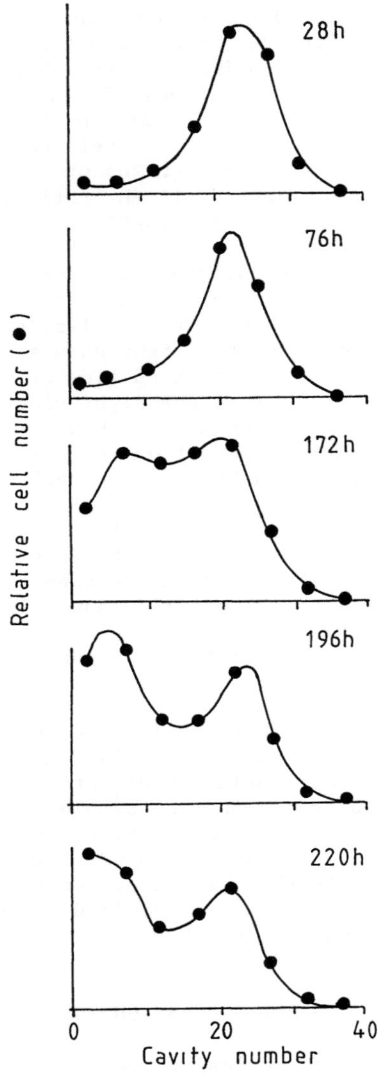

Fig. 1 *Countercurrent distribution of K-562 cells at different stages of culture growth (28, 76, 172, 196 and 220 hr). Cells were subjected to a 40 transfer countercurrent distribution at 4-5°C in a charge-sensitive phase system of 5% Dextran T500, 4% PEG 8000, 220 mosM sodium phosphate buffer (pH 7.4), 46 mosM NaCl and 5% heat-inactivated foetal bovine serum. The peak around cavity number 20 does not alter its position with time of culture and contains better than 90% viable cells. The left hand peak which developed with time in culture is composed primarily of dead cells. Redrawn from [11]*

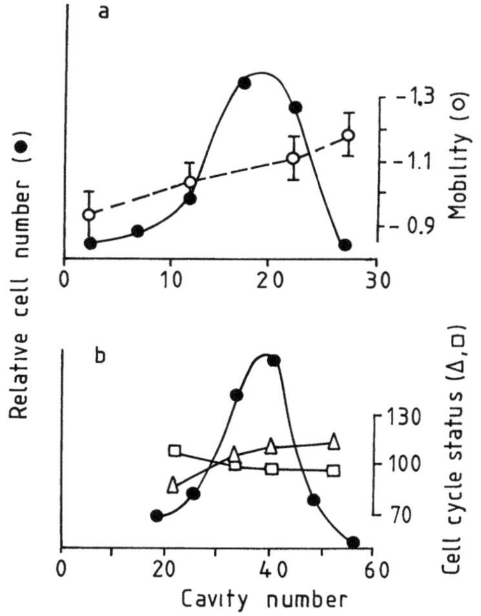

Fig. 2 *Countercurrent distribution of K-562 cells in a charge-sensitive phase system (see Fig. 1 for composition). (a) Heterogeneity of cell population is revealed by the increase in electrophoretic mobility (○ ums^{-1}V^{-1}cm; mean ± SD, n=10) through the profile. 40 transfers were performed. (b) Heterogeneity of the cell population is revealed by the differences in cell cycle status of cells through the profile. Results of flow cytometry (FACS) on cells in selected fractions are expressed as cells in G_o + G_1 early S (□) and cells in late S + G_2 + M (Δ); all expressed as a percentage in the unfractionated cells. 60 transfers were performed. Redrawn from [11]*

The peak of viable cells was shown to be heterogeneous by three criteria:

1) Cells pooled from each end of the CCD peak were subjected to a second CCD; cells from the left of the original peak distributed to the left of cells from the right of the original peak.

2) The electrophoretic mobilities of cells from different parts of the peak were different, increasing with increasing partition ratio (Fig. 2).

3) Cells under different parts of the the peak had dissimilar ratios of cells in different parts of the cell cycle; there was a tendency for cells in G_o + G_1 + early S to decrease and for those in late S + G_2 + M to increase with increasing partition ratio (Fig. 2).

It therefore appears that the cell heterogeneity under the CCD curve of K-562 cells is related to cells in different phases of the cell cycle.

THE EFFICIENCY OF CELL FRACTIONATION AND THE SPEED OF PHASE SEPARATION

The partitioning of cells arises from cell- and phase-dependent attachment of cells to phase droplets that form in mixed phase systems and the time-dependent delivery of the cells on the droplets to the bulk interface as phase separation proceeds [1,6,13]. This dynamic feature of cell partitioning results in partition ratios that show variation with time and the phase separation conditions and leads to the conclusion that cell separations and subfractionations by CCD depend on non-equilibrium conditions [13]. From these mechanistic studies of cell partitioning it was predicted that variation in settling time in CCD may improve cell fractionation and detect subtle differences in cell surface properties [13]. Recent studies by Walter and Krob [14] have demonstrated an effect of the rapidity of phase separation on the separability of cell populations by CCD; speeding up phase separation reduced the efficiency of cell separation. These studies have important implications for the design of CCD apparatus.

Walter makes an important distinction between the conditions of phase separation required for separation of macromolecules and for the separation of particulates [14]. Aqueous two-phase systems are considerably slower in phase separating than aqueous-organic systems [1] because the density difference between the two aqueous phases is small and because the polymer-containing phases are relatively viscous. Macromolecules partition between the bulk phases only; speeding up phase separation by use of a thin layer design in the CCD apparatus [15], reduces the time needed for separations. By contrast, rapid phase separation can decrease the opportunity for the cells to interact adequately with phase droplets and thereby reduce the efficiency of cell separation. Consequently "thick" layers of phase, as in the classical Craig CCD unit [16], may be more efficient. However, this improved efficiency would need to be balanced against the increased time required for CCD to be performed under such slower phase separation conditions.

This blend of the theoretical aspects of partitioning with experimental systems for examining population heterogeneity should lead to improvements in the resolution and applicability of cell separations by phase partitioning. New methodology for liquid-liquid chromatographic and countercurrent distribution techniques are described in Chapter 9 of this volume.

REFERENCES

1 D.E. Brooks, K.A. Sharp and D. Fisher, Theoretical aspects of partitioning, in: "Partitioning in Aqueous Two-Phase Systems. Theory, Methods, Uses, and Applications to Biotechnology," H. Walter, D.E. Brooks and D. Fisher, eds., Academic Press, Orlando (1985)

2 H. Walter, Surface properties of cells reflected by partitioning: red blood cells as a model, in: "Partitioning in Aqueous Two-Phase Systems. Theory, Methods, Uses, and Applications to Biotechnology," H. Walter, D.E. Brooks and D. Fisher, eds., Academic Press, Orlando (1985)
3 H. Walter and D. Fisher, Separation and subfractionation of selected mammalian cell populations, in: "Partitioning in Aqueous Two-Phase Systems. Theory, Methods, Uses, and Applications to Biotechnology. H. Walter, D.E. Brooks and D. Fisher, eds., Academic Press, Orlando (1985)
4 L.J. Karr, J.M. Van Alstine, R.S. Snyder, S.G. Schaffer and J.M. Harris, Cell separation by immunoaffinity partition in aqueous polymer two-phase systems, in: "Separations Using Aqueous Phase Systems: Applications in Cell Biology and Biotechnology," D. Fisher and I.A. Sutherland, eds., Plenum, New York (1989)
5 J. Stocks and D.E. Brooks, Second immunoaffinity ligands for cell separation, in: "Separations Using Aqueous Phase Systems: Applications in Cell Biology and Biotechnology," D. Fisher and I.A. Sutherland, eds., Plenum, New York (1989)
6 B.N. Youens, W.D. Cooper, F.D. Raymond, P.S. Gascoine and D. Fisher, Mechanism of cell partitioning; equilibrium contact angle measurements predict relative, kinetic partitioning behaviour, in: "Separations Using Aqueous Phase Systems: Applications in Cell Biology and Biotechnology," D. Fisher and I.A. Sutherland, eds., Plenum, New York (1989)
7 P. Eggleton, I.A. Sutherland and D. Fisher, Rapid partitioning of neutrophils by CCD, in: "Separations Using Aqueous Phase Systems: Applications in Cell Biology and Biotechnology," D. Fisher and I.A. Sutherland, eds., Plenum, New York (1989)
8 P. Eggleton, N. Crawford and D. Fisher, Functional heterogeneity in human neutrophils detected by continuous flow electrophoresis (CFE) and countercurrent distribution (CCD) in aqueous polymer two-phase systems, Biochem. Transactions, in press (1989)
9 H. Walter and E.J. Krob, Detection of surface differences between two closely related cell populations by partitioning isotopically labeled mixed cell populations in two-polymer aqueous phases. I. Human red blood cell subpopulations, Cell Biophys. 5:205 (1983)
10 H. Walter and E.J. Krob, Detection of surface differences between two closely related cell populations by partitioning isotopically labeled mixed cell populations in two-polymer aqueous phases. II. A correction, Cell Biophys. 5:301 (1983)
11 H. Walter and F.A. AL-Romaihi, The partitioning behaviour of K-562 cells associated with different stages of culture growth, Biochim. Biophys. Acta 924:249 (1987)
12 H. Walter, F.A. AL-Romaihi, E.J. Krob and G.V.F. Seaman, Fractionation of K-562 cells on the basis of their surface properties by partitioning in two-polymer aqueous-phase systems, Cell Biophysics 10:217 (1987)
13 D. Fisher and H. Walter, Cell separations and subfractionations by countercurrent distribution in two-polymer aqueous phase systems depends on non-equilibrium conditions, Biochim. Biophys. Acta 801:106 (1984)
14 H. Walter and E.J. Krob, Effect of rapidity of phase separation on the efficiency of cell fractionation by partitioning in aqueous two-phase systems, Biochim. Biophys. Acta 966: 65 (1988)
15 P.-Å. Albertsson, Thin layer counter-current distribution, Anal. Biochem. 11:121 (1965)
16 L.C. Craig and D. Craig, Laboratory extraction and countercurrent distribution, in: "Techniques of Organic Chemistry" A. Weissberger, ed., 2nd. ed., vol. 3, Part 1, Wiley (Interscience), New York (1956)

SEPARATION OF CHONDROCYTES IN AQUEOUS TWO-PHASE SYSTEMS

Claes-Goran Axelsson

Department of Clinical Chemistry
University of Umeå
S-901 85 Umeå, Sweden

Chondrocytes were isolated from bovine articular cartilage and put into a suspension culture. By using CCD the partition behaviour of the chondrocytes as followed at various times in culture. Freshly isolated cells are partitioned into one peak consisting of single cells. During culture two new peaks appeared in the CCD both of which contained cell aggregates of different kinds. By modifying the chemical composition of the cell surface it is demonstrated that the partition is regulated by the presence of proteoglycans at the cell surface. Besides affecting the charge density of the cells, increased concentration of proteoglycans also produces an increased cell aggregation.

INTRODUCTION

The chondrocyte is the only type of cell found in adult articular cartilage [1]. However, on the basis of cell morphology and chemical composition of its surrounding region, adult articular cartilage can be divided into three layers [1,2]. In the superficial layer the chondrocytes are numerous with a more flattened shape. The structure of the intercellular region in the deeper layer is organised into various regions due to the formation of collagen-proteoglycan networks with different orientation and diameter of the collagen fibrills [3]. In the superficial layer no such network is seen. Each cell is surrounded by a pericellular region being thickest in the deeper layer. This pericellular region consists mainly of various aggregates of proteoglycans. In spite of these variations, subpopulations of chondrocytes have not been detected when isolated chondrocytes are cultured in vitro.

The aim of this paper is to see if these ultrastructural variations around and on the cell surface of the chondrocyte can be regained in cells cultured in vitro and as a consequence be used for separation of chondrocytes in aqueous polymeric two-phase systems. Since cell associated proteoglycans are of major importance for the formation of the pericellular region and are highly sulphated, they contribute to a high negative charge density of the cell surface. Therefore a charge sensitive two-phase system was used in this study and the cells were separated with the countercurrent distribution (CCD) technique. With increasing time in suspension culture, three subpopulations of chondrocytes from bovine articular cartilage could be demonstrated in the

CCD. These subpopulations have been characterised by enzymatic modification of the cell surface and by scanning electron microscopy.

EXPERIMENTAL

Knee cartilage from freshly slaughtered one year old calves was taken and chondrocytes were isolated with sequential enzymatic digestion using hyaluronidase, trypsin and collagenase [4]. The cells were then cultured in spinner bottles at a cell density of 0.5 x 10^6 cells/ml. Dulbecco's modified Eagle's medium (DMEM) was supplemented with 10% fetal calf serum, antibiotics, Na-bicarbonate and L-glutamin [5]. At various times the chondrocytes were washed and resuspended in 0.4 ml of 0.5M sucrose and taken for CCD. Usually 3-5 x 10^6 cells were analysed by 29 transfers in the CCD at 4°C. Each fraction was analysed by its light scattering at 500 nm or by measuring the content of incorporated 35-sulphate. In some experiments the peak fractions were pooled and taken for microscopical examination. The viability of cells were judged by the trypan blue exlusion method, after isolation and before each CCD run, and was always >95%.

Two-phase System

A phase composition of 5% (w/w) Dextran T500, 3.5% (w/w) PEG 6000 (poly-(ethylene glycol)) with 20% (w/w) consisting of TMA-PEG, (trimethylamino-PEG 6000), 30 mM NaCl, 5 mM Na-phosphate buffer pH 7.3 and 0.25 M sucrose was used in this study. The chondrocytes are partitioned between upper phase and interface and in the CCD the upper phase is transferred leaving the whole interface and lower phase. The choice of phase composition was initially based on the partition of purified proteoglycans. It was then found that proteoglycans behaved like nucleic acid in the phase system, that is a very steep change in partition occurred with only small changes in the concentration of added salts or TMA-PEG [6]. To reduce the risk of variation in partition due to small changes in electrolytes between different experiments, a relatively high concentration of both NaCl and TMA-PEG was chosen.

RESULTS AND DISCUSSION

Freshly isolated chondrocytes distribute into one major peak, here called peak A, shown in Fig. 1. With increasing time in culture, two new peaks were observed, peaks B and C. During the first 36 hours of culturing the partition coefficient for peak A and C is rather constant while peak B seemed to increase its partition with increasing time in culture. After 4 days in culture, peak B could not be seen in the CCD, probably having moved so near peak C that no separation is obtained (not shown here). Also peak A decreased its partition so that an incomplete separation is obtained between peak A and C after 9 days in culture (Fig. 1). When chondrocytes are kept in suspension culture for 21 days, only peak C is observed in the CCD. The results presented so far are based on measurements of light scattering at 500 nm, which is a simple and common way to detect cells in a CCD. To confirm that the peaks shown in Fig. 1 represent chondrocytes, the experiments were repeated with the exception that cells were incubated with 40 μCi/ml of carrier free 35-sulphate for 4 hours and then washed extensively before being run in the CCD. 35-sulphate is incorporated with high specificity into the chondroitin chains of the proteoglycans [7]. In Fig. 2 it is shown that the partition behaviour of chondrocytes measured by radioactivity is almost the same as that measured by light scattering in Fig. 1. However, the position of peak A is now unchanged with time. One explanation for this is that there is some leaking of proteoglycans from the cells during CCD, and since proteoglycans are distributed in the CCD into one single peak around tube number 25, peak A here could represent both cell

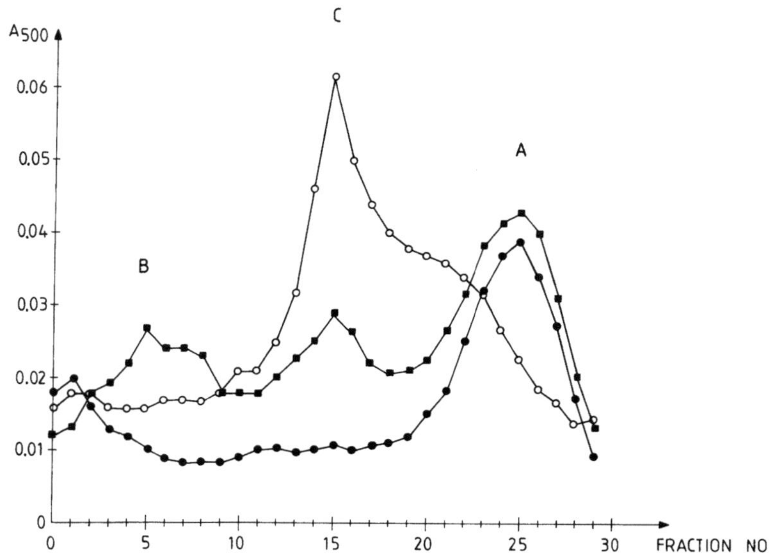

Fig. 1 *CCD of chondrocytes at various times of cell culturing. Freshly isolated cells (●), 36 hours (■) and 9 days (○) in suspension culture*

bound and free proteoglycans. It can also be demonstrated that peak B appeared in the CCD already after 4 hours in culture and with an amplitude that is almost the same as for peak A. Comparing with the light scattering results after 4 hours, the amplitude of peak B was less than one half of the amplitude of peak A (not shown here). Thus the relative 35-sulphate incorporation per cell appears to be higher in peak B during the first hours of culturing. This indicates that cells in peak B contain a high amount of newly sysnthesised proteoglycans.

Scanning Electron Microscopy

Microscopy of cells taken from the three peaks was performed and great differences between the cell fractions could be demonstrated. The technique and preparation of cells for the scanning electron microscopy was conventional. Cells from peak A, Fig. 3a, are seen as solitary cells having small buds or villi on their cell surface as well as areas with a smooth appearance. In peak B the majority of cells are found in large aggregates, some of which contain up to 100 cells, Fig. 3b. The cells are in close contact with each other and the pericellular region looks thicker and more uniform than for cells in peak A. For most of the cells in the aggregates, the pericellular region became confluent with that of the surrounding cells and gives the impression of a thin film of extracellular material that cover the cells. In the material from peak C, we find aggregates of 5-20 cells where the cells seem to be more apart from each other compared to cells in peak B, Fig. 3c. The intercellular space is filled with some material having a more porous or thread like appearance. This is most obvious when looking at chondrocytes from peak C cultured for 3 weeks, Fig. 3d.

The appearance of these peaks during cell culturing is proposed to represent subpopulations of chondrocytes. Since cell divisions do not occur to any great extent during the first weeks of culturing, as determined by constancy of cell number and a rather low rate of ^{3}H-thymidine incorporation [8,9], these three peaks found in the CCD are most unlikely to represent

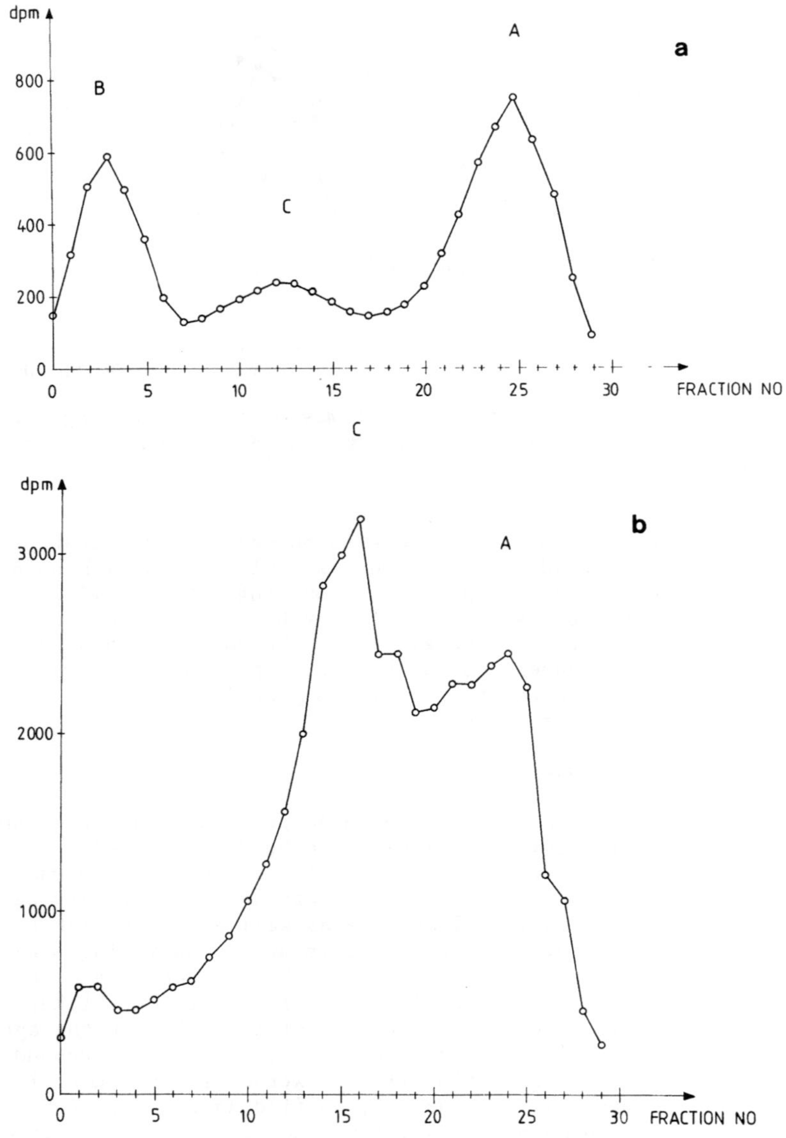

Fig. 2 *The same experiments as in Fig. 1 except that cells were incubated with 35-sulphate during 4 hours before run in CCD. Freshly isolated cells (a) and after 9 days (b) in culture*

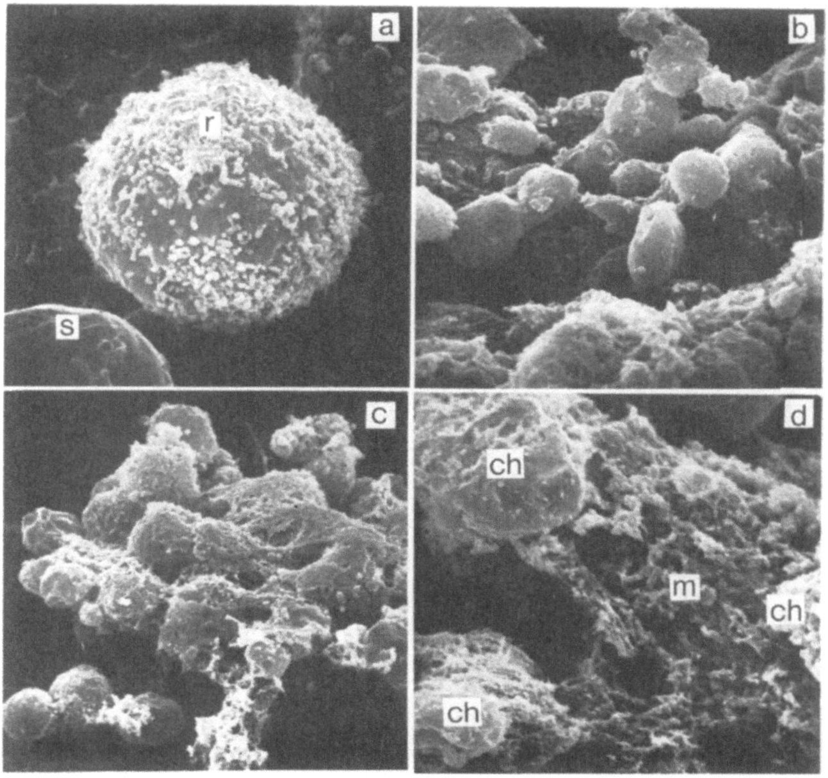

Fig. 3 *Scanning electron microscopy of chondrocytes taken from the three peaks in the CCD. Cells cultured for 2 days (a-c) and 3 weeks (d). Cells from peak A (a) showing both smooth (s) and rough (r) cell surfaces. Aggregates from peak B (b). Cell aggregates from peak C (c+d) with extracellular material (m) between the chondrocytes (ch)*

cells at various stages in the cell cycle. Instead the demonstration of subpopulations of chondrocytes can be explained in terms of both anabolic and catabolic events at the cell surface. Due to the isolation procedure, cells in peak A have lost some of their pericellular region, seen as the appearance of both rough and smooth areas on the cell surface. With increasing time of culturing the cell surface became reconstituted. The newly synthesised cell surface components, most probably aggregates of proteoglycans, then promote cell aggregation and thus form the type of cells seen in peak B. The final result of suspension culturing of chondrocytes is the type of cells seen in peak C where some intercellular material both separates and connects the cells producing a microenvironment similar to that found in vivo.

Treatment of the Cell Surface

To further characterise the cell surface of the three cell fractions, cells which had been cultured for 2 days were subsequently treated with either trypsin, chondroitinase or hyaluronidase. Both peaks B and C are absent in the CCD of trypsinised cells and only peak A with the same partition coefficient as for freshly isolated chondrocytes remains, Fig. 4a. Thus peak A represents cells with a cell surface composition insensitive to trypsin digestion while cells in both peaks B and C have reconstituted their cell surface with trypsin releasable material. However, if cells cultured for more than 4 days were trypsinised in the same way, not shown here, the

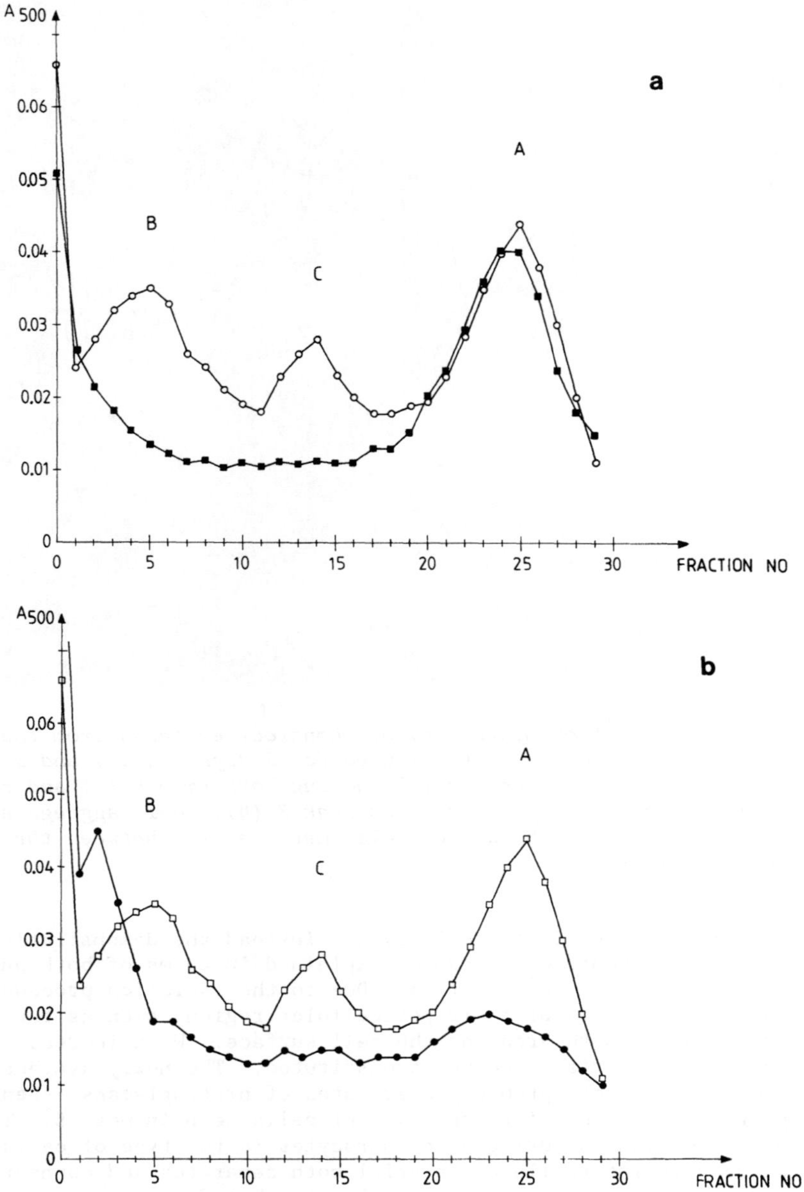

Fig. 4 *CCD of chondrocytes cultured for 2 days, with our without enzymatic treatment (a-c) of the cells before being run in the CCD, or cells cultured for 2 days with hyaluronic acid added to the culture medium (d). (a): 0.125% trypsin in 20 min (■), untreated cells (○). (b): 0.2 units chondroitinase ABC in 4 hours (●), untreated cells (○). (c): 0.05% hyaluronidase in 15 min (●), untreated cells (○). (d): hyaluronic acid (1 mg/ml) added to the medium (●), medium without acid (○)*

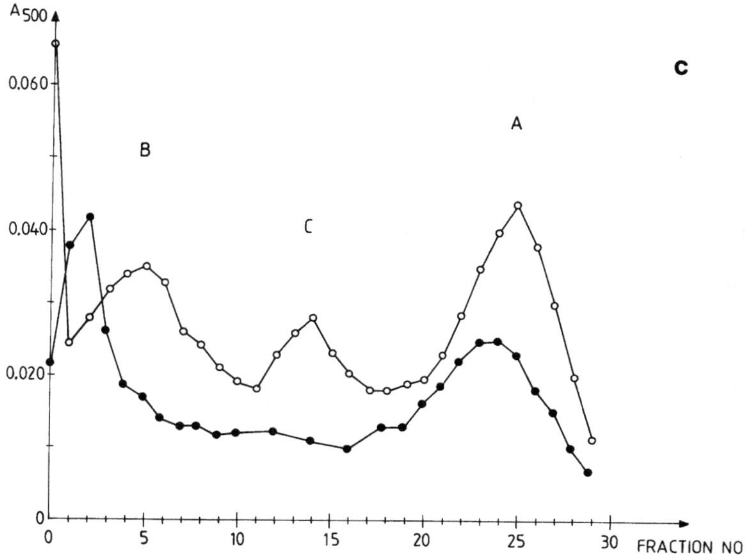
A500
0.060
0.040
0.020
0
B
C
A
c
0
5
10
15
20
25
30
FRACTION NO

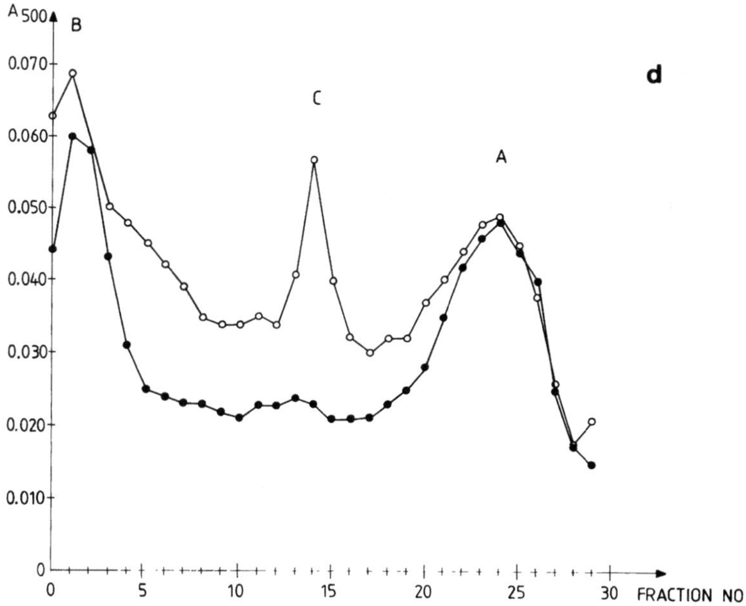
A500
B
0.070
0.060
0.050
0.040
0.030
0.020
0.010
0
C
A
d
0
5
10
15
20
25
30
FRACTION NO

partition of peak A is also affected. This is in agreement with the earlier proposition that also cells in peak A change their cell surface with increasing time of culturing, resulting in a continuous decrease in the partition coefficient of peak A.

Chondroitinase which splits off chondroitin sulphate from proteoglycans at the cell surface affects the partition of all three types of cell population by reducing the peak amplitude, Fig. 4b. The digested cells having a decreased charge density are partitioned preferably towards the interface.

A similar partition behaviour is seen with cells treated with hyaluronidase which is believed to affect the cell surface by releasing proteoglycans bound to the cell associated form of hyaluronic acid [10]. Thus, all subpopulations of chondrocytes have hyaluronic acid at their cell surface. If hyaluronic acid is added to the culture medium it competes with the membrane associated acid for the binding of proteoglycans released from the cell. In Fig. 4d the results are shown of culturing chondrocytes for 2 days with or without hyaluronic acid added to the medium. In the presence of hyaluronic acid, peak C disappears from the CCD of such cells. Thus, by inhibiting the binding of proteoglycans to the chondrocytes the type of aggregates seen in peak C cannot be formed. On the other hand, peaks A and B are not affected, suggesting a less efficient competition between added hyaluronic acid and newly synthesised proteoglycans from their corresponding cells.

CONCLUSIONS

The results presented here indicate that the partition behaviour of chondrocytes is regulated by the presence of proteoglycans at the cell surface. Besides affecting the charge density of the cells, increased concentration of proteoglycans at the cell surface also produces an increased cell aggregation. These two factors appear to determine the partition of chondrocytes in the charge sensitive two-phase system used in this study. Thus, freshly isolated cells behave as a rather homogenous cell fraction (peak A) having lost most of their pericellular region. In suspension culture they reconstitute their cell surface and accumulate some extracellular material forming a matrix structure (peak C). This takes place by initially forming cell aggregates (peak B).

Acknowledgements

The scanning electron microscopy was skillfully made by Dr P. Hörstedt, Department of Pathology, University of Umeå. Encouraging advice from Dr V. Shanbhag, Department of Biochemistry, University of Umeå, is greatly appreciated.

REFERENCES

1 R.A. Stockwell, "Biology of Cartilage Cells," Cambridge University Press, Cambridge (1979)
2 C.A. Poole, M.H. Flint and Beaumont, Morphological and functional interrelationships of articular cartilage matrices, J. Anat. 138:113 (1984)
3 A.R. Poole, A.R. Pidoux and L. Rosenberg, An immunoelectron microscope study of the organisation of proteoglycan monomer, link protein and collagen in the matrix of articular cartilage, J. Cell. Biol. 93:921 (1982)

4 W.T. Green, Behaviour of articular cartilage in cell culture, Clin. Orthop. Rel. Res. 75:248 (1971)
5 L.J. Jones, A. Klämfeldt and M.B. McGuire, Enhanced breakdown of bovine articular cartilage proteoglycan by conditioned synovial medium, Scand. J. Rheum. 11:41 (1982)
6 P.-Å. Albertsson, "Partition of Cell Particles and Macromolecules," Almqvist and Wiksell, Stockholm (1971)
7 A. Maroudas, Physicochemical basis for turnover studies, in: "Adult Articular Cartilage," M.A.R. Freeman, ed., Pitman Medical, London (1979)
8 S. Deqing, M.B. Aydelotte, B. Maldonado, K.E. Kuettner and J.H. Kimura, Clonal analysis of the population of chondrocytes from the swarm rat chondrosarcoma in agarose culture, J. Orth. Res. 4:427 (1986)
9 K.E. Kuettner, B.U. Pauli, G. Gall, V.A. Memoli and R.K. Schenk, Synthesis of cartilage matrix by mammalian chondrocytes in vitro, J. Cell. Biol. 93:743 (1982)
10 Y. Sommarin and D. Heinegård, Specific interaction between cartilage proteoglycans and hyaluronic acid at the chondrocyte cell surface, Biochem. J. 214:777 (1983)

PROBLEMS OF WHITE CELL SEPARATION TECHNIQUES IN RELATION TO THE PHASE PARTITIONING OF NEUTROPHILS

Paul Eggleton *, Neville Crawford + and Derek Fisher *

* Biochemistry Department, Royal Free Hospital
School of Medicine, University of London, London NW3 2PF

+ Biochemistry Department, Royal College of Surgeons of England
Lincoln's Inn Fields, London WC2 3PN, UK

The method selected for the isolation of neutrophils from peripheral blood influences the partition behaviour of the cells obtained. The phase forming polymers themselves can, to some extent, stimulate and alter the oxidative responsiveness of neutrophils. Care in the selection of batches of polymers for phase systems is clearly indicated.

INTRODUCTION

Neutrophils undergo marked morphological and functional changes in defence of the host against invading micro-organisms [1-3]. These changes involve modification of the cell surface membrane and cytoskeleton [4-5]. Quantitation of functional or biochemical characteristics of neutrophils has been limited by the fact that standard isolation techniques measure the mean activity of a bulk population of cells. Cell separation techniques such as counter current distribution (CCD) in charge-sensitive phase system and continuous flow electrophoresis (CFE) allows cell populations to be subfractionated on the basis of differences in surface charge.

We are undertaking a comparative study of the functional and biochemical heterogeneity of the circulating neutrophil population revealed by CCD and CFE.

As a preliminary to the CCD studies [6] we have examined the effect of neutrophil isolation methods on the partitioning behaviour of the cells. The most common methods of isolating neutrophils use commercial preparations of ficoll or dextran solutions, which may be harmful to the recovered cells [7-10]. In this study we have isolated neutrophils by three different methods, one of which does not employ these polymer solutions, and have assessed the purity, yield, viability and partition behaviour of the isolated neutrophils.

We have also screened a number of phase-forming polymers for their lack of ability to stimulate the oxidative burst in neutrophils using a histochemical method and a biochemical method to detect the formation of superoxide anion.

MATERIALS AND METHODS

Phase-forming polymers

Dextran T500 (lot KL-02434) and Dextran T40 (lot HB-24512) were obtained from Pharmacia, Dextran (average molecular weight 81600) was obtained from Sigma, Dextran PL 500 VC was a gift from Pfeiffer and Langen, Dormargen, West Germany, Aquaphase was a gift from Perstop AB, Sweden. PEG 8000 (lot 62806) was obtained from Union Carbide, PEG 6000 (lots 9844790 and 9159110), PEG 4000 (lot 629550) and PEG 20000 (lot 9971940f) were obtained from BDH (Poole, UK).

Neutrophil preparations

EDTA anticoagulated blood from volunteers was used to isolate neutrophils. For comparative studies of preparative methods neutrophils were isolated by dextran sedimentation, Lymphoprep centrifugation/dextran sedimentation and by a modified Dioguardi method [11] using ammonium chloride lysis followed by differential centrifugation in Hank's balanced salt solution (HBSS). Routinely neutrophils were prepared by the latter method.

Single tube partitioning

Two charge-sensitive phase systems (5% dextran-5% PEG; 4.5% dextran-4.5% PEG) in 270 mM sodium phosphate buffer, pH 7.4 were prepared on a w/w basis from a 40% w/w stock aqueous solution of PEG 6000 (lot 9844790) and 22.9% w/w stock aqueous solution of Dextran T500 and 808 mosM sodium phosphate buffer. Phase systems were equilibrated at 4°C for 15 hr. Top and bottom phases were removed and stored at 4°C. One ml aliquots of dextran-rich lower phase were pipetted in partition tubes. Approximately 1 x 10^6 white cells, isolated by each technique, were added to 1.05ml of top phase. 1ml of this suspension was added to the tubes containing the bottom phase. Tubes were then inverted 20 times, allowed to stand for 5 mins at 4°C to ensure temperature equilibration, remixed and allowed to phase separate for 30 mins at 4°C. 50 μl samples from the upper phase were then withdrawn and the number of neutrophils present counted electronically using a Coulter Counter.

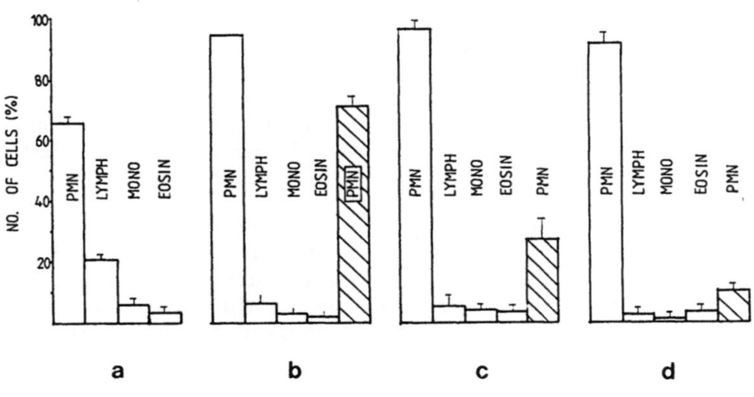

Fig. 1 *Purity and yields of neutrophils: the white cell composition of (a) whole blood and preparations of neutrophils isolated by (b) NH_4Cl lysis and differential centrifugation, (c) dextran sedimentation and (d) Lymphoprep centrifugation/dextran sedimentation are shown together with the yields of neutrophils obtained by each method (hatched areas)*

Partition coefficients were expressed as the number of neutrophils present in the top phase as a percentage of the total number of cells.

Polymer treatments and the oxidative responsiveness of neutrophils

The effect of pre-treatment of neutrophils in whole blood and isolated neutrophils with various phase reagents was assessed by the cells' ability to reduce nitroblue tetrazolium (NBT) dye [12] and ferricytochrome C [13] respectively.

For the NBT test 50 μl of whole blood were mixed with 150 μl of NBT dye solution (1mg/ml in HBSS) containing PEGs (5% w/w), dextrans (8% w/w) or Aquaphase (8% w/w) and incubated at 37°C for 15 min. The white cells, isolated as a buffy coat, were resuspended in 200 μl. 100 μl were cytocentrifuged and the resulting smear stained with Prodiff. 100 cells were scored for NBT positivity; cells classified as NBT-positive contained dense black formazan deposits.

For the cytochrome C reductase assay, 100 μl of isolated neutrophils (approx. 3 x 10^6 cells) were mixed with 1 ml solutions of dextrans (8% w/w, Aquaphase 8% w/w) or PEGs (5% w/w) prepared in HBSS. After 90 min at room temperature (21°C) the samples were centrifuged at 4000 rpm for 5 min and the cells resuspended in HBSS (500 μl). Superoxide anion was measured by ferricytochrome C reduction, the addition of superoxide dismutase being used to identify the specific contribution of superoxide anion [13]. Ferricytochrome C reduction was measured spectrophotometrically and expressed as the mean $\pm$ standard deviation of nanomoles of cytochrome C reduced in triplicate experiments.

RESULTS

Yield, purity and viability of neutrophils isolated by various methods

Neutrophils were isolated by each of the three methods described. Yield, purity and differential counts are shown in Fig. 1. The differential centrifugation method allowed the greatest recovery of neutrophils (72% $\pm$ 4.6), while dextran sedimentation and combined Lymphoprep centrifugation/ dextran sedimentation resulted in much lower yields of neutrophils (28.5% $\pm$ 2.0 and 17.2 $\pm$ 3.2 respectively). Neutrophil viability and purity was greater than 90% by all methods.

Table 1 *Effect of white cell isolation methods on the partitioning behaviour of neutrophils*

Isolation Method	Phase System[a]	Cells in top phase[b]
Modified Dioguardi Method	4.5% Dx/4.5% PEG	91.0 $\pm$ 8.1 (5)
	5.0% Dx/5.0% PEG	85.0 $\pm$ 12.1 (10)
Dextran sedimentation	4.5% Dx/4.5% PEG	82.0 $\pm$ 14.9 (5)
	5.0% Dx/5.0% PEG	22.0 $\pm$ 2.6 (10)
Lymphoprep centrifugation/ dextran sedimentation	4.5% Dx/4.5% PEG	61.0 $\pm$ 9.3 (9)
	5.0% Dx/5.0% PEG	39.0 $\pm$ 17.7 (10)

[a] Charge sensitive phase systems of Dextran T500 and PEG 6000 in 270 mOsm sodium phosphate, pH 7.4.
[b] Mean $\pm$ SD of the number of experiments shown in parenthesis.

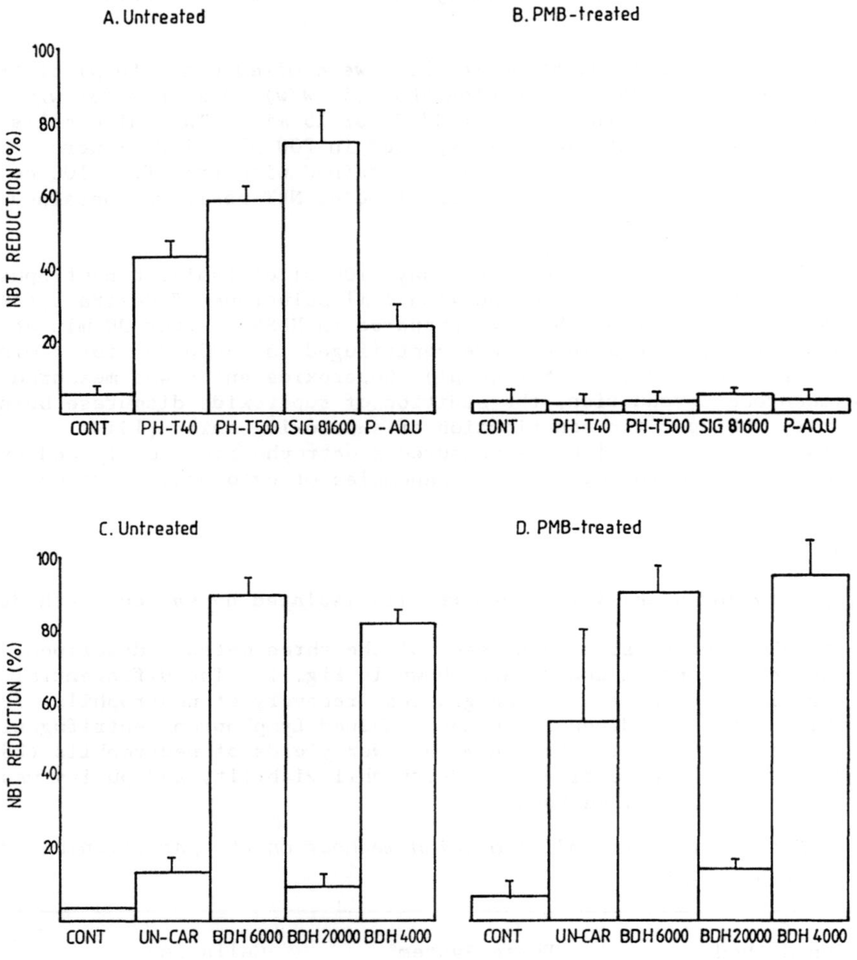

Fig. 2 *The effect of exposure of neutrophils to phase forming polymers on their oxidative responsiveness measured by the reduction of nitroblue tetrazolium (NBT) dye: (A) 8% solutions of dextrans and Aquaphase (P-AQU); (B) as A but also treated with polymyxin B (PMB); (C) 5% solutions of PEGs; (D) as C but also treated with polymyxin B*

Effect of isolation methods on the partition behaviour of neutrophils

The partitioning behaviour of isolated neutrophils was assessed using two different charge-sensitive two-phase systems (Table 1). Neutrophils isolated from whole blood using Lymphoprep and/or dextran resulted in the neutrophils having top phase partition coefficients (PC) of 39.0% ± 17.7 and 22.0% ± 2.6 respectively. These were significantly lower (P< 0.001) than the partition coefficient of neutrophils isolated by differential centrifugation (85.0% ± 12.1).

Stimulatory effect of phase-forming polymers on neutrophils in whole blood

The stimulatory effect of phase-forming polymers (dextrans, PEGs and Aquaphase) on neutrophils in whole blood at room temperature as detected by increased NBT reduction by neutrophils is shown in Fig. 2. Exposure of neutrophils to the polymer solutions led to an increased number of neutrophils reducing NBT above the baseline level of less than 10% observed in control cells (Figs. 2A and 2C). Pre-treatment of dextran and Aquaphase solutions with Polymyxin B (an endotoxin-inactivating antibiotic) reduced the level of NBT reduction to baseline levels (Fig. 2B). Pretreatment of polyethylene glycol solutions with the antibiotic, did not prevent the stimulatory effect of PEG on neutrophils (Fig. 2D).

Stimulatory effect of phase-forming polymers on isolated neutrophils

Stimulatory effects of the phase-forming polymers on isolated neutrophils were also observed. Exposure of isolated neutrophils to various dextrans (5% w/w) and PEGS (8% w/w) for 90 mins at 21°C led to increased reduction of cytochrome C above levels obtained with non-treated control cells (Fig. 3a).

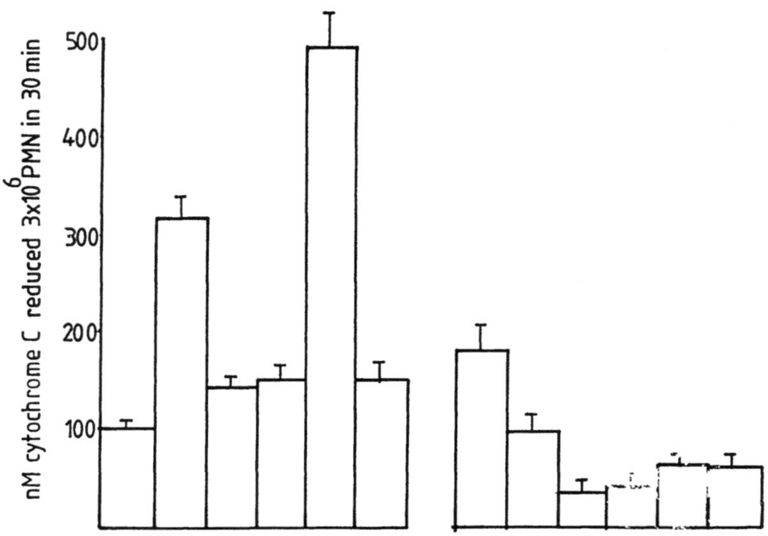

Fig. 3 *Effects of phase-forming polymers on the production of superoxide anion by isolated neutrophils: Neutrophils were exposed to polymer solutions, washed with HBSS and superoxide formation measured before (left hand diagram) and after stimulation with S. Aureus (right hand diagram). Mean ± SD of three experiments. (1) Control (HBSS); (2) 8% Aquaphase; (3) 8% Dextran T500 (lot MI 02434); (4) 8% Dextran PL 500 VC; (5) 5% PEG 6000 (lot 9159110); (6) 5% PEG 6000 (lot 9844790)*

Moreover, once neutrophils had been exposed to phase reagents for 90 minutes, their capacity to generate superoxide (i.e. reduce cytochrome C) in the presence of a bacterial stimulus was reduced.

DISCUSSION

Partitioning of cells in aqueous two-phase systems has been shown to be an effective method of separating cells on the basis of differences in surface charge and/or hydrophobic characteristics [14]. Yet very little work has been undertaken to evaluate the possibility that the phase reagents themselves may alter the surface and functional properties of the cells under test [15].

Since we wished to study the heterogeneity of the neutrophil population it was important to isolate the cells from blood in high yield to avoid the selective loss of cells during isolation. Although the three isolation methods compared (Fig. 1) gave adequate purity (90-95%) the recoveries of neutrophils from blood differed considerably. Dextran sedimentation of whole blood, resulted in low yields of neutrophils (30-40%) and the cells were susceptible to clumping and aggregation. Lymphoprep centrifugation followed by dextran sedimentation required a long centrifugation step. This resulted in many of the neutrophils becoming lodged between red blood cell aggregates, leading to the loss of a large number of neutrophils (up to 80%). Simple lysis of red cells in isotonic ammonium chloride followed by low speed differential centrifugation in HBSS, allowed high yields of neutrophils (65-80%) to be recovered in good morphological order.

The partitioning behaviour of neutrophils isolated by the three techniques also differed (Table 1). Cells isolated by dextran sedimentation or using Lymphoprep were observed to have lower top phase partition coefficients in fully charged phase systems than cells isolated by simple differential centrifugation. Brooks and coworkers [15] have reported similar findings, which they explain "by the ability of dextran to adsorb, in part irreversibly to cells. The cells carrying dextran on their surfaces then tend to partition with the dextran into the lower phase". Clearly an additional advantage of the differential centrifugation isolation method is that it avoids the possibility of polymer adsorption during isolation. It remains to be determined to what degree polymer adsorption occurs during cell partitioning in polymer two-phase systems, particulary during the relatively long exposure to phase systems in countercurrent distribution. Isolating cell without exposure to polymer may influence the heterogeneity revealed by CCD.

All of the phase reagents tested were stimulatory to some extent. Stimulation of neutrophils by dextrans could be prevented in the presence of polymyxin B (an endotoxin-inactivating antibiotic). This suggested that some batches of dextran contained bacterial endotoxin (LPS), the presence of which may have been directly responsible for the stimulatory effects observed.

In contrast, pre-treatment of PEG solutions with Polymyxin B did not remove their stimulatory properties. Attention has been given to impurities in PEGs with respect to the fusogenic property of relatively concentrated aqueous PEG solutions [16]. Antioxidants are present in commercial preparations and peroxides, aldehydes, aldehyde precursors and acidic impurities have been reported. These contaminants or the PEGs themselves may be the source of the stimulation. However, the variations in the degree of stimulation observed with PEGs, particularly that seen for different samples of the same molecular weight range, suggests that the polymers per se are unlikely candidates.

Exposure of the neutrophils to the phase-forming polymers reduced the ability of the neutrophils to respond subsequently to a defined bacterial stimulus. It therefore follows that unless precautions are taken to reduce the stimulation of the cells to a minimum, the heterogeneity revealed by CCD may simply reflect varying degrees of stimulation incurred during isolation and CCD and not the natural heterogeneity of the neutrophil population.

This study has shown that the isolation of neutrophils from peripheral blood influences the partition behaviour of the isolated cells. In addition, the phase reagents themselves appear to some extent to be responsible for stimulation and alteration of neutrophil functions. The mechanism of these effects remains to be determined, but it is likely that other surface-related phenomena are also affected by exposure of neutrophils to dextran and polyethylene glycols. Having demonstrated that most of the stimulation of cells by dextran can be removed by treatment of the dextran with Polymyxin B, raises the concern that bacterial endotoxin may be a contaminant of commercially supplied dextrans commonly used in phase partitioning systems and/or may develop significantly on storage. It is obviously important to standardise conditions used to partition all types of cells, but care is required when working with leukocytes which are particularly sensitive to exposure to endotoxin. For partitioning work with neutrophils an initial confirmation that polymer batches are not stimulatory and a routine screening of phase solutions prepared for partitioning are clearly indicated.

ACKNOWLEDGEMENT

We thank the S.E.R.C. for a Research Studentship to P.E. and the Peter Samuel Royal Free Fund.

REFERENCES

1 C.E. McCall, D.A. Bass, L.R. DeChatelet, A.R. Link, and M. ann, *In vitro* responses of human neutrophils to N-formyl-methionyl-leucyl-phenylalanine: correlation with effects of acutebacterial infection, J. Infect. Dis. 140:277 (1979)
2 J.E. Repine, C.C. Clawson, and F.C. Goetz, Bactericidal function of neutrophils from patients with acute bacterial infections and from diabetics, J. Infect. Dis. 142:869 (1980)
3 F.R Venezio, G.O. Westenfelder, and J.P. Phair, The adherence of polymorphonuclear leukocytes in patients with sepsis, J. Infect. Dis. 145:351 (1982)
4 R.L. Baehner, and L.A. Boxer, Disorders of polymorphonuclear leukocyte function related to alterations in the integrated reactions of cytoplasmoic constituents with the plasma membrane. Seminars in Hematol. 16:148 (1979)
5 F.S. Southwick, and T.P. Stossel, Contractile proteins in leukocyte function. Seminars in Hematol. 20:305(1983)
6 P. Eggleton, I.A. Sutherland and D.Fisher, Rapid partition of human neutrophils by CCD, in: "Separations Using Aqueous Phase Systems: Applications in Cell Biology and Biotechnology," D. Fisher and I.A. Sutherland, eds., Plenum, New York (1989)
7 C. Jarstrand, G. Lahnborg and L. Berhem, Human granulocyte function during dextran infusion, Acta Chir. Scand. 489:279 (1979)
8 R.D. Nelson, P.G. Quie and R.L. Simmons, Chemotaxis under agarose: A new and simple method for measuring chemotaxis and spontaneous migration of human polymorphonuclear leukocytes and monocytes, J. Immunol. 115:1650 (1975)
9 L. Poston, R.B. Jones and P.J.Hilton, Sodium transport of polymorphonuclear leukocytes; effects of isolation by the Ficoll/Triosil method, Clin. Sci. 62:563 (1982)
10 J.A. Roth and M.L. Kaeberle, Isolation of neutrophils and eosinophils from the peripheral blood of cattle and comparison of their functional activities, J. Immunol Methods. 45:153 (1981)

11 N. Dioguardi, A. Agostini, G. Fiorelli and B. Lomanto, Characterisation of lactic dehydrogenase of normal human granulocytes. J. Lab. Clin. Med. 61: 713 (1963).

12 P. Eggleton, Rapid techniques for the measurement of non-specific host responses to bacterial infection, M.Phil. Thesis, University of London (1987)

13 B. Babior, R.S. Kipnes and J.T. Curnutte, The production by leukocytes of superoxide, a potential bacterial agent, J. Clin. Invest. 52:741 (1973)

14 H. Walter, D.E. Brooks, and D.Fisher, "Partitioning in Aqueous Two-Phase Systems. Theory, Methods, Uses and Applications to Biotechnology," Academic Press Inc., (1985).

15 R.S. Hill, R. Norris-Jones, B. Still and D.E. Brooks, Surface charge and hydrophobic properties of fresh and cyropreserved blood phagocytes as determined by partitioning in two-phase aqueous polymer systems, Am. J. Hematol. 21:249 (1986)

16 D. Fisher, Mechanisms of cell fusion induced by polyethylene glycol, in: "Microorganisms in Aqueous Two-Phase Systems," IVA-rapport 330, Royal Swedish Academy of Engineering Sciences, Stockholm (1986)

T CELL HETEROGENEITY AND LYMPHOID CELL POPULATIONS IN NORMAL HUMANS AND RHEUMATOID ARTHRITIS PROBED BY PARTITIONING

J. Michalski *, C.C. McCombs * and H. Walter⁺

* Department of Medicine, Louisiana State University School of Medicine, New Orleans, Louisiana 70112, USA
⁺ Laboratory of Chemical Biology, Veterans Administrtion Medical Center, Long Beach, California 90822 and Department of Physiology and Biophysics University of California, Irvine, California

INTRODUCTION

A major advance in our understanding of the immune system has resulted from our ability to subdivide immune cells into several populations based on surface markers and functional characteristics. In the case of lymphocytes, T cells regulate antibody production and mediate "cellular immunity", B cells produce antibody, and "null" cells (not B or T) some of which act as "killer" cells against a variety of target cells without the requirement for prior immunization. Further understanding has come from the development of monoclonal antibodies that distinguish helper/inducer T cells from those with suppressor/cytotoxic activity [1]. More recently, these T cell subpopulations have been further subdivided into subsets with newer antibodies. Partitioning in two-polymer aqueous phase systems has been useful both to analyse lymphocyte heterogeneity and to subfractionate and partially purify lymphocyte subpopulations based on surface properties [2]. In the case of lymphocytes, studies using charge-sensitive phase systems show that humans [3] rats [4] and mice [5] have the same basic partitioning behaviour of the major subpopulations: B cells have a low partition ratio ($\underline{P}$), T cells have an intermediate $\underline{P}$ value and the null cells have the highest ratio.

In the studies reported here we used monoclonal antibodies to detect T cell subsets in lymphocytes subfractionated by partitioning in charge-sensitive phase systems. In the two systems used, null cells and a sub-set of suppressor/cytotoxic T cells were found in a small right-hand peak. We also briefly describe our preliminary findings concerning this cell population in patients with rheumatoid arthritis (RA).

METHODS

Partitioning was carried out in 2 slightly different phase systems: (I) 5% (w/w) Dextran T500, 4% (w/w) poly(ethylene glycol) 8000, 200 mosM sodium phosphate buffer, pH 7.4, 70 mosM NaCl and 5% (w/w) fetal bovine serum (normal T cell study) or (II) a system prepared identically except that the phosphate buffer was 150 mosM and the NaCl was 130 mosM (RA study).

Multiple extraction by countercurrent distribution (CCD) was performed on a thin-layer CCD apparatus as previously outlined by Walter [2]. CD4 bearing T cells (helper/inducer) were detected with OKT4 [7] or Leu 3 [8] monoclonal antibodies and CD8+ (suppressor/cytotoxic) with OKT8 or Leu 2 by direct immunofluorescence staining on a cell sorter (Becton Dickenson). Natural killer cell markers were detected with anti-Leu 7 [9] or HNK1 [10].

RESULTS AND DISCUSSION

A typical experiment in which human lymphocytes are subfractionated in these phase systems is shown in Fig. 1. The solid line indicates the distribution of mononuclear cells. There is a major left hand peak having a low partition ratio and a minor peak of high P value. The latter constitutes

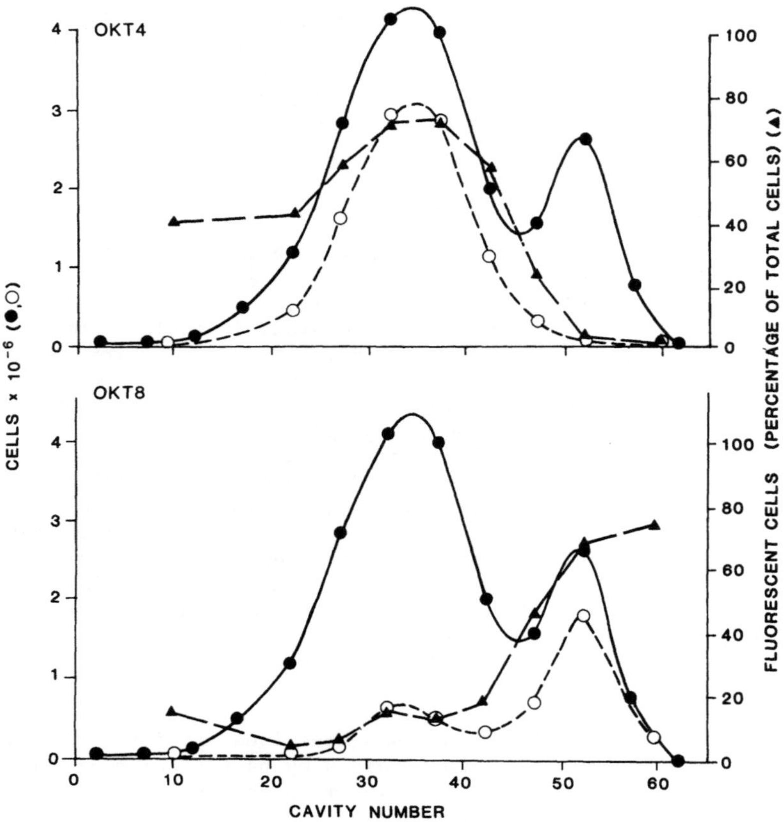

Fig. 1 *Countercurrent distribution (CCD) pattern of human peripheral blood mononuclear cells in a charge-sensitive dextran-poly(ethylene glycol) aqueous two-phase system. Distribution of total mononuclear cells is given in cell counts (●). In addition cells were analysed, using fluorescent monoclonal antibodies, for the OKT4 (top) and OKT8 (bottom) markers. Percentage of these cells in different cavities along the extraction train is indicated (▲) as is the distribution, in terms of counts, of cell bearing the specific marker indicated (○). From reference 11*

8-33% of lymphocytes in normal individuals and is enriched with null cells [12] and killer activity [13]. The helper/inducer T cell subset is found almost exclusively in the left hand peak (top of figure) whereas the suppressor/cytotoxic cells have a bimodal distribution and in this individual are found predominantly in the right hand peak (bottom part of figure). In five experiments, the proportion of CD8 positive cells found in the high P value peak ranged from 26-72%.

In three experiments we determined the proportion of cells having the Leu 7 marker (associated with natural killer activity). Virtually all of these cells have a high P value and a distribution that overlaps the right hand peak of cells and the distribution of CD8 positive cells with a high P value (data not shown).

The CD8 bearing T cell subpopulation encompasses lymphocytes having two functions: cellular cytotoxicity and suppression of antibody formation. This latter function has recently been shown to be mediated by a subpopulation of CD8 positive cells that co-expresses CD8 and Leu 7 as well as another marker, Leu 15 [14]. Since nearly all Leu 7 bearing cells are in the right hand peak, it seemed inescapable that the CD8 subset coexpressing Leu 7 should also be cells with a high partition ratio. We showed that this was the case by double labelling cells with a fluorescein labelled anti-CD8 (green) and a phycoerythrin labelled anti-HNK1 (red). Virtually all of the double labelled cells (CD8+, HNK1+) were in the right hand peak and most CD8+ cells with a high P value co-expressed the NK marker. In contrast, very few of the CD8+ cells with a low P value express HNK1. Partitioning in our charge-sensitive system has thus apparently resulted in a separation of the "suppressor" CD8+ cells described by Clement and coworkers [14] from the majority of CD8 bearing cells that lack this function. We are presently working on a series of experiments designed to examine the functional characteristics of the two populations of CD8+ cells.

Lymphocytes that express the Leu 7/HNK1 antigens also have a distinctive morphology; they are large and have granules stained by the Giemsa method [15]. Such cells have been called large granular lymphocytes (LGL) and are highly enriched in the right hand peak [12]. Marked increases in LGL are seen in rheumatoid arthritis (RA) patients and may be associated with severe leukopenia [16]. To determine if this severe increase in LGL in some patients might reflect a qualitative or quantitative change in the high P value of lymphocytes of most RA patients, we studied the partitioning behaviour of lymphocytes from 13 such patients compared with 32 control subjects. Our findings indicate that rheumatoid arthritis is associated with a modest but statistically significant increase in cells with high P value. Moreover, the expression of receptors for the F_c portion of IgG is greatly increased in the cells of the right hand peak suggesting that there are both quantitative and qualitative alterations of the high P value cells in RA.

Acknowledgement: This work was supported in part by a grant AM34223 from the National Institutes of Health and by the Medical Research Service of the Veterans Administration.

REFERENCES

1 J.D. Stobo, Lymphocyte I. T cells, in: "Basic and Clinical Immunology," D.P. Sites, J.D. Stobo, H.H. Fudenberg and J.V. Wells, eds., Lange Medical Publications, Los Altos (1984)
2 H. Walter, in: "Partitioning in Aqueous Two-Phase Systems: Theory, Methods, Uses, and Applications to Biotechnology," H. Walter, D.E. Brooks and D. Fisher, eds., Academic Press, Orlando (1985)

3 H. Walter, T.J. Webber, J.P. Michalski et al, Subfractionation of human peripheral blood lymphocytes on the basis of their surface properties by partitioning in two-polymer aqueous phase system, J. Immunol. 123:1687 (1979)
4 K. Nelson, P. Malmström, Å. Jönsson, et al, Separation of rat leukocytes by counter-current distribution in aqueous two-phase systems. II. Subpopulations which mediate selective and non-selective lysis of normal and colon carcinoma target cells in vitro, Cell. Immunol. 37:422 (1978)
5 J.P. Michalski, M. Razandi, C.C. McCombs, et al, Surface properties of lymphocyte subpopulations in autoimmune NZB/NZW F1 hybrid mice: alterations correlated with the immunodeficiency of ageing, Clin. Immunol. Immunopath. 29:15 (1983)
6 P.-Å. Albertsson, Separation of cells and cell particles by countercurrent distribution, Sci. Tool, 17:53 (1970)
7 P.C. Kung, G. Goldstein, E.L. Reinherz, et al, Monoclonal antibodies defining distinctive human T cell surface antigens, Science 206:347 (1979)
8 E.G. Engleman, C.T. Benike, E. Glichman, et al, Antibodies to membrane structures that distinguish suppressor/cytotoxic and helper T lymphocyte subpopulations block the mixed lymphocyte reaction in man, J. Exp. Med. 154:193 (1981)
9 T. Abo and C.M. Balch, A differentiation antigen of human NK and K cells identified by a monoclonal antibody, J. Immunol. 127:1024 (1981)
10 T. Abo, M.D. Cooper and C.M. Balch, Characterisation of HNK1+ (Leu 7) human lymphocytes. 1. Two distinct phenotypes of human NK cells with different cytotoxic capability, J. Immunol. 129:1752 (1982)
11 J.P. Michalski, B. Bozelka and H. Walter, Subfractionation of human T lymphocyte subpopulations by partitioning; surface heterogeneity of the OKT8/Leu 2+ T cell subset, Clin. Exp. Immunol. 67:565 (1987)
12 J.P. Michalski, S. Zanki, J.L. Anderson, et al, Subfractionation of human peripheral lymphocytes by partitioning in two-polymer aqueous phase systems. Characteristics of a small subpopulation of cells with high partition coefficient, J. Clin Lab. Immunol. 21:43 (1986)
13 E.M. Levy, S. Zanki and H. Walter, Countercurrent distribution of human peripheral blood lymphocytes; isolation of a sub-population enriched with natural killer and K cells, Eur. J. Immunol. 11:952 (1981)
14 L.T. Clement, C.E. Grossi and G.L. Gartland, Morphology and phenotypic features of the subpopulation of Leu2+ cells that suppresses B cell differentiation, J. Immunol. 133:2461 (1984)
15 T. Timeon, J.R. Ortaldo and R.B. Herberman, Characteristics of human large granular lymphocytes and relationship to natural killer and K cells, J. Exp. Med. 153:569 (1981)
16 J.C. Barton, E.F. Prasthofer, M.L. Egan, et al, Rheumatoid arthritis associated with expanded populations of granular lymphocytes, Ann. Intern. Med. 104:314 (1986)

SURFACE CHARACTERIZATION OF COLLOIDAL DRUG CARRIERS FOR DRUG TARGETING BY AQUEOUS TWO-PHASE PARTITIONING

Rainer H. Müller*, Stanley S. Davis and Eric Mak

Department of Pharmaceutical Sciences
University of Nottingham
Nottingham NG7 2RD, United Kingdom

* To whom correspondence should be addressed

Surface properties play an important role in the organ distribution of intravenously administered colloidal drug carriers (polymeric particles, parenteral fat emulsions). To avoid the clearance of the particles from the blood stream by the macrophage system of the body, they need to possess surface properties which do not lead to recognition as a foreign particle by the reticuloendothelial system (RES). Carriers with hydrophilic surfaces show a reduced opsonisation in the blood and less clearance by the RES. A hydrophilic surface can be created by coating polymeric particles with polymers or by selecting appropriate emulsifiers for emulsion systems. For the *in vitro* selection of suitable polymers and emulsifiers, aqueous two-phase partitioning has been employed to characterise the carrier surface.

INTRODUCTION

Drug targeting is the controlled delivery of drugs to their desired site of action. The selective accumulation of the drug in the desired tissue reduces toxic side effects in the rest of the body and thus enables that the amount of administered drug can be reduced. However, for anticancer drugs (cytotoxics), the amount administered is limited by the toxicity of the drug. In such cases, targeting would allow the application of a higher dose due to the reduction of toxic side effects.

One approach to achieve drug targetting is the incorporation of drugs in colloidal carriers and their administration by intravenous injection [1]. Carriers such as biodegradable polymeric particles [2], liposomes [3] or fat emulsions [4] can be used. The major obstacle for these systems is their recognition by the body as being foreign and their subsequent removal from the blood circulation by the macrophages of the recticuloendothelial system (RES) (mainly by the Kupffer cells in the liver and the spleen macrophages). Within the first 5 minutes after intravenous injection, up to 90% of the administered dose is removed from the blood by the liver and spleen [5]. However, clearance by the RES and the body distribution of the carriers depends, amongst other factors, on their surface properties [6]. In general, opsonisation of the carriers in the blood and clearance by the RES decrease the less hydrophobic and the less charged the carriers are. Determination of carrier charge [7] and surface hydrophobicity [8] are therefore of paramount importance for the characterisation of the systems and the estimation of their *in vivo* behaviour.

Hydrophilic carrier surfaces can be created by coating hydrophobic polymeric carriers with block co-polymers or phospholipids or by the preparation of fat emulsions using these compounds as emulsifiers and emulsion stabilisers. The co-polymers form a hydrophilic coating layer (particles) or interfacial film (emulsions) which acts simultaneously as a steric stabiliser. The steric stabilising layer protects against the uptake by macrophages [5,9] and keeps the carriers in the blood circulation. Compared with uncoated carriers (e.g. polystyrene particles), a reduction of liver/spleen uptake from 80-90% to approximately 25-30% could be achieved (measured with gamma-scintigraphy using radioactive labelled carriers [5]). Taking into account that 25% of the measured carrier concentration in the liver/spleen region is due to carriers circulating in the blood (blood pool of the liver), the clearance by the liver/spleen macrophages is reduced to only a few percent of the administered dose.

The selection of coating materials for *in vivo* studies was based on the data obtained with a range of characterisation methods. The thickness of the coating layer was determined by Photon Correlation Spectroscopy (PCS) [10,11] and the particle charge of coated carriers by electrophoresis measurements using Laser Doppler Anemometry (LDA) [11,12]. The surface hydrophobicity was assessed by adsorption measurements with Rose Bengal [8] and by Hydrophobic Interaction Chromatography (HIC) [13]. The protection of the coating layers against phagocytosis by macrophages was tested in cell cultures of mouse peritoneal macrophages [14].

Aqueous two-phase partitioning seems to be a promising technique which can be used to complete the spectrum of surface characterisation techniques. It is a sensitive technique which can be applied to detect changes in the surface properties of polymeric or emulsion carriers by comparing the partitioning behaviour of uncoated and coated carriers or modified emulsions. For the reduction of RES clearance it is important that the polymer covers the carrier surface perfectly without bare patches or exposing hydrophobic parts of the polymer to the carrier surface. For emulsion carriers, phospholipids (natural lecithins or synthetic phospholipids) can be selected which provide the greatest hydrophilicity. The RES clearance of commercial fat emulsions for parenteral nutrition seems to differ in dependence on the supplier. Incorporation of drugs in carriers can change their surface properties if the drug is exposed at the surface of a polymeric carrier or incorporated in the interfacial film of emulsions due to the amphiphilic character of the drug. The effect of drug incorporation on the *in vivo* fate might be estimated by measuring the change in surface properties using aqueous two-phase partitioning. This assumption has yet to be confirmed by comparing partitioning data with *in vivo* experiments.

EXPERIMENTAL

The composition of the aqueous two-phase systems was 16.0% Dextran 20, 12.0% PEG 2000, 0.01 M phosphate buffer and 0.05 M NaCl. The systems were allowed to settle by gravity and readings were taken after 2 or 4 hours' settling time. The carrier concentration was determined by absorption measurements at 400 nm.

Particle sizes and the thickness of adsorption layers were determined by Photon Correlation Spectroscopy (Spectrometer RR102 in connection with K 7025 Multibit Correlator, Malvern Instruments, UK).

Table 1 *Standard latex particles used as model carriers. The binding constants K were obtained from adsorption measurements with Rose Bengal and are used to place the latex in order of increasing hydrophobicity*

polystyrene particle (code)	size (um)	surface group	binding constant (K) (ml/ug)
PS-OH	0.25	hydroxyl	0.041
PSF	0.19	fluorescent label	0.120
PS-AR-NH_2	0.18	aromatic NH_2	0.135
PS-AL-NH_2	0.19	aliphatic NH_2	0.227
PS-1.0	1.00	-	0.279
PS-0.17	0.17	-	0.289
PS-0.06	0.06	-	0.400

MATERIALS

Dextran 20 was purchased from Sigma (Poole, UK) and PEG 2000 and PEG 35,000 were kindly supplied by Hoechst (Frankfurt, West Germany). Latex particles, differing in size and surface properties were bought from Polysciences (Northampton, UK) and Poloxamers and the Poloxamine 908 from Ugine Kuhlmann (Bolton, UK). Fat emulsions were obtained from Braun Melsungen AG (Lipofundin S, Lipofundin MCT, Endolipide - Melsungen, West Germany), KabiVitrium Ltd (Intralipid - Uxbridge, UK) and Morishita Pharmaceuticals (Venolipid - Japan). The phospholipids were kindly supplied by Nattermann AG (Cologne, West Germany).

RESULTS AND DISCUSSION

To investigate the partitioning of uncoated polymeric carriers and to study the influence of different surface groups on the partition coefficient, a range of non-biodegradable model carriers (polystyrene latex particles) were used. These latices are used as standard particles and possess defined surface properties (Table 1).

The unmodified polystyrene particles possess carboxyl and sulphate groups on their surface; on the modified latex, additional surface groups are introduced by chemical reaction. The surface hydrophobicity of such particles was determined previously [6,8] by measuring the adsorption of the hydrophobic dye Rose Bengal on the surface. From the adsorption isotherms obtained, the binding constant K was calculated and the particles could then be placed in order of increasing hydrophobicity. Distinct differences were apparent between the most hydrophobic latex ($K = 0.40$ ml/ug) and the most hydrophilic particles (hydroxlated latex, $K = 0.04$ ml/ug) (Table 1).

In the partitioning experiments (Table 2), no distinct reproducible differences could be found. The method seems to be unsuitable to detect differences in surface hydrophobicity between the latex particles. It is thought that even the relatively hydrophilic PS-OH particles are still too hydrophobic to have significantly different partitioning properties from the other latex (e.g. PS-0.17).

Furthermore, aggregation of the particles in the PEG phase was seen; this could well influence the partitioning process and would interfere to some extent with the absorption assay. The interactions of the particles

Table 2 *Partitioning of uncoated latex particles with different sizes and surface groups (surface hydrophobicity) in a Dextran 20/PEG 2000 two-phase system*

polystyrene particles (code)	Top phase (%)	Interface (%)	Bottom phase (%)
PS-OH	86.1	13.9	0
PSF	92.1	7.9	0
$PS-AR-NH_2$	97.3	2.7	0
PS-1.0	96.8	3.2	0
PS-0.17	87.5	12.5	0
PS-0.06	95.0	5.0	0

with the polymers making up the two phase system and the time dependent tendency to form aggregates were investigated using Photon Correlation Spectroscopy (PCS). Table 3 shows the increase in size of polystyrene latex particles dispersed in the PEG rich top phase with time. To investigate which polymer is responsible for the observed aggregation, latex particles (PS-0.06) were mixed with PEG and with dextran solutions at different polymer concentrations and measured directly with PCS. To make an adsorption more easily detectable, a high molecular weight PEG (MW 35.000) was used. The adsorption of PEG could be measured above polymer concentrations of 0.001% and adsorption layers of 40 to 60 Å were found (Table 4). The polydispersity index (a measure for the polydispersity of the sample) was almost constant (0.08) within the whole concentration range making it unlikely that PEG causes the aggregation. In dextran solutions with a polymer concentration of 0.005%, no adsorption could be detected for the molecular weights 20,000, 500,000 and 700,000. A low polydispersity index of 0.07 was found. However mixing with dextran MW 2,000,000 led to an increase in size of the latex by 35 nm and in the polydispersity index from 0.07 to 0.14 thus indicating particle aggregation. Mixing of latex with isolated dextran rich bottom phase led to immediate formation of aggregates which were too large to be measured by PCS (>> 5 um).

Coating of PS-0.17 particles with hydrophilic block co-polymers from the Poloxamer and Poloxamine series did not alter their partitioning. This is thought to be due to a specific interaction between the PEG-rich top phase

Table 3 *Size of uncoated and coated latex in top phase (PEG rich) in dependence on time (PD-Index - Polydispersity Index, approx. 0.06 for monodisperse latex particles*

time (h)	uncoated latex 145 nm		Poloxamine 908 coated latex	
	Size (nm)	PD-Index	Size (nm)	PD-Index
0.25	161	0.10	180	0.07
0.5	174	0.13	-	-
1.0	175	0.15	181	0.08
2.0	182	0.14	184	0.07
5.0	188	0.24	183	0.07
9.0	205	0.24	186	0.05

Table 4 *Adsorption of PEG 35,000 onto latex particles*

PEG 35,000 conc. (%)	layer thickness (A)	Polydispersity Index
0.00001	5	0.08
0.0001	4	0.07
0.001	44	0.08
0.05	52	0.08
0.1	41	0.09
1.0	62	0.12

and the ethylene oxide chains comprising the Poloxamer and Poloxamine structure. For particles coated with these polymers, a PEG-free two-phase system is required. In contrast to naked particles, the coated latex did not form aggregates in the two-phase system (Table 3) probably due to the steric stabilisation effect provided by the polymers [15].

More hydrophilic particles, or those coated with hydrophilic polymers cannot be characterised by the Rose Bengal probe method due to the lack of adsorption of the hydrophobic dye. For these particles the aqueous two-phase partitioning method was found to be more suitable. Coating of the model latex with a range of phospholipids altered their partitioning behaviour (Table 5). The percentage of particles in the top phase could be reduced from 80-100% to 40% and with two coating materials (Infusol, NC 95), no particles remained in the top phase. As a comparison, the phospholipids on their own in the form of dispersions were partitioned. Theoretically, if the particles are coated perfectly with a phospholipid, they should partition in a similar way to the phospholip dispersion. For all the phospholipids, nothing was found in the top phase. This indicates that the phospholipids did not provide an intact coat for the particles still left in the top phase. However, two coatings (Infusol, NC 95) performed well, as indicated by a similar partitioning of coated particles and the pure phospholipid dispersion. Thus the aqueous two-phase partitioning could be a suitable test for the quality (extent of surface coverage) of coating films. It should be noted that in contrast to the used two-phase system, phospholipid vesicles behave differently in other systems (e.g. Dextran 500/PEG 6000).

Table 5 *Partitioning of pure phospholipid dispersions and phospholipid coated model carriers (latex)*

	Partitioning of					
	pure phospholipid (%)			phospholipid coated carrier (%)		
phospholipid	Top	Bottom	Inter	Top	Bottom	Inter
Soya-PE-H	-	28.5	71.5	80.7	6.8	12.5
Egg-PC-H	-	32.1	67.9	86.1	9.9	4.0
Egg-PG	-	31.1	68.9	39.8	7.9	52.3
DPPC	-	28.8	71.2	45.7	24.9	29.4
NC 95	-	12.6	87.4	-	44.6	55.4
Infusol	-	16.3	83.7	-	24.4	75.6

Table 6 *Partitioning of parenteral fat emulsions in an aqueous two-phase system Dextran 20/PEG 2000*

Fat Emulsion	Top Phase (%)	Interface (%)	Bottom Phase (%)
Lipofundin S10%	0	49.0	51.0
Lipofundin S20%	0	50.0	50.0
Lipofundin MCT10%	0	41.0	58.9
Endolipide 10%	0	50.4	49.6
Intralipid 20%	0	65.5	34.7
Venolipid 20%	0	65.1	34.9

An emulsion drug carrier in the form of a fat emulsion, intended for parenteral nutrition has also been investigated. The particles have a hydrophilic surface provided by the phospholipid (lecithin) emulsifying agent. These systems are of particular interest because of their low clearance by the reticuloendothelial system following intravenous injection. Differences in the partitioning of different fat emulsions obtained from various suppliers were found (Table 6). These differences are attributed to variations in the nature of the natural lecithins used as stabilising agents. The incorporation of drugs into fat emulsions can change the surface hydrophilicity of oil droplets and subsequently might well alter the *in vivo* organ distribution. Quantification of these effects should be possible by two-phase partitioning. Further applications of the methodology to drug carriers include the assessment of detergent residues (from the particle production process) in the surface of polymeric drug carriers and the effect of sterilisation (gamma-irradiation) on the surface hydrophobicity of the polymer particles.

CONCLUSIONS

Aqueous two-phase partitioning is not suitable to characterise hydrophobic polymeric drug carriers due to only small differences in partitioning and interaction with the polymers (dextran) leading to the formation of aggregates. Coating of the carriers with sterically stabilising polymers (Poloxamer, Poloxamine) avoids this aggregation. Poloxamer coated latex could therefore be used as a standard particle for two-phase partitioning systems.

However this method allows the characterisation of more hydrophilic carriers such as coated polymeric particles and emulsions. The change in partitioning after modification of the carriers (coating, drug incorporation, sterilisation etc) compared with the distribution of the unmodified carrier is an indication of the extent of the change in surface properties (e.g. extent of polymer coating). To characterise Poloxamer and Poloxamine coated particles, PEG-free systems are required to avoid the interaction between the PEG in the top phase and the ethylene oxide chains comprising the structure of these block co-polymers.

REFERENCES

1 S.S. Davis and L. Illum, The targeting of drugs using polymeric microspheres, Brit. Polymer J. 15:160 (1983)

2 F. Koosha, R.H. Müller and C. Washington, Production of polyhydroxybutyrate (PHB) nanoparticles for drug targeting, to be published in J. Pharm. Pharmacol. Suppl. (1987)

3 G. Gregoriadis, Targeting of drugs, Nature 265:407 (1977)

4 S.S. Davis, Emulsion systems for the delivery of drugs by the parenteral route, in: "Optimisation of Drug Delivery", H. Bundgaard, A. Bagger-Hansen and H. Kofod, eds., Munksgaard, Copenhagen (1982)

5 L. Illum, S.S. Davis, R.H. Müller, E. Mak and P.E. West, The organ distribution and circulation time of intravenously injected colloidal carriers sterically stabilised with a blockcopolymer - Poloxamine 908, Life Sciences 40:367 (1987)

6 S.S. Davis, S. Douglas, L. Illum, P.D.E Jones, E. Mak and R.H. Müller, Targeting of colloidal carriers and the role of surface properties, in: "Targeting of Drugs with Synthetic Systems", G. Gregoriadis, J. Senior and G. Poste, eds., Plenum, New York (1986)

7 R.H. Müller, B.W. Müller and S.S. Davis, Amplitude Weighted Phase Structuration - a new method for simultaneous charge and size determination of colloidal drug carriers, Proc. 4th International Conference on Pharmaceutical Technology, 3-5 June, Paris, France 1:205 (1986)

8 R.H. Müller, S.S. Davis, L. Illum and E. Mak, Particle charge and surface hydrophobicity of colloidal drug carriers, in: "Targeting of Drugs with Synthetic Systems", G. Gregoriadis, J. Senior and J. Poste, eds., Plenum, New York (1986)

9 L. Illum and S.S. Davis, Effect of the nonionic surfactant Poloxamer 338 on the fate and deposition of polystyrene microspheres following intravenous administration, J. Pharm. Sc. 72:1086 (1983)

10 H.Z. Cummins and E.R. Pike, eds., Photon Correlation and Light Beating Spectroscopy, Nato Advanced Study Institute Series B, Plenum, New York (1973)

11 E.O. Schulz-DuBois, eds., Photon Correlation Techniques, Springer, Berlin (1983)

12 H.Z. Cummins and E.R. Pike, eds., Photon Correlation Spectroscopy and Velocimetry, Nato Advanced Study Institute Series B, Plenum, New York (1976)

13 R.H. Müller, S.S. Davis, L. Illum and E. Mak, Surface characterisation of colloidal drug carriers coated with polymers, Macromolecular Preprints 161 (1986)

14 L. Illum, L.O. Jacobsen, R.H. Müller, E. Mak and S.S. Davis, The importance of surface characteristics in determining the uptake of polystyrene microspheres by peritoneal macrophages, Biomaterials 8:113 (1987)

15 D.H. Napper, Polymeric stabilisation of colloidal dispersions, Academic Press, London (1983)

USE OF PARTITIONING TO STUDY DIFFERENTIATING CELLS OF *DICTYOSTELIUM DISCOIDEUM*

D.J. Watts, Gillian M. Knight, J. Lawry * and P.T. Sharpe

Department of Biochemistry and Department of Virology *
University of Sheffield
Sheffield, S10 2TN, UK

Partitioning in a dextran-poly(ethylene glycol), two-phase system in a thin-layer, countercurrent distribution (TLCCD) apparatus separates amoebae of *D.discoideum* at 10-11 h development into two populations. This method takes at least 10 h to complete. A much quicker method for separating the two populations has now been developed which makes use of two, different phase systems. The two populations separated by the rapid method are similar to those separated by TLCCD in comprising amoebae with different DNA contents. This indicates that the difference in DNA content between amoebae in the two populations separated by TLCCD does not arise during the long time needed to complete the countercurrent distribution.

INTRODUCTION

During growth, cells of the cellular slime mould *Dictyostelium discoideum* remain amoeboid and solitary. Wild-type strains have to make use of bacteria as a food source but mutant strains have been isolated that can grow axenically in a semi-defined medium as well as with a bacterial substrate. When deprived of nutrients, the amoebae enter the developmental phase of the life cycle. The amoebae then collect synchronously into aggregates, each containing about 10^5 cells, and the aggregates eventually transform into fruiting bodies. Each fruiting body comprises a spore mass supported by a tall, cellular stalk and, during development, amoebae have therefore to differentiate into stalk cells and spores. Because differentiation in *D.discoideum* gives rise to only these two cell types, it is much simpler than, for example, differentiation in developing embryos. *D.discoideum* is therefore widely studied as a "model system" of development and differentiation.

In standard, laboratory conditions, *D.discoideum* amoebae start to aggregate about 8h after initiation of development and formation of fruiting bodies is completed after a further 16-20 h. Studies of the initiation of *D.discoideum* differentiation have been hampered by difficulties in separating the stalk and spore precursor cells during early development (6-10 h) but we have found that separation can be achieved by partitioning the amoebae in an essentially "zero-potential" dextran-poly(ethylene glycol), two-phase system in a thin-layer, countercurrent distribution (TLCCD) apparatus. This method separates cells differing in cell surface hydrophobicity. It was found that amoebae harvested at any time up to 7 h development distributed as a single peak during TLCCD but, at 8 h development, the distribution became very broad and indicated that the amoebae had become very heterogeneous in cell surface properties. Between 9 h and 11 h development, it was apparent that partitioning separated the amoebae into two populations and it was found that it was the population of amoebae with the more hydrophilic cell surface properties that would eventually form spores, whereas stalk cells were formed from the population of amoebae having more hydrophobic cell surface properties [1,2].

The two populations of amoebae, isolated by TLCCD at 10 h development, were found to differ, not only in cell surface properties, but also in polypeptide composition [3] and in DNA content [2]. However, TLCCD takes at least 10 h and it is therefore possible that the differences in properties between the two populations of amoebae separated by TLCCD arise, not during development, but during the long period of partitioning. Some evidence supporting this criticism has been obtained. Amoebae separated by TLCCD at 10 h development and having hydrophilic cell surface properties were found to have a DNA content twice that found in amoebae having more hydrophobic cell surface properties [2]. However, flow cytometry, which makes use of a fluorescence activated cell sorter (FACS), can be used to determine the DNA content of amoebae that have not been subjected to partitioning, and it was then found that, at any stage of development, the amoebae all had the same DNA content [4]. Thus it appeared possible that the differences in DNA content between populations of amoebae separated by TLCCD were being created during partitioning. More definite evidence that changes occur in amoebae during TLCCD has been obtained from investigations of the cells' mRNA content. mRNA coding for a low molecular weight polypeptide that is synthesised by amoebae after 9h development [5] was found in amoebae at 11 h development but was lost from these amoebae during TLCCD (B.D. Hames, P.T. Sharpe and D.J. Watts - unpublished results). This loss of mRNA was not unexpected since, although developmentally-regulated mRNA has a half-life of 3-4 h in undisturbed aggregates, the half-life decreases to 30-40 min when the aggregates are dissociated [6].

In order to investigate control of the initial events of cell differentiation in *D.discoideum*, it would be helpful to be able to identify a mRNA species that is synthesised during early development exclusively in the spore precursor cells or exclusively in the stalk precursor cells. Studies of control of the cell-type-specific transcription of this mRNA should then lead to a better understanding of the mechanisms determining whether an amoeba will differentiate into a stalk cell or a spore. Thus there was a need for a partitioning technique that could be used to separate the stalk and spore precursor cells in less than an hour so that the cells would retain intact mRNA molecules. A new method of cell separation has therefore been developed that makes use of two different phase systems and which can be completed in about 40 min.

EXPERIMENTAL

Materials - Dextran T500 and poly(ethylene glycol) 4000 (PEG) were obtained from Pharmacia and BDH Chemicals respectively. Mithramycin and pure *Clostridium perfringens* DNA were from Sigma, and Hoechst 33258 was from Calbiochem. Plastic tubes were purchased from Sterilin.

Methods - Amoebae of *D.discoideum* strain Ax-2 were grown axenically in HL5 glucose medium [7] and were harvested during exponential growth. The amoebae were washed with distilled water at 5°C and left to initiate development on filters at 22°C [7]. After 10-11 h development, aggregates were washed off the filters in distilled water at 5°C. The suspensions were centrifuged and the aggregates were resuspended in distilled water at 5°C and dissociated into single cells by agitation on a Vortex mixer.

The phase systems comprised 5.1% (w/w) dextran and 5.1% (w/w) PEG, or 5.5% (w/w) dextran and 5.5% (w/w) PEG, to which had been added, per 20 g final mixture, 1.0 ml 1M NaCl, 1.0 ml 0.2 M Na_2SO_4 and 0.1 ml 0.2 M phosphate (KH_2PO_4 K_2HPO_4) buffer pH 7.8. These mixtures give essentially "zero-potential", two-phase systems. Phase systems were also made up containing the polymers and salts at slightly higher concentrations so that, when 4.5 g phase system was diluted by addition of 0.5 ml cell suspension, the phase systems would have the same polymer and salt concentrations as the 5.1% and 5.5% systems. The phase systems were kept at 5°C.

0.5 ml cell suspension (approximately $3x10^8$ amoebae) was added to 4.5 g "concentrated" phase system at 5°C in a 15 ml tube for the first partitioning step and the capped tube was inverted 30 times to obtain mixing. The tube was left horizontally at 5°C for 6 min for phase separation to occur. The upper phase could then be carefully removed by suction into a Pasteur pipette and added to fresh lower phase for the second partitioning step. Alternatively, fresh upper phase could be added to the lower phase and interface obtained from the first partitioning step. Amoebae recovered from the second partitioning were washed with 50 mM phosphate (NaH_2PO_4/K_2HPO_4) buffer pH 6.5.

The DNA content of amoebae was determined in a fluorimetric assay, as previously described [2]. Amoebae, suspended at known density, were sonicated and samples of the extract were added to the fluorochrome Hoechst 33258 which binds specifically to DNA to give a fluorescent complex. The increase in fluorescence on addition of cell extract to the Hoechst 33258 was measured in a fluorimeter and each assay was calibrated by addition of a known amount of *Cl.perfringens* DNA.

DNA content was measured in individual amoebae by flow cytometry as described by Durston et al [4]. Amoebae were fixed for at least 30 min in 70% ethanol at 5°C and were then resuspended at 10^6 cells ml^{-1} in mithramycin (100 $\mu g\ ml^{-1}$ mithramycin, 30 mM $MgCl_2$) which also binds specifically to DNA to give a fluorescent complex. A FACS 420 flow cytometer and accessory computer CONSORT 30 (Becton-Dickinson) were used to analyse 50,000 mithramycin-stained amoebae from each sample. The 4 W argon laser was tuned to an excitation wavelength of 457.9 nm at 100 mW and fluorescence emission was measured through a 530 nm band pass filter of 30 nm band width. Fluorescence intensity, measured on a linear axis, was considered to be proportional to the DNA content of individual amoebae stained with mithramycin.

RESULTS AND DISCUSSION

Rapid Cell Separation - Previously TLCCD has been used to separate amoebae at 10-11h development into two populations differing in cell surface properties[1], but this involved partitioning the amoebae 59 times in a dextran-PEG, two-phase system and took 10h. Because TLCCD also showed that the amoebae were very heterogeneous in cell surface properties, it was apparent that it would not be possible to separate amoebae into the two populations in a shorter time by using many fewer partitioning steps. It therefore appeared that rapid separation of amoebae by partitioning would have to involve use of two different phase systems. One phase system would then be used to isolate amoebae having the most hydrophobic cell surface properties, whereas the second phase system would be used to isolate amoebae having the most hydrophilic cell surface properties; amoebae with intermediate cell surface properties would have to be discarded.

It could be predicted that amoebae having the most hydrophilic cell surface properties would partition into the lower phase plus interface of a two-phase system containing low concentrations of dextran and PEG, whereas amoebae having the most hydrophobic cell surface properties would partition into the upper phase of a two-phase system containing dextran and PEG at a higher concentration. Studies of the distribution of amoebae at 11h development in phase systems containing various concentrations of dextran and PEG showed that 10-15% of the amoebae partitioned into the upper phase of a two-phase system comprising 5.5% dextran, 5.5% PEG and it was assumed that these were the amoebae with the most hydrophobic cell surface properties. When, in order to repeat partitioning, the upper phase with its complement of amoebae was added to fresh lower phase (from a 5.5% dextran, 5.5% PEG two-phase system), it was found that virtually all the amoebae repartitioned into the upper phase. It therefore appeared that amoebae recovered from the upper phase after the second partitioning step would be highly enriched in amoebae with hydrophobic cell surface properties. Only about 5% of the amoebae originally taken for partitioning were recovered from the upper phase of the second step but, by using two tubes of the 5.5% dextran, 5.5% PEG two-phase system for each step, it was possible to obtain approximately twice the number of amoebae having hydrophobic cell surface properties that could be recovered from a TLCCD run on a single rotor of a "Bioshef" TLCCD apparatus.

It was originally intended that a two-phase system, in which no more than 20% of the added amoebae would partition into the lower phase plus interface, should be used to isolate amoebae having the most hydrophilic cell surface properties. However, it was found that, in order to obtain a dextran-PEG, two-phase system at 5°C, the polymers had to be at a minimum concentration of 5.0%, and that about 50% of the amoebae added to this system then distributed into the lower phase plus interface. This phase system was also inconvenient to use and a 5.1% dextran, 5.1% PEG two-phase system has been used instead. 50-60% of the amoebae added to this system partitioned into the lower phase plus interface, but a considerable proportion of the amoebae at the interface was lost when all the upper phase was sucked off into a Pasteur pipette. After partitioning had been repeated, by addition of fresh upper phase (from a 5.1% dextran, 5.1% PEG, two-phase system) to the lower phase, about 5% of the amoebae originally taken for partitioning were recovered from the lower phase plus interface.

It took about 40 min to separate and wash the amoebae.

DNA Content of Amoebae - A fluorimetric DNA assay based on use of the fluorochrome Hoechst 33258 was used to determine the DNA content of amoebae isolated by the rapid separation technique at 10-11 h development. This assay showed (Table 1) that the DNA content in amoebae with hydrophilic cell surface properties ("hydrophilic" amoebae) was twice that in amoebae having hydrophobic cell surface properties ("hydrophobic" amoebae). Unseparated amoebae had, on average, an intermediate DNA content, as would be expected. However, the DNA content of the unseparated amoebae was biased towards the low DNA content of the hydrophobic amoebae and this was consistent with previous studies [1], making use of TLCCD, which indicated that hydrophobic amoebae predominate in a population of amoebae at 10-11 h development.

Overall, the results were virtually identical with those obtained previously [2] when the DNA content of amoebae separated by TLCCD was determined (Table 1) and this would imply that the rapid separation technique can be as effective as TLCCD in giving good cell separations. This was unexpected since partitioning of the amoebae in the 5.1% dextran. 5.1% PEG phase system had suggested that the hydrophilic amoebae recovered from the

Table 1 *DNA Content of Amoebae Separated by Partitioning at 10-11 h Development*

	DNA Content (pg/cell)[a]	
	Amoebae Separated by the Rapid Technique	Amoebae Separated by TLCCD[b]
Hydrophilic amoebae	0.38±0.006(4)	0.40±0.005(3)
Hydrophobic amoebae	0.19±0.008(4)	0.21±0.006(3)
Unseparated amoebae	0.23±0.012(4)	0.23±0.026(3)

[a] Results are given as means ± S.E.M. with the numbers of observations in parentheses

[b] Results for amoebae separated by TLCCD are for peak I (hydrophilic amoebae) and peak II (hydrophobic amoebae) of the TLCCD profile [1] and are reprinted by permission from Biochemical Journal 217:839, copyright (c)19, The Biochemical Society London

lower phase plus interface of this system would be heavily contaminated with hydrophobic amoebae. However, the amoebae obtained from this phase system were mainly those partitioning into the lower phase because amoebae partitioning at the interface tended to be discarded along with the amoebae in the upper phase. It would be expected that amoebae partitioning into the lower phase would have more hydrophilic cell surface properties than amoebae partitioning at the interface, and it is possible that it was because of the failure to recover most of the amoebae from the interface that the population of hydrophilic amoebae was less contamined than expected with hydrophobic amoebae.

Because the results for the DNA content of amoebae separated by the rapid technique and by TLCCD were so similar (Table 1), it is also possible to conclude that the difference in DNA content between the hydrophilic and hydrophobic amoebae separated by TLCCD does not arise during the long time needed to complete the TLCCD.

Flow cytometry - Flow cytometry can be used to determine the DNA content of individual cells if the cells have been stained with a fluorochrome such as mithramycin that binds specifically to DNA to give a fluorescent complex. It is therefore possible to use this technique to determine whether a population of cells contains individual cells differing in DNA content. When flow cytometry was employed to investigate the DNA content of amoebae during development, it was found that the amoebae all appeared to have the same DNA content [4]. This result was confirmed for amoebae at 10-11 h development (Fig. 1A) and was clearly not in agreement with results obtained previously when the fluorimetric assay had been used to determine the DNA content of amoebae separated by TLCCD [2]. To investigate this discrepancy, flow cytometry was used to determine the DNA content of every sample of hydrophilic and hydrophobic amoebae separated by the rapid partitioning technique. As shown in Figs. 1B and 1C, flow cytometry indicated that the hydrophilic and hydrophobic amoebae had the same DNA content, although the fluorimetric assay, using Hoechst 33258, suggested that the two populations results were not obtained because amoebae were stained with mithramycin for

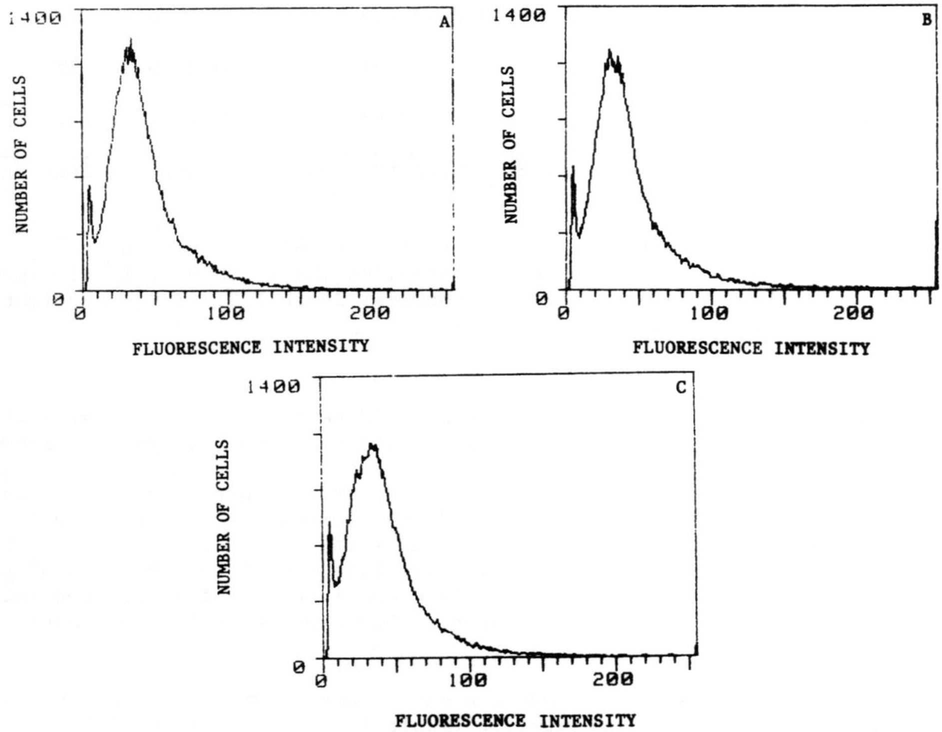

Fig. 1 *Fluorescence intensity distributions for amoebae, at 10-11 h development, stained with mithramycin. (A) Unseparated amoebae; (B) hydrophilic amoebae; (C) hydrophobic amoebae. The fluorimetric assay based on use of Hoechst 33258 indicated that the DNA contents of amoebae taken from the same populations were 0.23 pg/cell (unseparated); 0.41 pg/cell (hydrophilic amoebae) and 0.18 pg/cell (hydrophobic amoebae)*

flow cytometry, whereas Hoechst 33258 was used in the fluorimetric assay; flow cytometry still indicated that the hydrophilic and hydrophobic amoebae had the same DNA content when the amoebae had been stained with Hoechst 33258.

Further Investigations of the Two Methods for Assaying DNA - Flow cytometry and the fluorimetric assay were both used to determine the DNA content of amoebae of various strains (both haploid and diploid) of *D.discoideum* at various stages of growth and development. However, it was found that, in each experiment, the two methods gave similar results. Thus these investigations have not led to any indication of the reasons for the two methods giving different results when used to assay the DNA content of amoebae separated by partitioning. Nor have control experiments suggested that the two assay methods give different results when amoebae have been in contact with a two-phase system.

CONCLUSIONS

A rapid cell separation technique has been developed that has been helpful in confirming that the differences in cell surface properties and DNA content, found when amoebae of *D.discoideum* at 10-11 h development are subjected to partitioning in a TLCCD apparatus, do not arise during the long time needed to complete the countercurrent distribution. Furthermore, it is hoped that future experiments will demonstrate that the rapid technique may also be used to separate *D.discoideum* amoebae which, because of the short time need to effect cell separation, will retain a considerable amount of undegraded, developmentally-regulated mRNA. However, there appears to be no reason why the rapid cell separation technique should be restricted to studies of *D.discoideum*. We would expect that, with suitable modification, the technique could be used generally to select cells with specified cell surface properties from out of a population of cells having a variety of surface properties.

It has not been possible to decide whether the DNA content of *D.discoideum* amoebae at 10-11 h development is correctly estimated by flow cytometry or by the fluorimetric assay that makes use of the fluorochrome Hoechst 33258. The latter assay method has given results which indicate that the population of amoebae at 10-11 h development contains amoebae in both the G1 and G2 phases of the cell cycle [2], whereas flow cytometry suggests that the amoebae are all in the same (G2) phase of the cell cycle [4,8]. There have also been investigations of the cell cycle during 1 *D.discoideum* development that were not dependent on estimations of amoebal DNA content and these have led to the conclusion [9] that, in a population of amoebae at 10-11 h development, there are no amoebae in mitosis or S phase, but there amoebae in both G1 and G2 phase. This conclusion would support the view that use of the fluorochrome Hoechst 33258 in a fluorimetric assay of the DNA content of amoebae separated by partitioning gives valid results, and it would therefore appear probable that partitioning does separate *D.discoideum* amoebae that are in different phases of the cell cycle.

REFERENCES

1 P.T. Sharpe, T.E. Treffry and D.J. Watts, Studies of early stages of differentiation of the cellular slime mould *Dictyostelium discoideum*, J. Embryol. Exp. Morph. 67:181 (1982)
2 P.T. Sharpe, G.M. Knight and D.J. Watts, Changes in the DNA content of amoebae of *Dictyostelium discoideum* during growth and development, Biochem. J. 217:839 (1984)

3 P.T. Sharpe, R.M. Sharrard and D.J. Watts, Polypeptide compositions of amoebae of the cellular slime mould *Dictyostelium discoideum* separated by partitioning during development, Bioscience Reports 5:121 (1985)
4 A.J. Durston, C.J. Weijer, J.F. Jongkind, A. Verkerk, A. Timmermans and W.T. Kulve, A flow fluorimetric analysis of the cell cycle during growth and differentiation in *Dictyostelium discoideum*, Roux's Arch. Dev. Biol. 194:18 (1984)
5 D.G. Wilkinson, J. Wilson and B.D. Hames, Synthesis of spore proteins during development of *Dictyostelium discoideum*, Biochem. J. 215:567 (1983)
6 J.G. Williams, C.J. Pears, K.A. Jermyn, D.M. Driscoll, H. Mahbubani and R.R. Kay, The control of gene expression during cellular differentiation of *Dictyostelium discoideum*, in: "Regulation of Gene Expression", I. Booth and C. Higgins, eds., Cambridge University Press, Cambridge (1986)
7 D.J. Watts and J.M. Ashworth, Growth of myxamoebae of the cellular slime mould *Dictyostelium discoideum* in axenic culture, Biochem. J. 119:171 (1970)
8 C.J. Weijer, G. Duschl and C.N. David, A revision of the *Dictyostelium discoideum* cell cycle, J. Cell Sci. 70:111 (1984)
9 I.M. Zada-Hames and J.M. Ashworth, The cell cycle and its relationship to development in *Dictyostelium discoideum*, Dev. Biol. 63:307 (1978)

PARTITION OF ALGAL CELLS OF VARIOUS STRAINS IN AQUEOUS TWO PHASE SYSTEMS

J. Burczyk [ab] and D. Szymakowska [a]

[a] Silesian Medical Academy, Department of Pharmacognosy and Phytochemistry, Katowice-Sosnowiec, Poland
[b] Institute of Zootechnics, Laboratory of Applied Biochemistry, PL-43400 Cieszyn-Gumna 43

Green algae belonging to the genus Chlorella/Chlorococcales represents one of the simplest and commonest group of plant organisms. They have been proposed for practical application in agriculture and biotechnology as protein-rich food or feed, for conversion of energy. This genus includes small, more or less spherical cells, lacking in suitable morphological properties useful for distinguishing strains valuable because of their high biotechnological capacity. This has prompted scientists to find useful criteria for characterising and distinguishing the various strains as well as classifying the unicellular green algae.

To date the list of chemical properties used includes pigment composition (especially the presence of secondary carotenoids), occurrence of sporopollenins, algal resistant biopolymers, presence of hydrogenase, base composition of DNA, ability to metabolise various substrates, ability to form definite products, acid tolerance, salt tolerance and antigenic composition. The behaviour of cells in two-phase system, i.e. the reactivity to pH changes, composition of phase system, presence of ions and ligands, depends upon definite composition of cell surface.

Studies of the partition of algal cells of 18 different Chlorella strains and 3 Scenedesmus strains in aqueous two-phase systems were aimed at finding additional characteristics useful for distinguishing them.

Algal cells were partitioned in two-phase system composed of: A) 5% w/w Dextran T500 (Pharmacia), 4% w/w PEG 6000 (Union Carbide), 0.5% w/w DEAE-Dextran, 50 mM NaCl and 12.5 mM citrate/phosphate buffer pH 2-8 and B) the same composition as system A) but with the replacement of DEAE-Dextran with 0.5% w/w sodium-dextran-sulphate.

Two algal groups that showed differences in reactivity to pH changes in the phase system were distinguished using single step partition. The majority of the strains in the first group reacting in System A by pushing about 60% of strains when increasing the pH value from the top phase to the interface.

The second group consisted of strains which did not react at all to the changes of pH in either system. Only one strain, Chlorella fusca (strain 211-8p) changed its partition from the top phase to the bottom in the range of pH tested. But cross-points described for cells of other organisms, e.g. Penicillium, were not observed.

The presence of sporopollenin-like hydrophobic polymers in the outer cell wall layer of some strains tested here did not lead to any distinctive behaviour of these strains. It seems that this fact can be explained by the presence of masking glycoproteins or polysaccharides on the algal surface which coat sporopollenin-like algal resistant polymer known for its special resistance against biological and chemical agents.

In conclusion, aqueous two-phase systems can be used for distinguishing strains belonging to the genus Chlorella.

ANTIBIOTIC - NEUTROPHIL INTERACTIONS

STUDIED BY PHASE PARTITIONING

Paul Eggleton*, **Derek Fisher*** **and Neville Crawford**[+]

* Biochemistry Department, Royal Free Hospital School of Medicine, University of London, London NW3 2PF, UK

[+] Biochemistry Department, Royal College of Surgeons of England, Lincoln's Inn Fields, London WC2A 3PN, UK

INTRODUCTION

In an investigation of phase-forming polymers concerning their stimulatory effect on neutrophils (PMNLs), a number of dextrans were found to stimulate neutrophil oxidative metabolism as detected by nitroblue tetrazolium (NBT) reduction [1]. This effect was removed by addition of polymyxin B sulphate - a cyclic polypeptide antibiotic which binds to lipid A [2] and neutralises many of the biological effects of bacterial endotoxin [3,4]. This indicated that contaminating bacterial lipopolysaccharide (LPS) is present in some batches of dextran, but can be removed by addition of polymyxin B. Unfortunately, we observed that polymyxin B also enhanced the phagocytic capacity of the neutrophil. Consequently its incorporation into the phase reagents as a means of neutralising the effects of LPS is precluded. These observations prompted us to examine the surface properties of neutrophils in the presence and absence of polymyxin B by single step partitioning in charge-sensitive and non-charge-sensitive phase systems and the contribution of these surface properties to the observed increase in neutrophil phagocytosis.

METHODS

Pretreatment of neutrophils with 100 and 500 units polymyxin B per 10^6 PMNLs for 30 minutes at 37°C led to increases in the number of bacteria phagocytosed, compared to control (untreated) cells. PMNL viability was checked by two methods: exclusion of trypan blue and by a fluorescent method employing the uptake of acridine orange and ethidium bromide. No difference in the viability of PMNLs before or after pretreatment with polymyxin B was detected.

Surface charge-related properties of polymyxin B-treated and untreated cells were investigated using a charge-sensitive phase system (4.5% w/w Dextran T500, 4.5% w/w PEG 6000 in 270 mosM sodium phosphate, pH 7.2) and cell electrophoresis. Non-charge sensitive properties were examined using a non-charge-sensitive phase system (4.15% w/w Dextran T500, 4.15% w/w PEG 6000 in 270 mosM NaCl and 30 mosM sodium phosphate, pH 7.2).

Table 1 Partitioning properties of polymyxin B treated and untreated neutrophils

Type of partitioning	Untreated PMNL		Polymyxin B treated PMNL	
	subject 1	subject 2	subject 1	subject 2
* Charge-sensitive partition coefficient (PC)	61.2±1.4	63.0±8.7	60.7±4.2	63.8±7.5
+ Non-charge-sensitive partition coefficient (PC)	26.3±11.4	21.8±1.3	45.4±14.4	43.0±6.1

Neutrophils were incubated for 30 minutes with and without 500 units polymyxin B per 10^6 cells before measuring their partition coefficients (PC).
* Mean ± SD (n=5) + Mean ± SD (n=6)

RESULTS AND DISCUSSION

As shown in Table 1, there were no significant differences in the charge-sensitive partition coefficients of treated and untreated cells, but the non-charge-sensitive partition coefficients of treated cells were significantly raised ($P<0.05$).

Two possible interpretations for the change in the non-charge-sensitive partitioning of the neutrophils are as follows. Direct binding of the antibiotic to the neutrophil could occur; polymyxin B rapidly attaches to cell surfaces probably to the negatively charged phospholipids of the membrane [5]. However, this might be expected to produce change in surface charge. Alternatively, polymyxin B may perturb the PMNL membrane resulting in conformational changes, such as pseudopod formation. Previous workers have shown that polymyxin affects membrane-related responses such as degranulation and chemotaxis [6]. What is clear is that cell partitioning is a useful method to analyse physicochemical differences in the cell surface induced by antibiotics. Future studies should provide information relating to the ability of antibiotics to influence the membrane and functional characteristics of the cells.

REFERENCES

1 P. Eggleton, D. Fisher and N. Crawford, Problems of white cell separation techniques in relation to phase partitioning of neutrophils, in: "Separations Using Aqueous Phase Systems: Applications in Cell Biology and Biotechnology," D. Fisher and I.A. Sutherland, eds., Plenum, New York (1989)

2 D.C. Morrison and D.M. Jacobs, Binding of polymyxin B to the lipid A portion of bacterial lipopolysaccharides, Immunochemistry 13:813 (1976)

3 J.J. Corrigan and B.M. Bell, Endotoxin induced intravascular coagulation; prevention with polymyxin B sulphate, J. Lab. Clin. Med. 77:802 (1971)

4 J.J. Corrigan, O.F. Sieber, H. Ratajczak, and B.B. Bennet, Modification of human neutrophil response to endotoxin with polymyxin B sulphate, J. Infect. Dis. 130:384 (1974)

5 C.M. Kunin, Binding of antibodies to tissue homogenates, J. Infect. Dis. 121:55 (1970)

6 P.H. Naccache, M.M. Molski and R.I. Sha'afi, Polymyxin B inhibits phorbol 12-myristate 13-ocetate, but not chemotactic factor induced effects in rabbit neutrophils, FEBS 193:227 (1985)

SURFACE PROPERTIES OF KLEBSIELLA AND E COLI:

ROLE OF CAPSULAR POLYSACCHARIDE IN PROTECTION AGAINST PHAGOCYTOSIS

D. Fisher*, P. Allen+, J.R. Saunders+ and C.A. Hart#

* Department of Biochemistry, Royal Free Hospital
School of Medicine, University of London
Rowland Hill Street, London NW3 2PF

Departments of + Microbiology and # Medical
Microbiology, University of Liverpool
P O Box 147, Liverpool L69 3BX, UK

INTRODUCTION

Klebsiellae are Gram negative capsulate bacteria which are important pathogens, causing bacteremia, pneumonia and wound infection. The K21 serotype of Klebsiella is not homogeneous but consists of strains producing one or other of two chemically distinct types of capsular polysaccharide (K21a and K21b); K21b is more commonly encountered among clinical isolates [1]. We have investigated the role of the capsule of *Klebsiella pneumoniae* (K21b) as a determinant of pathogenicity by examining the resistance to phagocytic uptake and killing by polymorphonuclear leucocytes (PMNLs) of the native bacteria and a spontaneously derived capsule-less mutant. In addition, we have investigated the role of the antigenically similar colanic acid capsule produced by E coli DH1 transformed by plasmid pLV213 which carries the rcs gene [2]. The capsules are hydrophilic, negatively charged polysaccharides whose influences upon the surface properties of the bacteria can be determined by phase partitioning [3]. The various strains of bacteria provide particulates with varying surface properties, and allow correlations between these physicochemical properties and the biological properties of resistance to phagocytosis to be examined. Avoidance of phagocytosis is a strong virulence factor, which can be associated with altered surface properties, such as those detected by phase partitioning [4].

PARTITIONING METHODS

Partitioning was performed in two phase systems: a non-charge sensitive system of 6.2% w/w Dextran T500 - 4% w/w PEG 6000 (BDH) in 0.03 M Tris buffer, pH 7.2 and a charge sensitive system of 6.2% w/w Dextran T500 - 4.4% w/w PEG 6000 (including 0.55% w/w TMA-PEG) in 0.03 M Tris buffer, pH 7.2.

RESULTS

Bacterial Physicochemical Surface Properties determined by Partitioning - Table 1 shows that Klebsiella capsule (K21b) is negatively charged and increases non-charge sensitive partitioning.

Table 1 *Effect of capsule on partitioning and phagocytosis*

Bacterial species and strain	Capsule	Top phase partitioning (%) * Non-charge sensitive	Charge-sensitive	Susceptibility to phagocytosis
K. aerogenes non-capsulate strain (1L919)	-	0.9 ± 0.1	4.3 ± 0.4	phagocytosed
K. aerogenes K21b native strain (1L918)	+	19.1 ± 1.8	88.6 ± 5.2	resistant
E. coli DH1 (non-capsulate)	-	3.0 ± 0.3	18.7 ± 0.8	phagocytosed
E. coli transformed to produce colanic acid capsule (2L223)	+	1.9 ± 0.2	66.3 ± 3.0	phagocytosed

* mean ± SEM of four replicate samples

Biological Consequences of Bacterial Surface Properties: Phagocytosis and Intracellular Killing - The non-capsulate strain of Klebsiella was readily phagocytosed and killed by PMNL, whereas the capsulate strain (K21b) was resistant. By contrast the colanic acid and capsule on transformed *E. coli* gave no protection against phagocytosis and both capsulate and non-capsulate strains were phagocytosed and killed (Table 1).

CONCLUSIONS

This comparison of two examples of capsules (K21b and colanic acid) which are structurally different shows:

1 Both capsules impart a negative charge to the bacterial surface.
2 The capsules give different degrees of protection against phagocytosis and intracellular killing (K21b protects whereas colanic acid does not).
3 Although the capsules cannot be distinguished immunologically, they can be distinguished by non-charge sensitive partitioning.
4 If physicochemical surface properties contribute to the inability of capsulated klebsiella to be phagocytosed, it is the non-charge properties which are important in this case and not the surface charge.

REFERENCES

1 P. Allen, J. Williams, C.A. Hart and J.R. Saunders, J. Gen. Microbiol. 133:1365 (1987)
2 P. Allen, C.A. Hart and J.R. Saunders, J. Gen. Microbiol. 133:331 (1987)
3 P.M. Allen, D. Fisher, J.R. Saunders and C.A. Hart, J. Med. Microbiol. 24, 363 (1987)
4 K.-E. Magnusson and O. Stendahl, in: "Partitioning in Aqueous Two-Phase Systems. Theory, Methods, Applications, and Uses in Biotechnology", H. Walter, D.E. Brooks and D. Fisher, eds., Academic Press, Orlando, Florida (1985)

FRACTIONATION BY COUNTERCURRENT DISTRIBUTION (CCD) OF RAT ERYTHROCYTES DURING ANIMAL DEVELOPMENT

M. Pinilla, M. Moreno, P. Jimeno and J. Luque

Departamento de Bioquimica y Biologia Molecular
Universidad de Alcalá de Henares
Alcalá de Henares (Madrid) Spain

INTRODUCTION

Preliminary studies have shown that red blood cells from new-born and adult rats may be fractionated in aqueous two-phase systems on the basis of differences in cell size and surface properties [1]. Our aim is to extend this work and to use CCD as an appropriate fractionation procedure for biochemical studies of red cell switching (from fetal to definitive erythrocytes) during animal development. Erythrocytes from 19 days-fetuses, new-born young animals (6,8,12 and 24 days of age) and adult Wistar rats were fractionated by CCD. Dextran T-500 (Pharmacia) and poly(ethylene glycol) (PEG)-6000 (Serva) were used for the preparation of 5% dextran/4% PEG two-phase systems, containing 0.09 M NaPB/0.03 M NaCl, pH 6.8 (charged systems) [1,2]. A CCD apparatus with 60-cavity rotors [1,2] was employed. Experimental CCD conditions were those previously used for the fractionation of erythrocytes from adult rats [1,3,4]: top/bottom phase volume ratio of 1.3 (0.9 ml or 0.75 ml of top phase and 0.7 ml or 0.55 ml of bottom phase, respectively); settling time, t=5 min. Specific activities (U/mgHb) of bisphosphoglycerate mutase (BPGM) (EC 5.4.2.4), the enzyme responsible for 2,3-bisphosphoglycerate (BPG) synthesis, and pyruvate kinase (PK) (EC 2.7.1.40), the glycolytic enzyme involved in the regulation of BPG levels in erythrocytes, have been determined in adjacent pooled CCD fractions from all of the above red cell populations.

RESULTS AND DISCUSSION

A progressive displacement towards the right-hand cavities of the CCD rotor (increasing G values) and an increasing homogeneity of CCD profiles is being observed for erythrocytes from new-born to adult rats [5]. Both the displacement and the increasing homogeneity of CCD profiles are a consequence of differences in several physicochemical parameters (e.g. cell size, sialic acid content and electrophoretic mobility) [1], because of the substitution of fetal (new-born rats) by definitive (adult rats) erythrocytes during animal development. As a confirmation of this behaviour, G values calculated from the two peaks observed after the fractionation of an artificial mixture of erythrocytes from new-born and adult rats (1:1), were similar to those obtained after independent fractionation of fetal and definitive erythrocytes.

An increase in BPGM activity and a decrease in PK activity (i.e. a decrease in PK/BPGM ratio) and, as a consequence, an increase of BPG levels, have been observed in erythrocytes during rabbit development [6,7]. Such an inverse relationship between BPG concentration and PK/BPGM ratio has also been found in rats, and has been used here to approach the identification of fetal (low BPG and BPGM, high PK) and definitive (high BPG and BPGM, low PK) erythrocytes:

a) As a consequence of the highest PK activity and the lowest BPGM activity in erythrocytes from new-born rats, a very high PK/BPGM ratio was found in all CCD fractions. It means that only fetal erythrocytes were present in this cell population.

b) A lower PK activity and a higher BPGM activity, i.e. a lower PK/BPGM ratio than in new-born rat erythrocytes, is observed in erythrocytes from 6 days old rats. However, because of the progressive decrease of PK activity and slight increase of BPGM activity, a decreasing PK/BPGM ratio is observed from left- to right-hand CCD fractions from 6 days old rats. This seems to be the first indication of the presence of fetal erythrocytes in the left-hand CCD fractions and of definitive erythrocytes in the right-hand CCD fractions.

c) The progressive decrease of PK activity and increase of BPGM activity from left- to right-hand CCD fractions from 8 and 12 days old rats, gives rise to a parallel decreasing PK/BPGM ratio. This is now a clear indication of the predominance of fetal erythrocytes in the left-hand CCD fractions and of definitive erythrocytes in the right-hand CCD fractions.

d) A decrease of PK activity and a constant BPGM activity is found from left- to right-hand CCD fractions from 24 days old rats. Since at this stage all red cells are definitive erythrocytes, the still decreasing PK/BPGM ratio being observed from left- to right-hand CCD fractions is now the first indication of the ageing process taking place in blood. This conclusion is also supported by results obtained for erythrocytes from adult rats [4].

Finally, the presence of two red cell populations in 8 days old rats was demonstrated by increasing the settling time to 20 min. This is a confirmation of the influence of increasing settling time to improve the fractionation by CCD of heterogeneous cell populations. After measurement of BPGM and PK activities in the CCD fractions from these two red cell populations, values of PK/BPGM ratio serve again to indicate that fetal erythrocytes become fractionated towards the left- (high PK/BPGM ratio) while definitive erythrocytes are fractionated towards the right-hand cavities (lower PK/BPGM ratio) of the CCD rotor. As a consequence, BPG levels are lower in fetal than in definitive erythrocytes.

REFERENCES

1 H. Walter, D.E. Brooks and D. Fisher, eds., Partitioning in aqueous two-phase systems, Academic Press, Orlando (1985)

2 P.-Å. Albertsson, Partition of Cell Particles and Macromolecules, 3rd ed. J. Wiley, New York (1986)

3 M. Pinilla, P. Rodriguez-Horche and J. Luque, Cell Biochemistry and Function 5:301 (1987)

4 J. Luque, M.D. Delgado, P. Rodrıguez-Horche, M.T. Company and M.

Pinilla, Bioscience Reports 7:113 (1987)
5 J. Luque, M.D. Delgado, E. Ferrer, M. Moreno, M. Pinilla and P. Sancho, NATO ASI Series, Vol H8, I.N. Rich, ed., Springer, Heidelberg (1987)
6 W. Jelkmann and C. Bauer, Pflügers Archiv. 389:61 (1980)
7 J.G. Gilman, Biochem. J. 192:335 (1980)

ATTEMPTS TO FRACTIONATE RAT BONE MARROW CELLS BY COUNTERCURRENT DISTRIBUTION (CCD) IN DEXTRAN/POLYETHYLENE GLYCOL TWO-PHASE SYSTEMS

P. Sancho, A.I. Garcia-Pérez and J. Luque

Department of Biochemistry and Molecular Biology
University of Alcála de Henares
Alcalá de Henares (Madrid) Spain

INTRODUCTION

Cell suspension of bone marrow contains a mixture of different haematopoietic cell types. This population then provides a suitable sample to be used as a model system to study the fractionation of heterogeneous cell populations [1-3]. Bone marrow (from femur and tibia) and erythrocytes were prepared from male adult Wistar rats [1]. Reticulocytes were from anaemic (phenylhidrazinic rats) [3]. Dextran T-500 (Pharmacia) and poly(ethylene-glycol) PEG-6000 (Serva) were used for the preparation of two-phase systems [4,5]. Various versions of the countercurrent distribution (CCD) apparatus with 60-cavity rotors [4,5], made at the Universities of Madrid, Umeå and Sheffield, were employed. G values were obtained from equation: G=i/n-i (n=55 transfers).

RESULTS AND DISCUSSION

The fractionation of rat bone marrow cells in a 5%dextran/4%PEG charged system (0.09 M sodium phosphate buffer (NaPB), pH 6.8, 0.03M NaCl) was first reported [6] by using those experimental CCD conditions (top/bottom phase volume ratio, L=1.3, settling time, t=5 min) applied to the fractionation of erythrocytes [5]. It was shown that erythroid cell development in bone marrow is accompanied by an increase in the partition ratio, G, of cells [5,6]. In our hands, under identical CCD conditions, this cell population became distributed towards the right-hand side cavities of the CCD rotor, giving rise to a single peak profile (G=6.13) [2]. This was a consequence of the high partition ratio (P=80%) observed for bone marrow cells in test-tube (individual partition) experiments [7]. Variations in experimental CCD conditions [4,5] were tested to increase the efficiency of fractionation of rat bone marrow cells.

Charge of the system - Fractionation by CCD in systems of decreasing charge (i.e. phosphate concentration) allows relating partitioning behaviour of cells with decreasing charge and increasing hydrophobicity of cell surface. Decrease of charge will tend to centre the distribution profiles around cavity 30 of the CCD rotor [4,5]. Fractionation by CCD (at L=1.3 and t=5 min) of rat bone marrow cells, reticulocytes and erythrocytes were carried out using 5% dextran/4%PEG highly charged (0.09M NaPB, 0.03M NaCl), middle-charged (0.05M NaPB, 0.09M NaCl) and uncharged (0.01M NaPB, 0.15M

NaCl) two-phase systems. A displacement towards the left and a broader cell subfractionation profile was only observed for bone marrow cells in middle-charged (peaks: G=0.67 and G=2.67) and uncharged (peaks: G=0.45 and G=2.44) systems. This is in agreement with results reported for human bone marrow cells in uncharged systems [8]. Such a behaviour can be explained on the basis of the decreasing partition values observed in test-tube experiments from highly charged (P=80%) to middle-charged (P=60%) and uncharged (P=25%) systems [7]. As expected for an homogeneous cell population, neither reticulocytes nor erythrocytes were able to be subfractionated. Experiments were then carried out under conditions designed to increase the low efficiency of fractionation observed for bone marrow cells in highly charged systems. Our aim was to decrease partitioning without affecting the charge of the system, i.e. to achieve a better resolution as a function of cell surface charge.

Settling time and/or top/bottom phase volume ratio - The subfractionation of heterogeneous bone marrow cell suspensions from human and rats was not achieved, using a high top/bottom phase volume ratio (L=0.9/0.7=1.3), when the settling time allowed for the phases to be separated after each mixing step was increased up to 20 min (G=6.13) [2]. Obviously, a subfractionation was neither observed for homogeneous erythrocyte population (G=6.13) [2]. However, using a lower top/bottom phase ratio (L=0.4/0.7=0.6), bone marrow gives rise to two broad subpopulations when settling time was increased from 5 to 20 min (peaks: G=0.14 and G=1.38) [2]. These findings suggested that settling time in combination with phase volume ratio (see following section), could be used to determine the efficiency of fractionation of heterogeneous cell populations. After detailed examination of CCD profiles from bone marrow cells [2,3], it was confirmed that contaminating erythrocytes in the original cell suspensions became distributed towards the right-hand side cavities (high G), whereas immature erythroid cells were mainly located in the cell population distributed towards the left-hand side cavities (lower G) [5,6].

Volume of bottom phase - The decrease of bottom phase volume used for the partition of cells gave rise to a decrease in the distribution coefficient, G [4,5]. Thus, when the volume of bottom phase in each cavity of the CCD rotor was decreased from 90% (0.7 ml) to 75% (0.55 ml) of total bottom phase capacity (0.79 ml), without the top/bottom phase volume ratio (L=1.3) or settling time (5 min) being varied, a displacement towards the left (lower G) and an enlargement of the CCD profile was observed (peaks: G=0.57 and G=1.75). The volume of bottom phase therefore is also an important parameter in cell fractionation [4,5].

Polymer concentration - Test-tube experiments have shown that the increase in polymer concentration from 5%dextran/4%PEG to 5.4%dextran/4.4%PEG decrease the partition values, P, of bone marrow cells from 80% to 40% [7]. However, the fractionation by CCD (L=1.3, t=5 min) with increasing polymer concentration did not favour the resolution of bone marrow cells (G=12.75). A partial subfractionation was only achieved when the volume of the bottom phase was kept at 0.55 ml (peaks: G=0.53 and G=1.39). Therefore, the bottom phase volume seems to be a more important parameter than the increase in polymer concentration for the fractionation of heterogeneous samples. Obviously, the fractionation by CCD of erythrocytes always showed a single peak profile (G=2.67).

In summary, the increase in resolution by CCD of heterogeneous cell populations depends not only on polymer concentration and charge of the system, but it is also related to some methodological CCD improvements (settling time, top/bottom phase volume ratio, volume of bottom phase) developed to influence the kinetics of cell partition during CCD.

REFERENCES

1 M. Pineda, M. Pinilla and J. Luque, Cell Biochem. Func. 2:254 (1984)

2 P. Sancho, M.D. Delgado, A.I. Garcia-Pérez and J. Luque, J. Chromatogr. 380:339 (1986)

3 J. Luque, M.D. Delgado, E. Ferrer, M. Moreno, M. Pinilla and P. Sancho, NATO ASI Series Vol H8, I.N. Rich, ed., Springer, Heidelberg (1987)

4 P.-A. Ålbertsson, Partition of Cell Particles and Macromolecules, Wiley, New York (1986)

5 H. Walter, D.E. Brooks and D. Fisher, eds., Partitioning in Aqueous Two-Phase Systems: Theory, Methods, Uses and Applications to Biotechnology, Academic Press, New York (1985)

6 H. Walter, E.J. Krob and G.S. Ascher, Exp. Cell. Res. 79:63 (1973)

7 A.I. Garcia-Pérez, M.N. Recio, P. Sancho and J. Luque, J. Chromatogr. 403:131 (1987)

8 O.M. Smith, S. Green, G. Francis and D. Fisher, Biochem. Soc. Trans. 14:752 (1986)

THE PARTITIONING OF LIPOSOMES OF DEFINED SIZE AND COMPOSITION

C.P.S. Tilcock *, T. Dempsey+, P.R. Cullis * and D. Fisher +

Departments of Biochemistry
* University of British Columbia
2146 Health Science Avenue
Vancouver BC, Canada V6T 1W5
+ Royal Free Hospital School of Medicine
Rowland Hill Street, London NW3 2PF, UK

Liposomes are a well-established experimental system in which to model biological membranes. Studies with multilamellar vesicles (MLVs) and small unilamellar vesicles (SUVs), prepared by sonication, by Eriksson and Ålbertsson [1] and by Sharpe [2,3], have demonstrated the importance of the polar head group in determining the partitioning of liposomes.

Recently a rapid method for the production of unilamellar vesicles of defined, uniform size by an extrusion technique (VETs: Vesicles by Extrusion Techniques) has been developed in the Biochemistry Department, University of British Columbia [4].

We have compared the partitioning behaviour of VETs extruded through filters with pore sizes from 50 nm to 600 nm (VET_{50} to VET_{600}) with that of MLVs in a charge-sensitive phase system (5% Dextran T500-5% PEG 6000 (BDH) in 0.11 M sodium phosphate buffer, pH 6.8. Two different negatively charged lipsome systems were used:

i)	egg phosphatidyl glycerol	(EPG)	40% w/w
	egg phosphatidyl choline	(EPC)	60% w/w
ii)	egg phosphatidic acid	(EPA)	40% w/w
	egg phosphatidyl choline	(EPC)	60% w/w

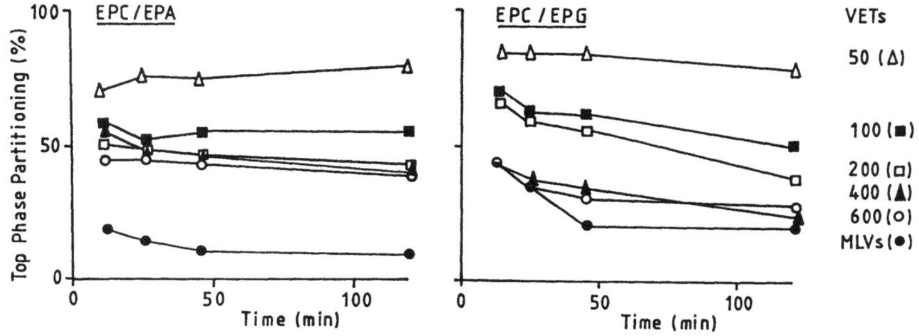

Fig. 1 *Time course of partitioning of VETs of varying sizes*

Partitions were sensitive to the size of the vesicle preparations, top phase partitioning increasing with decreasing size.

Using preparations of a defined size (VET_{100}), the effect of the liposome composition on the partitioning in phase systems of 5% dextran-5% PEG that had low electrostatic potential difference between the bulk phases, $\Delta\psi$ (0.15 M NaCl in 0.01 M phosphate buffer), intermediate $\Delta\psi$(0.075 M NaCl in 0.068 M phosphate buffer) and high $\Delta\psi$ (0.11 M phosphate buffer) was examined.

The lipid compositions used were as follows:

neutral:	egg phosphatidyl choline (EPC)
negatively charged:	EPA-EPC (40% w/w) EPG-EPC (40% w/w) Phosphatidyl inositol (PI)-EPC (40% w/w) Brain ganglioside-EPC (40% w/w)
positively charged:	stearylamine-EPC (20% w/w)

Vesicles of EPC partitioned predominantly to the top phase (70%) with 15% in the bottom phase and at the interface. Partitioning was virtually independent of the $\Delta\psi$, a result consistent with the liposomes having no net charge.

Incorporation of the negatively charged lipids into the EPC vesicles rendered partitioning sensitive to $\Delta\psi$, top phase partitioning increased as $\Delta\psi$ increased. This arose from two contrasting features. Firstly, in the low $\Delta\psi$ phase incorporation of the negatively charged lipid caused increased partition to the bottom phase. This increased wetting by the lower phase has been confirmed by contact angle measurements on EPG-EPC mixtures. Secondly, as $\Delta\psi$ increased partition moved in favour of the interface and then the top phase.

The partition of the positively charged vesicles was similar to that of EPC in the low $\Delta\psi$ phase, but was very sensitive to increasing $\Delta\psi$, showing increased partition to the lower phase.

The observations indicate that liposomes will prove to be an acceptable system in which to determine the molecular features of non-charge sensitive partitioning, charge sensitive partitioning, and their inter-relationship. Furthermore, partitioning should provide useful information on surface properties of lipsomes related to their use as drug delivery vehicles.

This work is supported by a NATO grant for international collaboration in research.

REFERENCES

1 E. Eriksson, and P.-A. Ålbertsson, The effect of lipid composition on the partition of liposomes in aqueous two-phase systems, Biochim. Biophys. Acta 507:425 (1978)
2 P.T. Sharpe and G.S. Warren, The incorporation of lipids with defined carbohydrate sequence into liposomes and the effects on partition in aqueous two-phase systems, Biochim. Biophys. Acta 772:176
3 P.T. Sharpe, The partition of charged liposomes in aqueous two-phase systems, Molec. Cell. Biochem. 68:151 (1985)
4 L.D. Mayer, M.J. Hope and P.R. Cullis, Vesicles of variable size produced by a rapid extrusion procedure, Biochim. Biophys. Acta 858:161 (1986)

Chapter 5

Immuno-affinity Partitioning

SECOND IMMUNOAFFINITY LIGANDS FOR CELL SEPARATION

S. Jill Stocks and Donald E. Brooks

Departments of Pathology and Chemistry
Department of Pathology, 2211 Wesbrook Mall
Vancouver BC, Canada V6T 1W5

Recent developments in immunoaffinity partition have included the use of a more general second affinity ligand and monoclonal antibodies. In this paper, poly(ethylene glycol) (PEG) and trypan-blue derivatised antibodies are used as second affinity ligands in two-phase systems containing PEG and dextran to immunospecifically increase the partition coefficient of human erythrocytes. In order to extend the technique towards future clinical applications, a more stringent separation problem involving transformed mouse lymphocytes is being approached. In this case a biotin-antibody ligand is employed. All the ligands were able to immunospecifically increase cell partition.

INTRODUCTION

There are many existing procedures to identify and isolate pure molecular species from complex solutions but methods to separate sub-populations of cells from mixed suspensions are less well developed. One of the methods which has been used to separate cells on the basis of surface antigen differences is immunoaffinity partition. This technique has been applied to the separation of erythrocytes of different species [1,2] by the use of polyethylene glycol (PEG) modified antibody and a two phase polymer system containing PEG and Dextran (Dx). Although a separation was obtained using a PEG-modified primary antibody, a different affinity ligand would be required for each separation problem. This disadvantage can be overcome by using a second, more general, derivatised antibody as the affinity ligand. The following discussion is based on three model separation problems, two using erythrocytes and one using lymphocytes. All of the separations use monoclonal antibodies rather than polyclonal antibodies as one of the ligands. Monoclonal antibodies have several advantages, in particular that they recognise one antigenic determinant only, they can be isolated as a pure molecular species and an infinite supply of identical antibody is available.

The first system examined consisted of human and rabbit erythrocytes, polyclonal rabbit anti-human NN glycophorin (rαNN glyc.), PEG 1900 derivatised monoclonal mouse anti-rabbit IgG, F_c fragment, IgG (mαrF_c) and a phase system containing PEG and Dx [3]. The schematic depiction of a series of binding reactions shown in Fig. 1 demonstrates the way in which human erythrocytes are immunospecifically coated with PEG. This will increase the partition of human erythrocytes into the upper, PEG-rich phase of the system

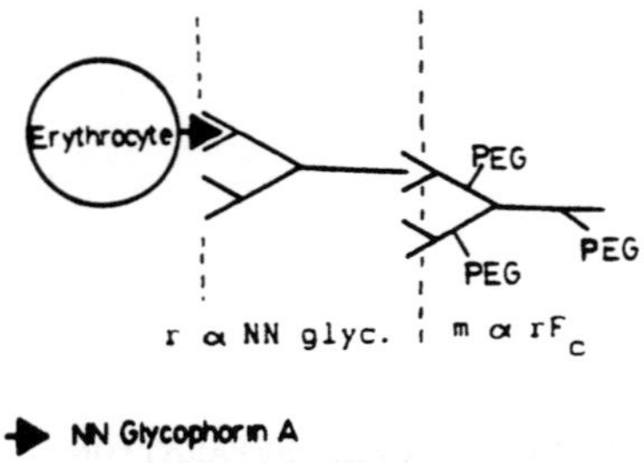

Fig. 1 *A schematic representation of PEG-modified mouse monoclonal anti-rabbit IgG, F_c fragment, IgG bound to an erythrocyte via rabbit anti-human NN glycophorin*

while the partition of cells not bearing the human NN glycophorin antigen will be unaffected as they will not be coated with PEG.

The second system also consisted of human and rabbit erythrocytes but the primary antibody was a monoclonal mouse anti-human NN glycophorin (mαNNglyc.) and the second affinity ligand was a polyclonal sheep anti-mouse IgG, F_c Fragment, IgG (sαmF_c) derivatised with trypan blue dye. Trypan blue has a partition coefficient (PC) of 10. Again the phase system contained PEG and Dextran.

The third system was developed primarily to examine the possibility of applying immunoaffinity partition to the cell separation problem of bone marrow cleansing for autologous bone marrow transplant as leukaemia therapy. This problem, and almost any clinical application, is a far more stringent separation than that of erythrocytes where the antibody recognises a high density antigen such as glycophorin. In the bone marrow purging situation, the technique would have to separate normal from malignant cells, possibly with very small surface antigen differences only distinguishable by monoclonal antibodies. The model separation being studied is of two sub-lines, MBL-2 (4.1) and MBL-2 (2.6) which were unknown prior to the development of the rat monoclonal antibody, YE1 48.10 [4,5] and are currently only separable by Fluorescent Activated Cell Sorting using this antibody. The approach taken here is to first estimate the number of binding sites for the antibody per cell type and the equilibrium free YE1 48.10 saturation concentration. Then a combination of PEG-derivatised avidin and biotinylated YE1 48.10 are used as affinity ligands with MBL-2 (4.1) and MBL-2 (2.6) in a phase system containing PEG and Dextran. A schematic description of the binding reactions is shown in Fig. 2.

MATERIALS AND METHODS

Antibody Preparations

Rabbit αNN glyc. and sαmF_c were purified from immune rabbit or sheep serum by ammonium sulphate precipitation followed by Fast Protein Liquid

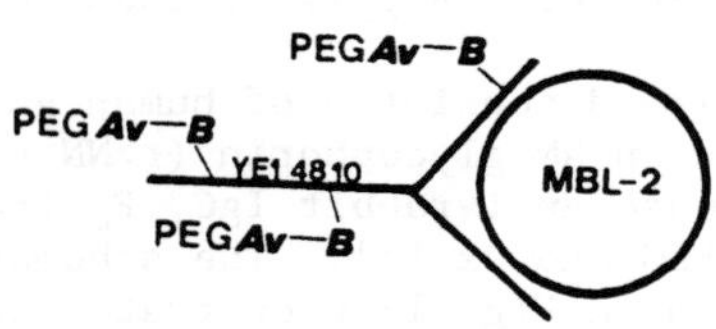

Fig. 2 *A schematic representation of PEG-modified avidin (PEG-Av) bound to an MBL-2 cell via biotinylated (B) YE1 48.10 antibody*

Chromatography (FPLC) on a mono Q column [3]. Monoclonal mαrF$_c$ hydridomas were made by standard fusion techniques, selected by an Enzyme-linked Immunosorbent Assay (ELISA) and recloned twice [6]. The antibody was purified from ascites or culture supernatant as described previously [3].

Rat monoclonal antibody YE1 48.10 hybridoma cells were a gift from Dr F. Takei of the Terry Fox Cancer Research Laboratory. The antibody was purified as described previously [3] from culture supernatant grown in roller bottles. All antibody preparations were assessed for purity by SDS-PAGE and activity by ELISA. All F$_c$ preparations were by papain digestion [7].

Protein Modifications

Cyanuric chloride-activated PEG 1900 [8] was added to IgG or avidin in sodium borate buffer (0.1M, pH 9) to give a 3:1 PEG: lysine molar ratio. After 40 minutes at room temperature the reaction mixture was purified by dialysis or gel filtration. The extent of protein modification was assayed using fluorescamine for IgG [9] and for avidin by the displacement of a dye [2-(4-hydroxy phenylazobenzoic acid)] (HABA) which binds to avidin, by biotin [10,11].

Trypan blue-modified IgG was made by mixing IgG (5mg/ml), trypan blue (10mg/ml) and 1-ethyl-3(3-diaminopropyl) carbodiimide-HCl (EDAC) (40 mg/ml) in sodium acetate buffer, pH 4.8 end over end for four hours at room temperature. The product was purified by extensive dialysis against 1M NaCl. The extent of protein modification was estimated by using radio labelled IgG and measuring the dye absorbtion.

Biotinylated IgG was made by mixing N-hydroxysuccinamido-biotin (BNHS, 20 mg/ml in DMF) with IgG, at the required mole ratio, end over end for four hours at room temperature, followed by dialysis.

Cell Preparations

Rabbit and human erythrocytes were obtained by venipuncture and washed three times with phosphate buffered saline (PBS) prior to use. MBL-2 (4.1) and MBL-2 (2.6) were also a gift from Dr Takei. They were maintained in tissue culture at 37°C, 5% CO_2 in RPMI 1640 medium containing 5% fetal bovine serum. They were washed three times with PBS and kept at 4°C for at least thirty minutes prior to use.

Phase System Preparations

Stock solutions of PEG 8000, Dx T500, sodium phosphate (0.3M, pH 7.2), sodium chloride (0.6M) and water were combined to form the required system. The first example used a 5% Dx, 3.4% PEG, 130 mM NaCl and 10 mM sodium phosphate combination. The second example used an almost identical system, the only difference being 4% PEG. The third case used 5% Dx, 4% PEG and 135 mM sodium phosphate.

Partitioning Experiments

All experiments were done in 2 ml of phase system at room temperature. The upper and lower phases and interface were sampled and the cells counted using a Particle Data Inc Electrozone Celloscope. In the case of radio-labelled proteins, the samples were counted on an LKB 1282 Compugamma.

Binding Studies

All experiments measured equilibrium binding of radio-labelled YE1 48.10 at 4°C in PBS containing sodium azide. The number of molecules of bound YE1 48.10 per cell was calculated from the free YE1 48.10 concentration before

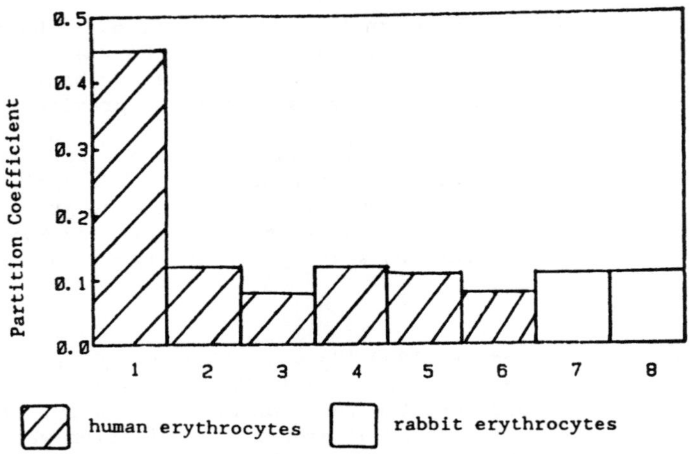

1. PEG -m α rF_c and rαNN glyc.
2. No antibody
3. rαNN glyc. only
4. PEG -mαrF_c only
5. Native mαrF_c and r α NN glyc
6. Native mαrF_c only
7. As in 1.
8. As in 2.

Fig. 3 *The effect of PEG-modified monoclonal mouse anti-rabbit IgG, F_c fragment, IgG and rabbit anti-human NN glycophorin on the partition coefficient of human and rabbit erythrocytes*

and after incubation at 4°C with MBL-2 (4.1) and MBL-2 (2.6) cells for one hour.

RESULTS AND DISCUSSION

The PEG-mαrF_c had a PC of 10 compared to 1 for the native mαrF_c and was modified by the attachment of 31-32 molecules of PEG 1900 per IgG molecule. This amount of bound PEG would, in theory, predict an even larger PC but apparent shielding of some PEG molecules by others prevents complete interaction of the bound PEG with the system. This has been previously observed [1]. The apparent binding activity of the mαrF_c was decreased significantly by PEG modification, in the order of a 50 fold decrease in binding activity [3] but it was still able to increase the PC of rαNN glyc. from 1 to 2.

The effect of the combination of PEG-mαrF_c (0.18 mg/ml) and rαNN glyc. (0.24 mg/ml) on the partition of human and rabbit enthrocytes is summarised in Fig. 3. The antibodies were able to increase the PC of human erythrocytes by a factor of approximately 5 compared to the controls (Fig. 3, 2-6) and rabbit enthrocytes (Fig. 3, 7-8), thus effecting a 5 fold enrichment of human erythrocytes over rabbit erythrocytes in the upper phase.

While a degree of separation of erythrocytes was obtained using the PEG-modified antibody, it was apparent that PEG modification caused a significant decrease in antibody binding. For this reason a smaller molecule, trypan blue (M=960), with a PC of 10 was used to modify sαmF_c. The conjugate had 28 trypan blue (TB) molecules per IgG molecule and a PC of 5. As shown in Table 1, the combination of trypan blue-sαmF_c and mαNN glyc. significantly increased the PC of human erythrocytes over that of rabbit erythrocytes. The smaller PC increase for the rabbit erythrocytes is probably due to non-specific binding of the TB-sαmF_c to the rabbit erythrocytes; a less likely explanation is cross reaction of the monoclonal mαNN glyc.

Table 1 *Results of partitioning experiments using trypan blue-modified sheep anti-mouse F_c IgG as a second affinity ligand*

Cell Type	1° Ligand	2° Ligand	PC (% U.P.)	Control PC (% U.P.)	Factor of PC Increase
-	-	TB-s mF_c	5(83%)	0.3(23%)	17
-	mIgG	TB-s mF_c	3.4(77%)	0.6(37%)	7
human erythrocytes	m NN glyc.	TB-s mF_c	2.6(70%)	0.1(18%)	26
rabbit erythrocytes	m NN glyc.	TB-s mF_c	0.25(20%)	0.09(8%)	3

Binding studies of MBL-2 (4.1) and MBL-2 (2.6) are summarised in the plots shown in Fig. 4. From the Scatchard plot (Fig. 4a), it was estimated that each MBL-2 (4.1) cell had 2 x 10^6 binding sites for YE1 48.10 and each MBL-2 (2.6) cell, 6.5 x 10^5 binding sites or 3 x 10^9 sites/cm^2 and 1 x 10^9 sites/cm^2 respectively, suggesting that a difference in surface antigen density of only a factor of three was the basis for separation. This is a very small difference considering that human erythrocytes have 5 x 10^{11} glycophorins per cm^2 [12] and rabbit erythrocytes have zero human glycophorin. The direct plot (Fig. 4b) indicates that saturation occurs at approximately 0.3 mg/ml free YE1 48.10 concentration. The two cells were indistinguishable by their native PC or with a PEG 18:2 ester ligand which dramatically increases cell partition into the upper phase. Biotinylated YE1 48.10 had a PC of 5 when 10 molecules of biotin were bound per IgG. As shown in Fig. 5, the addition of PEG-Av to the phase system rapidly moved essentially all the remaining biotin - YE1 48.10 into the upper phase, while avidin moved all of the biotin YE1 48.10 to the lower phase.

Biotinylated YE1 48.10 was able to immunospecifically alter the PC of MBL-2 (4.1) and MBL-2 (2.6) cells as shown in Table 2.

The data in Table 2 demonstrates that the partition of these cells may be immunospecifically altered using biotinylated YE1 48.10. However, due to the difficulty of producing sufficient quantities of pure YE1 48.10, only concentrations below saturation have been tried.

It is anticipated that the PC difference between the two cells can be improved by optimising the number of biotins per IgG and increasing the concentration of biotin-YE1 48.10.

The addition of PEG-avidin to the system actually had the effect of decreasing the partition with increasing PEG-avidin concentration, while increasing concentrations of avidin had no effect. This may be attributable to the increase in interfacial tension of the system with increasing PEG, making it harder for cells to desorb from the interface. It is not due to the biotin-avidin reaction blocking the binding of biotin YE1 48.10 to the cell as avidin would also have this effect and avidin has no effect on the cell partition. The theory that the increasing interfacial tension keeps the cells at the interface is supported by the fact that at very low PEG-avidin concentration the P.C. of the cells is actually slightly increased.

Table 2 *Results of partitioning experiments using biotinylated YE1 48.10 and MBL-2 cells*

Cell Type	YE1 48.10 Concen-tration mg/ml	biotins per IgG (molec./molec.)	PC of Biotin-YE1 48.10	% Cells in Upper Phase	PC of Cells	Factor of Cell Increase
MBL-2 (2.6)	0.20	0	1	45%	0.8	
MBL-2 (4.1)	0.20	0	1	45%	0.8	
MBL-2 (2.6)	0.20	17	4	48%	0.9	1.1
MBL-2 (4.1)	0.20	17	4	63%	1.7	2.1
MBL-2 (2.6)	0.20	10	2	50%	1.0	1.2
MBL-2 (4.1)	0.20	10	2	62%	1.6	2.0
MBL-2 (2.6)	0.15	0	1	30%	0.4	
MBL-2 (4.1)	0.15	0	1	31%	0.4	
MBL-2 (2.6)	0.15	27	5	27%	0.4	1.0
MBL-2 (4.1)	0.15	27	5	46%	0.85	2.1

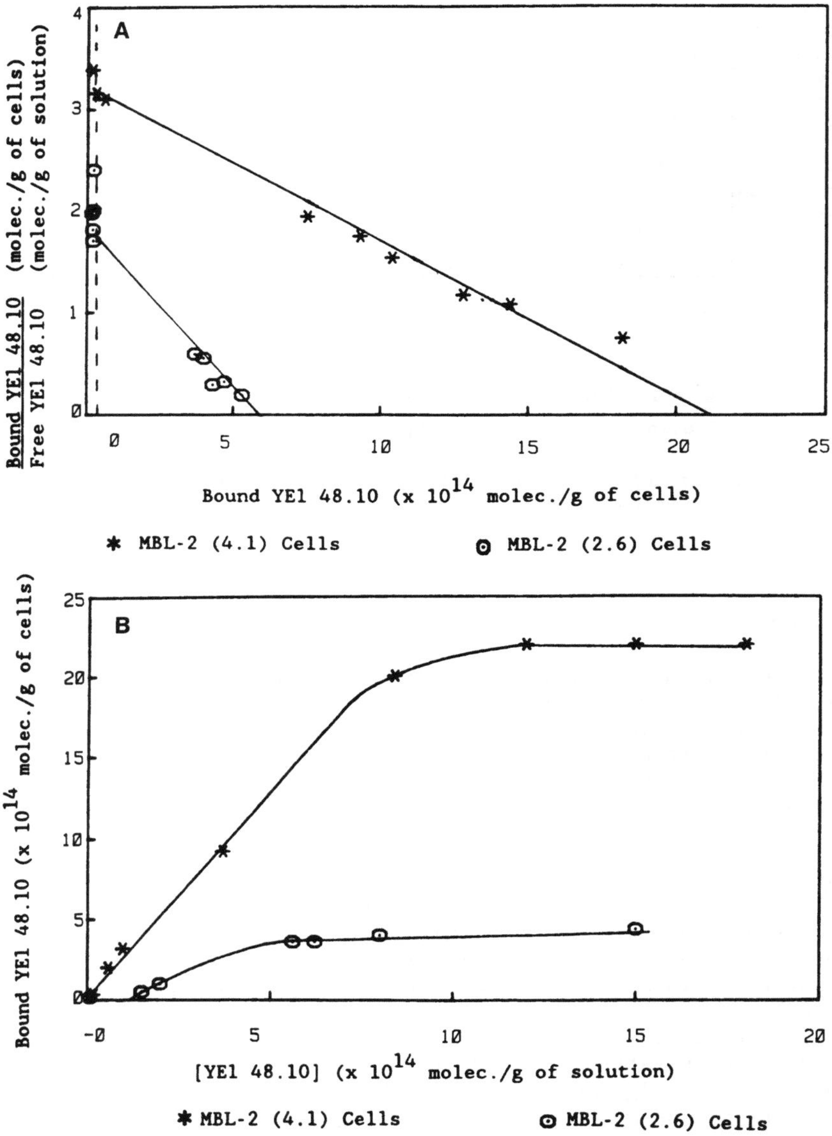

Fig. 4 *a) A Scatchard Plot of the binding of YE1 48.10 antibody to MBL-2 (4.1) and MBL-2 (2.6) cells and b) A binding isotherm of YE1 48.10 antibody with MBL-2 (4.1) and MBL-2 (2.6) cells*

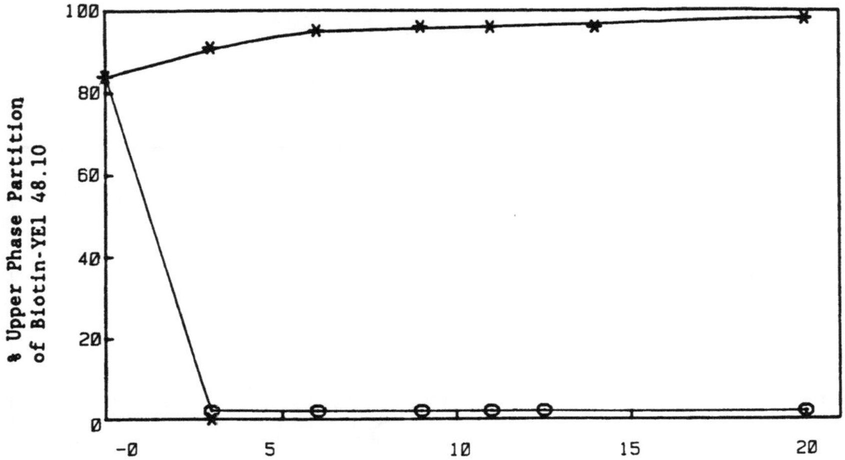

Fig. 5 *The effect of PEG-avidin and avidin on the partition of biotin-YE1 48.10*

CONCLUSION

PEG and trypan blue modified antibodies were found to be effective second affinity ligands and were capable of effecting an immunospecific increase in the partition coefficient of human erythrocytes. The use of a biotinylated antibody as an affinity ligand for the separation of closely related cell types has had some preliminary success and further research should be undertaken.

REFERENCES

1 K.A. Sharp, M. Yalpani, S. Howard and D.E. Brooks, Synthesis and application of a poly(ethylene glycol)-antibody affinity ligand for cell separations in aqueous polymer two-phase systems, Anal. Biochem. 154:110 (1986)

2 L.J. Karr, S.G. Shafer, J.M. Harris, J.M. Van Alstine and R.S. Snyder, Immuno-affinity partition of cells in aqueous polymer two-phase systems, J. Chromatog. 354:269 (1986)

3 S.J. Stocks, Development of a General Ligand for Immunoaffinity Partitioning in Two Phase Aqueous Systems, Anal. Biochem., 172, in press (1988)

4 F.J. Takei, Two surface antigens expressed on proliferating mouse T lymphocytes defined by rat monoclonal antibodies, Immunol. 130:6:2794 (1983)

5 P. Chan and F.J. Takei, Expression of a T cell receptor-like molecule on normal and malignant murine T cells detected by rat monoclonal antibodies to nonclonotypic determinants, Immunol. 136:4:1346 (1986)

6 J.W. Goding, in: Monoclonal Antibodies: Principles and Practice, Academic Press, New York (1983)

7 R.R. Porter, The hydrolysis of rabbit δ-globulin and antibodies with crystalline papain Biochem. J. 73:119 (1959)

8 A. Abuchowski, T. Van Es, N.C. Palczuk and F.D. Davis, Effect of covalent attachment of poly(ethylene glycol) on immunogenicity and circulating life of bovine liver catalase, J. Biol. Chem. 252:3582 (1977)
9 S.J. Stocks, A.J.M. Jones, C.W. Ramey and D.E. Brooks, A fluorometric assay of the degree of modification of protein primary amines with poly(ethylene glycol), Anal. Biochem. 154:232 (1986)
10 N.M. Green, A spectrophotometric assay for avidin and biotin based on binding of dyes by avidin, Biochem. J. 94:23C (1965)
11 N.M. Green, Spectrophotometric determination of avidin and biotin, Meth. Enzymol. 18A:418 (1970)
12 R.E. Lovrien and R.A. Anderson, Stoichiometry of wheat germ agglutinin as a morphology controlling agents and as a morphology protective agent for the human erythrocyte, J. Cell. Biol. 85:534 (1980)

CELL SEPARATION BY IMMUNOAFFINITY PARTITION IN AQUEOUS POLYMER TWO-PHASE SYSTEMS

Laurel J. Karr[1], James M. Van Alstine[1], Robert S. Snyder[1], Steven G. Shafer[2] and J. Milton Harris[2]

[1] Space Science Laboratory, NASA/Marshall Space Flight Center, Huntsville, AL 35812

[2] Department of Chemistry, University of Alabama in Huntsville, Huntsville, AL 35899

Two methods for immunoaffinity partitioning are described. In the first, poly(ethylene glycol) (PEG) was covalently coupled to immunoglobulin G antibody preparations directed toward surface antigens unique to either sheep or human erythrocytes. The modified antibody preparations exhibited a reduction in ability to agglutinate cells and increased affinity for the PEG-rich upper phase in Dextran-PEG aqueous two-phase systems. When either preparation was incubated with a mixture of target and control erythrocytes, the partition of the erythrocyte population bearing the complimentary antigen was selectively increased into the upper phase while the control cells remained unaffected. Effective separation of cell populations was obtained in 2 hours using 30-transfer countercurrent distribution.

The second method describes the use of PEG-modified Protein A as a general immunoaffinity ligand to complex with cells and unmodified antibody and selectively increase their partition to the PEG-rich phase. Effective cell separations were also obtained with this method, which may eliminate the need to prepare a different PEG-modified antibody for each cell type to be purified.

INTRODUCTION

Affinity phase partitioning has proven useful in both preparative and analytical separations of macromolecules and cells [1-7]. The present article describes an extension of affinity partitioning to include use of antibody affinity ligands, either covalently bound to PEG (PEG-Ab) [3,4] or as a composite ligand complexed with PEG-modified Staphylococcal protein A (PEG-SpA) [7].

Several steps are necessary to determine the feasibility of using PEG-modified antibodies (PEG-Abs) as immunoaffinity ligands: (1) PEG must be linked to the antibody; (2) PEG-Ab must retain antigen binding activity; and (3) PEG-Ab must itself partition into the PEG-rich phase, and must also cause a specific increase in partition of target cells.

Human and sheep red blood cells (RBCs) were chosen as a model system. Native antibodies to RBC stroma are commercially available and are directed toward antigens present in large numbers on the cell surface. Thus any negative results would not be due to insufficient binding sites. RBCs are easily quantified, their partitioning characteristics have been extensively studied [8,9], and one aspect of antibody activity can be measured by haemagglutination assays. However, it is possible that PEG-modification could block one binding site, making a molecule incapable of agglutination but still able to bind RBCs. The effects of PEG molecular weight and the degree of modification upon antibody activity and its affinity for the upper phase are described, as are the effects of varying phase system compositions.

Staphylococcal protein A (SpA) binds with the Fc portion of most immunoglobulin G molecules, leaving the Fab regions free for antigen recognition, a phenomenon proved useful in a wide variety of immunological methods [10-13]. To eliminate the need to prepare a separate PEG-Ab derivative for each cell type to be purified, SpA was derivatised with PEG and used with unmodified antibody to cause a selective shift in target cell partition which is similar to that seen when using directly modified antibody.

IMMUNOAFFINITY PARTITION STUDIES USING ANTIBODIES MODIFIED DIRECTLY WITH PEG

Protein Derivatisation with PEG

The derivatisation of antibodies and other proteins with PEG is described in detail elsewhere [3]. Briefly, PEG-5000 or PEG-1900 mono-methyl ether is activated with cyanuric chloride and specific quantities are allowed to react with protein for 1 hour at 4°C, followed by removal of unattached PEG by diafiltration. Degree of modification can be analysed based on the reaction of trinitrobenzenesulfonic acid or flouorescamine [14] with free lysine groups.

The effect of PEG-derivatisation upon protein partitioning is seen in Table 1. A dramatic partition shift results when substituting BSA with PEG. Although PEG-Ab did not show such a large shift, it was still sufficient to indicate feasibility of the immunoaffinity method.

Table 1 *Effects of PEG Substitution on Partitioning of BSA and Native Antibody in Dextran-PEG Two-Phase Systems*

Protein modified	Lysines modified (%)	Partition (%)[a]
BSA[b]	0	31 ± 3
BSA	10	89 ± 3
BSA	39	96 ± 1
BSA	47	96 ± 1
BSA	76	98
Native-Ab[c]	0	50 ± 3
Native-Ab	47	87 ± 3

[a] Percentage of total protein partitioning into the upper phase: BSA (mean ±SD, n = 2); native-Ab (mean ± SD, n = 4)

[b] Bovine serum albumin in a two-phase system consisting of 5.0% Dextran T500, 4.0% PEG 8000, and 0.02% sodium azide in 150 mM NaCl, 7.3 mM Na_2HPO_4, 2.3 mM NaH_2PO_4, pH 7.2 (buffer V)

[c] Sheep IgG anti-human RBCs in a system consisting of 4.6% dextran, 3.9% PEG, and 0.02% sodium azide in buffer V

Table 2 *Effects of PEG Substitution on Antibody Induced Cell Partition and Aggregation*

PEG M_w [a]	Lysines modified (%)	Minimal hemagglutination concentration (g/ml) [b]	Partition (%) [c]
Control	0	<0.2	35 ± 2
5000	27	23 ± 8	92 ± 2
5000	40	19 ± 8	84 ± 4
5000	51	5 ± 2	80 ± 3
1900	45	1 ± 0	57 ± 1
1900	70	3 ± 1	61 ± 2

[a] Molecular weight of PEG used for substitution of sheep IgG anti-human RBCs

[b] Lowest antibody concentration exhibiting appreciable hemagglutination in microtitre assay (mean ± SD, n = 8)

[c] Percentage of total cells partitioning into the upper phase (mean ± SD, n = 2) in a two-phase system consisting of 4.6% Dextran T500, 3.9% PEG 8000 and 0.02% sodium azide in 150 mM NaCl, 7.3 mM Na_2HPO_4, 2.3 mM NaH_2PO_4, pH 7.2 [(4.6, 3.9) V]. Control value is in absence of PEG-Ab

Enzymes substituted with PEG have been shown to lose varying degrees of activity [15]. Similarly, the ability of PEG-Ab to agglutinate red cells is diminished as compared to that of native antibody (Table 2). Direct microtitre hemagglutination assays were performed as described previously [3]. Deactivation appears to depend on molecular weight, with PEG 5000 causing more deactivation than PEG-1900. However, the relationship between percent substitution and antibody activity is less clear, perhaps due to the somewhat approximate nature of these measurements (a discussion of the effects of PEG-substitution on enzyme activity using more precise techniques appears elsewhere in this volume [15]). Control experiments showed no agglutination of sheep RBCs with either native or PEG-modified antibodies to human RBCs.

Single-Tube Partition Experiments

It is interesting that the PEG-Ab with the greatest deactivation also gave the highest percentage partition (Table 2), due at least in part to sedimentation of cell clumps with the more active PEG-Ab. In single-tube experiments, there was a direct relationship between the concentration of PEG-Ab added (mg/ml) and the log of the partition coefficient K (defined as the ratio of cells in the top phase to the remainder of cells) [3]. This relationship plateaus at log K = 1.0 (91% of cells in upper phase). Linear regression analysis of initial points below log K = 1.0 gives a slope of 1.35×10^{-2}, with a y (log K) intercept of -1.60 (r = 0.99, n = 7). A shift in target RBC partition to the PEG-rich top phase was not seen with unmodified Ab to either cell type, either because of the necessity for PEG attachment or because of extensive agglutination and sedimentation. Other control experiments showed that partitioning of sheep RBCs was not affected by anti-human PEG-Ab, nor was partitioning of human RBCs affected by anti-sheep PEG-Ab.

Several aspects of phase system composition were investigated to determine their effects on immunoaffinity phase partitioning, using human RBCs and PEG-Ab directed against them. Results of these experiments are presented in Table 3. Shifts in cell partition caused by addition of PEG-Ab

Table 3 *Effects of Phase System Composition on PEG-Ab Induced Shifts in Cell Partition*

Two-phase system[a]	Interfacial tension (μN/m)[b]	Electrostatic bulk phase potential (mV)[b]	Cell location[c]	Partition (%)[d] no PEG-Ab	Partition (%)[d] PEG-Ab	K_B/K_A[e]
(5, 4) I	11.0	2.2	Upper	14 ± 1	58 ± 4	9
(5, 4) II	≈ 8	1.0	Upper	6 ± 1	64 ± 1	28
(5, 3.5) V	0.7	0.2	Upper	2 ± 1	77 ± 1	
(4.6, 3.9) V	≈ 3	0.2	Upper	2 ± 1	69 ± 3	
(5, 4) V	4.9	0.3	Upper	0 ± 1	30 ± 1	43
(7, 4) V	17.2	0.2	Upper	0 ± 1	4 ± 2	
(7, 5) V[a]	5.0	0.2	Upper	0 ± 1	1 ± 1	
			Interface	45 ± 3	97 ± 3	
			Lower	55 ± 3	3 ± 1	

[a] All systems compounded with PEG 8000 and Dextran T500 of the same lots except (7, 5) V which was prepared with Dextran T40. (5, 4) I, 5% dextran, 4% PEG, 109 mM Na_2HPO_4, 35 mM NaH_2PO_4, pH 7.2 (buffer I); (5, 4) II, 5% dextran, 4% PEG, 50 mM NaCl, 73 mM Na_2HPO_4, 23 mM NaH_2PO_4 pH 7.2 (buffer II); (5, 3.5) V, 5% dextran, 3.5% PEG, 150 mM NaCl, 7.3 mM Na_2HPO_4, 2.3 mM NaH_2PO_4, pH 7.2 (buffer V); (4.6, 3.9) V, 4.6% dextran, 3.9% PEG, buffer V; (5, 4) V, 5% dextran, 4% PEG, buffer V; (7, 4) V, 7% dextran, 4% PEG, buffer V; (7, 5) V, 7% Dextran T40, 5% PEG, buffer V

[b] Taken from reference 16. Approximate values interpolated for purposes of discussion

[c] Referring to the upper or lower phase or to the interface between them

[d] Percentage of total cells. Mean ± SD of two independent determinations using 643 μg/ml of PEG-Ab (from rabbit)

[e] Relative effect of ligand addition, where K is the ratio of cells in the upper phase to the remainder of cells. B is with the presence of PEG-Ab and A is without.

become more pronounced as the interfacial tension of the phase systems decreases (compare systems (7, 4) V, (5, 4) V, and (5, 3.5) V). The effects of changes in electrostatic bulk phase potential are difficult to determine since increasing potential (by salt variation) is accompanied by an increase in interfacial tension (compare (5, 4) V, (5, 4) II, and (5, 4) I). However, with increasing potential of the phase system the cell partition coefficient in the absence of ligand is also increased, thus decreasing the relative effect (K_B/K_A) when ligand is present.

Dextran T40 (molecular weight 40,000) was used to prepare the (7, 5) V system shown as the final entry in Table 3. RBCs normally partition fairly equally between the interface and the lower phase in this system [8,9]. It is interesting that incubation with PEG-Ab shifts partition almost totally to the interface with no increase in partition to the upper, PEG-rich phase.

An apparent discrepancy should be pointed out. The entries for the (4.6, 3.9) V system without ligands vary considerably between Table 2 and Table 3. The difference seen is due to use of different lots of PEG, an observation not uncommon in phase partitioning [16].

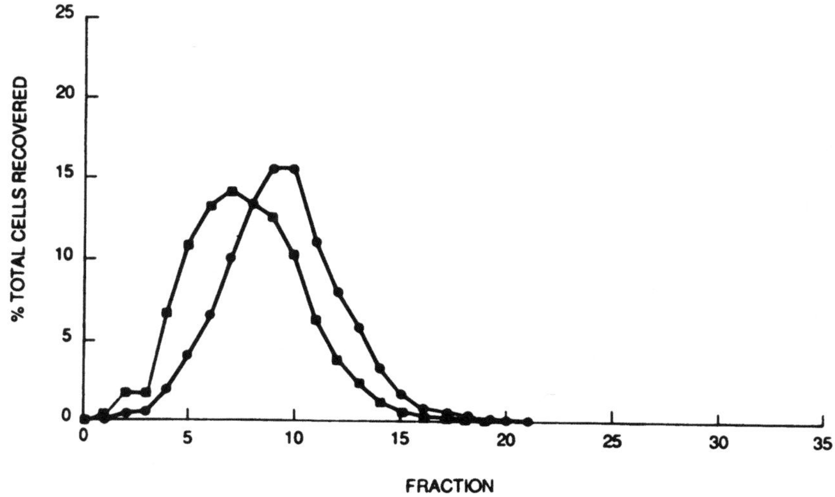

Fig. 1 *30-transfer CCD of a mixture of 3 x 10^7 untreated human (●) and sheep (■) RBCs in a (4.6, 3.9) V two-phase system*

Countercurrent Distribution (CCD) Experiments

Having demonstrated that PEG-Ab can selectively shift target cell partitioning to the PEG-rich upper phase in single-tube experiments, it remained to be determined if complete separation of two cell types could be effected by CCD following a single incubation procedure. A Biosheff MK2 apparatus was utilised to perform 30-transfer CCD experiments in the (4.6, 3.9) V phase system described above, with a 3 min settling time and a 12 second shaking time.

As seen in Fig. 1, CCD profiles of untreated sheep and human RBCs overlap considerably. Fig. 2A shows the results of a typical experiment when CCD was performed on a mixture of 3 x 10^7 human and sheep RBCs following incubation (at 37° for 15 min) with PEG-Ab directed against human RBCs at an incubation concentration of 1.3 mg/ml. Positions of cell populations could be reversed when the mixture was instead incubated with PEG-Ab directed against sheep RBCs, Fig. 2B. Again nearly complete resolution of the two cell populations was seen.

IMMUNOAFFINITY PARTITION STUDIES USING A COMPOSITE LIGAND OF PEG-SPA AND UNMODIFIED ANTIBODY

Single-Tube Partition Experiments

Although PEG-Ab works well as an immunoaffinity ligand in an erythrocyte model system, a more general reagent is desirable to preclude the necessity of a separate antibody derivatisation for each cell type to be purified. To this end, protein A was derivatised with PEG-5000, with 60% of the available lysine groups being modified [7]. The same model cells and phase system were used so that direct comparisons could be made.

Cell agglutination upon incubation with PEG-SpA and unmodified antibody was of concern. There is no deactivation of the antibody since it is not directly derivatised with PEG. Moreover, protein A is known to form soluble complexes with rabbit IgG which vary in composition and ability to agglutinate cells depending on the relative proportions of Protein A and IgG [17]. Qualitative single-tube partition experiments were performed to

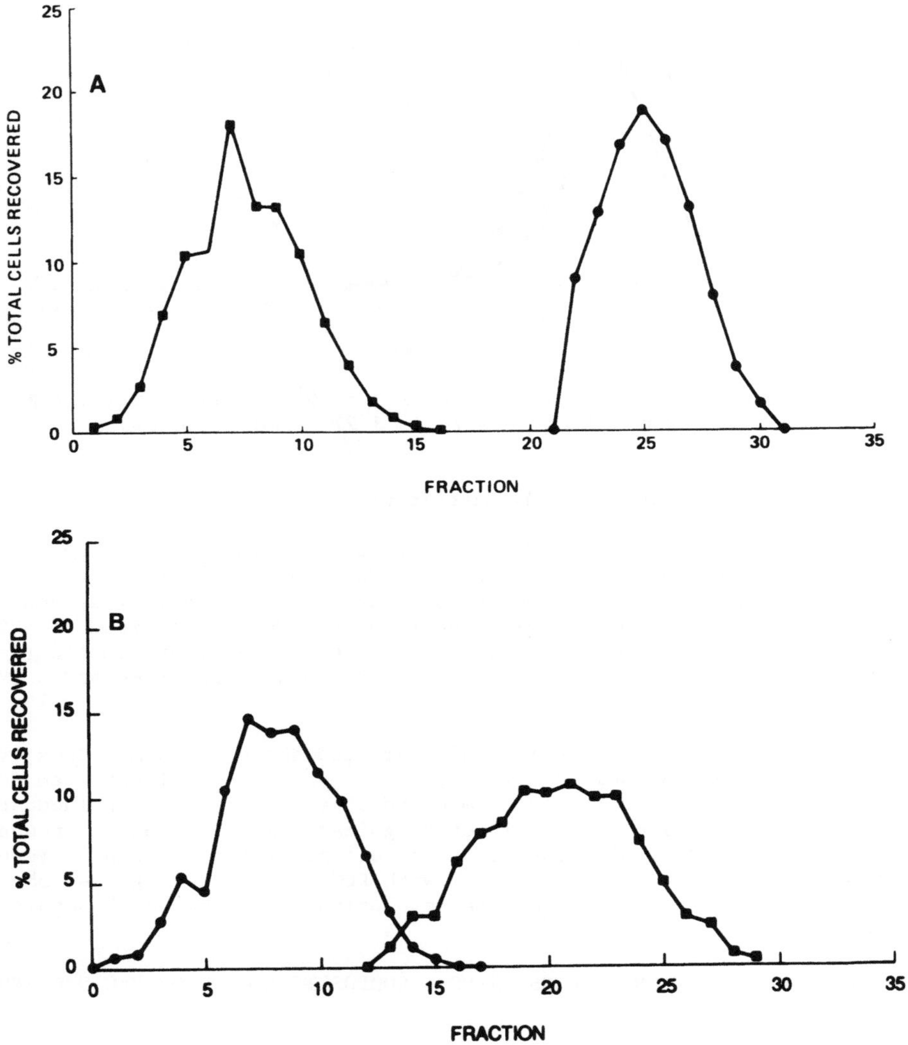

Fig. 2 *30-transfer CCD of mixtures containing 3 x 10^7 human (●) and sheep (■) RBCs in a (4.6, 3.9) V two-phase system. (A) RBC mixture was incubated with PEG-Ab directed against human RBCs and washed with upper phase prior to CCD; (B) identical to A except RBCs were incubated instead with PEG-Ab directed against sheep RBCs; (C) RBC mixture was incubated with PEG-SpA and unmodified antibody directed against human RBCs and washed with upper phase prior to CCD; (D) identical to C except RBCs were incubated with PEG-SpA and unmodified antibody directed against sheep RBCs*

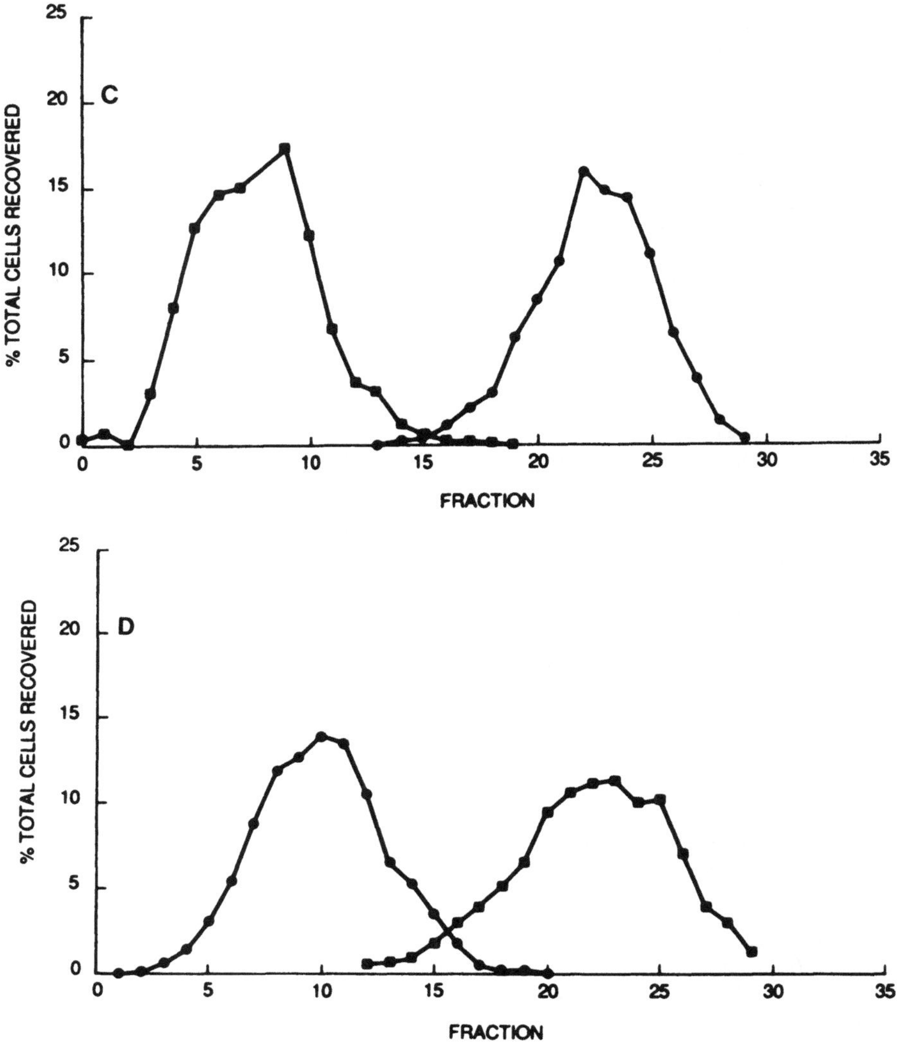
C
D
% TOTAL CELLS RECOVERED
FRACTION
0
5
10
15
20
25
30
35

optimise incubation parameters, such as concentration of reagents and their order of addition, as well as length of time and temperature of incubation.

The order in which reagents were combined greatly affected the extent of cell agglutination. It was not practical to add unmodified antibody to cells prior to adding PEG-SpA, as agglutination was caused even by very dilute antibody preparations. At antibody concentrations low enough to prevent agglutination, no significant increase in cell partition was seen. Unmodified antibody was also mixed first with PEG-SpA for 60 minutes at room temperature followed by addition of cells and a further incubation for 15 minutes at 37°. This worked well in single-tube partition experiments but cells later agglutinated during CCD. Although not further pursued, additional experiments using this order of addition but varying relative concentrations of PEG-SpA and antibody might prove beneficial. We did find, however, that little agglutination occurred when PEG-SpA was first added to pelleted cells and mixed, followed by addition of unmodified antibody.

Both temperature and duration of incubation procedures affected the maximum increase in target cell partition. There was little difference between incubation at 37° or at room temperature, but at 0°, attachment of PEG-SpA appeared to be considerably diminished (as evidenced by only slight shifts in cell partition). A comparison of incubation times of 15, 30 and 60 minutes showed progressively higher partition shifts. Incubation times greater than 60 minutes were not tested.

Also affecting the magnitude of partition shift and cell agglutination was the proportion of reagents used in incubation procedures. Best results were obtained when approximately 25-35 times more PEG-SpA was added than was necessary to bind all IgG molecules on a molar basis; this excess serving to lessen the likelihood of formation of IgM-like soluble complexes [17].

To summarise, incubation conditions we found to be optimal are: (1) PEG-SpA (at a molar concentration 25-35 times higher than IgG) is added to pelleted cells and mixed; (2) unmodified antibody is then added and incubation proceeds for 60 minutes at room temperature with occasional agitation; (3) cells are washed once with upper phase prior to partition. Although these conditions appeared to be optimal for the erythrocyte model cells used, it should be stressed that different conditions may prove necessary for other cell types or for IgG antibodies from other species or differing subclasses.

In single-tube partition experiments, human RBCs incubated as described with PEG-SpA (0.80 mg/ml) and unmodified anti-human RBC antibody (0.12 mg/ml), partitioned an average of 70% into the upper phase, whereas the same cells without treatment partitioned only 7.5% into the upper phase. Similarly, following incubation of sheep RBCs with PEG-SpA (0.80 mg/ml) and anti-sheep RBCs antibody (0.08 mg/ml) partition increased from 3% to 65%. There was no increase in partition seen for either cell type when incubated with PEG-SpA alone.

Countercurrent Distribution Experiments

Fig. 2C and 2D show typical results of cell separations using PEG-SpA as a composite immunoaffinity ligand with unmodified antibodies. Mixtures of 3×10^7 human and sheep RBCs were incubated with PEG-SpA and unmodified antibody using the same conditions and concentrations as for single-tube partition experiments. Depending on the antibody preparation used, the partition of either cell type was selectively increased and effective separation obtained with 30-transfer CCD. Separations are comparable to those obtained using directly modified antibodies and use less than one-tenth the antibody concentration.

CONCLUSIONS

Immunoaffinity partition in Dextran-PEG aqueous two-phase systems has been demonstrated using both PEG-modified antibodies and PEG-modified Protein A with unmodified antibodies. Both methods allow for the effective isolation of specific cell populations using CCD. The PEG-SpA method should be of more general use, with the PEG-Ab method being used where undue cell agglutination or lack of antibody reaction with Protein A presents a problem. In terms of cost-effectiveness, gentleness, capacity, and recovery of cells, immunoaffinity partitioning may offer advantages over fluorescence activated cell sorting and other immunoaffinity techniques.

ACKNOWLEDGEMENTS

The authors thank the National Aeronatuics and Space Administration and the Universities Space Research Association (J.M.V.A.) for the financial support of this research.

REFERENCES

1 G. Kopperschläger, G. Lorenz and E. Usbeck, Application of affinity partitioning in an aqueous two-phase system to the investigation of triazine dye-enzyme interactions, J. Chromatogr. 259:97 (1983)
2 G. Johansson, M. Joelsson, B. Olde and V.P. Shanbhag, Affinity partitioning of biopolymers and membranes in ficoll-Dextran aqueous two-phase systems, J. Chromatogr. 331:11 (1985)
3 L.J. Karr, S.G. Shafer, J.M. Harris, J.M. Van Alstine and R.S. Snyder, Immuno-affinity partition of cells in aqueous polymer two-phase systems, J. Chromatogr. 354:269 (1986)
4 K.A. Sharp, M. Yalpani, S.J. Howard and D.E. Brooks, Synthesis and application of a poly(ethylene glycol)-antibody affinity ligand for cell separations in aqueous polymer two-phase systems, Anal. Biochem. 154:110 (1986)
5 S.D. Flanagan, G. Johansson, B. Yost, Y. Ito and I.A. Sutherland, Toroidal coil countercurrent chromatography in the affinity partitioning of nicotinic cholinergic receptor enriched membranes, J. Liq. Chromatogr. 7:385 (1984)
6 M.K. Pangburn and H. Walter, Specific deposition of complement protein C3b on abnormal PNH erythrocytes permits their separation by partitioning. Possible general approach for isolation of specific cell populations, Biochimica et Biophysica Acta 902:278 (1987)
7 L.J. Karr, J.M Van Alstine, R.S. Snyder, S.G. Shafer and J.M. Harris, Cell separation by immunoaffinity partitioning with poly(ethylene glycol)-modified Protein A, in aqueous polymer two phase systems, J. Chromatogr. 442:219 (1988)
8 P.-Å. Albertsson, in: "Partition of Cell Particles and Macromolecules," Wiley Interscience, New York (1986)
9 H. Walter, D.E. Brooks and D. Fisher, in: "Partitioning in Aqueous Two-Phase Systems," Academic Press, Orlando (1985)
10 J.J. Langone, Use of labeled Protein A in quantitative immunochemical analysis of antigens and antibodies, J. Immunol. Methods 51:3 (1982)
11 J.J. Langone, Applications of immobilised Protein A in immunochemical techniques, J. Immunol. Methods 55:277 (1982)
12 T.G.I. Ling and B. Mattiasson, A general study of the binding and separation in partition affinity ligand assay. Immunoassay of β_2-microglobulin, J. Immunol. Methods 59:327 (1983)

13 T.M. Phillips, W.D. Queen, N.S. More and A.M. Thompson, Universal support medium for high-performance immunoaffinity chromatography, J. Chromatogr. 327:213 (1985)
14 S.J. Stocks, A.J.M. Jones, C.W. Ramey and D.E. Brooks, A fluorometric assay of the degree of modification of protein primary amines with poly(ethylene glycol), Anal. Biochem. 154:232 (1986)
15 J.M. Harris, K. Yoshinaga, M.S. Paley and M.R. Herati, New activated PEG derivatives for affinity partitioning, in: "Separations Using Aqueous Phase Systems: Applications in Cell Biology and Biotechnology," D. Fisher and I.A. Sutherland, eds., Plenum, New York (1989)
16 S. Bamberger, D.E. Brooks, K.A. Sharp, J.M. Van Alstine and T.J. Webber, Preparation of phase systems and measurement of their physicochemical properties, in: "Partitioning in Aqueous Two-Phase Systems," H. Walter, D.E. Brooks and D. Fisher, eds., Academic Press, Orlando (1985)
17 J.J. Langone, M.D.P. Boyle and T. Borsos, Studies on the interaction between Protein A and immunoglobulin G. II. Composition and activity of complexes formed between Protein A and IgG, J. Immunol. 121:333 (1978)

NEW ACTIVATED PEG DERIVATIVES FOR AFFINITY PARTITIONING

J.M. Harris, K. Yoshinaga, M.S. Paley and M.R. Herati

Department of Chemistry
University of Alabama in Huntsville
Huntsville, AL 35899, USA

If affinity partitioning is to be applied industrially, it is likely that the chemistry of polymer-ligand coupling will be critical to commercial viability. Factors to be considered include reagent cost, ease of handling, loss of ligand activity, reactivity of intermediates, and linkage stability. Consequently, a great deal of effort has been invested in PEG chemistry [1,2]. This same chemistry is pertinent to other areas such as catalysis, biosensors, and coatings for control of wetting, electroosmosis, and protein adsorption [3,4]. In this article we describe recent work in three areas: (1) effects of known methods for PEG-protein coupling on protein activity and stability, (2) synthesis of heterofunctional PEG derivatives, and (3) preparation and utility of PEG aldehyde.

EFFECTS OF PEG-PROTEIN COUPLING ON PROTEIN ACTIVITY

In this section we examine the effects on protein activity of substitution extent, PEG molecular weight (MW), PEG functionality (i.e. monofunctional or difunctional), and coupling method, and we examine the effect of PEG substitution on protein response to pH and heat. All these experiments have been conducted on alkaline phosphatase as a model protein. This model was chosen because of the ease and accuracy of assay (p-nitrophenylphosphate hydrolysis), ready availability, and extensive literature on this enzyme [5].

Effects of PEG MW and extent of modification

First we examine the effects of PEG MW and extent of substitution on alkaline phosphatase activity using the cyanuric-chloride route to PEG-protein coupling (note that substitution takes place primarily on the nucleophilic lysine amino groups of the protein) [6,7]. In a previous work with PEG-substituted antibody we tentatively noted a decrease in antibody activity with increasing PEG MW and extent of substitution, but firm conclusions were prevented because of the limited accuracy of the assay method (cell agglutination) [8]. A series of experiments are presented in Table 1 for coupling of the monomethyl ethers of PEG 1900 and 5000 (M-PEG) to alkaline phosphatase.

From Table 1 we see that increasing percent substitution and MW give a general decrease in enzyme activity. Note, however, that extensive substitution must occur with M-PEG 1900 before appreciable activity loss is evident. It is also noteworthy that even enzyme with essentially complete substitution (we could not substitute more than 87% of the lysines available to trinitrobenzenesulfonic acid, our assay method) retains almost half the activity of unsubstituted enzyme. Examination of Table 1 shows that 23% modification with M-PEG-5000 gives deactivation equivalent to that from 75% substitution with M-PEG-1900. Since there are more bound lysines with the 75% M-PEG and roughly equal masses of M-PEG attached for 75% M-PEG 1900 and 23% M-PEG 5000, it appears that the key to deactivation is bound PEG mass. We believe that the gradual loss of enzyme activity results primarily from conformational restriction of the enzymes and possibly from interference with substrate approach to the active site.

Table 1 *Activities of alkaline phosphatase modified by M-PEG1900 and 5000 [7]*

M-PEG MW	% Mod.	Activity (mol/min/mg)
(unmod.)	0	12.0
1900	5	11.6
	29	11.6
	41	11.8
	58	9.9
	74	8.0
	86	6.6
5000	23	7.9
	38	6.6
	48	6.8
	82	5.3

The most surprising result from this study is the response of enzyme activity to substitution with difunctional PEG 4000, 8000 and 20000. These polymers have two active terminae and thus the possibility exists for formation of "loops" in which both ends of the polymer are attached to a single protein chain; crosslinking between proteins did not to occur (as shown by HPLC), presumably because an excess of PEG was used in the substitution reaction. If these loops are formed, one might suppose that they would lead to significant loss of enzyme activity because of the consequent loss of protein chain mobility. However, as shown in Table 2, substitution with difunctional PEG gives quite active enzyme and little response to degree of substitution. It is also interesting that varying the MW of the difunctional PEGs from 4000, 8000 to 20000 has little effect on enzyme activity; in fact, the most active conjugate is obtained with the highest PEG MW (20000). Possibly, these results are due to stabilisation of a preferred enzyme conformation by large loops of PEG.

Table 2 *Activities of alkaline phosphatase modified with cyanuric-chloride activated PEG*

PEG MW	% Mod.	Activity (mol/min/mg)
4000	41	8.6
	77	8.6
8000	40	8.4
	66	8.4
	72	8.9
20000	54	10.1
	63	9.9
	84	9.6

Table 3 *Effects of coupling method on activity of alkaline phosphatase modified with M-PEG-5000*

Activation	% Mod.	% Native Activity
none	0	100
cyanuric chloride	88	33
	62	67
tresyl chloride	77	86
	73	82
carbonyl diimid.	78	77
	56	91
succinimidyl	79	93
	61	98

Effect of coupling method

It has been observed previously that there can be substantial loss of enzyme activity upon PEG coupling and that the extent of activity loss can depend on coupling method [9,10]. In Table 3 we show the results on alkaline phosphatase activity of coupling M-PEG 5000 using activated M-PEG prepared by reaction with four different reagents: (1) cyanuric chloride [6,7]; (2) tresyl chloride [11]; (3) carbonyl diimidazole [10,12] and (4) succinimidyl chloride [9].

The most active PEG-protein was obtained by use of the succinimidyl derivative, which gave a protein conjugate of comparable activity to the native enzyme, even at a high modification level of 79%. The tresylate and imidazole gave a slightly less active conjugate than the succinimidyl derivative. On the other hand, modification with the cyanuric-chloride derivative, while satisfactory at moderate levels of modification, gives extensive deactivation at high degrees of modification. This loss of activity with cyanuric-chloride modification has been observed with several other proteins [9,10,13] and thus is presumably a general phenomenon. However, the cyanuric-chloride procedure remains an attractive one because of the ease of preparing the activated PEG and the demonstrated utility of proteins modified by this procedure [8,9]. The succinimide procedure does give highly active enzyme, but this activated PEG is the most difficult to prepare. The tresylate and diimidazole procedures are technically straightforward, and they also give highly active products. A possible drawback to the tresylate procedure is that tresyl chloride is a rather expensive starting reagent.

The cyanuric-chloride derivative has two replaceable chlorines at its active site, so it is plausible to attribute the increased deactivation with this derivative to intramolecular crosslinking of the protein as earlier suggested by Boccu et al [13]. This crosslinking, however, differs from the loop formation potentially occurring with difunctional PEGs (above) in that the two active chlorines of the cyanuric-chloride derivative are quite close together so that extreme, rigid deformation of the protein chain could result. On the other hand, the two active sites of activated PEG (X-PEG-X, where X is the active site) are separated by the long polymer chain so that crosslinking, if it occurs, would introduce little deformation or rigidity into the protein chain.

PEG substitution and enzyme stability

Other workers have noted the increased serum lifetime of PEG-modified proteins [1,9,10,13] which presumably results from PEG's ability to exclude deactivating antibodies and enzymes from the covalently bound protein. Similarly, it has also been claimed that PEG attachment enhances the stability of enzymes in a sterile environment. The experiments we have described thus far were done with two forms of alkaline phosphatase, one from calf intestine (a glycoprotein of MW 140,000) and one from E. coli (a nonglycosylated protein of MW 80,000). Interestingly, the mammalian enzyme is relatively unstable in sterile medium, losing half its activity after four days' storage at room temperature. The bacterial enzyme, on the other hand, loses little activity after weeks under the same conditions. To test the claim that PEG substitution increases the thermal stability of the enzyme we examined the thermal stability, under sterile conditions, of calf intestine alkaline phosphatase in which M-PEG-5000 is attached by the four methods described above, cyanuric chloride (88% modification), tresyl chloride (77% modification), carbonyl diimidazole (78% modification), and succinimide (79% modification). Interestingly, the last three modifications showed practically identical rates of activity loss as the unmodified enzyme (half life four days at room temperature). The cyanuric-chloride modified

Table 4 *Effects of pH on alkaline phosphatase activity upon modification with M-PEG-5000 cyanuric chloride at 30°C*

		Activity (mol/min/mg enzyme)		
% MOD.	pH	8.5	9.8	10.5
0		11.3	15.7	5.10
		(72%)	100%)	(32%)
35		7.06	11.1	4.47
		(64%)	(100%)	(40%)
51		4.75	7.37	3.13
		(62%)	(100%)	(42%)

enzyme lost activity more slowly, with a half life of seven days. Possibly the crosslinking attributed above to this derivative slows the degradation process.

Hydrolytic deactivation of a protein in a sterile environment presumably proceeds by slow denaturation. One would suspect, as shown by our experiments, that covalent binding of M-PEG would do little to slow this process. This result is in sharp contrast to the greatly retarded rate of enzyme deactivation observed in vivo, and reinforces the idea that in vivo deactivation is slowed by PEG's ability to exclude deactivating proteins such as antibodies and enzymes.

We have also briefly examined the effects of pH and temperature on modified alkaline phosphatase activity, and find only subtle effects. Alkaline phosphatase modified to varying extents with M-PEG 5000 showed linear Arrhenius plots up to 40°C with only slight changes in relative rates (activation parameters discussed in next section). Similarly, activities were maximum at pH 9.8 for unmodified and modified alkaline phosphatase, Table 4. Thus it is apparent that enzyme modification with PEG has no effect on the pH and temperature response of the enzyme.

Mechanistic study of PEG-modified alkaline phosphatase

We have also examined the effects of altering temperature and substrate concentration on the kinetics of p-nitrophenylphosphate hydrolysis catalysed by M-PEG-5000 modified alkaline phosphatase with the aim of testing the hypothesis presented above that the cyanuric-chloride method of PEG attachment causes activity loss by reducing protein mobility.

Reaction rates for cyanuric-chloride modified alkaline phosphatase were determined at 20, 25, 30, 35 and 40°C for 31% and 58% modification. Treatment of these data with the appropriate kinetic expressions provides a more detailed look at the activity loss accompanying modification, Table 5.

The data show that the loss of activity (increasing free energy of activation) accompanying increasing modification results from an increasingly negative entropy of activation. Although it is easy to over-interpret this sort of trend (because of probable solvent contributions to entropy of activation and because of the possibility of multiple competitive rate-determining steps) an increasingly negative entropy of activation can be interpreted as indicating a highly ordered (rigid) activation process.

Analysis of the kinetics in terms of the Michaelis-Menten approach is informative, Table 6. Modification with M-PEG by the tresylate route gives

Table 5 *Activation parameters (kcal/mol) for M-PEG-5000 substitution on alkaline phosphatase by the cyanuric-chloride method*

% Mod.	*% Native Act.*	$G^{\#}$	$H^{\#}$	$-T S^{\#}$	$S^{\#}$ (e.u.)
0	100	18.7	7.36	11.3	-37.8
31	81.3	18.8	6.99	11.8	-39.4
58	65.9	18.9	6.54	12.4	-41.4

slight increase in enzyme activity resulting from slight improvements in both V_{max} and K_m. On the other hand, modification by the cyanuric-chloride route gives a loss of activity resulting from a decrease in V_{max}; the binding constant is essentially unchanged. This result indicates that substrate binding to the enzyme is unaltered by M-PEG modification and that loss of activity results from a loss of turnover ability of the catalyst. Again, such a result could well be due to te aforementioned loss of enzyme conformational flexibility upon M-PEG-cyanuric-chloride binding. Recall that the rate-determining step for this enzyme is said to be a conformational change in the protein [5,7].

SYNTHESIS OF HETEROFUNCTIONAL PEG'S

PEG derivatives with different terminae (such as methoxy and hydroxyl) have many potential uses. For example, the ether terminus of the commonly used monomethyl ether of PEG (M-PEG) is relatively inert, so activation of the hydroxyl terminus permits coupling of the polymer to various ligands and surfaces through a single covalent linkage. Unfortunately M-PEG, its derivatives, and PEG-based detergents are the only commercially available heterofunctional PEGs. Consequently we have examined methods for ethylene-oxide polymerisation that permit synthesis of useful heterofunctional PEGs at atmospheric pressure with equipment readily available in most organic laboratories.

We have found that anionic polymerisation initiated by alkoxides and promoted by crown ethers is effective, equation (1) (EO is ethylene oxide). This method is based on previous work of Shalati and Overberger [14] and Eisenbach [15] who showed that the crown ether functions to keep the oxyanion at the terminus of the growing chain soluble and reactive.

$$RO^-Na^+ + EO \xrightarrow[\text{DMF, 1 atm.}]{\text{15-crown-5}} RO\text{-}(CH_2CH_2O)_n\text{-}H \quad = \quad \mathbf{A} \qquad \text{-----} \ (1)$$

One heterofunctional PEG synthesised via equation (1), which we have found to be very useful is the benzyl ether of **A** (where R = $PhCH_2$). The hydroxyl terminus of this compound can be activated in a conventional manner (preceding section) and attached to some substrate. The benzyl group can then be removed by reaction by trimethylsilyl iodide to reveal a new hydroxyl group (equation 2).

$$PhCH_2O\text{-}(CH_2CH_2O)_nH \xrightarrow[\text{step 2. couple, } RNH_2 \text{; step 3. remove benzyl}]{\text{step 1. activate}} HO\text{-}(CH_2CH_2O)_n\text{-}NHR = \mathbf{B} \ \text{---} \ (2)$$

Table 6 *Michaelis-Menten parameters for hydrolysis of p-nitrophenylphosphate catalysed by M-PEG-5000 modified alkaline phosphatase (at 30°C)*

Mod. Method	% Mod.	V_{max} (mol/min/mg)	K_m (mM)	Activity (1.67 mM) (mol/min/mg)
none	0	20	0.49	16.0
tresylate	41	23	0.60	16.5
cyanuric chl.	51	13	0.43	10.3

This new hydroxyl group can in turn be activated and coupled to other substrates. We have used this procedure to couple various molecules to surfaces via a PEG linker.

Another polymer we have synthesised via equation 1 is the monoamine $_2HN-(CH_2CH_2O)_nH$; in this case the alkoxide of equation 1 is replaced by sodium amide. As with the benzyl derivative, this polymer can be attached to a surface or soluble molecule through one terminus (the amino in this case) to produce a polymer tethered at one end and with a free hydroxyl group on the other end.

PEG ALDEHYDE

We have had an interest for some time in the possible utility of PEG aldehyde for reductive amination with amines [16]. For example, reductive amination of lysine amino groups with PEG aldehyde could provide a ready means of coupling PEG to proteins through a highly stable amine linkage.

$$\text{PEG-CHO} + \text{R-NH}_2 \xrightarrow{\text{NaCNBH}_3} \text{PEG-CH}_2\text{-NH-R} \quad ----- \quad (3)$$

Several routes to synthesis of the aldehyde have been reported [1,17] but in a recent investigation of these methods [17] we had difficulty obtaining good yields in a reproducible fashion with any of the published procedures. Consequently we applied six of the most promising routes to synthesis of the oligomeric compound methoxyethoxyacetaldehyde (**C**). This compound was chosen because its small size permits application of a large number of analytical techniques which either could not be used or could only be used with difficulty with the polymeric analogue. The six routes examined were: (1-4) direct oxidation of the alcohol with MnO_2, ceric ammonium nitrate, dimethylsulfoxide/acetic anhydride, and pyridium chlorochromate; (5) synthesis of the diethyl acetal from the diethyl acetal of 2-chloroacetaldehyde, followed by hydrolysis; and (6) synthesis of the diol followed by cleavage with potassium periodate.

Methods 5 and 6 proved to be effective routes to synthesis of **C**, while the other four routes gave poor yields at best. Once **C** was in hand, it became apparent why previous work with the polymeric analogue had met with mixed success: the aldehyde is sensitive to bases (even very weak ones), decomposing rapidly to give a polymeric degradation product (apparently by aldol condensation). Thus reaction with amines to give reductive amination occurs in competition with a base-catalysed decomposition reaction.

Because of the potential utility of reductive amination we also examined the ease of preparation, stability and reductive amination of the benzlaldehyde derivative of PEG. This compound proved to be stable in the presence of base and easy to prepare (by the published route of Bayer) [18].

Reaction with amines gives the imine, which is stable itself and which can be reduced to the coupled amine with sodium barohydride. We are now exploring uses of the polymeric benzaldehyde derivative for coupling to surfaces and proteins.

REFERENCES

1 J.M. Harris, Synthesis of poly(ethylene glycol) derivatives, J. Macromol. Sci.-Rev. Macromol. Chem. Phys. C25:325 (1985)
2 J.M. Harris and M. Yalpani, Polymer-ligands used in biospecific affinity partitioning and their synthesis, in: "Partitioning in Aqueous Two Phase Systems," H. Walter, D.E. Brooks and D. Fisher, eds., Academic Press, New York (1985)
3 B.J. Herren, S.G. Shafer, J.M. Van Alstine, J.M. Harris and R.S. Snyder, Control of electroosmosis in coated quartz capillaries, J. Colloid Interface Sci. 51:46 (1987)
4 J.M. Van Alstine, J.M. Harris, R.S. Snyder, P.A. Curreri, S. Bamberger and D.E. Brooks, Separation of aqueous two-phase polymer systems in microgravity, Eur. Space Agency SP-222:309 (1984)
5 R.B. McComb, G.N. Bowers, Jr. and S. Posen, "Alkaline Phosphatase," Plenum, New York (1979)
6 S.G. Shafer and J.M. Harris, Preparation of cyanuric-chloride activated poly(ethyleneglycol), J. Polym. Sci. Polym. Chem. Edn. 24:375 (1986)
7 K. Yoshinaga, S.G. Shafer and J.M. Harris, Effects of poly(ethyleneglycol) substitution on enzyme activity, J. Bioact. Compatible Polym. 2:49 (1987)
8 L.J. Karr, S.G. Shafer, J.M. Harris, J.M. Van Alstine and R.S. Snyder, Immuno-affinity partition of cells in aqueous polymer two-phase systems, J. Chromatography 354:269 (1986)
9 Y.K. Park, A. Abuchowski, S. Davis and F. Davis, Pharmacology of Escherichia Coli-I-Asparaginase poly(ethyleneglycol) adduct, Anticancer Research 1:373 (1981)
10 C.O. Beauchamp, S.L. Gonias, D.P. Manapace and S.V. Pizzo, A new procedure for the synthesis of poly(ethyleneglycol)-protein adducts, Anal. Biochem. 131:25 (1983)
11 K. Nilsson and K. Mosbach, Immobilisation of ligands with organic sulfonyl chlorides, Methods in Enzymology 104:56 (1984)
12 P. Ferruti, M.C. Tanzi, L. Rusconi and R. Cecchi, Succinic half-esters of PEGs and their derivatives as oligomeric drug-binding matrices, Makromol. Chem. 182:2183 (1981)
13 E. Boccu, G.P. Velo and F.M. Veronese, Pharmocokinetic properties of poly(ethyleneglycol) derivatised superoxide dismutase, Pharmacol. Res. Commun. 14:113 (1982)
14 M.D. Shalati and C.G. Overberger, Grafting of living poly(ethylene oxide) onto polystyrene via aromatic nucleophilic displacement of activated nitro groups, J. Polym. Sci., Polym. Chem. Edn. 21:3425 (1983)
15 C.D. Eisenbach, M. Peuscher, G. Wegner and M. Weiss, Synthesis of poly(ethylene oxide) with unimodal and narrow molecular weight distribution, Makromol. Chem. 184:2313 (1983)
16 J.M. Harris, M. Yalpani, J.M. Van Alstine, E.C. Struck, M.G. Case, M.S. Paley and D.E. Brooks, Synthesis of poly(ethyleneglycol) derivatives, J. Poly. Sci., Poly. Chem. Ed. 22:341 (1984)
17 M.S. Paley and J.M. Harris, Synthesis of the aldehyde of oligomeric polyoxyethylene, J. Polym. Sci. Polym. Chem. Edn. 25:2447 (1987)
18 E. Bayer, H. Zheng and K. Geckeler, Functionalisation of soluble polymers, Polymer Bulletin 8:585 (1982)

COUPLING OF PEG TO PROTEINS BY ACTIVATION WITH TRESYL CHLORIDE. APPLICATIONS IN IMMUNOAFFINITY CELL PARTITIONING

C. Delgado, G.E. Francis* and D. Fisher

Department of Biochemistry and
*Department of Haematology, Royal Free Hospital
School of Medicine, University of London
Rowland Hill Street, London NW3 2PF, UK

Cell partitioning in aqueous two-phase systems formed by dextran and poly(ethylene glycol) (PEG) is a developing technique for cell fractionation using countercurrent distribution, which has been used to separate a wide range of cell types on the basis of surface properties such as charge, hydrophobicity and ligand binding [1,2]. Nevertheless, the method has lacked the specificity required for most biological separation problems. Recently, a dramatic increase in the selectivity of partitioning has been achieved by combining phase partitioning and affinity methods. Cells expressing antigen can be specifically pulled into the top PEG-rich phase of the system by treating the cells with the corresponding antibody coupled to PEG [3,4]. Coupling of PEG to antibodies has been accomplished by activation of the polymer with cyanuric chloride. The subsequent coupling of the activated polymer to the protein requires pH and ionic strength which are non-physiological. In addition, the PEG-antibody is isolated from the unreacted activated PEG before the incubation with the cell mixture.

We have developed a very simple method for the coupling of PEG to proteins under near physiological conditions [5]. Monomethoxy-PEG-5000 (MPEG-5000, Union Carbide, USA) is activated with tresyl chloride (2,2,2-trifluorethanesulfonyl chloride, Fluka FRG) essentially under the conditions described by Nilsson and Mosbach [6]. The coupling is then carried out in 0.05 M sodium phosphate buffer pH 7.5 containing 0.125 M sodium chloride (0.344 osmolar) and is achieved within 2 h of reaction at room temperature.

After establishing that the morphology and volume of human red blood cells (RBC) were not affected by such medium (30 min at 37°C) we have attempted to develop an immunoaffinity partitioning method without any re-isolation of the complex MPEG-immunoglobulin (MPEG-Ig).

As a model system we chose human RBC coated with human anti-Rh_0(D) and we used antihuman immunoglobulin coupled to MPEG as the immunoaffinity ligand.

Tresylated MPEG was coupled to antihuman immunoglobulin in the buffer described above by using a molar ratio MPEG:Lys groups of 20:1. Coupling of MPEG to the immunoglobulin was demonstrated by the increase in the partition coefficient of the protein in a dextran-PEG phase system (4.75% w/w Dx T-500, Pharmacia Sweden: 4.75% w/w PEG-6000, BDH, UK) in 0.15 M NaCl buffered with 0.01 M sodium phosphate, pH 6.8.

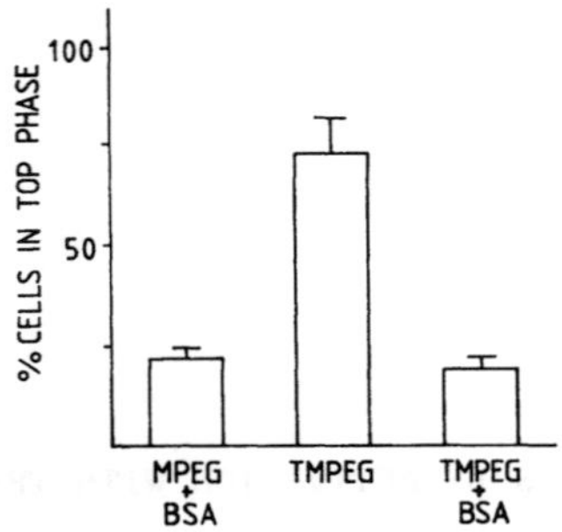

Fig. 1 *The linkage of tresylated PEG (TMPEG) to red cells can be prevented by the presence of albumin (BSA)*

We found that tresylated MPEG can be linked directly to the surface of the RBC during incubation with the result that the cells showed increased partition into the top phase (Fig. 1). Therefore, before introducing the cells in the medium containing the MPEG-modified immunoglobulin the excess tresylated MPEG had to be inactivated. This was attempted by adding BSA to the reaction medium. To demonstrate inactivation of tresylated MPEG, the partition coefficient of RBC introduced in the medium containing the polymer was compared with that of cells incubated in the presence of an equivalent amount of MPEG. Total inactivation of the excess tresylated MPEG was achieved by adding BSA to provide a molar ratio of lysine groups in BSA to tresylated MPEG of 4:1.

Immunoaffinity partitioning was then attempted by introducing cells coated with anti-Rh_o(D) into the medium containing MPEG-Ig in which excess tresylated MPEG had been inactivated with BSA. The concentrations of Ig ($F(ab')_2$ fragment) used were 0.12 and 1 mg/ml. The concentrations of cells were 5 x 10^7 and 8 x 10^6 cells/ml, respectively. After 30 min of incubation at 37°C cells were spun down and introduced in a two-phase system formed by 4.6% w/w PEG-6000 and 4.6% w/w Dextran T-500 prepared in 0.15 M NaCl buffered with 0.01M sodium phosphate, pH 6.8. In no case was the partition coefficient of the cells significantly increased with respect to the partition coefficient of the cells incubated in buffer. As the selected PEG-dextran two-phase system seemed to be suitable for affinity partitioning (the partition coefficient of the cells increased when these are directly linked to MPEG, see above), these results indicate that this procedure was unable to produce immunoaffinity partitioning where the expression of the antigen on the surface of the target cells was relatively weak (60,000 antigen sites/cell) [7].

REFERENCES

1 P.-Å. Albertsson, Partition of Cell Particles and Macromolecules, 3rd edition, Wiley, Interscience, New York (1985)

2 H. Walter, D.E. Brooks and D. Fisher, eds., in: "Partitioning in Aqueous Two-Phase Systems. Theory, Methods, Uses and Applications to Biotechnology," Academic Press, New York (1985)

3 L.J. Karr, S.G. Shafer, J.M. Harris, J.M. Van Alstine and R.S. Snyder, J. Chromatogr. 354:269 (1986)

4 K.A. Sharp, M. Yalpani, S.J. Howard and D.E. Brooks, Anal. Biochem. 154:110 (1986)

5 C. Delgado, G.E. Francis and D. Fisher, in preparation.
6 K. Nilsson and K. Mosbach, in: "Methods in Enzymology,", W.D. Jakoby, ed., Academic Press, New York (1984)
7 C. Bloy, D.B. Blanchard, P. Laubin, D. Goosseus, P. Rouger, C. Salmon and J.P. Cartron, Blood 69:1491 (1987)

Chapter 6

Theoretical Aspects of Phase Partitioning

AN OVERVIEW OF THEORETICAL DEVELOPMENTS FOR PROTEIN PARTITIONING IN AQUEOUS TWO-PHASE POLYMER SYSTEMS

Jesse N. Baskir and T. Alan Hatton

Department of Chemical Engineering
Massachusetts Institute of Technology
Cambridge, Massachusetts 02139, USA

Theories of protein partitioning in two-phase systems which account for the effects of different aspects of system composition - such as the choice of materials, protein size, polymer molecular weight, polymer concentration, salt concentration, and affinity ligands - are reviewed. Although the present theories provide some information about specific aspects of partitioning, a comprehensive and fundamental theory which can be used to predict protein partitioning behaviour has not yet been developed. Some recommendations for future work are given.

INTRODUCTION

The tendency for two different polymers in a common solvent to separate into two distinct phases is a well known phenomenon that has been demonstrated to be the general rule for most polymer/polymer/solvent systems [1]. The phases in such systems are made up primarily of solvent, with the remainder of the phase consisting mostly of one of the two phase forming polymers.

The discovery by Albertsson [1] in the late 1950's of biomaterial partitioning in two-phase systems made from aqueous solutions of dextran and poly(ethylene glycol) heralded the beginning of a new and powerful technique for the separation and analysis of biological particles. Over the last 3 decades many groups [1-3] have used variations of this technique for laboratory scale separations of proteins, cells, cell organelles, viruses, cell membrane fragments, and other biological materials. In addition, the technique has been applied by Albertsson [4], Walter [5], Zaslavsky [6] and others as a way of determining biomaterial surface properties such as charge and hydrophobicity.

The recent expansion of the biochemicals industry has greatly increased the demand for inexpensive large scale separations methods. Kula [7] has demonstrated the effectiveness of two-phase aqueous extraction for the large scale separation of several proteins. However, the lack of a fundamental understanding of how biomaterial partitioning depends on the system composition remains a major obstacle to the widespread industrial application of aqueous two-phase extraction. In this paper we will briefly describe the primary efforts which have been made to model and explain protein partitioning behaviour in aqueous two-phase polymer systems, and will recommend future directions for investigations into the theory of such partitioning. A more complete review of these theories will be published elsewhere [8].

Factors governing partition of proteins in aqueous two-phase systems

The partitioning of proteins in a two-phase system is described by the partition coefficient K, defined by

$$K = C_T/C_B \tag{1}$$

where C_T and C_B are the concentrations of the protein in the top and bottom phases, respectively. In general, the dependence of the partition coefficient can be divided into several roughly independent factors. Therefore, K may be written as [4]

$$\ln K = \ln K_{polymer} + \ln K_{salt} + \ln K_{ligand} + \ln K_{other} \tag{2}$$

where $K_{polymer}$, K_{salt}, K_{ligand} and K_{other} are the contributions to protein partitioning from the phase forming polymers, salts, affinity ligands, and other system factors. Before examining the theory behind protein partitioning in two-phase systems, it is helpful to first examine the empirical evidence regarding the effect of the different system parameters on protein partitioning.

Protein related factors

Although protein partitioning is a complex function of the system composition, a few generalisations can be made. Protein partitioning has been found to be more one-sided for large particles than for small ones. Small molecules, such as glucose, partition almost evenly between the two phases, whereas proteins have partition coefficients on the order of 0.1 or 10 [9]. Very large particles, such as cells, commonly partition between the interface and one of the two phases, with no cells found in the second phase [5].

Qualitatively, Bronsted [10,11] expressed the relationship between the particle partition coefficient K and the size of the particle as

$$K = \exp(\lambda M/kT) \tag{3}$$

where M is the molecular weight of the particle, and λ is a factor which depends in some form on the system composition. In general, this relationship has been shown to be true (see Fig. 1). Globular proteins, however, are usually tightly folded molecules whose internal structure is virtually inaccessible to the solution. Consequently, protein partitioning is related more directly to the protein's surface properties, in particular the protein surface area (which is a function of the protein's molecular weight) and surface charge [1,4,7-10].

Effect of the phase polymers

Protein partitioning also depends on the choice of phase-forming polymers, their molecular weight, and their concentration. The choice of phase forming polymer will determine the properties of the phase such as its surface tension and hydrophobicity, which will, in turn, affect protein partitioning [4-6,12]. In general, if the molecular weight of one of the phase polymers is increased, protein partitioning favours the other phase [1], although the effect approaches an asymptotic limit at high molecular weights. Protein partitioning also becomes more one-sided as the overall

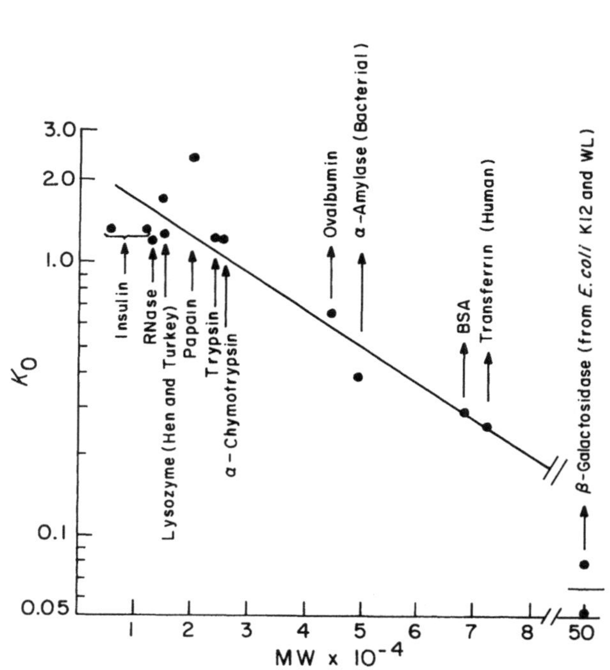

Fig. 1 *Relationship between K at the cross-point, K_o, and the molecular weight of some proteins. System: 7% Dextran 500, 4.4% PEG 8000, 0.1 M NaCl or 0.05 M Na_2SO_4, 10 mM phosphate or glycine buffer. Temperature, 20°C. Reprinted by permission from Sasakawa and Walter [12]*

concentration of the phase polymers in the system is increased (i.e. for systems further from the critical point of the phase diagram).

Salt effects

Numerous studies have indicated that protein partitioning is a complex function of the salts added to the system. At relatively low salt concentrations (<0.5 M), the influence of salts on protein partitioning is independent of the ionic strength of the salt and is related primarily to the salt composition in the system [13,14]. Many salts have been found to distribute unevenly between the two phases, due to their unequal affinity for the two phase-forming polymers [15]. This creates an electric potential difference of about 0-7 mV between the two phases in such systems [14]. The partition of proteins with a net charge has been found to depend on the magnitude of the protein charge and on the distribution potential between the two phases [14]. The partition of proteins at their isoelectric point is roughly independent of the salt composition of the system [16].

At salt concentrations above 0.5 M, two-phase systems can be produced from ternary mixtures of PEG, salt and water [1]. In such systems, one phase contains most of the polymer, and the salt concentration in the two phases can differ by as much as 1 M or more. Protein partitioning in such systems depends strongly on the salt ionic strength as well as the polymer concentration.

Affinity ligands

Although the phase polymers normally interact nonspecifically with biomaterials, ligands which bind to specific sites on a given protein can be attached covalently to one of the polymers to enhance the partitioning of the protein into that phase [17]. Only a few percent of the total polymer need be modified to optimally augment the separation. The enhancement of partitioning depends on the number of binding sites on the protein, and increases asymptotically with the concentration of polymer-ligands [13]. Partition also depends on the molecular weight of the polymer tail attached to the ligand [13].

Protein Partitioning Models: Polymer Effects

The virial expansion model. In a multi-phase system, the thermodynamic criterion for equilibrium is that, for all species, the chemical potentials of the given species are equal in all phases. If the chemical potentials are expressed in terms of the molal concentrations of the components in each phase, then the equilibrium concentrations of each component can be calculated. For ideal mixtures, the chemical potential is related simply to the natural logarithm of the mole fraction of the species. However, polymer-polymer two phase systems are highly non-ideal. To account for the non-idealities, Ogston [18] developed general thermodynamic expressions for the chemical potential of each species based on a virial expansion of the chemical potential in terms of the molalities of all components in the system. Edmond and Ogston [19] subsequently used forms of these virial expansions which contain only lower order terms to derive a system of equations for the phase equilibrium of PEG/dextran systems. The coefficients of the terms in the equations are related to the second virial coefficients for the components and to the interaction coefficients between components.

King et al [20] have recently shown that it is possible to determine these coefficients for given components from light scattering experiments. They have used these values to predict phase diagrams for dextran/PEG systems and are presently trying to extend this to predict protein partitioning in two-phase systems. The model gives the appropriate qualitative relationship between the partition coefficient and the polymer concentration, but does not account explicitly for the polymer molecular weight. However, the virial coefficients in the model can be related to the volume "excluded" by each component in solution [19,21] to give an effective hard sphere volume for each. This volume increases with the molecular weight of the material. Correlation of the virial coefficients with the effective molecular radii of dextran and PEG of various molecular weights has given reasonable agreement with independent measurements made from intrinsic viscosity experiments.

Excluded volume models based on the virial expansion have been used to study the precipitation of proteins from aqueous solutions of PEG. These models assume that the polymer "excludes" other components from a fraction of the water, thereby concentrating them in the remainder of the solution. The solubility of the protein is, therefore, a function of its concentration in the "free" solution, not on its overall concentration. Juckes [22] used these arguments to develop a general equation relating the solubility of the protein exponentially to the polymer concentration in solution. Although the model correctly predicts the linear relationship between the natural logarithm of the protein solubility and polymer concentration, it incorrectly predicts that the slope of the solubility curve should increase with increasing molecular weight, when, in fact, the opposite is observed [23].

The lattice model. The adsorption of polymer molecules to surfaces has long been of interest to people working in the area of colloidal stabilisation and flocculation. Recent lattice based theories of polymer adsorption to infinite flat plates have been developed which allow the calculation of the polymer segment density profile near the surface and the interfacial energy. The most sophisticated theories also provide detailed information concerning polymer chain conformations near the surface [24,25].

In the lattice model, each site in the lattice is occupied by either a solvent molecule or a polymer chain segment. Solvent molecules occupy only a single site, while the polymer chains are made up of a connected series of segments. Energetic interactions in the lattice occur only between unlike nearest-neighbour pairs. Using the lattice formalism, Scheutjens and Fleer [24,25] were able to develop expressions for the system energy (the sum of unlike nearest-neighbour interaction energies) and the combinatorial entropy of the polymer chains. From statistical mechanics, they used these expressions to determine the equilibrium set of polymer chain conformations in the system.

Baskir et al [26] have extended the theory of Scheutjens and Fleer to account for particles of finite dimensions and shapes. The system consists of a protein particle at the centre of a curved lattice whose shape is determined by the shape of the particle (see Fig. 2). The protein surface is assumed to be homogeneous, so that the energetic interaction of the polymer segments or solvent with the protein surface are the same at every point on the surface. Using this model, Baskir et al have been able to calculate the Gibbs' energy for proteins in aqeuous polymer solutions (and, therefore, the distribution coefficient for proteins in two-phase systems) as a function of the polymer chain length and concentration, the size and shape of the protein, and the polymer/water and polymer/protein interaction energies. All parameters for the model may be determined *a priori* except for the polymer/protein interaction energy, which thus far has been used as a semi-adjustable parameter. Baskir et al have shown that this interaction is slightly attractive and does not vary much for a given polymer with different proteins.

The model indicates that the phase-forming polymers influence protein partitioning through two primary mechanisms: energetic interactions between the polymer chain segments and the protein surface, and the entropic effect of steric exclusion of the polymer chains from near the protein. Calculations using this modified lattice model have demonstrated very good agreement with data on protein partitioning as a function of polymer molecular weight (see Fig. 3) and protein size. The model has also reproduced the correct trends and given good agreement for the solubility of proteins in aqueous PEG solutions as a function of polymer molecular weight and protein size [26].

Protein Partition Models: Salt Effects

Polymer/polymer two-phase systems (low salt concentration). The type of systems most frequently described in the literature are those in which the system consists of water, two phase polymers, and salts at an overall system concentration of roughly 0.1 - 0.5 M. In these systems the distribution of salts does not create a significant concentration gradient between the two phases [1]. If the salt ions have different affinities for the two phases, a small electrical potential, known as the distribution potential, is created. A general thermodynamic theory for the partition of charged particles in two phase systems with a distribution potential has been developed by Albertsson [1] and elaborated by Brooks [27]. In this development, the electrochemical potential, which contains an extra term that accounts for electrostatic interactions, is used, so that

$$\mu_p = \mu_p{}^\circ + RT\ln a_p + z_p F\Psi \qquad (4)$$

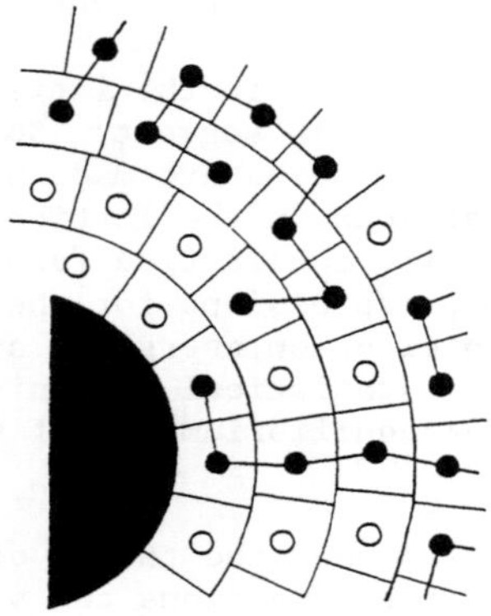

Fig. 2 *Lattice model for a spherical particle in an aqueous polymer solution. The lattice is shown here in two dimensions; however, the actual lattice would be three dimensional. Only part of the lattice is shown. From Baskir et al [26]*

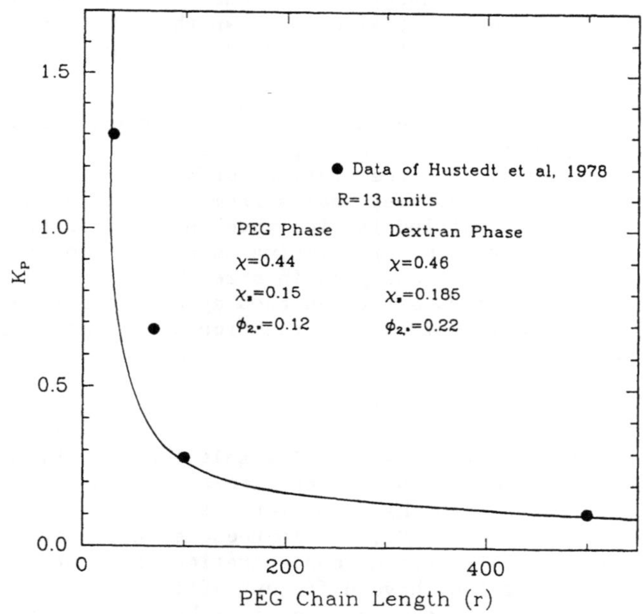

Fig. 3 *Distribution coefficient K_p for pullulanase in a 12% PEG 6000, 1% Dextran T500 biphasic system. Lattice model calculations (solid line) are compared to the data of H. Hustedt, K.H. Kroner, W. Stach and M.R. Kula, Biotech. Bioeng. 20:1989 (1978). From Baskir et al [26]*

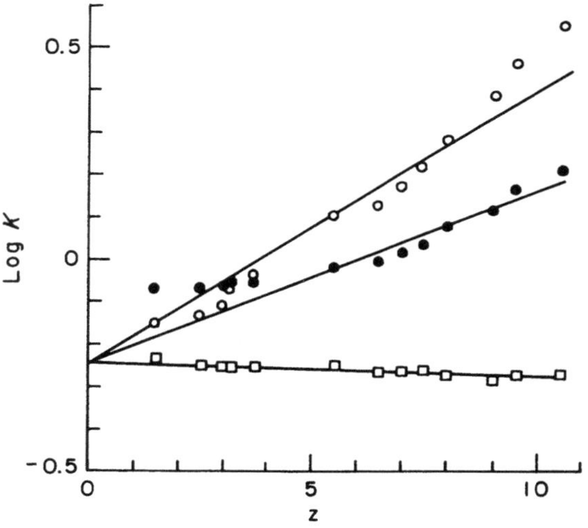

Fig. 4 *Relationship between partition coefficient, K, and net charge, z, of ribonuclease A. System: 8.4% Dextran 500, 5.8% PEG 8000. Temperature, 20°C. From Johansson [13]*

where μ_p is the electrochemical potential of the protein in the phase, $\mu_p{}^\circ$ is the standard state chemical potential for the protein, a_p is the activity of the protein in the phase, z_p is the net protein charge, F is the Faraday constant, and Ψ is the electrostatic potential of the phase. By setting the electrochemical potential for the protein in the two phases equal, the natural logarithm of the distribution coefficient is found to be

$$\ln K_p = \ln K_p{}^\circ + z_p F \Delta\Psi / RT \quad (5)$$

Provided the protein concentration is low compared to that of the salt, the distribution potential is essentially independent of protein partitioning.

A direct result of this theory is that, as long as the presence of the salt does not alter the polymer composition of the two phases, the partitioning of a protein in two identical sets of phase systems made using different salts should be equal when the net protein charge is zero. This fact has been used by Albertsson to estimate the isoelectric points of several proteins using "cross partitioning" [16].

The theory also indicates that protein partitioning should be very sensitive to protein charge. Although this has been seen experimentally, protein partitioning data do not show the expected linear relationship between $\ln K_p$ and z_p (see Fig. 4). Brooks has pointed out, however, that this relationship will only be true if $K_p{}^\circ$ is only a function of the polymer composition of the system, which is not always the case. Deligny and Gelsema [28] and Brooks [27] have developed an alternative method for determining the potential difference between the two phases by comparing protein partition in two sets of systems in which two different monovalent salts with a common cation are used.

Polymer/Salt Two-Phase Systems (High Salt Concentration)

When the two-phase systems are made from combinations of water, polymer and an inorganic salt, protein partitioning is strongly influenced by the salt concentration in each phase. A general theory for the solubility of proteins in aqueous salt solutions has been developed by Melander and Horvath [29]. The model describes the effect of the salt on both the electrostatic and hydrophobic interactions with the protein. The Gibbs' energy of the particle is calculated as the sum of electrostatic interactions of the protein with the salt and the energy required to create a cavity in the solution large enough to hold the protein. The electrostatic terms are related to the salt concentration through a combination of a Debye-Huckel term (appropriate at low salt concentrations) and a Kirkwood term (good at higher salt concentrations) [29]. The cavity term is the product of the surface area of the protein and the surface tension of the solution. For many inorganic salts, the surface tension of the solution is proportional to the salt concentration in the solution, the proportionality constant being dependent on the type of salt. Consequently, the Gibbs' energy of the protein can be written in terms of the molal concentration of the salt. The general form of the protein solubility equation derived from this theory agrees well with experiments, although the exact values of the coefficients must be derived empirically.

Kim [30] has developed a model of protein partitioning in PEG/salt systems which combines the virial expansion/exluded volume model for polymer/protein interactions with the salt/protein interaction model developed by Melander and Horvath. In Kim's model, the polymer in the polymer-rich phase is assumed to exclude the protein and salt from a fraction of the water in the phase, so that they are effectively concentrated in the remainder of the phase. The volume of the phase available to the protein and the salt is calculated as a function of the polymer concentration, and the protein solubility in the polymer phase is then calculated as a function of the effective salt concentration using the Melander and Horvath theory. The protein solubility in the salt phase is calculated from the Melander and Horvath theory without any excluded volume adjustment. The coefficients in the equation, which determine the molar excluded volume of the polymer for the salt and the protein and the salting out constant of the salt for the protein must be determined through solubility experiments.

Affinity Ligand Partitioning of Proteins

The specificity of protein separations in two-phase polymer systems can be enhanced by the covalent binding of affinity ligands to one of the phase polymers. Flanagan and Barondes [17] have developed a general thermodynamic model to describe the influence of polymer-ligands on partitioning. The distribution coefficient of the protein-ligand complex is related to the difference of the Gibbs' energy of the complex in the two phases. The Gibbs' energy difference can be broken down into separate parts by visualising the protein transfer between the two phases as a multistep process. First, the protein-ligand complex in one phase dissociates into a protein and a polymer-ligand molecule. Each molecule then transfers into the second phase, and, finally, the protein-ligand complex reforms in the second phase. The distribution coefficient can, therefore, be expressed in terms of the distribution coefficients for the free protein and free polymer-ligand, and the equilibirum association constants for the protein and polymer ligand in each of the phases. If the protein contains N independent and equivalent binding sites, then the distribution coefficient of the protein is exponentially related to the number of binding sites.

Experimentally, it has been found that the dependence of the protein partitioning on the number of binding sites is less than that predicted by the theory of Flanagan and Barondes. This theory assumes that ligand binding is independent of the phase composition and the number of previously bound ligands. In general, it cannot be assumed that the binding constants in the two phases are equal [27]. Neither can it be assumed that binding of multiple ligand molecules to a protein are independent and equal since ligand binding can alter the protein conformation. In addition, the long polymer tails of bound polymer-ligands interfere with each other sterically. No theories have been developed yet which explicity account for the dependence of protein partition or ligand binding on the length of the polymer tail.

Future Directions

Several areas remain for further investigation. First, the modelling of protein-polymer interactions at present still relies on a highly idealised picture of the protein molecule. A model which provides a more realistic description of the protein surface is needed. As a first approximation, the difference between "hydrophobic" and "hydrophilic" regions of the protein surface should be taken into account. Such an approach may be possible using a suitable modification of the lattice model, or may require the use of Monte Carlo simulation.

In addition, further work is needed to elucidate the effects of affinity ligands on protein partitioning. At the present time, the theory is unable to account explicitly for the influence of the length of the polymer tail attached to the ligand, particularly for the case where several ligands are bound to the biomaterial. A molecular level understanding of the distribution of the polymer tails is needed to clarify how the polymer tail affects ligand binding to the protein and whether polymer-ligands interfere with each other. This problem may be suitable for lattice modelling. Models of polymer chains attached to a surface by a terminal end have been developed [24,25], although these treat only pure polymer phases, or isolated chains. Such models should be able to provide detailed information regarding the conformation of polymer tails of polymer-ligands bound to proteins.

Finally, a model for polymer-ion interactions is needed both to help determine salt partition in two-phase polymer systems, and to allow prediction of electrostatic potentials generated by the addition of different salts. At present, the virial expansion models appear the most promising for this area. However, such models shed little light on the mechanisms that determine differences in polymer-ion interactions. Models which account explicitly for electrostatic and dipole interactions between salts, the phase polymers, and water are desireable. A more complete understanding of all of these factors will be critical to the future use of this technique for laboratory and commercial applications.

ACKNOWLEDGEMENT

The authors would like to thank the Biotechnology Process Engineering Center at M.I.T. for the generous support of this work.

REFERENCES

1 P.-Å. Albertsson, "Partition of Cell Particles and Macromolecules", Wiley, New York (1971)
2 D. Fisher, The separation of cells and organelles by partitioning in two-polymer aqueous phases, Biochem. J. 196:1 (1981)
3 H. Walter, D.E. Brooks and D. Fisher, eds., "Partitioning in Aqueous Two-Phase Systems," Academic Press, New York (1985)

4 P.-Å. Albertsson, Partition between polymer phases, J. Chromatogr. 159 :111 (1978)
5 H. Walter, Partition of cells in two-polymer aqueous phases: A surface affinity method for cell separation, Methods Cell Separation 1:307 (1977)
6 B.Y. Zaslavsky, N.M. Mestechkina and S.V. Rogozhin, Characteristics of protein-aqeuous medium interactions measured by partition in aqueous ficoll-dextran biphasic system, J. Chromatogr. 260:329 (1983)
7 M.R. Kula, K.H. Kroner and H. Hustedt, Purification of enzymes by liquid-liquid extraction, Adv. Biochem. Eng. 24:73 (1982)
8 J.N. Baskir, T.A. Hatton and U.W. Suter, in preparation
9 P.-Å. Albertsson, Interaction between biomolecules studied by phase partition, Methods Biochem. Anal. 29:1 (1983)
10 J.N. Bronsted, Z. phys. Chem., A (Bodenstein-Festband):257 (1931)
11 J.N. Bronsted and E. Warming, Z. phys. Chem., A 155:343 (1931)
12 S. Sasakawa and H. Walter, Partition behaviour of native proteins in aqueous dextran-poly(ethylene glycol)-phase systems, Biochemistry. 14:2760 (1972)
13 G. Johansson, Partitioning of proteins, in: "Partitioning in Aqueous Two-Phase Systems," H. Walter, D.E. Brooks and D. Fisher, eds., Academic Press, New York (1985)
14 G. Johansson, Effects of salts on the partition of proteins in aqueous polymeric biphasic systems, Acta Chem. Scand. B 28:873 (1974)
15 G. Johansson, Partition of salts and their effects on the partition of proteins in a dextran-poly(ethylene glycol)-water two-phase system, Biochim. Biophys. Acta 221:387 (1970)
16 P.-Å. Albertsson, Cross partition and isoelectric points of proteins, Nature 228:1329 (1970)
17 S.D. Flanagan and S.H. Barondes, Affinity partitioning: A method for purification of proteins using specific polymer-ligands in aqueous polymer two-phase systems, J. Biol. Chem. 250:1484 (1975)
18 A.G. Ogston, Some thermodynamic relationships in ternary systems, with special reference to the properties of systems containing hyaluronic acid and protein, Arch. Biochem. Biophys. Suppl. no. 1:39 (1962)
19 E. Edmond and A.G. Ogston, An approach to the study of phase separation in ternary aqueous systems, Biochem. J. 109:569 (1968)
20 R. King, H.W. Blanch and J.W. Prausnitz, Thermodynamics of aqueous polymer-polymer two-phase systems, presented at ACS National Convention in Anaheim, California, September 10 (1986)
21 A.G. Ogston and C.F. Phelps, The partition of solutes between buffer solutions and solutions containing hyaluronic acid, Biochem. J. 78:827 (1960)
22 I.R.M. Juckes, Fractionation of proteins and viruses with poly(ethylene glycol), Biochim. Biophys. Acta 229:535 (1971)
23 D.H. Atha and K.C. Ingham, Mechanism of precipitation of proteins by poly(ethylene glycol)s: Analysis in terms of excluded volumes, J. Biol. Chem. 256:12108 (1981)
24 J.M.H.M. Scheutjens and G.J. Fleer, Statistical theory of the adsorption of interacting chain molecules. 1. Partition function, segment density distribution, and adsorption isotherms, J. Phys. Chem. 83:1619 (1979)
25 J.M.H.M. Scheutjens and G.J. Fleer, Statistical theory of the adsorption of interacting chain molecules. 2. Train, loop and tail size distribution, J. Phys. Chem. 84:178 (1980)
26 J.N. Baskir, T.A. Hatton and U.W. Suter, Thermodynamics of the partition of biomaterials in two phase aqueous polymer systems: Effect of the phase forming polymers, Macromol. 20:1300 (1987)
27 D.E. Brooks, K.A. Sharp and D. Fisher, Theoretical aspects of partitioning, in: "Partitioning in Aqueous Two-Phase Systems,"

H. Walter, D.E. Brooks and D. Fisher, eds., Academic Press, New York (1985)
28 C.L. DeLigny and W.J. Gelsema, On the influence of pH and salt composition on the partition of polyelectrolytes in aqueous polymer two-phase systems, Sep. Sci. Technol. 17:200 (1982)
29 W. Melander and C. Horvath, Salt effects on hydrophobic interactions in the precipitation and chromatography of proteins: An interpretation of the lyotropic series, Arch. Biochem. Biophys. 183:200 (1977)
30 C.W. Kim, Interfacial condensation of biologicals in aqueous two phase systems, PhD Thesis, Massachusetts Institute of Technology, Cambridge, Massachusetts

PROTEIN TRANSPORT PROCESSES IN THE WATER-WATER INTERFACE IN INCOMPATIBLE TWO PHASE SYSTEMS

M.P. Tombs and S.E. Harding

University of Nottingham
Department of Applied Biochemistry and Food Science
Sutton Bonington, Loughborough, UK

Proteins, such as ovalbumin, albumin, cytochrome C or a chromobacter lipase were added to either the upper or lower phase of a PEG 6000, Dextran T500 system, and the phases placed carefully in contact. Diffusion processes and interfacial accumulation were then observed by using the optical scanning system of an MSE Centriscan ultracentrifuge run at low speeds. The overall process was one of simple diffusion, though in such complex systems movement up the protein concentration gradient can occur and was observed. Interfacial accumulation roughly in accord with expectation based on a simple interfacial tension theory was also seen.

INTRODUCTION

The phenomenon of incompatible phase separation is known and largely understood. Tompa for example gave a full theoretical description in 1956 [1].

Since then Ogston and coworkers over a long period applied the concepts of coexclusion to biological macromolecules, particularly hyaluronic acid and related carbohydrates. This resulted in theoretical developments [2,3] and an awareness of the role that coexclusion can play in the functional properties of biological macromolecules [e.g. 4].

More recently it led to an interesting insight into the way in which coexclusion can affect the diffusion of macromolecules in matrices [5,6]. Rather unexpectedly diffusion can be greater in the presence of a matrix than in its absence in properly chosen conditions [7].

In a distinct line of investigation, Albertsson and coworkers [8] used multiple aqueous phase systems for the purification and isolation of both macromolecules and cell fragments.

Despite all this there has been almost no work on the events at the water-water interface. This is surprising since it would appear to offer excellent opportunities to model biologically significant systems.

While it is unlikely that any biological situation occurs where two phases are produced by coexclusion (with the possible exception of mobilised seed protein bodies[9]), membranes almost always separate two aqueous phases. Thus the water-water interface provides a locus, at thermodynamic equilibrium, in which synthetic membranes could be examined.

Numerous processes of biotechnological interest involve the movement of proteins towards, into and through matrices. The production of immobilised enzymes, biosensors, and the use of multiple phase separation itself all involve a diffusion step. Moreover, there is often a matrix, or at least a concentrated solution to be penetrated.

There are few previous studies bearing on this. England and Nazarian [10] measured the diffusion rate of ribonuclease and concluded that the interface was freely penetrated near the critical point composition.

Shanbhag [11] made some transfer rate measurement on proteins in PEG-dextran systems in stirred cells, and found a relationship between partition coefficients and transfer rate. Meares [12] has recently analysed interphase diffusion in terms of stagnant layers. Wells [6] and Laurent et al [7] have developed ideas about diffusion to account for events at unstable layers, while Linde [13] has considered two immiscible fluids.

EXPERIMENTAL

We measured the concentration profile by using the scanning optical system of an MSE Centriscan ultracentrifuge. We loaded 0.22 ml of lower phase followed by the same volume of top phase into the 10 mm cell. At the position of the interface the cross sectional area was 25.9 mm^2. The rotor was run at 4000 rpm at 25° which should produce no significant sedimentation. Appropriate blanks were included, mainly the two phases with no added protein.

Most experiments were done in the two phases produced by mixing 5 g of PEG 6000 (BDH Biochemical grade, with low 280 nm absorbance) 5.3 g of Dextran T500 (Pharmacia) with 100 ml of 50 mm sodium phosphate pH 7. After equilibration phases were separated and protein added to the appropriate phase. Proteins were obtained from Sigma, FITC derivatives from Molecular Probes Ltd, except lipases and FITC lipases, which were prepared as described elsewhere [16].

Surface tension measurement (we thank Dr J. Mead) by the spinning drop method was $0.09 \pm 0.03 \times \Delta\rho$ (about 0.03) so that the interfacial energy was 0.0027 - 0.0036 erg. cm^{-2}.

RESULTS

Transient water fluxes

Wells [6] has recently proposed a theory to describe disequilibria due to diffusion in moderately concentrated polymer solutions. Although this was intended for layered solutions, it can also be applied to quaternary two-phase systems.

Considerable transient water flux would be expected and, as is shown in Fig. 1, complex steps in the concentration gradients formed. These gradually

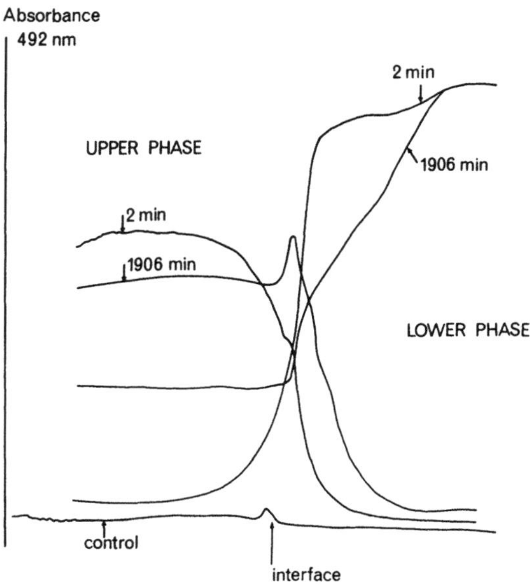

Fig. 1 *Concentration profiles measured in three cells run simultaneously. The lowest, control trace, contained no protein. The others are for fluorescein labelled bovine serum albumin at 1 mg/ml in either the upper or lower phases*

disappeared, and after about 3 days equilibrium was reached, and it was possible to assess the magnitude of the interfacial accumulation.

The initial transient fluxes were asymmetric, and were greater when protein was initially in the lower phase than when it was in the upper phase. Predialysis almost completely eliminated them: in this, both phases, containing protein were inside the sac, and both phases with no protein were outside the sac. After dialysis to equilibrium, which would result in a small hydrostatic pressure difference, the protein-containing phases from inside the sac were placed in contact with the corresponding phase from outside.

Simplification of the interfacial accumulation theory described by Albertsson [8] leads to a relationship

$$\frac{C_i}{C_1} = e^{6.8\,\tau_{12}}$$

where C_i is the interfacial concentration, C_1 the bulk phase concentration at equilibrium, and τ_{12} the interfacial tension in the absence of protein.

This predicts values of 3-4% for the interfacial excess concentration, as found for albumin and ovalbumin. Cytochrome C and the lipase gave higher values, but of the correct order of magnitude.

CONCLUSIONS

Interfacial turbulence and accumulation are important in the practical application of phase separation, particularly where it is hoped to separate the phases without the use of centrifuges. These results show that very complex initial transient fluxes occur and have implication for the correct choice of phase and systems.

REFERENCES

1 H. Tompa, Polymer solutions, Butterworths, London (1956)
2 E. Edmond and A.G. Ogston, Biochem. J. 109:569 (1968)
3 A.G. Ogston, B.N. Preston and J.D. Wells, Proc. Roy. Soc. A. 333:297 (1973)
4 T.C. Laurent, Chemistry and physiology of polysaccharides, Proc. Milan Symp. Churchill, London (1968)
5 A.G. Ogston, J. Phys. Chem. 74:668 (1970)
6 J.D. Wells, J. Phys. Chem. 90:2433 (1968)
7 T.C. Laurent, B.N. Preston and L.O. Sundeloff, Nature 279:60 (1979)
8 P.-Å. Albertsson, Partition of Cell Particles and Macromolecules, 3rd edition, Wiley, New York (1985)
9 M.P. Tombs, B. Newsom and P. Wilding, Int. J. Prot. and Peptide Res. 6:253 (1974)
10 J. England and G.M. Nazarian, Biochemistry 9:1553 (1970)
11 V.P. Shanbhag, Biochem. Biophys. Acta 320:517 (1973)
12 P. Meares, Farad. Discussions Chem. Soc. 77:7 (1984)
13 H. Linde, Farad. Discussions Chem. Soc. 77:181 (1984)
14 T.C. Laurent and B. Preston, J. Phys. Chem. 87:648 (1983)
15 J.D. Wells, K. Edsman, T.C. Laurent and L.O. Sundeloff, J. Phys. Chem. 90:2425 (1986)
16 G.A. Roberts and M.P. Tombs, Biochem. Biophys. Acta 902:327 (1987)

WATER: THE THIRD COMPONENT IN POLYMER TWO-PHASE SYSTEMS

T.E. Treffry [1], T.H. Lilley [2] and P.J. Cheek [2]

Biochemistry [1] and Chemistry Departments [2]
University of Sheffield, Sheffield S10 2TN, UK

Phase systems which are close to the critical point can be changed from biphase to monophase by increasing temperature or adding urea or water. Lower temperature or tert. Butanol have the opposite effect. These observations imply that water structure plays an important part in phase separation. Flow calorimetry of dextran and PEG solutions shows that these polymers are strongly solvated and that solute:solvent:solvent interactions are predominant in these systems. In mixing experiments the critical point appears thermochemically trivial and it is suggested that like polymer hydates form micro-phases at concentrations below the critical point.

The standard approach to an understanding phase formation by polymers in solution is the statistical mechanical treatment produced by Flory and Huggins. This analysis, originally developed to explain anomalous vapour pressures over a polymer in solution in its monomer, leads Brooks et al [1] and Sjöberg et al [2] to conclude that polymer:solvent interactions are not important in determining phase separation and that only polymer:polymer interactions are involved. We believe that this conclusion is unsatisfactory in that it neglects the unique properties of water as a solvent and is moreover contrary to a great deal of evidence from studies of mono- and oligosaccharides. A variety of experimental observations and some new, preliminary, thermodynamic approaches are presented in which phase formation is discussed in terms of competitive solvation, water structure and the optimisation of hydrogen bonding in the system as a whole.

A number of PEG6000/Dextran T500 systems were prepared at compositions close to the binodal. These "poised" systems can be monophasic or biphasic depending on experimental conditions. Equal weights of top and bottom phase from a 4% PEG : 5% dextran system were taken and diluted with water at room temperature until monophasic, i.e. they remain optically clear when vortexed. For a 13 g system this required 1.2 ml water or 145 mg urea (0.9 M), or 116 mg guanidine hydrochloride (0.46 M), but guanidine thiocyanate, up to 1.3 M was ineffective. Conversely, the system made monophasic by dilution becomes biphasic on the addition of TBA (tert. butanol) or lower molecular weight PEGs. The effects of urea and water in producing a mono-phase appear additive.

In a different type of experiment (Fig. 1), a 3% PEG 5.5% dextran system was used. Equal aliquots (0.5 ml) of phases equilibrated at 4°C were found to show a transition from monophase to biphase at 12°C. Addition of 10 μl

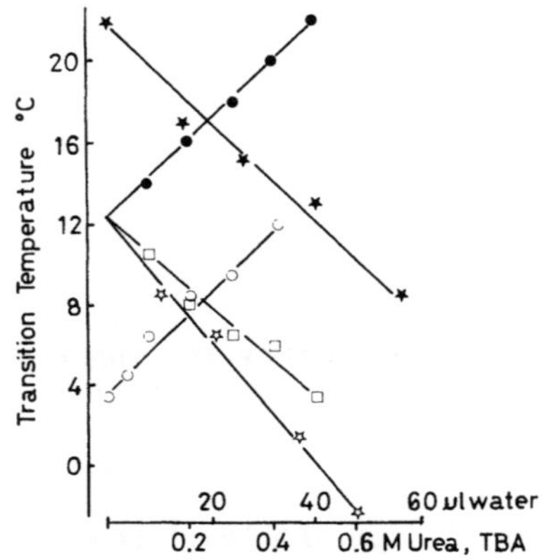

Fig. 1 *The effect of additions of water (□), urea (✶) or tert. butanol (TBA) (●) to 1.0 ml of a phase system with a transition temperature of 12.5°C and of TBA* _after_ *final addition of water (○), and of urea after TBA (☆)*

TBA raised this temperature to 14°C. Urea on the other hand reduced the transition temperature; 3 mg to 10°C, 20 mg to 5.5°C. This latter result was mimicked by 40 µl water and the further addition of 20 µl TBA raised the transition temperature 10°C. Some interesting conclusions can be drawn from these data. Adding 0.33 mmol urea is "equivalent" to adding 2 mmol water; adding 0.2 mmol TBA seems to "withdraw" this water.

We have also investigated the enthalpies of mixing PEG and dextran solutions using a flow micro-calorimeter [3] at concentrations where phase

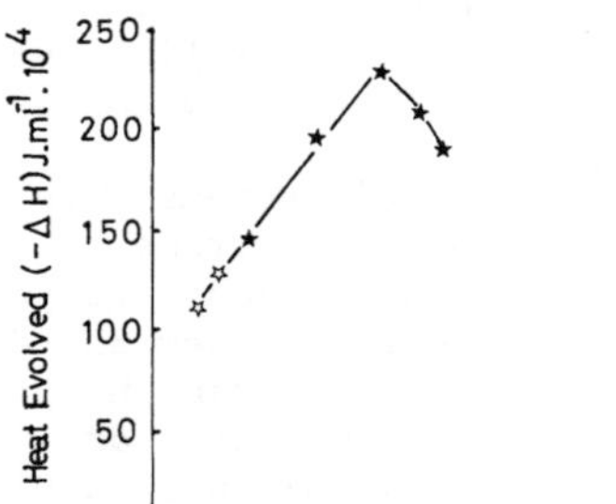

Fig. 2 *Heat evolved in a flow micro-calorimetry experiment mixing 8% PEG6000 and 12% Dextran T500 as indicated*
Monophase (☆) Biphase (✶)

Table 1 *Enthalpic virial coefficients (h_{pp}) for the homotactic interaction (per sub-unit) of PEG in water at 25°C*

Mol. Wt. $x10^{-3}$	20	12.5	8.8	3.4	1.0	0.46	0.2
h_{pp} $J.Kg.mol^{-2}$	44	49	56	67	100	128	156

separation occurs and when it does not. In this calorimeter two solutions can be mixed at a controlled mixing ratio. In an experiment using 8% PEG and 12% dextran at 25°C we start with a mixing ratio of 1 to 4 where these solutions are miscible. As the proportion of PEG is increased (Fig. 2) the heat of mixing increases steadily. There is no thermal discontinuity associated with the monophase-biphase transition but a maximum exothermy is reached at a value corresponding to 5.3%:PEG 4% dextran where reference to a phase diagram prepared for the polymers used in these experiments [4] indicates the furthest departure from the binodal. The most interesting feature of this experiment is that phase separation appears to be thermochemically trivial.

In separate experiments it was shown that the dilution of dextran is slightly exothermic but the values obtained were close to the limits of detection for the apparatus used. The enthalpies of dilution of a range of PEGs with different molecular weights were also determined [5] and are shown as the enthalpic terms for subunit:subunit interactions [6] (Table 1). From these values coefficients for homotactic [7] interactions for PEG6000 in water were interpolated. Values for the homotactic and heterotactic [7] interactions (Table 2) of dextran and PEG were calculated from mixing experiments using 5% solutions of each polymer which are miscible in all combinations.

Descriptions of phase separation in aqueous polymer two-phase systems based on Flory-Huggins have been praised for their utility [1,2] but I (T.T.) have never found them intellectually satisfying. These descriptions conjure up a vision of giant polymer molecules dominating a system in which the tiny solvent molecules are reduced to an entirely passive role. The polymers are seen as tumbling in a void interacting only with each other. This construct entirely disregards the unique properties of water as a solvent arising from its ability to form H-bonds both with itself and solute molecules. It does not explain our data using poised systems. That the solvent plays an important role in the system is indicated by our enthalpic measurements. The qualitative conclusion that can be drawn from our data is that what might be called "thermochemical repulsion" takes place between these solutes in water. This is in accord with the fairly extensive information which is available on saccharidic molecules in water [7] which has been interpreted both from a qualitative [8], and rigorously quantitative [9] viewpoint in terms of extensive solvation of the solutes. It has been shown recently [9] that the occurrence of a positive enthalpic virial coefficient (which corresponds to an exothermic enthalpy of dilution) necessarily means that solvent:solute interactions are dominant and that they prevail over solute:solute interactions. In a formal sense the enthalpic coefficient represents the balance between these two components and can be expressed as:

$$h_{AB} = X_{AB} - (X_{AW} + X_{BW}) \tag{1}$$

Table 2 *Enthalpic virial coefficients, h, for the interactions indicated by the subscripts. D, Dextran. P, PEG*

	h Polymer $J.Kg.mol^{-2}$	h Sub-unit $J.Kg.mol^{-2}$
h_{DD}	5.0×10^{-9}	640
h_{PP}	1.1×10^{-6}	60
h_{DP}	4.8×10^{-6}	13

where X is the enthalpic term for the interactions of the polymers (A, B) and water (W) as indicated. Because interactions do occur, each will be negative, but clearly the enthalpic change observed must be positive, if the parenthetical term is more negative than the first X term.

At a molecular level the explanation for this [8,9] seems to be that, at least for saccharides, the solute molecules can be accommodation within the water structure and are thus in a "comfortable" environment and prefer to be surrounded by water molecules rather than to interact with each other. For the PEG systems the picture may be somewhat different in that we envisage relatively strong directional H-bonds between the ether linkages and the water molecules and at least vestigial hydrophobic-solvation about the apolar groups.

The systems we are describing are of course complicated and almost certainly contain a variety of components but, having established that solvation is a major factor, an explanation for the behaviour of aqueous-polymer two-phase systems which initially appealed to us was based on competitive hydration.

$$\text{"Polymeric" Water} \underset{B}{\overset{A}{\rightleftharpoons}} \text{"Monomeric" Water} \qquad (2)$$

"Polymeric" water includes structured water and water in polymer hydrates. Higher temperature, urea and dilution are "structure breakers" A. Lower temperature, TBA or added polymer are "structure makers" B.

PEG appears to bind two water molecules per repeating unit [10] and dextran appears to bind six [8]. As solute concentrations are increased, monomeric water becomes limiting, like molecules are attracted because their bound water has the same steric arrangement and can be "shared". Urea or increased temperature would promote a 2-phase to 1-phase transition by making more water "available" (favouring A). Conversely TBA or lower temperature promotes phase separation (favouring B). In a 100 g system, 5 g PEG and 5 g dextran could between them bind up to 6 g water. In such a system, which is bi-phasic, there is still a lot of water unaccounted for. It is tempting to think of water as a third polymer but that approach is difficult to sustain and still leaves our results with TBA and our heat of mixing experiments unexplained.

The molecular structure of urea is such that it can participate in a maximum of six direct hydrogen bonds but as these lie more or less in the same plane it is very disruptive of water structure. Quantitatively the relationship noted between the addition of urea or water to a poised system is remarkably close in terms of the numbers of water molecules involved. TBA in solution is surrounded by a water clathrate of 30 molecules but there is evidence that these clathrates associate to involve fewer, ca 9, water molecules per TBA molecule [11]. These findings refer to the amine rather than the alcohol but its behaviour, which again matches our observations (Fig. 1), would not be expected to differ.

The indication that TBA:water clathrates are close-packed rather than scattered suggests a more productive version of proposition 2 based, not on competitive hydration, but on the maximisation of the number of undistorted H-bonds in the system.

$$\text{More H-bonds} \underset{B}{\overset{A}{\rightleftharpoons}} \text{Fewer H-bonds} \qquad (3)$$

In the general case we posit 3 states of water; bonded water, that close to the polymer molecule ("the polymer hydrate") but distinguishable from the bulk water, and the bulk water. Each polymer makes hydrogen bonds with neighbouring water molecules and in doing so can distort the water structure for a considerable distance. The nature of this perturbation will vary with the steric arrangement of the bonding groups on the polymer.

Where one of the polymers is dextran which rather than perturbing may stabilise water structure, there are only two states of water; that associated with PEG and the remainder. This would explain why in dextran:PEG two phase systems very little of the dextran is found in the PEG-rich phase. A much higher proportion of the PEG is found in the dextran-rich phase.

It is suggested that the system is arranged to minimise discontinuities which limit bond formation. Below the critical point there is bulk water between unlike polymer hydrates and there is no sharp discontinuity between them. As polymer concentration is increased bulk water is subsumed into the

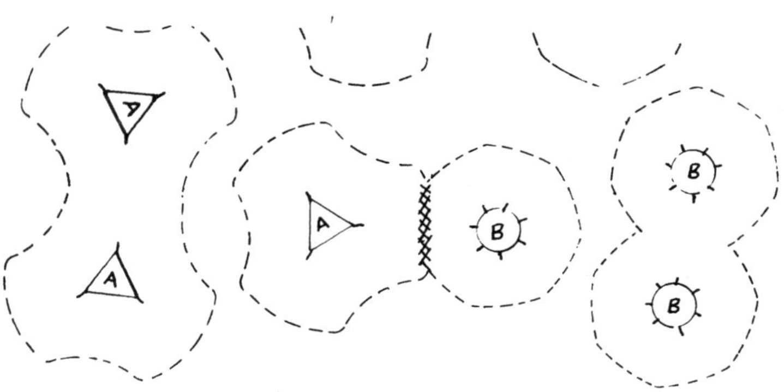

Fig. 3 *A and B represent polymer molecules surrounded by shells of solvent molecules structured according to the steric arrangement of bonding sites on the polymer which may or may not match structure in the bulk solvent. Like shells may merge producing micro aggregates. When unlike shells are forced together structural discontinuity leads to turbidity and phase separation*

polymer hydrates and unlike hydrates form separate phases to minimise H-bond distortion. This transition is dramatic to the eye but not to the calorimeter. An important inference from this is that even at low polymer concentrations aggregations of like polymer hydrates are favoured (Fig. 3). Phase separation is merely the point where the boundaries of pre-existing microphases sharpen (producing turbidity) and coalesce to form macrophases. Such factors as temperature, urea and TBA, or, more generally, structure breakers and structure makers, modify the conditions under which phase separation occurs through their effects on the H-bonding propensities of the system as a whole. Johansson [12] has also suggested, on the basis of light scattering measurements, the existence of micro-aggregates below the critical point.

Our present tentative hypothesis fits the data obtained thus far and makes two testable assertions; that H-bonds within the solvent and between solvent and solute control the behaviour of aqueous polymer two-phase systems, and, more remarkably, that phase separation is the coming together of pre-existing micro-aggregates rather than initiation of the aggregation of like polymers. It also appears more satisfying than approaches derived from the classic work of Flory and Huggins which was initially applied to entirely different systems where the atomic constitution of the polymer and the solvent were similar. We do not presume that our analysis emphasising the primacy of polymer solvation and nett hydrogen bonding is definitive, but we do suggest that any detailed attempt to describe the behaviour of aqueous polymer systems must consider the properties of each component of the system. Descriptions which neglect the solvent are seriously flawed.

REFERENCES

1 D.E. Brooks, K.A. Sharp and D. Fisher. Theoretical aspects of partitioning, in: "Partitioning in Aqueous Two-Phase Systems," H. Walter, D.E. Brooks and D. Fisher, eds., Academic Press, London (1985)
2 A Sjöberg, H. Wennerstrom, F. Tjerneld, Microorganisms in aqueous two-phase systems, in: "Microbial Physiology for Biotechnological Innovation," IVA Rapport 330, Swedish Academy of Engineering Sciences, Stockholm (1987)
3 P.J. Cheek, K.G. Davis and T.H. Lilley, J. Chem. Soc. Faraday Trans. in press
4 T. Treffry and B. Sheth, in: "Separations Using Aqueous Phase Systems: Applications in Cell Biology and Biotechnology," D. Fisher and I.A. Sutherland, eds., Plenum, New York (1989)
5 S.H. Gaffney, E. Haslam, T.H. Lilley and T.R. Ward, Homotactic and heterotactic interactions in aqueous solutions containing some saccharides. Experimental results and an empirical relationship between saccharide solvation and solute-solute interactions, J. Chem. Soc. Faraday Trans. 1:84 (1988), in press
6 G.M. Blackburn, T.H. Lilley and E.W. Walmsley, J. Chem. Soc. Faraday Trans. 76:915 (1980)
7 D.A. Cesaro, in: "Thermodynamics for Biochemistry and Biotechnology," H.J. Hiaz, ed., Springer Verlag, Berlin (1986)
8 S.H. Gaffney, E. Haslam, T.H. Lilley and T.R. Ward, Abs. L604, 8th Int. Symp. on Solute-Solute-Solvent Interactions, Regensburg (1987)
9 T.H. Lilley, Chapter 1, in: "Biochemical Thermodynamics," M.N. Jones, ed., Elsevier (1987)
10 P. Molyneux, "Water Soluble Synthetic Polymers: Properties and Behaviour," CRC Press, Boca Raton (1981)
11 D.W. Davidson, Clathrate hydrates, in: "Water: a Comprehensive Treatise," F. Franks, ed., Plenum, New York & London (1973)
12 G. Johansson, unpublished observations after presentation of this paper

CONTACT ANGLES AS AN ANALYTICAL TOOL FOR INVESTIGATING TWO-PHASE INTERACTIONS WITH BIOLOGICAL SURFACES: A REVIEW

John F. Boyce [1] and Donald E. Brooks [1,2]

[1] Department of Pathology, The University of British Columbia
2211 Wesbrook Mall, Vancouver, B.C., Canada V6T 1W5

[2] Department of Chemistry, The University of British Columbia
Vancouver, B.C., Canada V6T 1W5

The contact angle approach offers several advantages for the interpretation of aqueous two-phase partition behaviour, and as an implement for investigating the physico-chemical nature of the cell surface. The development and evolution of the technique, and its applications are reviewed here.

INTRODUCTION AND THEORY

Over the last twenty five years aqueous two-phase partition has proved to be a useful and sensitive method for separating biological cells and particles. While the technique is good at detecting subtle differences between cell surfaces, its utility is lost when details concerning the chemical and physical attributes of the cell surface are sought. Aqueous two-phase partition of larger particles is limited from an analytical perspective because it is a dynamic, non-equilibrium process complicated by the effects of phase convection and particle sedimentation.

For molecules and tiny particles of molecular dimensions, the assumption that the distribution will be driven by random thermal motion is not unreasonable. The partition coefficient, K (concentration of material in the upper phase)/(concentration of material in the bottom phase) may then be determined by the standard state Helmholtz free energy difference, $\Delta F°$, according to the Boltzman distribution:

$$K = \exp\ (\Delta F° \ /\ k\ T) \qquad (1)$$

The work required to transfer a particle from the bottom phase to the top phase of the system will simply be the product of the particle surface area (A) and the interfacial free energy difference for the particle in both phases, i.e.:

$$\Delta F° = F°_{Final} - F°_{Initial} = A\ (\gamma_{P,T} - \gamma_{P,B}) \qquad (2)$$

where: γ is the interfacial free energy and the subscripts P, T and B represent the particle, top and bottom phases respectively. This expression immediately suggests that the equilibrium of a small particle between two liquid phases is thermodynamically identical to spreading a droplet of one phase in the presence of another over a substrate composed of the same material as the particle.

As illustrated in Fig. 1, the forces acting on the three-phase contact line at the surface can be resolved to yield an equation named after Thomas Young who described it qualitatively in 1805 [1]:

$$\gamma_{P,B} - \gamma_{P,T} = \Delta\gamma = -\gamma_{T,B} \cdot \cos\theta \qquad (3)$$

where $\Delta\gamma$ is the interfacial free energy difference and is the contact angle. If Eq. (3) and Eq. (2) are substituted into the original expression for K the close relationship between the contact angle and partition coefficient is revealed:

$$K = \exp(-A \cdot \gamma_{T,B} \cdot \cos\theta / kT) \qquad (4)$$

This expression describes a molecular partition between the top and bottom phases where the contact angle will measure zero for a lower phase partition and 180° for a partition into the upper phase.

However, for larger particles such as cells, the partition typically takes place between one phase and the interface. For a partition between the top phase and the interface the work of transfer becomes:

$$\Delta F^\circ = \left[\frac{A\,(\gamma_{P,B} - \gamma_{P,T} - \gamma_{T,B})^2}{4\gamma_{T,B}}\right] \qquad (5)$$

as originally derived by Albertsson [2]. Replacing $\gamma_{P,T} - \gamma_{P,B}$ in Eq. (5) with the right side of Young's equation and substituting the result into Eq. (1) yields the intimate relationship between the partition coefficient and the contact angle,

$$K = \exp\left[\frac{-A \cdot \gamma_{T,B}(1 + \cos\theta)^2}{4kT}\right] \qquad (6)$$

An almost identical expression was published by Brooks and co-workers in 1986 [3], the single difference being the use of the complementary contact angle (180° $-\theta$) in their derivation.

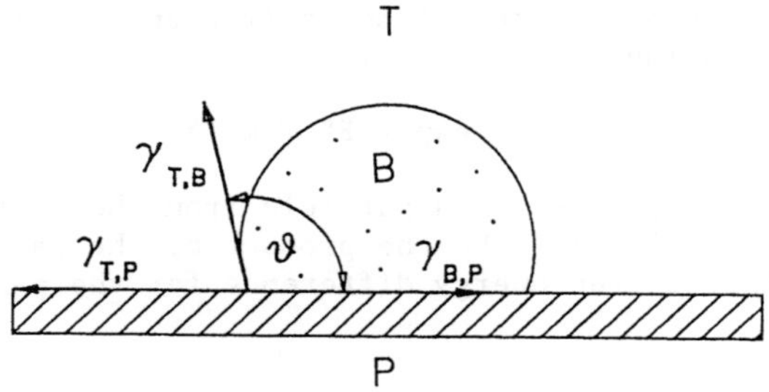

Fig. 1 *Force components of the Young Equation*

The partition coefficient calculated with Eq. (6) represents an ideal, equilibrium situation where thermal energies are assumed to be the sole driving force behind the particle distribution. The equation does not, as Sharp [4] has shown, accurately predict the experimentally determined partition coefficient, possibly because large gravitationally induced randomising energies are superimposed upon the thermodynamic considerations. An approximate exponential dependence of K on $[\gamma_{T,B}(1+\cos\theta)^2]$ is observed, however. Nevertheless, Eq. (6) is useful for making qualitative predictions about a cell population's partitioning behaviour from contact angle measurements. For example, Eq. (6) clearly shows that, as the contact angle increases, the partition of cells into the upper phase also increases until θ reaches a maximum of 180°. At this point, the partition of cells into the upper phase reaches 100%.

The real value of the contact angle approach, however, lies not in its ability to predict partitioning behaviour but in its usefulness as a simple analytical tool for investigating physical and chemical properties of the cell surface. Since the method is an equilibrium technique, the thermodynamic aspects of a cell's interaction with its physical environment can be examined independently from the rheological factors which complicate the analysis of phase partitioning. Cell surface properties are important in several biological phenomena including cellular adhesion, cell motility and phagocytosis. It is in these areas that the contact angle approach has found its most useful application.

EARLY DEVELOPMENTS

The association between contact angles and two-phase polymer systems evolved in 1980 from work done at the University of Western Ontario in London, Canada, by a group investigating the surface properties of various biological materials, including cells. Until that time, the contact angle technique was used to estimate surface energies of materials with a variety of oil/water test systems [5]. While these systems could be used with an equation of state [6] to calculate an "absolute" interfacial free energy, the large interfacial tension between the oil test droplet and the bulk aqueous medium (10-30 $mJ\cdot m^{-2}$) rendered them unsuitable for detecting subtle changes of a cell surface that may have profound biological effects. There were also concerns about the effects of oil on the hydrophobic components of the cell membrane. Aqueous two-phase polymer systems were seen as a means of avoiding these difficulties since their inherently low interfacial tensions (typically 10^{-2}-$10^{-3} mJ\cdot m^{-2}$) impart a high degree of sensitivity, and their aqueous nature provides a safe and sustaining environment for cell homeostasis.

Among the first papers to appear was a series published by Gerson in 1980 shortly after he arrived from Western Ontario at the Basel Institute for Immunology. The first of these papers [7] attempted to establish a quantitative relationship between cell surface energies as estimated with the contact angle technique and two-phase partition. Gerson's derivation pivots around a treatment of suspended cells as a molecular species in solution with equal chemical potentials in each of the two phases at equilibrium. The application of equilibrium thermodynamics for the analysis of a temporary suspension of cells is certainly incorrect, and this work can be criticised for describing cells with parameters that are only defined for molecular solutes, e.g. standard state chemical potentials [3]. Nevertheless, the final form of Gerson's equation bears resemblance to the equation derived by Brooks et al [3] and Eq. (6):

$$K_{Eq} = \exp\,(\alpha\cdot\gamma_{T,B}\cdot\cos\theta + \beta) \qquad (7)$$

$$= \exp\,-(\alpha\cdot\Delta\gamma+\beta)$$

where: α and β are empirically derived constants, and $\Delta\gamma$ (from Eq. (3)) is the interfacial free energy difference between the substrate/top and the substrate/bottom phase boundaries. By varying the polymer concentration to produce systems of different interfacial tensions, Gerson was able to test Eq. (7) and show the existence of a linear relationship between the logarithm of the partition coefficient and $\gamma_{T,B}\cdot\cos\theta$ for several lymphocytic cell lines. Gerson's work confirmed with an independent technique the importance of the cell surface free energy as a determinant of cell partitioning behaviour.

A similar study was performed on various types of bacteria in the next paper of the series [8], but the components of the interfacial energy difference, $\gamma_{P,B}$ and $\gamma_{P,T}$ were determined independently at the air/cell interface using Neumann's equation of state [6,9]. Unfortunately, the behaviour of aqueous polymer mixtures departs significantly from the empirical data used in the derivation of the equation, and it therefore yields incorrect results when applied to these systems [10]. It should also be noted that a comparison of cells in two different physical states is made here: The interfacial free energy difference ($\Delta\gamma$) was obtained from contact angles measured on cell layers that had been "allowed to dry in air", while the partitioned cells would obviously remain well hydrated in the aqueous environment of the polymer system. Any conclusion about the dependence of K on $\Delta\gamma$ under these circumstances is ambiguous because different surfaces are compared.

Some time later, Gerson [11] introduced an amended version of the equation of state that corrected for a discontinuity in the original function for high surface tensions. This modification, however, had no effect on the inadequacies of the equation when describing low tension aqueous systems.

PHYSICO-CHEMICAL CHARACTERISATION OF THE CELL SURFACE

About the same time as Gerson's papers appeared, an extensive study, investigating the surface properties of various cell types in aqueous phase mixtures, was published by Schürch and collaborators from Western Ontario [12]. Appreciating the limitations of the equation of state when applied to aqueous polymers, Schürch et al adapted a technique, introduced by Fox and Zisman [13] and later modified by Good and Girifalco [14], to calculate the critical interfacial tension of spreading, γ_C, for a cell surface. While not thermodynamically identical to the interfacial free energy, the critical interfacial tension approach provides an empirical characterisation of a surface's physical chemistry. (γ_C is determined by extrapolation from a plot of $\gamma_{T,B}{}^{-\frac{1}{2}}$ *vs.* $\cos\theta$, and is defined as the critical interfacial tension of a drop for complete spreading on a specific substrate. At this point the interfacial tension of the droplet approximates the interfacial free energy of the substrate and the substrate/drop interfacial free energy reaches a minimum or zero).

With this contact angle based technique, Schürch was able to reproducibly characterise the surface properties of various cell types (erythrocytes, neutrophils, macrophages and lymphocytes) and demonstrate direct surface effects from the actions of several biochemical agents, including concanavalin A, bacterial lipopolysaccharide and calcium ionophore.

Shortly after their original study, McIver and Schürch [10] published a second investigation that explored the physical nature of the cell surface in the context of its role in biological adhesion. The critical interfacial tension approach and the equation of state with oil/water test systems were used as parallel techniques for investigating synthetic and native erythrocyte surfaces. McIver and Schürch's results indicate that the effective interfacial free energy of the cell surface can vary by as much as three orders of magnitude depending on where the interfacial boundary is positioned in the surface. The outer-most region of the cell, the carbohydrate-rich glycocalyx, is associated with the lowest interfacial energy and the hydrophilic interactions that predominate there make it well matched to the sensitivities of the aqueous phase system. A low glycocalx/water interfacial free energy ($\approx 1 \times 10^{-3}$ $mJ \cdot m^{-2}$) implies that any adhesional contacts formed will be of a reversible nature, a scheme well suited for such transient phenomena as cell motility and chemotaxis. The higher interfacial energies sensed with the oil/water systems may pertain to less reversible phenomena associated with the lipid bilayer, for example phagocytosis and cell fusion.

Some time after the above study was published, Schürch and McIver collaborated with others to report *in vitro* experiments that reproduced some of the results previously observed with model membrane systems [15]. Cell growth under hypoxic conditions is known to alter features of the cell membrane, including enzyme activity, lipid and protein composition, and membrane morphology. To exploit this phenomenon, fibroblasts were cultured in an environment depleted of oxygen and their surface properties monitored through contact angle measurements. Higher contact angles were observed on hypoxic cells which the authors suggest is consistent with a decline in the molecular weight components on the membrane surface. A biochemical analysis of membrane constituents supports this interpretation by showing that the size (molecular weight) and number of membrane glycoproteins isolated from hypoxic cells were significantly less than those isolated from cells cultured under normal conditions.

CELL SURFACE ALTERATIONS - IMPLICATIONS FOR PHAGOCYTOSIS AND ADHESION

After their original study demonstrated that contact angles could be used to detect differences in the surfaces of several cell types, Schürch and McIver specifically concerned themselves with alveolar macrophages where adhesion and phagocytosis are important physiological functions of the cell. They were able to show that, in addition to the release of lysosomal enzymes and superoxide (O_2^-), macrophage activation is accompanied by a sharp drop in the contact angle [16,17]. In fact, these three parameters became the main criteria for judging macrophage activation when the cells were exposed to particulate and soluble stimulii. Both serum-treated yeast particles (zymosan) and phorbol myristate acetate (PMA) dissolved in the medium elicited potent activation responses. While leukotrienes as a class in general had minimal effects on enzyme or superoxide release, leukotriene D_4 did produce a marked change in the macrophage surface. The consistency of the contact angle measurements showing this effect was such that a concentration dependence could be established for stimulation of the cells. In this study the contact angle method revealed an aspect of macrophage stimulatory response which would probably have been missed with conventional testing procedures.

McIver and Schürch [18] have recently reported studies similar to those above where the nature of neutrophil stimulation by the adhesion and aggregation trigger N-formyl methionyl leucyl phenylalanine (fMLP) was investigated. Like leukotriene D_4, a concentration dependence for fMLP was also established. The same paper described an interesting study where contact angles were used to follow the clinical course of a disorder known as Adult Respiratory Disease Syndrome (ARDS). ARDS is characterised by massive intravascular neutrophil adhesion, and neutrophils taken from patients suffering with the disorder show substantially different surface properties compared to controls. Contact angles on cells from recovering patients show a steady increase towards control values; however, those from patients who eventually died from the disorder continued to decline. The authors suggest that these profound surface effects are probably the result of *in vivo* activation of the cells by chemoattractants, based on the response of neutrophils stimulated by fMLP in the laboratory.

From a thermodynamic perspective, adhesion of cells (or particles) to other cells or substrates is favoured if the change in the free energy for the process of attachment is negative, i.e. if the total energy of the exposed surface is decreased. The change is, per unit area,

$$\Delta F^{\circ}_{Adh} = \gamma_{C,S} - \gamma_{C,L} - \gamma_{S,L} \tag{8}$$

where γ is the interfacial free energy and the subscripts: P, S, and L, represent the cell (or particle), substrate and bulk liquid phases respectively. Thus adhesion of a particle to a substrate is favoured for relatively great particle/liquid and substrate/liquid interfacial energies, i.e.

$$(\gamma_{C,L} + \gamma_{S,L}) > \gamma_{C,S} \tag{9}$$

Since $\gamma_{C,S}$ cannot be determined, many of the studies already discussed have concentrated on evaluating adhesion phenomena by considering changes in $\gamma_{C,L}$ produced by chemical or biological stimulation of the cell. However, in some situations $\gamma_{C,L}$ remains constant and it is the substrate interfacial free energy that changes to promote adhesion. The following two studies are investigations of pathological processes that fall into the later category.

Platelet adhesion to the blood vessel wall is an important first step in thrombus formation and may be involved with the initiation of atherosclerosis. Platelets normally do not stick to the healthy endothelial lining of arteries, but adhere rapidly when injury exposes the subendothelial material. To explore whether this attachment process could be explained with a thermodynamic mechanism, contact angles were used to assess the relative free energy of an intact arterial surface compared to a similar surface denuded of endothelium [19]. The results show a sharp drop in the contact angle from approximately 90° to 20° when the endothelial layer is removed from the arterial wall, thus implying that the exposed underlayer is of significantly higher interfacial free energy. Adhesion would likely be highly favoured in this situation because the sum $(\gamma_{C,L} + \gamma_{S,L})$ in Eq. 9 would now exceed the cell/substrate interfacial free energy.

Bacterial infection of the bladder provides a second example where the physics implied by the contact angle approach can be used to explain pathological problems. Bladder surface mucin has been shown to be a non-specific inhibitor of adhesion for a wide variety of bacterial species, whether they are living or dead. Since mucin is composed primarily of highly hydrated glycosaminoglycans, Boyce and colleagues [20] postulated that bacterial attachment to the bladder surface may be inhibited thermodynamically through the maintenance of a low energy interface with the urine environment (i.e. keeping $\gamma_{S,L}$ small in Eq. (9)). This hypothesis was tested by using contact angles to estimate the *relative* interfacial free energy of the mucin layer before and after it had been degraded with a 0.3 M HCl acid treatment. The results and conclusions were similar to those obtained on arterial surfaces: the intact mucin layer may inhibit adhesion because it represents the lowest free energy configuration of the system, while the exposed epithelial surface cannot prevent bacterial attachment to its higher energy interface.

However, it must be emphasised that interpretation of *in vivo* biological behaviour from contact angle results is not straightforward. The substrates are tested with systems that consist of polymers known to adsorb to biological materials and this will no doubt have an effect on the contact angle. The analysis is further complicated by the fact that a contact angle change can result from an alteration in the bulk/substrate interfacial energy, the drop/substrate interfacial energy, or both, and with the two-phase polymer system neither of these two parameters can be controlled or manipulated independently. Nevertheless, experience and knowledge of the *in vivo* chemical environment does permit inference and focussed speculation to be made from results obtained with phase systems.

ELECTROSTATIC EFFECTS AND LIGAND INTERACTIONS

Perhaps more than anyone, Sharp [4] has made extensive use of the contact angle approach to investigate various factors affecting partition phenomena. Sharp's method differs significantly from the methods of other investigators in that his measurements are made on individual cells immobilised with a micro-pipette aspiration technique. A hypotonic buffer is used to swell the cells to a spherical shape so contact angles may be accurately calculated. Since the partition of cells is often manipulated by adding phosphate to create an electrical potential difference in phase systems, Sharp used contact angles to study electrostatic effects on cell surface interactions with the two-phase polymer system. His results indicate that the interfacial free energy difference is a linear function of the electrostatic potential difference between the polymeric phases. While the slope of the plot should correspond to the erythrocyte surface charge density, it reflects only about 5% of the glycocalyx charge density determined by chemical assay. Sharp concluded that the majority of the charge was inaccessible to the phase system possibly because the phase polymers are excluded from the glycocalyx.

Contact angles were also used by Sharp [4] to test a theory developed to describe the effect of affinity ligand binding on two-phase interaction with the cell surface.

$$\Delta\gamma_L - \Delta\gamma_O = n\ k\ T \quad \ln\left[\frac{(1 + c^B k^B)}{(1 + c^T k^T)}\right] \qquad (10)$$

Here, $\Delta\gamma_L$ and $\Delta\gamma_O$ are the interfacial free energy differences in the presence and absence of affinity ligand, n is the number of binding sites on the cell per unit area, k, the Boltzman constant, T, the absolute temperature, and c and k are the concentration and binding association constant for the top (T) and bottom (B) phases respectively. The binding of PEG-palmitate to erythrocytes was found to follow a simple Langmuir-type isotherm in each phase. The magnitude of the surface free energy difference, while proportional to the number of ligand molecules bound, was very small, i.e. less than 1% of the calculated binding energies. Sharp suggests this is because the head group of the ligand molecule is partially hidden from the phases within the glycocalyx, thus lending further support for the concept of polymer exclusion from the cell's carbohydrate-rich surface coat. More recently, Youens *et al* have confirmed Sharp's experimental findings, which correlate modifications of the cell surface through ligand binding with partition behaviour.

While the contact angle approach has been useful for shedding new light on the physical nature of the cell surface, many questions still remain. Since it is fundamentally a thermodynamic method, the technique is good at detecting subtle changes in cell surface properties, but the factors responsible for the changes are not easily elucidated. For example, the change in the contact angle may be individually or collectively attributable to: a decrease in the macromolecule surface density, an increase in carbohydrate molecular weight, changing polymer interactions or a myriad of other factors. Hope for resolving some of these uncertainties lies in careful control studies on model cell surfaces where variables may be examined in isolation and the complexity of the substrate increased as our knowledge expands.

ACKNOWLEDGEMENTS

This work was supported by a Canadian Heart Fellowship to J.F.B. and an M.R.C. (Canada) term grant to D.E.B.).

REFERENCES

1 T. Young, Philos. Trans. R. Soc., London 95:65 (1805)

2 P.-Å. Albertsson, in : "Partition of Cell Particles and Macromolecules," 3rd ed., Wiley-Interscience, New York (1986)

3 D.E. Brooks, K.A. Sharp and D. Fisher, Theoretical aspects of partitioning, in: "Partitioning in Aqueous Two Phase Systems," H. Walter, D.E. Brooks and D. Fisher, eds., Academic Press, Orlando (1985)

4 K.A. Sharp, Theoretical and experimental studies on erythrocyte partition in aqueous polymer two-phase systems, Ph.D. Thesis, University of British Columbia, Vancouver, 1985

5 J.F. Boyce, S. Schürch and D.J.L. McIver, Interfacial tensions in healthy and atherosclerotic rabbit aortae: Higher values on lesion surfaces, Atherosclerosis 37:361 (1980)

6 A.W. Neumann, R.J. Good, C.J. Hope and M. Sejpal, An equation of state approach to determine surface tensions of low energy solids from contact angles, J. Colloid Interface Science 49:291 (1974)

7 D.F. Gerson, Cell surface energy, contact angles and phase partition, I. Lymphocytic cell lines in biphasic aqueous mixtures, Biochim. Biophys. Acta 602:269 (1980)

8 D.F. Gerson and J. Akit, Cell surface energy, contact angles and phase partition. II Bacterial cells in biphasic aqueous mixtures, Biochim. Biophys. Acta 602:281 (1980)

9 C.J. Van Oss, C.F. Gillman and A.W. Neumann, "Phagocytic Engulfment and Cell Adhesiveness as Cellular Surface Phenomena," Marcel Dekker, New York (1975)

10 D.J.L. McIver and S. Schürch, Interfacial free energies of intact and reconstituted erythrocyte surfaces - implications for biological adhesion, Biochim. Biophys. Acta 691:52 (1982)

11 D.F. Gerson, Interfacial free energies of cells and polymers in aqueous media, in: "Physiochemical Aspects of Polymer Surfaces," Vol. 1, K.L. Mittel, ed., Plenum, New York (1983)

12 S. Schüurch, D.F. Gerson and D.J.L. McIver, Determination of cell medium interfacial tensions from contact angles in aqueous polymer systems, Biochim. Biophys. Acta 640:557 (1981)

13 H.W. Fox and W.A. Zisman, The spreading of liquids on low energy surfaces. II Modified tetrafluorethylene polymers, J. Colloid Sci. 7:109 (1952)

14 R.J. Good and L.A. Girifalco, A theory for estimation of surface and interfacial energies. III Estimation of surface energies of solids from contact angle data, J. Phys. Chem. 64:561 (1960)

15 S.A. Johnstone, S. Schürch, D.J.L. McIver, E.A. Jacobson and E.R. Tustanoff, Membrane glycoprotein and surface free energy changes in hypoxic fibroblast cells, Biochim. Biophys. Acta 815:159 (1985)

16 N.A.M. Paterson, D.J.L. McIver and S. Schüurch, Zymosan enhances leukotriene D_4 metabolism by porcine alveolar macrophages, Immunology 56:153 (1985)

17 N.A.M. Paterson, D.J.L. McIver and S. Schürch, The effect of leukotrienes on porcine alveolar macrophage function, Prostaglandins, Leukotrienes Med. 25:147 (1986)

18 D.J.L. McIver and S. Schürch, Polymer mixing and the thermodynamics of cell adhesion at fluid interfaces, J. Adhesion 22:253 (1987)
19 J.F. Boyce, P.C. Wong, S. Schürch and M.R. Roach, Rabbit arterial endothelium and subendothelium: A change in interfacial free energy that may promote initial platelet adhesion, Circ. Res. 53:372 (1983)
20 J.F. Boyce, R.B. Podesta, S. Schürch and M.R. Roach, Rabbit bladder-surface mucin: A thermodynamic mechanism for inhibiting bacterial adhesion, Urol. Res. 11:93 (1983)
21 B.N. Youens, A.W. Foulds, W.D. Cooper and D. Fisher, Interfacial properties of cells partitioning in aqueous polymer two-phase systems, Biochem. Soc. Trans. 14:751 (1986)

MACROPHAGE SURFACE AFFINITY CHANGES DURING PHAGOCYTOSIS

Samuel Schürch[1], Nigel A.M. Paterson[2,3] and Donald J.L. McIver[3,4]

Faculty of Medicine[1], University of Calgary
Calgary, Alberta, Canada T2N 4N1, and
Departments of Medicine[2], Pharmacology and Toxicology[3]
and Medical Biophysics[4], University of Western Ontario
London, Ontario, Canada N6A 5C1

We have used a combination of surface thermodynamic, spectro-photometric and morphometric techniques to investigate the mechanism of particle ingestion (phagocytosis) by isolated alveolar macrophages. We studied the time and dose-dependence of the effects on macrophages of both structurally specific particles (serum complement-opsonised zymosan) and non-specific particles (non-opsonised silica). Particle exposure increased the cell affinity for the dextran phase of 4% Dextran 2M/4% PEG 20K phases in a dose- and opsonin-dependent manner which correlated well with biochemical and morphological assessment of cell activation. We do not yet know the structural basis of the cell surface changes which are responsible for the alterations in polymer phase affinity, but likely candidates are phagocytosis-induced changes in the amount, conformation or molecular weight of membrane glycoproteins.

INTRODUCTION

Phagocytosis or the engulfment and intracellular digestion of foreign particles is phylogenetically one of the oldest cellular activities, yet it remains of central importance in such diverse roles as host defence mechanisms, cell surface receptor regulation and physiological and pathological responses to foreign chemicals. Both physical and chemical mechanisms are clearly involved in phagocytosis but exactly how are these physicochemical forces coordinated to drive the changes in the cell surface which occur during particle ingestion?

A minimal description of particle engulfment by phagocytic cells would regard the phagocytic process as an equilibrium solubility phenomenon in which the particles are "dissolved" in the phagocyte, and which is driven simply by changes in the interfacial free energies at the particle-medium, particle-phagocyte and phagocyte-medium interfaces. The idea is not new: in one form or another it has been around since at least the 1920's [1,2]. What we call the "solubility" model of phagocytosis is illustrated schematically in Fig. 1, where the particle phase, the extracellular phase and intracellular phases are denoted as 1, 2 and 3, and the interfacial free energies have their usual meanings. Quantitatively, this model - which is

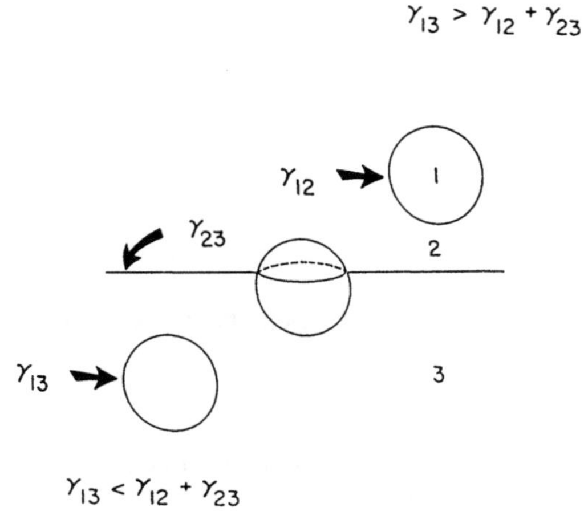

Fig. 1 *Phase partition model of the distribution of a particle (1) between the extracellular phase (2) and the intracellular phase (3)*

simply a phase partition process itself - predicts that endocytosis is thermodynamically favoured when

$$\gamma_{13} < \gamma_{12} + \gamma_{23}$$

while the thermodynamic condition for exocytosis is

$$\gamma_{13} > \gamma_{12} + \gamma_{23}$$

Testing these predictions requires obtaining numerical estimates of the three interfacial free energies.

Even if direct measurements of these energies for living cells were to become available - and for the time being they are not - one should still ask how far can we reasonably expect this type of simple equilibrium model to apply to a complex vital process such as phagocytosis? Is it not naive in the extreme to expect that measurements based on macroscopically time- and space-averaged static surface chemistry will describe the dynamic microscopic discrimination displayed by phagocytic cells? The point at which the simple model will cease to be useful is not yet clear, not least because we do not yet have the means of rigorously testing even the simple model. Certainly, more complex models can (and have been) devised to include, for example, additional energy terms (both equilibrium and non-equilibrium) such as membrane curvature, electrostatic potential changes and cytoplasmic force-discplacement work, but it seems that such proliferation of components should only be resorted to when simpler measures have proved inadequate.

One of the most challenging demands of a model of the physics and chemistry of phagocytosis will be to explain how chemically specific processes, such as opsonisation by complement or immunoglobulins which affect phagocytosis, can do so by altering the physical forces which must ultimately be responsible for particle translation from the environment into the cell. Almost as demanding, however, is explaining how the same forces operate in the cellular ingestion of structurally non-specific materials such as particles of rocks, hydrocarbons or fluorocarbons. The majority of studies of phagocytosis (e.g. 3) has concentrated on the specific aspects of phagocytosis; much less is known about the non-specific aspects of phagocytosis [4]. A more detailed understanding of the relationships between these two apparently different phenomena might well help in answering such important questions as: why are some forms of particles (e.g. silica) much more toxic than others?

A frontal attack on such a complex problem is bound to fail. We have therefore concentrated on the first part of the question: can interfacial free energy changes during phagocytosis be measured under biologically non-perturbing conditions (such as are believed to exist in aqueous phase systems) and do these biophysical measurements correlate in a time, dose and stimulus specific manner with parallel measurements of the biochemical changes associated with phagocytosis?

METHODS

The methods have been described in detail elsewhere [5-7]. Briefly, alveolar macrophages were isolated by saline lavage of the lungs of anaesthetised pigs, and purified if necessary by density gradient centrifugation to yield macrophage populations of at least 99% purity and viability. The cells were incubated in Tyrode's buffer at 37°C at a concentration of 1×10^7 cells per ml in the presence or absence of varying dose- and time-exposures to zymosan or silica ("opsonised" or "non-opsonised"). At the end of the incubation, the cells were centrifuged onto flat transparent substrates (glass or polystyrene cover slips) at a density which gave complete monolayer coverage of the substrate by the cells. The supernatant was analysed for superoxide production or the release of the lysosomal enzymes N-acetyl beta glucosaminidase or arylsulfatase, and the cell monolayer and substrate were immersed in the PEG phase of a phase separated 4% dextran/4% PEG phase system as previously described [5-7]. Final advancing contact angles [8] on the cell layers were measured using drops of the dextran phase ranging between 0.1 and 1 mm in diameter, and therefore sampling between 100 and 1000 cells per drop. After measuring the contact angles (usually 20 to 40 replications per sample) the cells were fixed with

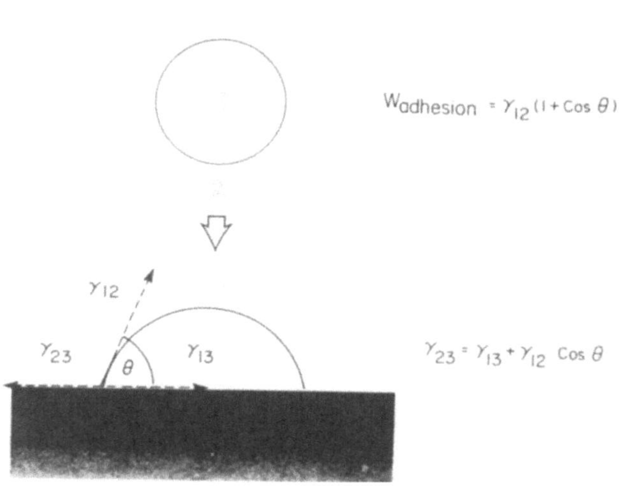

Fig. 2 *Ideal (equilibrium) relations between the contact angle θ, the work of adhesion W_{adh} and the interfacial free energies γ, at a solid/liquid/liquid interface*

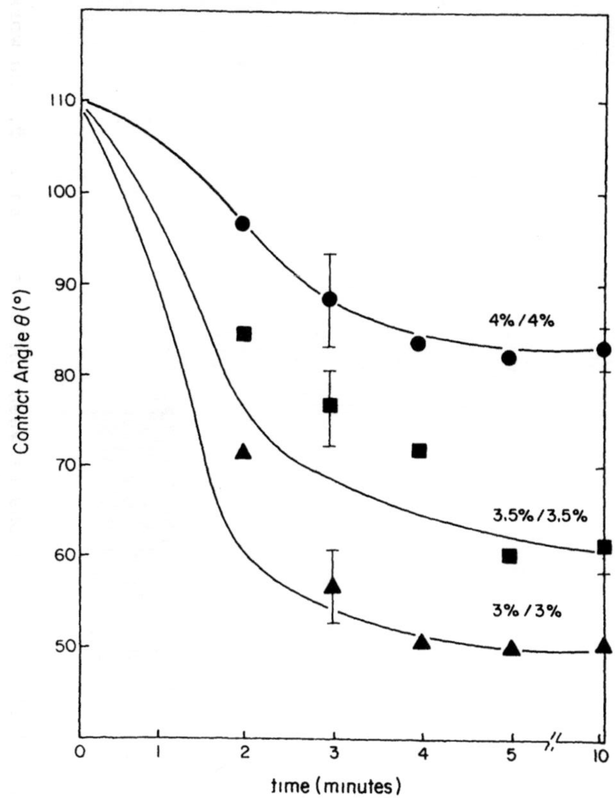

Fig. 3A *Kinetics of wetting of macrophage layers by three different dextran 2M/PEG 20K phase systems. Each point represents the mean ± S.E. mean of 20-40 separate contact angle measurements.* (●) *4%/4%* (■) *3.5%/3.5%* (▲) *3%/3%*

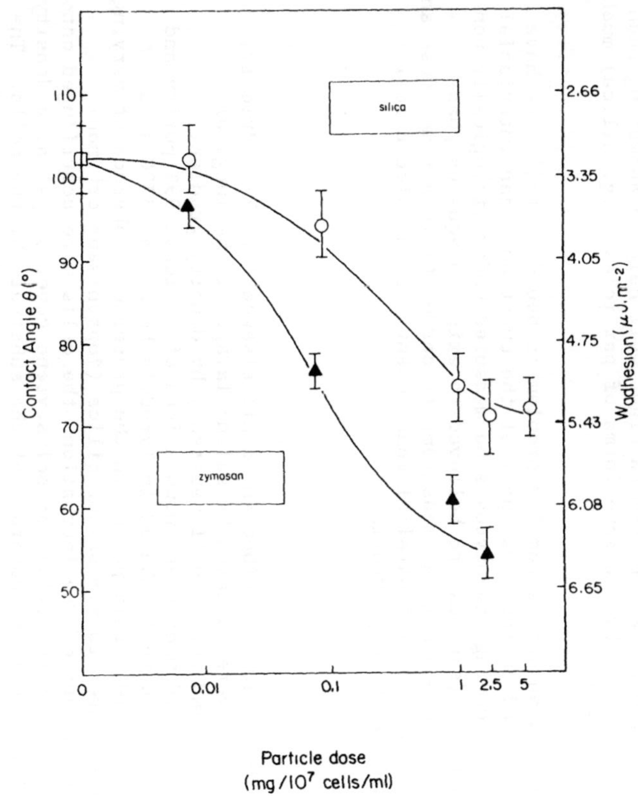

Fig. 3B *Particle dose-response curves for macrophage surface wetting after exposure to non-serum-treated silica* (○) *or to serum treated zymosan* (▲). *Each point represents the mean ± S.E. mean of 30-40 separate measurements*

2% glutaraldehyde and examined under differential interference contrast light microscopy (D.I.C. Nomarski type) for morphometric studies. In this case, the sole purpose of the fixative was to eliminate possible morphological changes in the cells subsequent to the time at which the contact angles were measured.

The principle of the contact angle measurement is shown in Fig. 2, where the dextran drop is denoted as phase 1, the PEG phase as 2 and the cell layer as phase 3. From measurements of the contact angle and the interfacial tension, the work of adhesion of the cells to the dextran phase in the PEG phase is calculated. The work of adhesion corresponds to the difference in the standard state free energy between the cells in phase 1 and phase 2 [9].

RESULTS AND DISCUSSION

The wetting of the cells by macroscopic phase droplets occurs over several minutes and varies with phase composition

Fig. 3A shows the time taken to reach the final advancing contact angle by dextran drops of three different phase compositions. The rate of wetting is greatest in the system of lowest polymer concentration, although the time to reach constant advancing contact angles is similar in all cases - about 3 minutes. This is in distinct contrast to the rates of micro-wetting shown in the following paper, where equilibrium wetting by similar phase systems is achieved in a few seconds; the magnitudes of the final advancing angles however, are similar in both the macroscopic and microscopic measurements, suggesting that the measured work of adhesion is indeed independent of the path of wetting.

W_{adh} of phagocytosing macrophages varies with the nature and dose of particle

Fig. 3B shows the particle dose-contact angle response curves for macrophages incubated with varying doses of either non serum-treated silica (circles) or serum-treated zymosan (triangles). In all cases the time of incubation was chosen from the plateau of the incubation time-response curve (usually about 20 minutes). Both particles increase the affinity (work of adhesion) of the cells for the dextran phase in a dose-dependent manner, and in both cases the responses appear to reach a plateau at high particle concentrations. However, the opsonised zymosan has a much greater maximum effect (W_{adh} max) on the macrophage affinity: W_{adh} max of 3.30 $\mu J.m^{-2}$ was measured for zymosan compared with 2.08 $\mu J.m^{-2}$ for silica. We attribute this difference to the activation of macrophage complement receptors by complement components adsorbed to the zymosan (see below). Fig. 4 shows photographs of the variation in contact angle with particle exposure.

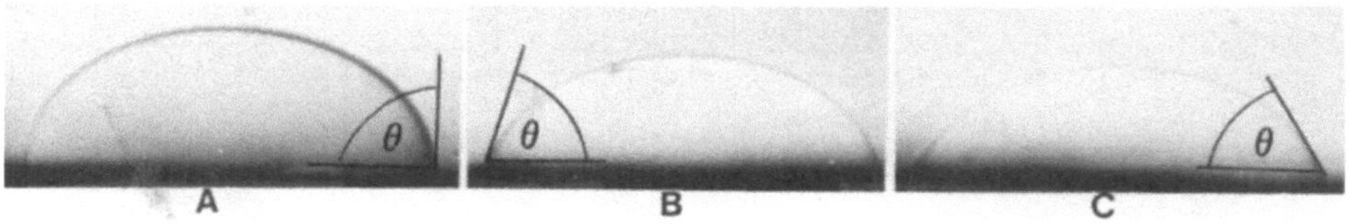

Fig. 4 *Photomicrographs of the final advancing contact angle formed on macrophage surfaces by the 4%/4% dextran/PEG phase system, before (A) or after exposure to non-serum treated silican (B) or serum-treated zymosan (C). The final particle dose in each case was 1 mg/ml·10⁷ cells. The scale bar is 100 μm*

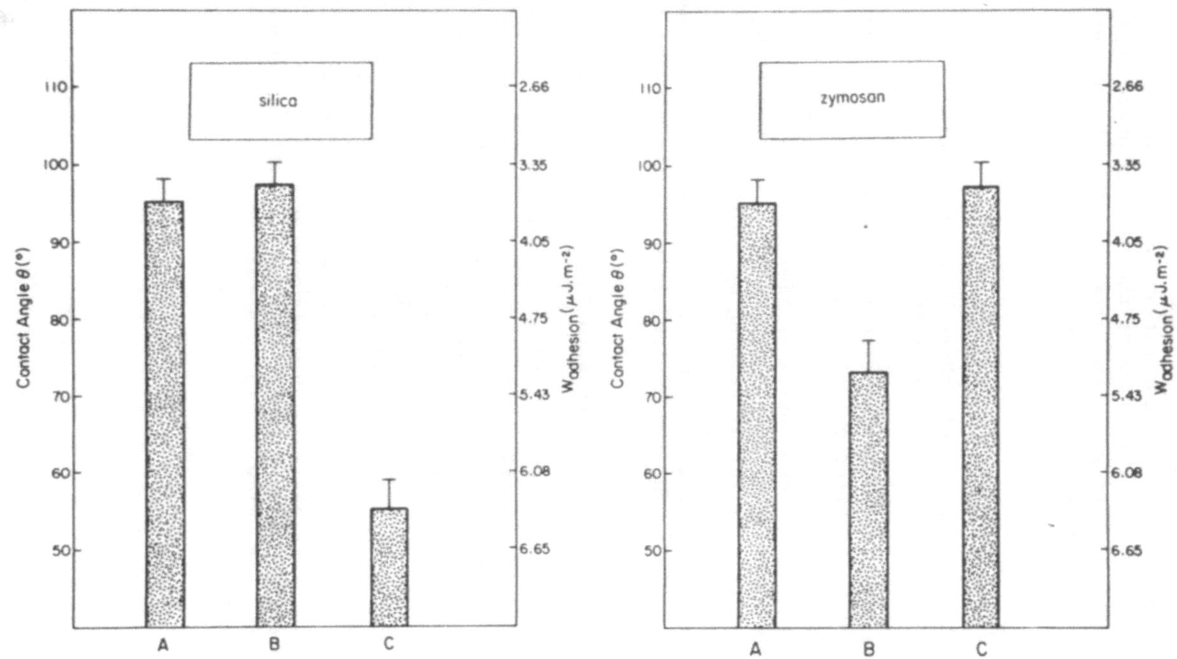

Fig. 5 *Effect of serum pre-treatment on surface wetting of macrophages in response to zymosan or silica exposure. A. Control; B. serum-treated particles; C. non-serum-treated particles. Each bar represents the mean ± S.E. mean of 15-20 separate measurements*

Serum pre-treatment enhances zymosan-induced changes in macrophage dextran phase affinity, but inhibits those due to silica

Fig. 5 shows that the changes in macrophage dextran phase affinity induced by maximal doses of particles differ according to whether or not the particles have been pre-treated with serum. Serum treatment "activates" the zymosan while it "passivates" the silica with respect to the ability of the particles to induce changes in the dextran phase affinity of the macrophage surfaces.

The nature of the substrate and the extent of stimulus-induced cell surface area change (spreading) has much less influence on W_{adh} than the nature of the ingested particle

Fig. 6 shows the results of planimeter tracings of D.I.C. micrographs of the cell surface areas when the cells were fixed after contact angle measurements were made on the cells sedimented onto two transparent substrate materials - glass or polystyrene microscope cover slips. The substrate alone does have a small effect on the dextran phase affinity: the cells spread more on the high energy glass surface than on the polystyrene and this spreading is associated with a small increase in dextran phase affinity. In both cases of particle exposure, however, the particle effects completely dominate the substrate effect. Silica ingestion is associated with a dramatic increase in cell surface area, which is independent of the

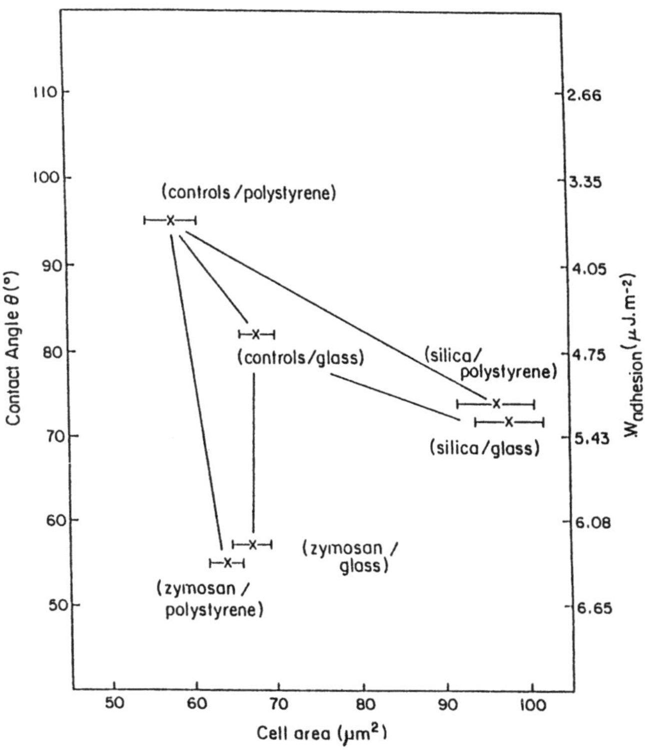

Fig. 6 *Effect of substrate and particle exposure on macrophage wetting and surface areas. Each point represents the mean± S.E. mean of 10-20 area determinations or wetting measurements*

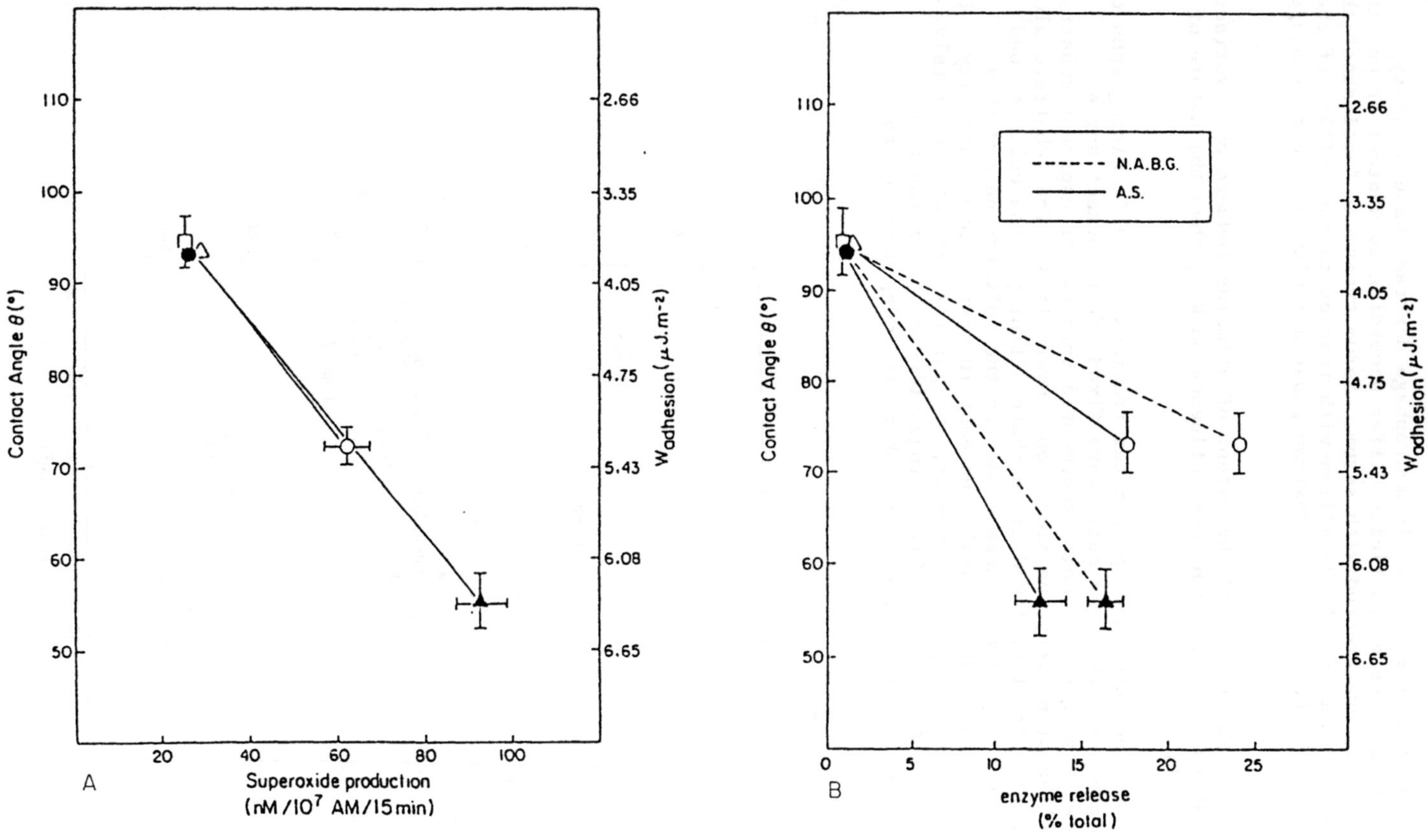

Fig. 7 *Relations between macrophage wetting and (A) superoxide anion production or (B) release of the lysosomal enzymes N-acetyl beta glucosaminidase (NABG) or aryl sulfatase (AS). Each point represents the mean ± S.E. mean of 6-8 separate enzyme determinations or wetting measurements. (●) control; (○) silica; (▲) zymosan. Serum treatment is shown by filled symbols and non-serum treatment by open symbols*

substrate, and a relatively small increase in dextran affinity, as noted above. The opposite trend is observed for zymosan: a large increase in dextran affinity is associated with a small increase in cell surface area; once again, the substrate has little effect.

Particle-induced macrophage phase affinity, superoxide production and lysosomal enzyme release vary directly with the dose and nature of the stimuli

Fig. 7A shows that the rate of superoxide production correlates directly with the change in dextran affinity for both serum-treated and non-serum-treated particles. The doses illustrated were the maximally effective doses shown in Fig. 3; lower doses (not shown) produced intermediate effects. Serum treated zymosan induced the maximum change in both dextran affinity and superoxide production; non-serum-treated silica was intermediate in effects on both dextran affinity and superoxide production. Serum-treated silica or non-serum-treated zymosan were without effect on either measurement.

A similar but slightly different trend was observed when particle effects on lysosomal enzyme release were studied. Fig. 7B shows the relation between the change in dextran affinity and the release of the two lysosomal marker enzymes N-acetyl beta glucosaminidase (NABG) and arylsulfatase (AS). Here silica induced a greater release of enzymes than zymosan, but a smaller change in dextran affinity. Once again, serum-treated silica or non-serum-treated zymosan were without effect.

CONCLUSIONS

The experiments described here are basically descriptive: they show that phagocytic stimuli induce stimulus-specific, dose-dependent changes in macrophage polymer phase affinity, which correlate directly with independent biochemical and morphological measurements of cell activation. The experiments do not, of course, prove that these correlations are causal, nor do they yet allow us to test the predictions or limitations of the "solubility" model of phagocytosis. Young's equation (Fig. 2) yields only the <u>difference</u> in surface free energies, i.e. the work of adhesion. Thus, the three interfacial free energies in Fig. 1 cannot be estimated from Young's equation without an independent relationship between the state functions (an equation of state) in the polymer phase/particle/cell surface system. Experimental and theoretical advances in the area of polymer-solvent-surface interactions (for example, [10]) may well yield such relationships; in the meantime, the present experiments do allow some interim conclusions.

Especially interesting are the opposing influences of serum pre-treatment on the effects of zymosan and silica on the macrophages. While the mechanism of serum passivation of silica is unknown (as indeed is the nature of the "non-specific macrophage particle receptor" [4]), the activation of zymosan by serum is due to adsorption of complement components (opsonisation). The fact that opsonisation changes the phase affinity of the macrophages indicates that opsonins may not only act as catalysts to increase the rates of particle-phagocyte interactions [3], but that they may also increase the thermodynamic driving force for phagocytosis by making the phagocytes stickier.

Unfortunately, the experiments give us only hints about what perhaps are the most interesting questions: what is the molecular basis of the changes in cell surface composition which cause the phase affinity changes and what is the relationships of these changes to the cell biology of phagocytosis?

A theoretical analysis of polymer phase partition, based on the Flory-Huggins' statistical mechanical theory of polymer interactions [11], indicates that the majority of polymer phase behaviour is attributable to polymer segment-polymer segment interactions, and thus that the biological basis of the phase affinity changes observed in the present experiments should be sought in changes in the amount and molecular weight of glycocalyx proteins [9]. This suggestion is quite consistent with our previous experimental findings [5,6,12,13]. An additional factor which is not directly addressed in the original Flory-Huggins' theory, but which may also play a role in cell surface-polymer phase interactions, is the role of polymer conformation. The polymer affinity is the resultant of all of these effects, and this is at the same time both the strength and weakness of the surface thermodynamic approach.

One of the main agendas in cell partitioning now is to relate the phase thermodynamics with the molecular structure of the cell surface - to unravel the unit molecular contributions to the thermodynamic potential functions. Only with a good understanding of these two approaches can we hope to fully realise what analytic aqueous polymer phase partitioning is uniquely suited to tell us: the relationships between the physics and chemistry of intermolecular and surface forces in living biological cells - intact, hydrated, unfixed and unstained; still breathing, in fact, as Fig. 7A shows.

Supported by grants from the Medical Research Council of Canada to S.S., N.A.M.P. and D.J.L.M., and from the Ontario Thoracic Society to D.J.L.M. K. Ellis and L. Goela provided expert technical assistance.

REFERENCES

1 W.O. Fenn, The theoretical response of living cells to contact with solid bodies, J. Gen. Physiol. 4:373 (1922)
2 C.J. Van Os, C.F. Gilman and A.W. Neumann, in: "Phagocytic Engulfment and Cell Adhesiveness as Surface Phenomena," Marcel Dekker, New York (1975)
3 P.J. Edelson, Monocytes and macrophages: aspects of their cell biology, in: "The Cell Biology of Inflammation," G. Weissmann, ed., Elsevier, New York (1980)
4 J. Aggeler and Z. Werb, Ultrastructural aspects of phagocytosis by macrophages, in: "Advances in Inflammation Research", Vol. 8," G. Weissmann, ed., Raven, New York (1984)
5 S. Schürch, D.F. Gerson and D.J.L. McIver, Determination of cell-medium interfacial tensions from contact angles in aqueous polymer systems, Biochim. Biophys. Acta 640:557 (1981)
6 D.J.L. McIver and S. Schürch, Interfacial free energies of intact and reconstituted erythrocyte surfaces; implications for biological adhesion, Biochem. Biophys. Acta 691:52 (1983)
7 N.A.M. Paterson, D.J.L. McIver and S. Schürch, Zymosan enhances leukotriene D_4 metabolism by porcine alveolar macrophages, Immunology 56:153 (1985)
8 A.W. Neumann, Contact angles and their temperature dependence: Thermodynamic status, measurement, interpretation and application, Adv. Colloid Interface Sci. 4:105 (1974)
9 D.E. Brooks, K.A. Sharp and D. Fisher, Theoretical aspects of partitioning, in: "Partitioning in Aqueous Two-Phase Systems. Theory, Methods, Uses and Applications to Biotechnology," H. Walter, D.E. Brooks and D. Fisher, eds., Academic Press, Orlando (1985)
10 E.A. Evans, Intrinsic colloidal attraction/repulsion between lipid bilayers and strong attraction induced by non-adsorbing polymers,

International Symposium on Molecular Mechanisms of Membrane Fusion, Buffalo, N.Y. (1987)
11 P.J. Flory, Thermodynamics of high polymer solutions. J. Chem. Phys. 9:660 (1941)
12 S. Schürch, K. Ellis and D.J.L. McIver, ^{14}C-labelled protein adsorption at a lipid-water interface investigated by using phase separated aqueous polymers. American Chemical Society, 57th Colloid and Surface Symposium, Toronto, A 217 (1983)
13 S.A. Johnstone, S. Schürch, D.J.L. McIver, E.A. Jacobson and E.R. Tustanoff, Membrane glycoprotein and surface free energy changes in hypoxic fibroblast cells, Biochim. Biophys. Acta 815:159 (1985)

A POLYMER MIXING APPROACH TO CHEMOATTRACTANT-STIMULATED NEUTROPHIL ADHESION

Donald J.L. McIver [1] and Samuel Schürch [2]

Departments of Pharmacology and Toxicology, Medical Biophysics, Medicine and The Robarts Research Institute [1]
University of Western Ontario, London, Ontario, Canada N6A 5C1
and Faculty of Medicine [2], University of Calgary
Calgary, Alberta, Canada T2N 1N4

We have used photometric and aqueous phase polymer affinity techniques to investigate the kinetics and surface thermodynamics of neutrophilic leukocyte responses to the soluble bacterial chemotaxin N-formyl methionyl leucyl phenylalanine (fMLP). Unlike the relatively slow responses of macrophages to particles, neutrophils respond to chemoattractants within seconds and display a complex series of time-dependent changes which in part reflect cellular adaptation to the chemotaxin by surface receptor internalisation [1,2]. The kinetics of these changes can be resolved by photometric methods, but not by the relatively slow techniques of macroscopic polymer phase wetting. We have therefore used the fungal metabolite cytochalasin B to prevent reversal of the rapid neutrophil responses to fMLP. Under these conditions, wetting of the neutrophils by dextran/PEG phases is essentially independent of time, and depends only on the dose of chemoattractant. The dose-response curves for fMLP induced cell aggregation and increased dextran phase affinity are similar, each with an ED_{50} of 2×10^{-8} M, suggesting that the two processes may have the same basis, for example in specific granule glycoproteins which become expressed on the cell surface during aggregation. By contrast, chemotactic cell surface ruffling (assessed either by digital image analysis of photomicrographs or by perpendicular light scattering from the cell suspensions) and oxygen radical-dependent chemiluminescence have dose-response curves with values of ED_{50} ten times lower and higher, respectively, than aggregation and phase affinity changes. Using a lattice model of the cell surface and the polymer phases, a Zisman-type of approach to estimate a critical phase composition for spreading, and an assumed molecular weight of 200 kdal for the hypothetical "adhesion molecules" we estimate that the molecules may attain micromolar concentrations on the cell surface during maximal adhesion. Preliminary experiments using the micropipette aspiration technique of E. Evans [3] have shown that wetting studies using this technique have ample kinetic sensitivity to resolve fast changes in cell surface thermodynamics. So far, however, we have not been able to carry out these experiments under conditions which are optimal for chemotaxin-induced adhesion.

INTRODUCTION

The blood-borne neutrophilic leukocytes (or "microphages" as they were originally called) have a number of similarities to macrophages, but also a number of differences. These phylogenetically younger, smaller, shorter lived cells are quicker to respond to stimuli (in keeping with their acute inflammatory function), have less developed phagocytic capabilities but are highly evolved for rapid responses to soluble chemical signals, especially those of bacterial origin.

When neutrophils are exposed to chemotactic stimuli, they rapidly remodel their surfaces and display a complex repertoire of biochemical and behavioural changes. If the cell concentration is sufficiently high to provide cell-cell contacts, a transient increase in adhesion is associated with microscopic aggregation, followed by disaggregation as the cells adopt the configuration associated with chemotactic migration [1].

What is the physicochemical basis of the changes in the cell surface aggregation, disaggregation and chemotaxis? Can aqueous phase wetting studies contribute to understanding the changes in surface affinity which drive the adhesive interactions between these cells? As with phagocytosis, both physical and chemical forces are clearly involved in stimulus-induced cell adhesion, but the relations between chemical and physical work in cell adhesion are only partly understood. The transience of the chemotaxin effects makes them difficult to study under conditions of equilibrium receptor occupancy and steady state adhesive responses.

When isolated purified neutrophils are studied *in vitro* by photometric techniques, rapid changes in the fluxes of scattered and transmitted light are observed following exposure to chemotaxins. While there is some controversy about the exact cellular origin of these fluxes [2,4] light scattering theory [5] attributes small angle scattering (axial scattering or transmittance) predominantly to "large" scale effects (for example, changes in whole cell shape or state of aggregation) and scattering at larger angles (for example at 90°) predominantly to "small" scale effects, including perhaps surface ruffling, degranulation or actin polymerisation [6]. In this context "large" and "small" are defined relative to the wavelength of the illuminating light. This controversy is still not settled, but from the point of view of the present study, we simply employ cell photometry as an empirical tool to define the kinetics of cell responses to chemotaxins (whatever in fact these kinetics may arise from), so that the rates of cell responses to chemotactic stimuli may be matched to the intrinsic rates of wetting of the cells by the polymer phases. Preliminary accounts of studies of the cellular origin of the luminous fluxes in neutrophil photometry have been presented elsewhere [7], and more detailed discussions are in preparation.

In the present study, we used two kinetic regimes (the presence or absence of the fungal metabolite cytochalasin B) to investigate the effects of chemotactic stimulation on neutrophil-polymer phase affinity, and we used continuous photometric recording of the stimulated neutrophils to provide the time scale of cell behaviour which is absent from the equilibrium thermodynamic measurements.

METHODS

The methods are quite similar to those described in the previous paper, and elsewhere in our earlier work [8-12]. Briefly, suspensions of purified human neutrophils were prepared by fractionating venous blood, using the dextran sedimentation/ficoll-hypaque density gradient/hypotonic lysis technique of Boyum [11], with minor modifications designed to reduce platelet contamination. The cells were >99% pure and viable and contained <1% platelets, and were suspended at a concentration of 1×10^7/ml in calcium- and magnesium-free Hanks balanced salt solution containing 1% bovine serum albumin. The cells were pre-incubated for 3 minutes at 37°C after the addition of calcium (1.2 mM) and magnesium (0.7 mM) and before the addition of stimuli. The incubation with various doses of fMLP, in the presence or absence of cytochalasin B, was carried out by stirring the cells at 37°C in a Payton model 1020 chemiluminescence photometer which was modified to record right angle light scattering in addition to axial light transmission and luminol-enhanced chemiluminescence. Contact angle measurements were carried out as described earlier; briefly, samples of the cell suspension sufficient to form a complete monolayer of cells were removed from the photometer cuvette at times determined by the light scattering signals, sedimented onto flat polystyrene supports, the aqueous phase was replaced by the PEG-rich phase of the polymer system and dextran droplets from the dextran-rich phase were allowed to settle onto the cells. The final advancing contact angles and dextran/PEG interfacial free energies were determined from digitised photomicrographs of the droplets, using the computer programme of Rotenburg, Boruvka and Neumann [12], which uses a Newton-Raphson technique to fit a LaPlacian curve to the drop profile, and which calculates the interfacial free energy, the drop volume and surface area, and the contact angle, using the drop dimensions, the gravitational acceleration vector and the density difference between the two phases as input data. Cell samples were also removed at various times for examination by differential interference contrast light microscopy, and the distributions of cell sizes and shapes were determined by digital image analysis, using a Dapple microcomputer based video digitising system, in collaboration with Dr B. Kaye, Laurentian University, Sudbury, Ontario. The cell samples were withdrawn and fixed through a light-proof cover on the photometer, which permitted the acquisition of morphological data without interrupting the ongoing photometric measurements; we estimated the time resolution of the fixation procedure to be between 10 and 20 seconds, so we have not attempted to report on changes in cell structure during the early phase of cell activation. Preliminary experiments using the cell micromanipulation procedure of E. Evans [3] were also carried out in collaboration with K. Sharp, D. Brooks and E. Evans at the University of British Columbia, Vancouver, B.C., to investigate the kinetics and surface thermodynamics of the responses of single neutrophils to stimuli.

RESULTS AND DISCUSSION

Neutrophil photometry and morphometry show rapid and complex kinetic changes arising from several underlying cellular processes. The photometric signals are composite signals arising from at least four underlying processes: cell shape change (ruffling), aggregation and disaggregation, degranulation and chemotactic polarisation. These conclusions are based on the analysis of the simultaneous measurement of light transmission, perpendicular light scattering, chemiluminescence and morphology, in the presence or absence of cytochalasin B. Fig. 1 shows the three simultaneous photometric signals recorded following a single dose (10^{-7} M) of fMLP. In the absence of the microfilament antagonist cytochalasin B the chemiluminescence has monophasic kinetics, with a single peak at 55 sec and a decline over the following 2 minutes. Light transmission has biphasic kinetics, with a peak at 2 minutes, followed by a reversal towards baseline with an inflection at 6 minutes, and a slower secondary increase in transmission over the following ten minutes. Perpendicular light scattering has triphasic kinetics: a rapid highly synchronous first peak occurs at 10 seconds, followed by a slower, less synchronised second peak between 1 and 2 minutes, followed by an inflection point and slower third phase of scattering which has identical kinetics to those of transmission. In the presence of cytochalasin B, the reveral of both light transmission and perpendicular light scattering is inhibited and after two minutes the light fluxes in the presence of this agent are essentially invariant with respect to time (curves marked by asterisks). "Typical" cell shapes (based on digitised profiles of at least 500 cells per

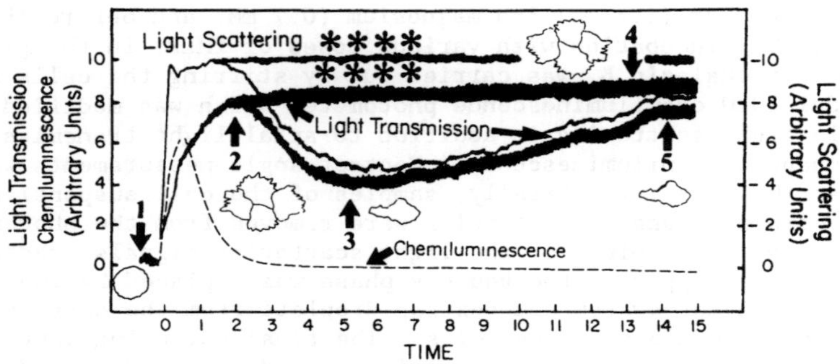

Fig. 1 *Kinetics and morphology of neutrophil suspension responses to 10^{-7} M fMLP in the absence or presence of cytochalasin B (curves marked with asterisks). In order to facilitate comparison, the sign of the light scattering curve has been reversed*

sample) determined in samples removed and fixed at the time points numbered in Fig. 1 are also shown. The salient morphological points are: in the absence of cytochalasin the initially spherical cells (1) ruffle and adhere into clumps (2) which are mostly dispersed by the time that light transmission and scattering reach the inflection point (3); at this time the cells have mostly adopted the polarised chemotactic configuration which persists for the rest of the recording (5). In the presence of cytochalasin B the aggregates do not disperse and the cells stay adherent and non-polarised (4).

In the absence of cytochalasin B small (10%) increases in dextran affinity are seen, while in the presence of cytochalasin B five times larger (50%) changes are seen. Fig. 2 shows the dose-response curves for macroscopic contact angles measured on sedimented layers of neutrophils, incubated for 15 minutes with various doses of fMLP, in the presence (solid circles) or absence (open circles) of cytochalasin B (i.e. removed from the cuvette at a point corresponding to number 5 in Fig. 1). In the absence of cytochalasin B, 10^{-7} M fMLP produces a small (10%) increase in the work of adhesion of the cells to the dextran phase. In the presence of cytochalasin B, much larger (50%) changes are seen in response to the same dose of fMLP. Microscopically, the populations of cells observed under the two conditions corresponding to these measurements were predominantly in the chemotactic or aggregated configurations respectively, so we can say that for this particular phase system 10% and 50% W_{adh} changes are associated with each of these cell states, when compared with the control state. Whether these associations are <u>causal</u> or not is another matter, discussed further below. Cytochalasin B alone in the concentration employed (1 µg/ml) had no effect on the control contact angles. The experiments on microscopic contact angles and divalent cation removal are discussed below (Sections 4 and 5).

In the presence of cytochalasin B, the fMLP contact angle dose-response curve is similar to the dose-response curve for the maximum increase in light transmittance, both having values of ED_{50} in the 2-5 x 10^{-8} M range. By contrast, the first peak of light scattering and chemiluminescence have ED_{50} ten times lower and higher, respectively. Fig. 3 shows the composite dose-response curve for the four photometric measurements (the first and second peaks of light scattering, the peak of light transmission and

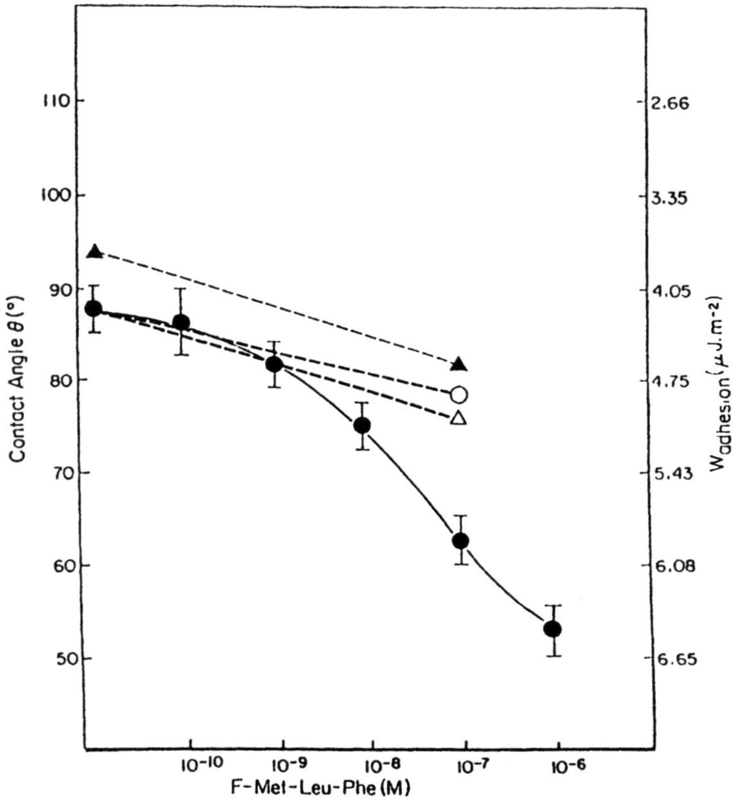

Fig. 2 *fMLP-dextran phase affinity dose-response curves for neutrophil contact angle measurements carried out under four different conditions: (▲) Micropipette aspiration technique; (○) macroscopic approach with 1 mM EDTA; (Δ) macroscopic approach with calcium and magnesium but no cytochalasin B; (●) macroscopic approach with calcium, magnesium and cytochalasin B*

chemiluminescence, all determined in the absence of cytochalasin B), and the single thermodynamic measurement, the work of adhesion response to fMLP in the presence of cytochalasin B. As has been shown previously, the first peak of light scattering has an ED_{50} in the nanomolar range [3,5], while chemiluminescence has an ED_{50} in the micromolar range. What has not been shown before, under conditions where kinetics presumably do not significantly confound the measurements, is the clear correlation between the dose-response curves for the increase in light transmission and the chemoattractant-induced increase in dextran phase affinity. Both curves have values of ED_{50} in the 2-5 x 10^{-8} M range, are parallel over their whole course and are statistically indistinguishable at any point. Thus, the cellular factors which are responsible for the increase in light transmission are also probably responsible for the increased dextran phase affinity in response to fMLP. It is more difficult to say if the dextran affinity increase is as well correlated with the second peak of light scattering, since this is a more difficult "peak" to locate, and since the scattered light signal is about 1000 times weaker in intensity than the transmitted signal, it is inherently much noisier. Thus, the apparent

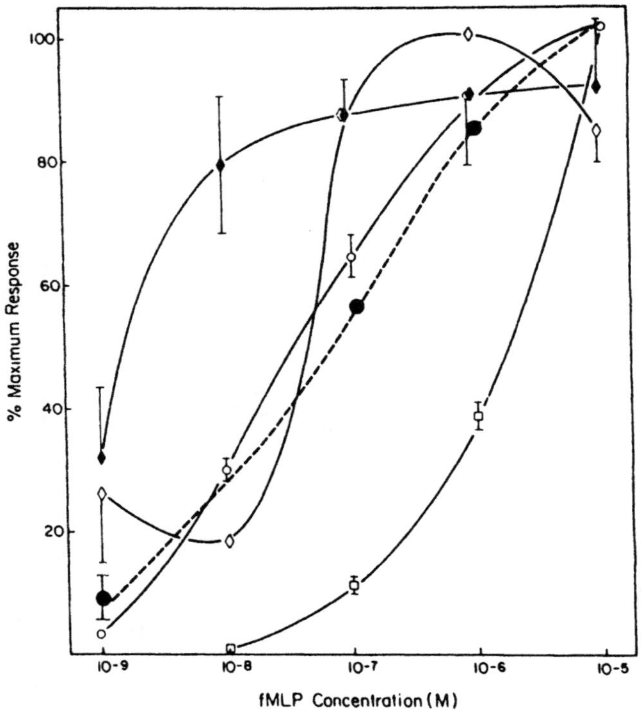

Fig. 3 *Composite dose-response curves (normalised to the maximum response as 100%) for the first (♦) and second (◊) peaks of perpendicular light scattering, light transmission (○), dextran phase affinity (●) and chemiluminescence (□)*

biphasic dose-response curve of the second peak of perpendicular light scattering may be artefactual; it may be that with more sensitive detection of scattered light, and an improved signal to noise ratio, that the second peak of scattering may also be found to correlate with dextran affinity. The obvious question that these observations raise is: what is the cellular origin of the light fluxes, and what is the relation of the dextran affinity changes to these processes?

Removing calcium from the medium or progressively adding EDTA reduces the light transmission change and the dextran phase affinity increase to 5% compared to control. The first peak of perpendicular light scattering is almost unaffected, but the second peak is substantially reduced. The effect of progressively lowering the medium calcium and magnesium provides part of the answer to the last question. As shown in Fig. 2, 0.1 mM EDTA for example almost abolishes the change in dextran phase affinity caused by fMLP. Fig. 4 shows that the change in light transmittance is almost abolished under the same conditions. In addition, quantitative aggregate size distribution counts (not shown) are reduced along with the transmittance changes, further strengthening the association between increased dextran phase affinity, enhanced light transmission and cell aggregation. On the other hand, the first peak of light scattering, which is thought to be associated with cell surface ruffling [2,3], is unaffected by conditions which profoundly affect the dextran phase affinity, indicating that surface ruffling *per se* likely does not contribute significantly to the macroscopic phase behaviour. This conclusion may also relate to the question "what is the effect of cell-surface roughness on the measured contact angles?" Obviously, this question

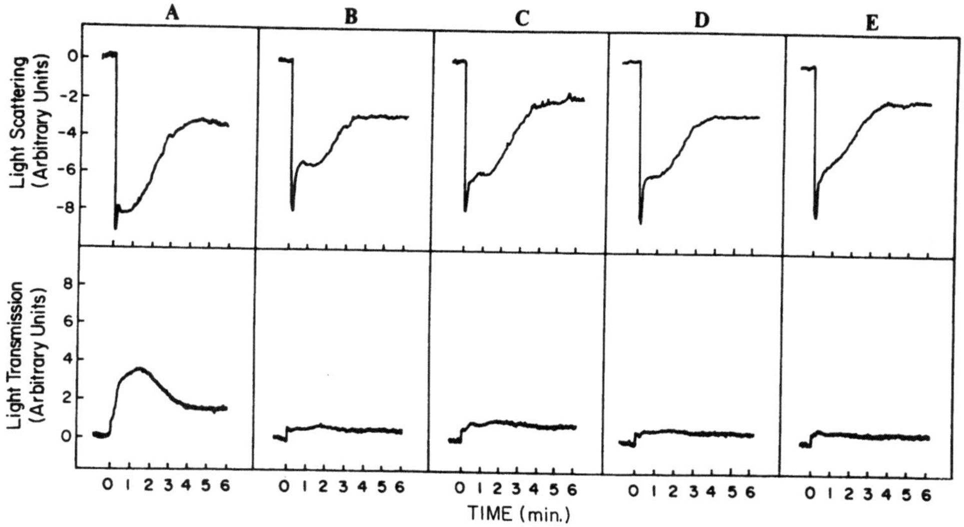

Fig. 4 *The effects of calcium removal and the progressive addition of EDTA on light transmission and perpendicular light scattering by neutrophil suspensions exposed to 10^{-7} M fMLP. A, Control response; B, No added calcium or magnesium; C, 0.1 mM EDTA; D, 1mM EDTA; E, 5mM EDTA*

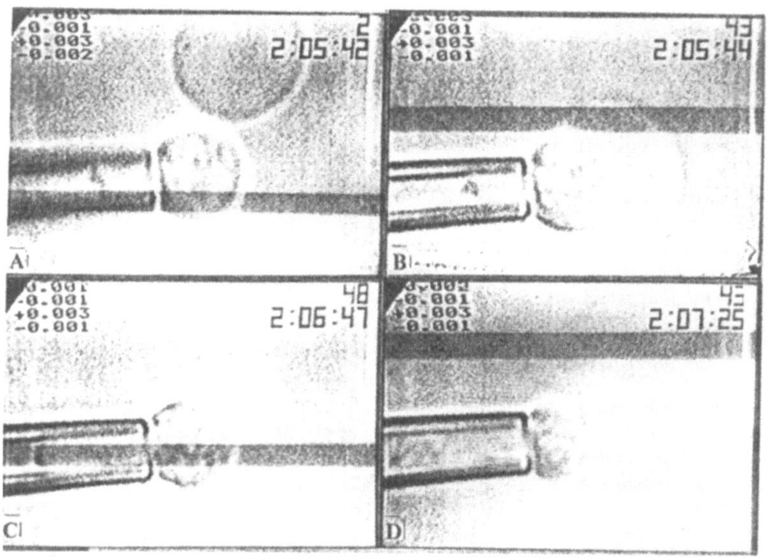

Fig. 5 *Video monitor photographs of single human neutrophils immersed in a 4%/4% dextran 2 M/PEG 20 K phase system comparable to that used in the preceding experiments. D.J.L. McIver, K.A. Sharp, D.E. Brooks and E.A. Evans [13]*

cannot be answered without specifying the drop size, or its ratio to the dimensions of the surface rugosities. For the drop sizes used here (0.1-1 mm) roughness on the scale of 0.1-1 μm range seems not to have an important influence on the contact angle. With much smaller drops (see below) this is probably not the case. The role of the second peak of light scattering is more difficult to analyse, partly because of the difficulty of deconvoluting the first and second peaks. However, the 50% reduction in the height of the second peak (possibly associated with reduction of degranulation) under the same conditions as inhibition of dextran phase affinity, light transmission increase and microscopic aggregation is consistent with the conclusion that all of these events may be related.

Preliminary micropipette aspiration studies on single neutrophils exposed to fMLP in dextran/PEG phases containing 0.1 mM EDTA demonstrate that under these conditions (1) cell wetting by the dextran phase occurs in about 2 seconds (2) that a 50 percent reduction in cell surface area has virtually no effect on the contact angle and (3) that 10^{-7} M fMLP has a small (5%) effect on the dextran affinity. Fig. 5 shows the results of preliminary experiments carried out in collaboration with Drs Kim Sharp, Don Brooks and Evan Evans [13], using the micropipette aspiration technique developed by Dr Evans [3]. In order to carry out these experiments without activating the neutrophils by contact with the micropipette, 0.1 mM EDTA is used in the suspending medium. This experimental constraint has an important bearing on the comparability of these experiments to some of the results reported above. Figs. 5A and 5B show that the time required to reach the final advancing contact angle (180° to 95°) is 2.41 seconds, in contrast to the three-five minutes required to reach equilibrium wetting in the macroscopic phase system shown in the previous paper; similar observations (not shown) have been made for the kinetics of neutrophil wetting by macroscopic polymer phase droplets. Figs. 5C and 5D show the results of an experiment in which the cell is aspirated into the pipette, sufficiently to cause a 50% reduction in the free area. If this manoeuvre were like compressing a monolyaer in a Wilhelmy balance (e.g. see 18), substantial changes in contact angle would be expected. In fact, no wetting changes are observed, suggesting that the effective surface concentration of the glycocalyx does not change under these conditions. Incidentally, this experiment (except for the presence of EDTA) probably models quite well what happens to a neutrophil as it gains egress from the vascular system by squeezing between the inter-endothelial slits. Teleologically, significant changes in adhesiveness during this process would probably be functionally detrimental. The effects of fMLP exposure on the microscopic contact angles were shown in Fig. 2: a 5% increase in the work of adhesion was observed with 10^{-7} M fMLP.

CONCLUSIONS

What do these experiments allow us to say about the mechanism of chemoattractant-induced cell adhesion? In particular, what can we say about the relations of this (these) process(es) to the changes in dextran phase affinity and to the underlying molecular events? The thermodynamic measurements are summarised in Fig. 2, which shows that the micropipette studies, and the macroscopic studies performed in the presence of EDTA or the absence of cytochalasin B show similar (small) trends of fMLP dependent increases in dextran phase affinity. By contrast, in the presence of cytochalasin B, a sigmoidal dose-response curve is seen for the change in contact angle with fMLP concentration, over the dose range which functional studies show produced marked alterations in cell behaviour. What are the functional changes which cause the dextran affinity changes? We suggest that they may be due to the addition of secretory granule membrane proteins to the cell surface following activation by an effective dose of fMLP, and that

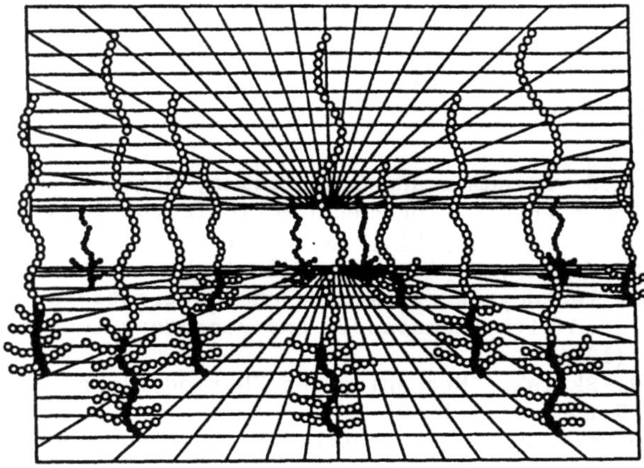

Fig. 6 *Schematic diagram of polymer mixing on the cell surface when the dextran contact angle approaches zero [12]. Reproduced with permission of Jonathan Wiley and Sons*

these proteins may be the molecular basis of both the increased dextran affinity and the cell-cell adhesion (aggregation). Degranulation and aggregation are both inhibited by EDTA; this may also be the explanation for the small phase affinity changes seen in the micropipette experiments. The aggregation process also spontaneously reverses in the absence of cytochalasin B; the affinity changes observed under these conditions are thus likely those associated with the state of the cell surface during chemotaxis. The aggregation process is stabilised by cytochalasin B, which presumably allows the initial dextran affinity changes to be observed under time-independent conditions.

Testing this proposal further requires the isolation, purification and reconstitution of the adhesion proteins into model membranes where their phase affinity behaviour can be studied under defined conditions [10]. In the meantime, some simple modelistic calculations may give some idea of what kind of phase behaviour to expect. If we carry out a Zisman-type of "critical spreading" [10] approach to the wetting of the neutrophil surface, we find that a maximally activated neutrophil (10^{-6} M fMLP in the presence of cytochalasin B) has a "critical spreading" polymer phase composition of 3.2%/3.2% dextran 2M/PEG 20K. Fig. 6 schematically illustrates the hypothesis that under these conditions, when the interfacial free energy between the cell surface and the dextran phase is very low and possibly zero, that the effective polymer concentrations in the two phases are equivalent. If this is correct, then the putative adhesion molecules must be present in quite high concentrations on the cell surface during maximal adhesions. Assuming a molecular weight in the 150-200 kdal range (based on our earlier studies of the relations between substrate adhesion, polymer phase affinity and membrane glycoprotein composition [11]) we reach the somewhat surprising conclusion that the adhesion molecules may attain micromolar concentrations on the cell surface during neutrophil adhesion. Or, given the high granule content of these cells (sometimes also known as "granulocytes"), and the possibility that aggregation and dextran phase affinity both result from the fusion of granule membrane proteins with the cell plasma membrane, perhaps this is not so surprising after all.

Supported by The Medical Research Council of Canada and The Heart and Stroke Foundation of Ontario.

REFERENCES

1 S.T. Hoffstein, R.S. Friedman and G. Weissman, Degranulation, membrane addition and shape change during chemotactic factor-induced aggregation of human neutrophils, J. Cell Biol. 95:234 (1982)

2 I. Yuli and R. Snyderman, Rapid changes in light scattering from human polymorphonuclear leukocytes exposed to chemoattractants. Discrete responses correlated with chemotactic and secretory functions, J. Clin. Invest. 73:1408 (1984)

3 E.A. Evans and R. Skalak, "Mechanics and Thermodynamics of Biomembranes," CRC Press, Boca Raton, Fl. (1980)

4 I. Yuli and R. Snyderman, Light scattering by polymorphonuclear leukocytes stimulated to aggregate under various pharmacological conditions, Blood 64:649 (1984)

5 P. Latimer, Light scattering and absorption as methods of studying cell population parameters, Ann. Rev. Biophys. Bioeng. 11:129 (1982)

6 L.A. Sklar, G.M. Omann and R.G. Painter, Relationship of actin polymerisation and depolymerisation to light scattering in human neutrophils: dependence on receptor occupancy and intracellular Ca^{++}, J. Cell Biol. 101:1161 (1985)

7 D.J.L. McIver and S. Schürch, Polymer mixing and the thermodynamics of cell adhesion at fluid interfaces, J. Adhesion, 22:253 (1987)

8 S. Schürch and D.J.L. McIver, Interfacial free energies at lipid-water interfaces. Comparison of equation-of-state predictions with direct experimental measurements, J. Colloid Interface Sci. 83:301 (1981)

9 S. Schürch, D.F. Gerson and D.J.L. McIver, Determination of cell-medium interfacial tensions from contact angles in aqueous polymer systems, Biochim. Biophys. Acta 640:557 (1981)

10 D.J.L. McIver and S. Schürch, Interfacial free energies of intact and reconstituted erythrocyte surfaces; implications for biological adhesion, Biochim. Biophys. Acta 691:52 (1983)

11 S.A. Johnstone, S. Schürch, D.J.L. McIver, E.A. Jacobson and E.R. Tustanoff, Membrane glycoprotein and surface free energy changes in hypoxic fibroblast cells, Biochim. Biophys. Acta 815:159 (1985)

12 D.J.L. McIver and S. Schürch, Free energies at the biosurface-water interface, in: "The Biophysics of Water," F. Franks and S. Mathias, eds., Jonathan Wiley, London (1983)

13 A. Boyum, Isolation of lymphocytes, granulocytes and macrophages, Scand. J. Immunol. 5, Supp. 5:9 (1976)

14 Y. Rotenburg, L. Boruvka and A.W. Neumann, Determination of surface tension and contact angle from the shapes of axisymmetric fluid interfaces, J. Colloid Int. Sci. 93:169 (1983)

15 D.J.L. McIver, K.A. Sharp, D.E. Brooks and E.A. Evans, unpublished results

MECHANISM OF CELL PARTITIONING; EQUILIBRIUM CONTACT ANGLE MEASUREMENTS PREDICT RELATIVE, KINETIC PARTITIONING BEHAVIOUR

Bernadette N. Youens*, W. David Cooper#, Frank D. Raymond*, Paul S Gascoine* and Derek Fisher*

*Department of Biochemistry, Royal Free Hospital School of Medicine, University of London, Rowland Hill Street London NW3 2PF, UK and # Shell Research Ltd, Thornton Research Centre, Chester CH1 1SH, UK

The relationships between cell partition, a dynamic process, and contact angle determinations, which are equilibrium thermodynamic measurements are investigated experimentally in biphasic systems of dextran and poly(ethylene glycol). Non-charge-sensitive partition, charge-sensitive partition and hydrophobic affinity cell partition all show correlations with contact angle measurements. Contact angles appear to provide a quantitative index of the degree of interaction of cells with the liquid-liquid interfaces of phase droplets which determine the kinetic features of partitioning.

INTRODUCTION

The partitioning of cells and other particulates (e.g. organelles and liposomes) in aqueous polymer two-phase systems, differs from that of macromolecules (e.g. proteins and nucleic acids) in that the bulk interface is involved; cells generally partition between one phase and the horizontal (bulk) interface.

Some years ago we made a number of observations which indicated that attention needed to be paid to the kinetics of partitioning [1-4]. The top phase partition of erythrocytes of different species in Dextran T500-PEG 6000 phase systems of low electrostatic potential difference between the phases (Dw) was time dependent. Although all the cells finally ended up at the bulk interface, which therefore represented their equilibrium position, the time courses (kinetics) of partitioning differed for the different cells; the relative order of top phase partition was:-

rat > turkey > chicken > human > pig > ox

Light microscopy studies led us to a qualitative explanation of these kinetics in terms of the association of cells with the surfaces of the phase droplets and streams (liquid-liquid microinterfaces) which form during the process of phase separation of the mixed system. Different cells were seen to differ in their ability to associate with the surfaces of the phase droplets. Strongly associating cells (e.g. pig erythrocytes) rapidly attached to the phase droplets and were rapidly delivered to the developing bulk interface in the early stages of phase separation. Their partition in the top phase was considerably lower at any time than that of weakly associating cells (e.g. rat erythrocytes) which were slow to attach to the droplets, finally attaching to the slowly moving small droplets that remained in the later stages of phase separation and were delivered slowly to the bulk interface. Cells attached to the outer surface of droplets present in the top phase and to the inner surface of droplets present in the bottom phase (Fig. 1); consequently any weakening of the association with the microinterfaces moved the cells into the PEG-rich bulk phase.

The association of the cells to the phase droplets was weakened by a number of factors:

(a) decrease in the concentration of the polymers in non-charge-sensitive phase systems
(b) increasing concentrations of PEG-palmitate (which interacts hydrophobically with the cell membrane)
(c) increase in the electrostatic potential difference between the phases ($\Delta\psi$)

which respectively increased top phase partition to provide:

(a) non-charge-sensitive partitioning
(b) hydrophobic affinity partitioning
(c) charge-sensitive partitioning

Subsequent studies have shown that these observations are not confined to the Dextran T500-PEG 6000 system; similar phase droplet-mediated kinetic features have been observed in the Aquaphase PPT-PEG 6000 [5] and the Dextran T40-PEG 6000 [6] phase systems.

In extending these studies we have considered that a quantitative index of the degree of interaction between solid and liquid interfaces may be provided by contact angle measurements [7]. A drop of dextran-rich bottom phase forms a contact angle (θ) on a layer of cells covered with PEG-rich top phase (Fig. 1). Therefore, we have examined experimentally the relationship between cell partitioning, a dynamic process, and contact angle determinations, which are equilibrium thermodynamic measurements, in a number of different partitioning conditions.

Top phase partition increased with increasing values of contact angles, showing a correlation that was consistent with that derived from theoretical treatments based on thermodynamic considerations by Sharp [8] and Gerson [9].

METHODS

PEG 8000 (Carbowax 8000, Union Carbide) was obtained from Unalco, London; PEG 6000 was obtained from BDH (Poole, UK); Dextran T500 was obtained from Pharmacia. PEG-palmitate was synthesised from PEG 6000 and palmitoyl chloride [10]. Phase systems were prepared on a w/w basis from stock aqueous solutions of 40% PEG, approximately 20% dextran (calibrated by polarimetry), 0.44M sodium phosphate buffer, pH 6.8, and 0.6M NaCl, and were allowed to phase separate at 25°C, the process being accelerated by centrifugation at 2500 g for 90 min. Top and bottom phases were separated and stored at 4°C.

Erythrocytes were isolated as described previously [1]. Cells were fixed by swirling in 0.15 M phosphate buffer, pH 7.4, containing 1% glutaraldehyde (70% v/v, double distilled, Ladd Chemical Company, Burlington Vermont).

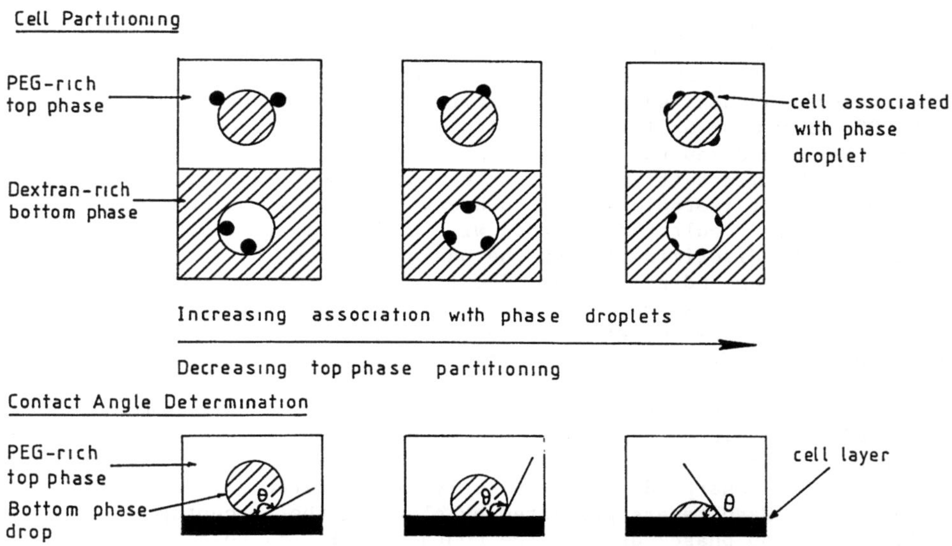

Fig. 1 *Association of cells with phase droplets during cell partitioning and contact angle determinations*

Cell partition was performed by adding 5 μl of packed cells to 5 x 1 cm glass tubes containing 1 g each of top and bottom phase. The contents were mixed for 1 min and the percentage of cells remaining in the top phase after a settling time of 30 min was determined electronically using a Coulter Counter.

Contact angle measurements were made on a lawn of cells produced by gently filtering the contents of the partition tube onto a 0.45 μm Millipore filter. A section was cut out, placed in a glass cuvette (5 x 1 cm) filled with top phase, and viewed horizontally in a Leitz microscope arranged with its stage in the vertical plane. Drops of the bottom phase were introduced with a capillary pipette (5 μl) and the angle of contact (Fig. 1) measured by means of the rotation of crosswires in the eyepiece. This method is far from being as refined as the procedure and apparatus used in the studies described by Boyce and Brooks [11] and must not be considered to provide an absolute measurement. However, it does provide a reproducible relative measure which is sufficient for the comparative studies described here.

Countercurrent distribution of mixtures of rat and human erythrocytes, and human and pig erythrocytes was performed in a Biosheff Mark 2 version [12] of Albertsson's thin layer apparatus comprising 60 chambers constructed by Dr T.E. Treffry and colleagues (Biochemistry Department, University of Sheffield). A phase system of intermediate $\Delta\psi$ of 5% Dextran T500 and 3.4% PEG 8000 in 0.075M NaCl buffered with 0.068M sodium phosphate, pH 6.8, equilibrated and separated at 4°C was used. 59 transfer cycles of 15 sec shaking and 7 min settling were performed at 4°C. Each fraction following the addition of isotonic saline to break the phases was counted electronically (Coulter Model ZBI) with a Channelyzer C1000 accessory, which permitted the discrimination on the basis of volume between rat and human cells, and human and pig cells. The fraction in the CCD profile that contained the highest number of cells of each type was noted (i) and used to provide an approximate value of the cells' partition ratio, G, by use of the relationship $G=i/(n-i)$ where n is the number of transfers [13].

Correlations of partition with contact angle (θ) were examined by performing the linear regression of log partition coefficient (K; ratio of the number of cells in the top phase to the remaining number of cells) on $(1 + \cos\theta)^2$. Sharp [8] has previously demonstrated such a linear dependence. Results are presented by plotting the percentage partition (% of cell in the top phase) against the contact angle (θ) together with the curve derived from fitting the data to the linear regression.

RESULTS

Background - Kinetics of cell partitioning

Fig. 2 gives examples of time courses observed for the non-charge-sensitive partition and for hydrophobic affinity cell partition of rat, human and pig erythrocytes. In both systems the relative top phase partition was

rat > human > pig

Fig. 2 also shows that charge-sensitive partition of human erythrocytes shows kinetic features.

Non-charge-sensitive cell partitioning and contact angle determinations

The partitioning and contact angle measurements of the erythrocytes (which had been fixed with glutaraldehyde) in a phase system of low $\Delta\psi$ were

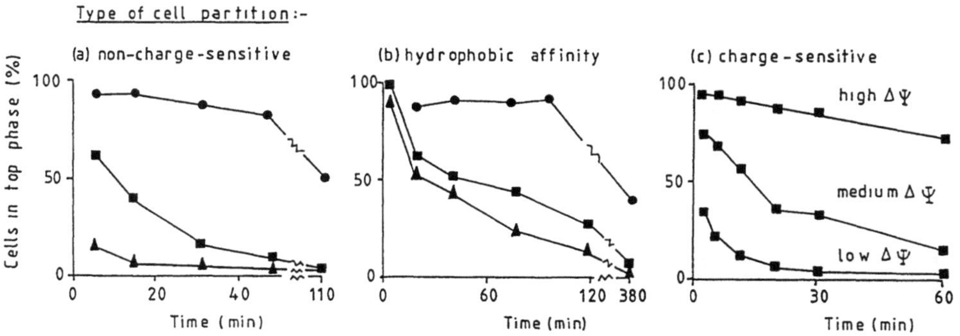

Fig. 2 *Time dependence of (a) non-charge-sensitive partitioning, (b) hydrophobic affinity partitioning and (c) charge sensitive partitioning of erythrocytes. Erythrocytes from rat (●), human (■) and pig (▲) were partitioned in (a) 5% Dextran T500-4.17% PEG 6000 in 0.15 M NaCl, 0.01M sodium phosphate, pH 6.8 (b) 5% Dextran T500-5% PEG 6000 in 0.15 M NaCl, 0.01 M sodium phosphate buffer, pH 6.8 containing 20 μg PEG-palmitate per 2 g of reconstituted phase system (c) 5% Dextran T500-5% PEG 6000 in 0.15 M NaCl, 0.01 M sodium phosphate (low Δψ), in 0.075 M NaCl-0.068 M sodium phosphate (medium Δψ), and in 0.03M NaCl-0.09M sodium phosphate (high Δψ)*

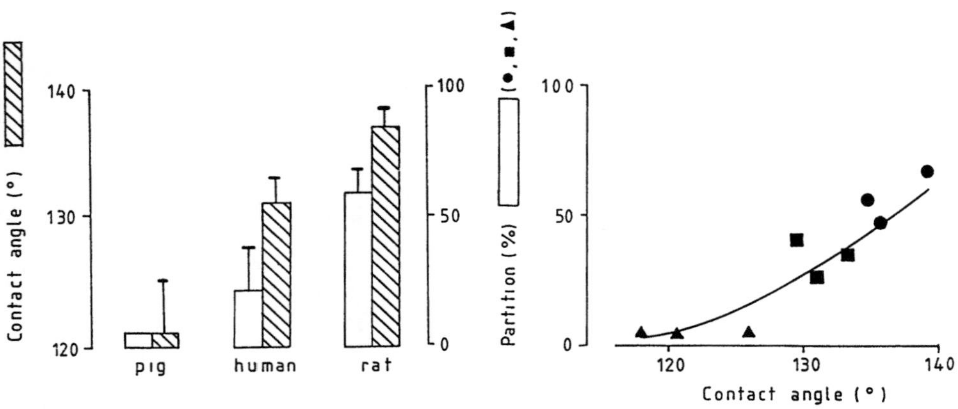

Fig. 3 *Correlation of non-charge sensitive partitioning of erythrocytes with contact angles. Fixed erythrocytes from rat (●), human (■) and pig (▲) were used in a phase system of 5% Dextran T500 and 4.2% PEG 6000 phase system in 0.15M NaCl containing 0.01M sodium phosphate buffer. The diagram on the left shows results of three experiments. On the right the curve shows the dependence of the percentage of cells in the top phase on contact angle calculated by fitting the data to a linear regression of log K on* $(1 + \cos\theta)^2$*. (r = 0.825; n = 9)*

compared (Fig. 3). The polymer concentrations were selected to give a phase system that was close to the critical point with significant top phase partitioning of cells at 20 min (5% Dextran T500, 4.2% PEG 6000 in 0.15M NaCl with 0.01M sodium phosphate, pH 6.8). The relative order of partition was:

rat > human > pig

as observed in the kinetic study in Fig. 2. Contact angles increased in the same order and correlated well with the partition ($r = 0.825$; $n = 9$; $P < 0.01$).

Charge sensitive partitioning and contact angle determinations

A phase system of 5% Dextran T500, 4.8 % PEG 6000 in 0.03M NaCl containing 0.09M phosphate buffer was used to provide charge-sensitive partitions of fixed erythrocytes from rat, human and pig that were markedly different (Fig. 4). Top phase partition and contact angles showed the same relative order:

rat > human > pig

and correlated well ($r = 0.93$, $n = 9$; $P < 0.001$).

The influence of contact angles on the separability of cells in mixtures was also examined. Mixtures of erythrocytes were separated by CCD in a charge-sensitive phase system (Fig. 5). The peak of rat erythrocytes appeared in fractions higher than the peak of human erythrocytes which in turn was in fractions higher than the peak of pig cells. This indicated an order of top phase partitioning:

rat > human > pig.

From the peak positions the effective partition ratio (G) was calculated. Fig. 5 shows that partition increased with the contact angles of the cells measured in the phase system used for the CCD separation.

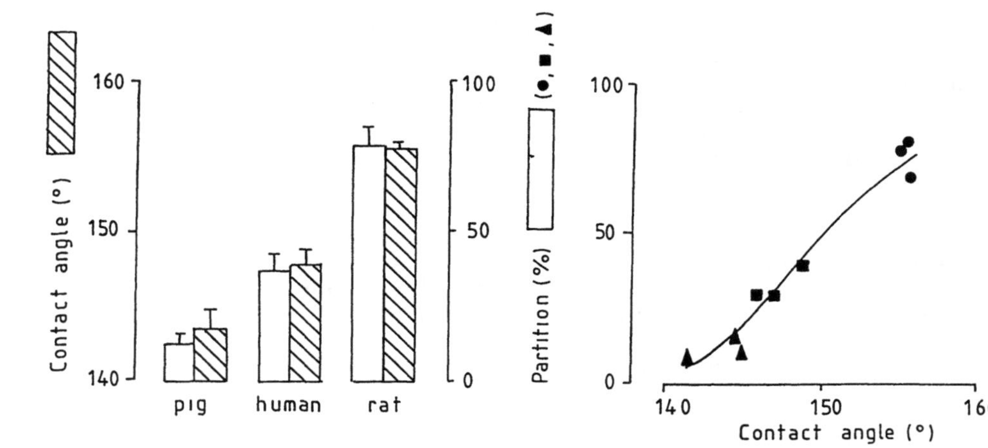

Fig. 4 *Correlation of charge-sensitive partitioning of erythrocytes with contact angles. Fixed erythrocytes from rat (●), human (■) and pig (▲) were used in a phase system of 5% Dextran T500 and 4.8% PEG 6000 phase system in 0.03M NaCl containing 0.09M sodium phosphate buffer. The diagram on the left shows results of three experiments. On the right the curve shows the dependence of the percentage of cells in the top phase on contact angle calculated by fitting the data to a linear regression of log K on* $(1 + \cos \theta)^2$. *(r = 0.93; n = 9)*

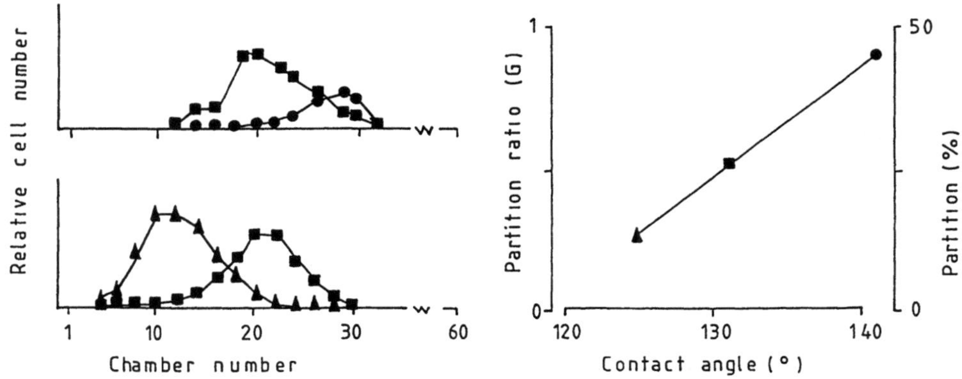

Fig. 5 *Correlation of the position of cells in countercurrent distribution with contact angles. Mixtures of erythrocytes from rat (●) and human (■), and pig (▲) were separated in a charge-sensitive phase system of 5% Dextran T500 and 3.4% PEG 8000 in 0.075M NaCl containing 0.068 M sodium phosphate buffer. On the right is shown the dependence of the partition ratio, G, calculated from peak position, on contact angle. The curve shows the dependence of G on contact angle calculated by fitting the data to a linear regression of log G on* $(1 + \cos \theta)^2$

Hydrophobic affinity cell partitioning and contact angle determinations

Cells were partitioned in a phase system of 5% Dextran T500 and 4.5% PEG 8000 containing PEG-palmitate. With increasing concentrations of the hydrophobic affinity ligand both the top phase partition and the contact angles increased. The partitioning of rat, human and pig erythrocytes differed markedly in sensitivity to PEG-palmitate, rat cells requiring much less than pig to take them completely into the top phase; human cells being intermediate in response. In Fig. 6 the dependence of partitioning on the contact angle in the presence of the same concentration of PEG-palmitate (20 μg added to a 2 g phase system) is shown. Because of the very different sensitivities of the cells' partitioning to PEG-palmitate there was only a narrow range of concentrations over which a comparison can be made i.e. when all three samples of erythrocytes show significant partitions without at least one sample being totally in the top phase.

The order of top phase partitioning and contact angles was:

rat> human > pig

and partitioning and θ correlated well ($r = 0.78$; $n = 9$; $P < 0.02$).

DISCUSSION

We find that equilibrium contact angle determinations predict the relative partitioning behaviour of cells. The top phase partition of the cells we have studied in a number of two-phase systems of PEG and dextran, sensitive to different surface properties, increased with increasing contact angles formed by drops of lower phase on layers of the cells covered by the top phase. We interpret this (Fig. 1) in terms of θ providing a measure of the association of cells to the microinterfaces of phase droplets; higher θ represents weaker association with phase droplets leading to higher top phase partition.

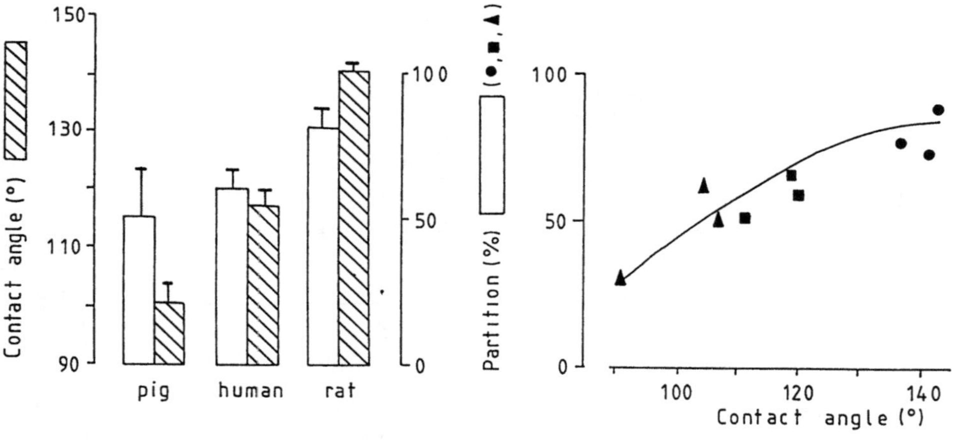

Fig. 6 *Correlation of hydrophobic affinity partitoning of cells with contact angles. Rat (●), human (■) and pig (▲) erythrocytes were partitioned in a non-charge-sensitive phase system of 5% Dextran T500 and 4.5% PEG 8000 in 0.15M NaCl and 0.01M sodium phosphate buffer containing PEG-palmitate. Contact angles were also determined. The diagram on the left shows results of three experiments. On the right the curve shows the dependence of the percentage of cells in the top phase on contact angle calculated by fitting the data to a linear regression of log K on* $(1 + \cos\theta)^2$ *($r = 0.78$; $n = 9$)*

McIver and Schürch [14] have used contact angle determinations to measure the wetting of cell surfaces by aqueous phase separated polymer solutions of dextran and PEG, expressed as the work of adhesion, W_{adh}, where

$$W_{adh} = \gamma_{12} (1 + \cos\theta)$$

where γ_{12} is the liquid-liquid interfacial tension.

A detailed thermodynamic theory for cell partitioning has been developed by Sharp, described in [8]. A close association between the predicted partition coefficient and the degree to which the cell is preferentially wetted by one of the phases, as described by contact angle measuements, is brought out; an exponential dependence of partition ratio on $(1 + \cos\theta)^2$ is predicted. Gerson has suggested a similar dependence [9]. When tested experimentally by Sharp, the form of this exponential dependence was confirmed, but the numerical coefficients were orders of magnitude different from those predicted; cells showed appreciable top phase partitioning when the theory predicted they should all be at the bulk interface.

The kinetic studies of cell partitioning identify two reasons for this [1-4, 8]. Firstly cells can be present at their equilibrium position at the liquid-liquid interface, but that corresponding to the surfaces of phase droplets as well as the bulk interface. If these phase droplets are small they, and their associated cells, will settle slowly to the bulk interface; cells will be present in the bulk top phase well after the bulk interface has formed and the apparent partition coefficient will be high. Secondly the association of cells to microinterfaces will be counteracted by the shear forces present as the phase streams and droplets move through the developing bulk phases. Weakly associating cells will find it more difficult to attach to the microinterfaces of fast moving streams and rapidly settling large droplets than will strongly associating cells. Attachment to slower moving droplets results in the weakly attaching cells finding their way to the bulk interface more slowly; their partition coefficient will be higher than predicted from thermodynamic considerations.

Despite the nonequilibrium conditions that pertain in cell partitioning our results are also consistent with the exponential dependence of partitioning on $(1 + \cos\theta)^2$ that thermodynamic considerations predict. This is not surprising since the theoretical considerations of Sharp are based on the thermodynamic measurments of the work required to remove the cell from the interface, assuming only thermal energies are responsible for the removal. The fact that there are other forces arising from the dymamics of partitioning which will influence this removal suggests, at a simple experimental level at least, that the thermodynamic relationships will act as a relative predictor of partitioning behaviour. However, the most important feature of the relation between partition and contact angle is its exponential nature; partitioning is a much more sensitive indicator of surface properties than the contact angle. Separation arising from small differences in contact angles are to be expected as we have shown. Nevertheless some caution needs to be expressed about the exact dependence on contact angles. In the range of angles encountered in this study (70°-140°); $(1+ \cos\theta)^2$, $(1 + \cos\theta)$ and θ are virtually linearly related and discrimination by correlations cannot be achieved.

Although contact angle measurement are not as sensitive or as easy to make as partition measurements, they offer an attractive experimental system with which to examine the membrane and surface components that influence the cell-phase interactions. The contact angle approach permits these interactions that contribute greatly to determining the partitioning behaviour of cells and other particulates to be studied free from the complicating kinetic features of partitioning.

This work has been supported by MRC, SERC (CASE Award to BYN), Shell Research Ltd, The Royal Society, the Royal Free Hospital School of Medicine (Junior Research Fellowship to FDR) and the Peter Samuel Royal Free Fund.

REFERENCES

1 F.D. Raymond and D. Fisher, Partition of rat erythrocytes in aqueous polymer two-phase systems, Biochim. Biophys. Acta 596:445 (1980)

2 F.D. Raymond and D. Fisher, Effect of poly(ethylene glycol)-palmitate on the partition of cells in aqueous polymer two-phase systems, Biochem. Soc. Trans. 8:118 (1980)

3 F.D. Raymond and D. Fisher, Cell surface properties and cell partition in two-polymer aqueous phase systems, in: "Cell Electrophoresis in Cancer and other Clinical Research," A.W. Preece and P.A. Light, eds., Elsevier North Holland Biomedical Press, Amsterdam (1981)

4 F.D. Raymond, The partition of cells in two polymer aqueous phase systems, PhD Thesis, University of London (1981)

5 B.N. Youens, Cell partition and interfacial properties in two-polymer aqueous phase systems, PhD Thesis, University of London (1986)

6 I. Shelton, The partition of cells in two-polymer aqueous phase systems, PhD Thesis, University of London (1984)

7 B.N. Youens, A.W. Foulds, W.D. Cooper and D. Fisher, Interfacial properties of cells partitioning in aqueous polymer two-phase systems, Biochem. Soc. Trans. 14:751 (1986)

8 D.E. Brooks, K.A. Sharp and D. Fisher, Theoretical Aspects of Partitioning, in: "Partitionining in Aqueous Two-Phase Systems. Theory, Methods, Uses, and Applications to Biotechnology," H. Walter, D.E. Brooks and D. Fisher, ed., Academic Press (1985)

9 D.F. Gerson, Cell surface energy, contact angles and phase partition I. Lymphocytic cell lines in biphasic aqueous mixtures, Biochim. Biophys. Acta 602: 269 (1980)

10 V.P. Shanbhag and G. Johansson, Specific extraction of human serum albumin by partition in aqueous biphasic systems containing pol(ethylene glycol)-bound ligand, Biochim. Biophys. Res. Commun. 61:1141 (1974)

11 J.F. Boyce and D.E. Brooks, Contact angles as an analytical tool for investigating two-phase interactions with biological surfaces: a review, in: "Separations Using Aqueous Phase Systems: Applications in Cell Biology and Biotechnology," D. Fisher and I.A. Sutherland, eds., Plenum, New York (1989)

12 T.E. Treffry, P.T. Sharpe, H. Walter and D.E Brooks, Thin-layer countercurrent distribution and apparatus, in: "Partitionining in Aqueous Two-Phase Systems. Theory, Methods, Uses, and Applications to Biotechnology," H. Walter, D.E. Brooks and D. Fisher ed., Academic Press, Orlando (1985)

13 H. Walter, D.E. Brooks and D. Fisher"Partitionining in Aqueous Two-Phase Systems. Theory, Methods, Uses, and Applications to Biotechnology," Academic Press, Orlando (1985)

14 D.J.L. McIver and S. Schürch, Polymer mixing and the thermodynamics of cell adhesion at fluid interfaces, J. Adhesion 22: 253 (1987)

DEMIXING KINETICS OF PHASE SEPARATED POLYMER SOLUTIONS IN MICROGRAVITY

S. Bamberger [1], D.E. Brooks [2], J.M. Van Alstine [1]
J.M. Harris [3] and R.S. Snyder [1]

[1] Space Science Laboratory, NASA/Marshall Space Flight Center, ES76, Alabama 35812
[2] Departments of Pathology and Chemistry, University of British Columbia, Vancouver, BC V6T 1W5, Canada
[3] Department of Chemistry, University of Alabama Huntsville, Alabama 35899

The efficiency of particle separation in polymer two-phase systems appears to be up to several orders of magnitude lower than theoretical estimations based on Boltzmann statistics. To test the hypothesis that this inefficiency is at least partially due to the convection and sedimentation that occur during gravity-driven demixing we have begun a series of experiments aimed at characterising particle partitions in space. The one-g demixing of systems possessing equal densities has been studied. We have also studied the demixing of PEG/dextran systems in low gravity on KC-135 aircraft and during one Shuttle mid-deck experiment. The demixing rate data from these experiments have been analysed by two different approaches. In both cases the results are consistent with coalescence being a dominant feature of the gravity-free demixing process.

The distribution of particles partitioned in PEG/dextran aqueous two-phase systems has been studied to test predictions made on the basis of Boltzmann equilibrium as determined only by the temperature, T, and the difference in free energy of a particle between either phase and the interface, $\Delta G°$ [1,2]. Under these assumptions the partition coefficient, K, (defined as the ratio of the number of particles free in the bulk phase to the number absorbed at the interface) can be expressed as a function of T and $\Delta G°$ by the Boltzmann equation:

$$K = \exp(-\Delta G°/kT) = \exp(-\Delta G'A/kT)$$

where $\Delta G'$ is the net free energy per unit area of the particle, A the surface area of the particle, and k the Boltzmann constant. Given a series of phase systems differing primarily in interfacial tension between the phases, τ, the relationship between log K and τ should be a straight line with a slope of -A/kT. Such a relationship has been verified experimentally and values of T determined from partition coefficient, particle surface area, and interfacial tensions. Theoretically calculated data differed widely from the experimental with the apparent values of kT larger by several orders of magnitude. These results suggest that the Boltzmann equilibrium for particles in two-phase systems is disturbed by some additional randomising force(s).

In the presence of gravity the phase separation process starts with a brief quiescent period where phase droplet domains grow as slowly coalescing droplet clusters. This period is followed, as the domains grow larger, by turbulent gravity-driven convective streaming of cylindrical phase regions past each other. This fluid movement creates a chaotic environment with shear forces likely to have a randomising effect on cell partitioning. In the final stage of demixing isolated phase droplets move to their respective bulk phase. In this stage particle-droplet interaction can strongly influence the final position of a particle depending on the droplet size [3,4]. Once the nature of the randomising force(s) is understood the separation procedure can be modified to minimise these effects. A first step towards this goal is a better knowledge of the influence of gravity on phase system demixing and phase droplet growth in the absence of particles or other material to be partitioned.

Factors likely to affect droplet growth and interaction are: interfacial tension, phase viscosities, bulk phase potential, density difference between the phases, phase volume ratio, wetting of the inner container walls by proximal phase droplets and, particularly at later stages, the container geometry [5]. Of these, turbulent fluid movement originating from the action of gravity is the least controllable. In the experiments presented here this force was suppressed in two ways: the gravitational field was virtually eliminated by performing experiments on Shuttle flight 51D and during short term (up to 30s) low gravity conditions on a KC-135 plane, and secondly phase density difference was balanced by adding small amounts of PEG to create an isopycnic dextran/Ficoll system. In all three types of experiments the

Table 1 *Composition and physical properties of phase systems with equal volumes of each phase containing the polymers dextran (DEX), poly(ethylene glycol) MW 8000 (PEG) and Ficoll 400 (FIC) (MW 400,000). Salt concentrations: 109 mM Na_2HPO_4, 35 mM NaH_2PO_4, pH 7.3 (I); 150 mM NaCl, 7.3 mM Na_2HPO_4, 2.3 mM NaH_2PO_4, pH 7.2 (II); 84.6 mM Na_2HPO_4, 25.4 mM NaH_2PO_4, pH 7.5 (III). All dextran fractions have MW 500,000 except the (7/5/0) system which contains a fraction with MW 40,000. (For experimental details see 5 and 6)*

Composition (% w/w) (DEX/PEG/FIC)	Salt	Viscosity (cP)		Interfacial Tension (μN/m)	Density (g/ml)		Tie Line Length (% w/w)
		Top	Bottom		Top	Bottom	
(5/3.5/0)[a]	I	3.36	24.6	5.24	1.0263	1.0563	9.41
(5/4/0)[a]	I	3.41	32.1	10.54	1.0281	1.0566	10.97
(7/5/0)[a,b]	II	4.96	12.2	6.00	1.0277	1.0578	11.69
(5.5/0.67/9.5)[c,d]	III	39.6	19.4	3.9	1.072	1.072	ND[e]
(7/0.29/12)[a,c,d]	III	ND	ND	16	1.085	1.085	ND

[a] System flown during space shuttle flight 51D.
[b] System flown on KC-135 aircraft.
[c] Isopycnic system at 1 g.
[d] The phase properties of the isopycnic systems are listed as follows: The Ficoll-rich phase in "Top" column, the dextran-rich phase in "Bottom" column.
[e] ND: not determined.

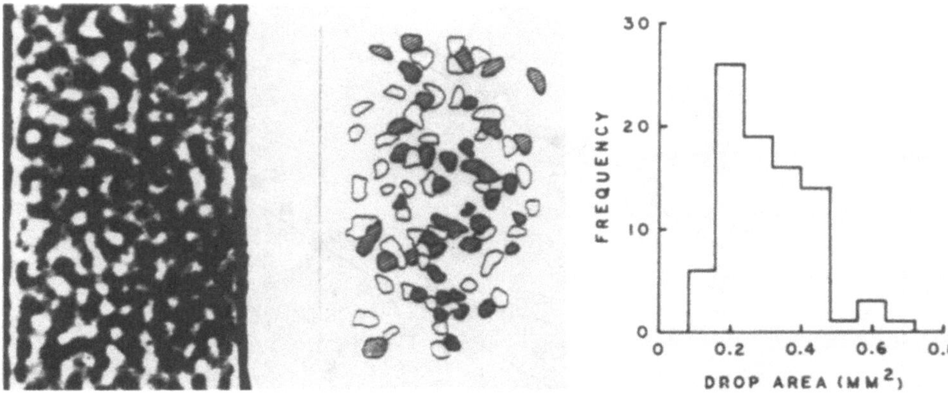

Fig. 1 *Determination of domain growth in dextran/PEG/Ficoll phase system (5.5/0.67/9.5) [See Table 1 for details]. Left: 90s after mixing. Middle: traced domains. Right: Histogram of traced domains*

systems were manually shaken and then photographed in a time sequence. To enhance the contrast between the phases 0.05 mg/ml of Trypan Blue dye was added to preferentially dye the PEG-rich or the Ficoll-rich phase. The dye did not measurably alter the physical properties of the systems. The composition of the systems and some of their physical properties [5,6] are listed in Table 1.

To assess gravity-independent demixing rates, photographs of demixing systems taken at defined time intervals were projected, and particularly well defined domains were traced. The resulting areas were digitised and the mean area for each time point calculated (Fig. 1). Table 2 presents the data reduced to the coefficients of the linear regression lines for a double logarithmic dependence. Fig. 2 demonstrates this dependence for two of the systems. In spite of the largely different phase viscosities and interfacial

Table 2 *Linear regression parameters of the logarithmic values of the mean radius (r [mm]) versus time (t [s]). CC = correlation coefficient*

System (See Table 1)	log r = A + B log t		
	A	B	CC
<u>Isopycnic 1g</u>			
(5.5/0.69/9.5)	-2.11	0.80	0.985
(7/0.29/12)	-1.23	0.55	0.970
<u>KC-135</u>			
(7/5/0)	-1.22	0.78	0.954
<u>Shuttle</u>			
(7/0.29/12)	-1.31	0.61	0.967
(7/5/0)	-0.89	0.66	0.989
(5/3/5/0)	-0.78	0.51	0.917

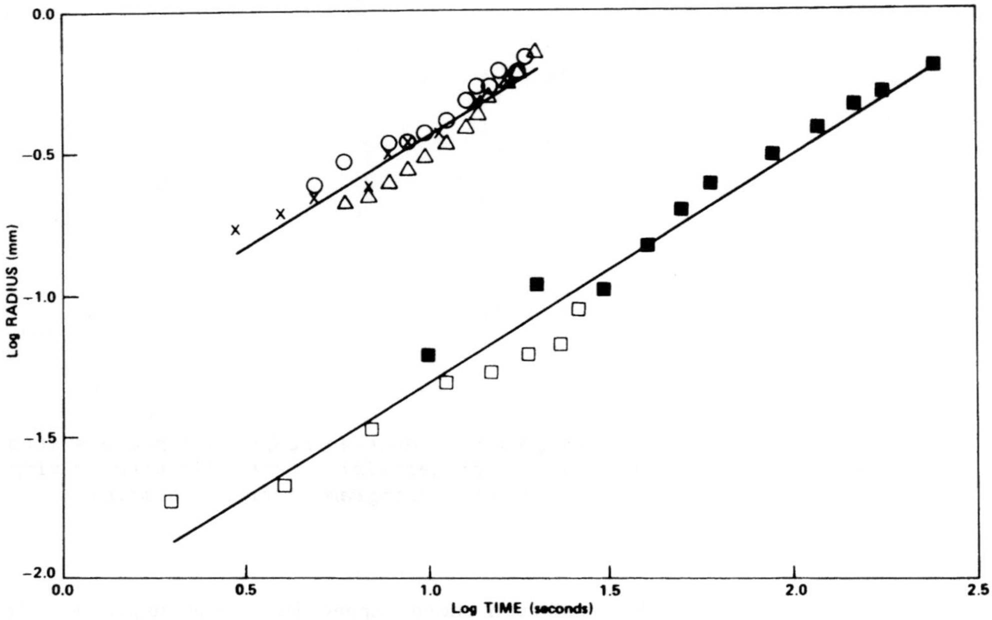

Fig. 2 *Growth of mean droplet radius with time in dextran, PEG, Ficoll phase system (7/5/0 - See Table 1 for details) on board of KC-135 (o, Δ, x), and in dextran, PEG, Ficoll phase system (5.5/0.67/9.5 - See Table 1 for details) at 1 g (□,■). Symbols represent individual experiments*

tensions the slopes of the regression lines average to 0.65 $\pm$ 0.11 suggesting a power law dependence between droplet radius and time.

The manual method of data analysis described above can be somewhat arbitrary in that domain recognition is dependent on the skill of the operator. We therefore analysed a control sample of the data by digitising the pictures using a Panasonic video camera, Model WV1800 with 12.5/75 mm zoom lens and TV Vision Frame Grabber (Image Technology, Woburn, Mass.) and subjecting the data to two-dimensional Fourier transformation [7-9]. The dominant frequency in both dimensions, originating from repetitive structures in the original picture, was determined and converted to its respective length. The example given in Fig. 3 indicates a mean repetitive structural distance of 0.80 mm. Considering that at a phase volume ratio of 1.0 the distance between adjacent domains of the same phase should be between one and two times the domain diameter, this is in good agreement with the mean domain diameter of 0.44 mm obtained by the first method. We are in the process of re-analysing all collected data with the improved method.

The results obtained have led to a number of conclusions concerning the demixing kinetics of polymer two-phase systems [5]. Gravity independent phase demixing proceeds at a more gentle rate than gravity dependent demixing, but at a pace adequate for Space Bioprocessing. Kinetic data from isopycnic systems and KC-135 based experiments are in agreement with data obtained in orbit. Phase droplet growth kinetics are more in keeping with coalescence processes rather than diffusion based mechanisms such as Ostwald ripening [5,6]. As expected, such kinetics appear to be strongly influenced by phase volume ratio, phase viscosities and interfacial tension. The amount of data is presently not sufficient to evaluate the dependence of the phase droplet growth on the parameters in question. More experimental data will be gathered on shuttle mission STS-26 projected for August 1988.

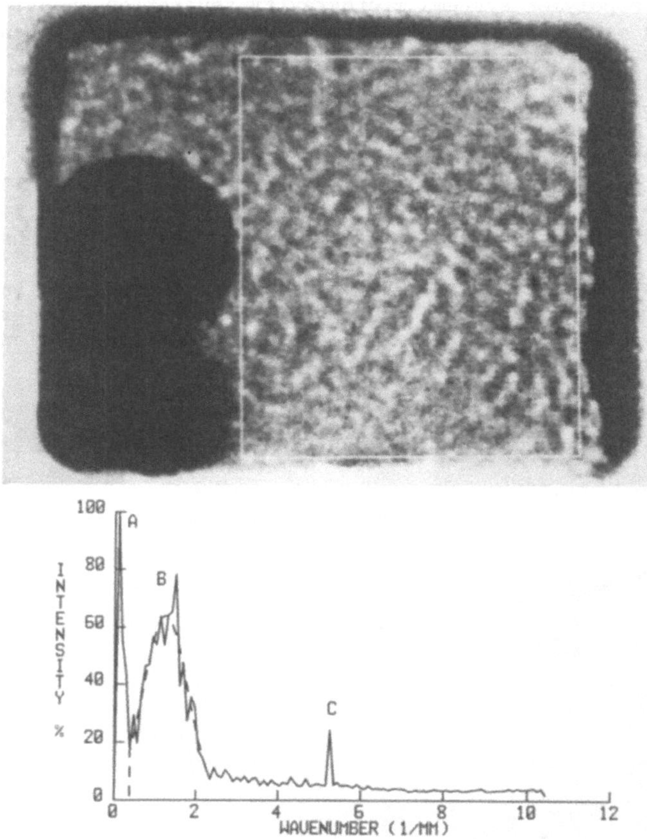

Fig. 3 *Determination of dominant frequency in (5.5/0.67/9.5). Top: chamber with phase system at 19 s after mixing. The two dark circular areas are the shadows of the mixing ball and the fill port. Bottom: Frequency distribution of the Fourier transformed area inside the white rectangle. Peak A results from inhomogeneities in domain size and intensity between areas close to the edge and at the centre. Peak B represents the frequency distribution of the dark and light regions. The dotted line is a Gauss distribution fitted to the values of peak B. Peak C is due to the grain size of the photograph*

ACKNOWLEDGEMENTS

S.B. presently holds a Research Associateship from the National Research Council. J.M.V.A. is supported by a fellowship from the University Space Research Association. This research is supported by a NASA grant (contract no. NAS8-35333).

REFERENCES

1 D.E. Brooks, K.A. Sharp and D. Fisher, Theoretical aspects of partitioning, in: "Partitioning in Aqueous Two-Phase Systems," H. Walter, D.E. Brooks, D. Fisher, eds., Academic Press, Orlando, Florida (1985)
2 D.E. Brooks, S.B. Bamberger, J.M. Harris and J.M. Van Alstine, Rationale for two phase polymer systems microgravity separation experiments, in: "Proceedings of the 5th European Symposium on Materials Sciences under Microgravity," European Space Agency (1984)
3 F.D. Raymond and D. Fisher, Effect of poly(ethylene glycol)-palmitate on the partition of cells in aqueous polymer two-phase systems, Biochem. Soc. Trans., 8:118 (1980)
4 F.D. Raymond and D. Fisher, Partition of rat erythrocytes in aqueous polymer two-phase systems, Biochim. Biophys. Acta 596:445 (1980)
5 S. Bamberger, J.M. Van Alstine, J.M. Harris, J.K. Baird, R.S. Snyder, J. Boyce and D.E. Brooks, Demixing of aqueous polymer two-phase systems in low gravity, Sep. Sci. Tech. 23:17 (1988)
6 J.M. Van Alstine, J.M. Harris, S. Snyder, P.A. Curreri, S. Bamberger and D.E. Brooks, Separation of aqueous polymer systems in microgravity, in: "Proceedings of the 5th European Symposium on Materials Sciences under Microgravity," European Space Agency (1984)
7 O.E. Brigham, "The Fast Fourier Transform", Prentice-Hall, Englewood Cliffs, N.J. (1974)
8 R.C. Singleton, On computing the fast fourier transform, Comm. ACM, 10:647 (1967)

DEXTRAN COATED GLASS SLIDES AS A MODEL FOR CELL SURFACE INTERACTIONS WITH AQUEOUS TWO PHASE POLYMER SYSTEMS

J.F. Boyce[1], J.M. Harris[2], J.M. Van Alstine[3] and D.E. Brooks[1]

[1] Department of Pathology, University of British Columbia
2211 Wesbrook Mall, Vancouver, B.C., Canada V6T 1W5
[2] Department of Chemistry, University of Alabama at Huntsville, Alabama, USA 35899
[3] USRA Space Sciences Lab, Marshall Space Flight Center Huntsville, Alabama, USA 35812

The significance of the external cell surface as a major determinant of cell partition behaviour is well established. Since the carbohydrate-rich glycocalyx is the site of many of the cell's primary interactions with its environment, glycocalyx structure and composition are important. Unfortunately, little is known about the physical structure of the glycocalyx such as its depth, density and mass distribution throughout the region.

As a first attempt to investigate some of these questions by surface energy methods, we have established a crude model of the glycocalyx using glass slides coated with terminally bound dextran polymers of different molecular weights. The quartz glass was first activated with aminopropylsilane before dextran was attached through reductive amination of its reducing end group in the presence of sodium cyanoborohydride [1]. Coatings prepared from narrow cut fractions of dextran with molecular weights of: 27.9k, 52k, 107k, 147.5k and 205k were examined using a sessile drop-contact angle technique. To compare the effects of different polymer sizes in the bulk solution, phase diagrams of low and high molecular weight dextrans with poly(ethylene glycol) (PEG) were constructed. From this information two systems, each having a tie line length of 14.7% (w/w) were chosen: 5% dextran 27.9k, 8.4% PEG 8000 and 5% dextran 205k, 5.25% PEG 8000.

For testing, samples of the coated glass (10 x 25 x 2 mm) were placed in a narrow viewing chamber and immersed in the PEG-rich upper phase. With the aid of a micromanipulator, small droplets (0.5 - 1 mm diameter) of the denser, dextran-rich, bottom phase were placed randomly over the glass surface and viewed from the side through a horizontally mounted Zeiss Tessovar microscope. The droplets were photographed 1 hr and 24 hr after being placed on the substrate to give some indication of the equilibration time with the surface. Enlarged tracings of the droplet profiles were made from the photographic negatives and these were digitised so that contact angles and interfacial tensions could be calculated by computer using Rotenberg's Axisymmetric Drop programme [2]. The results are summarised in Fig. 1.

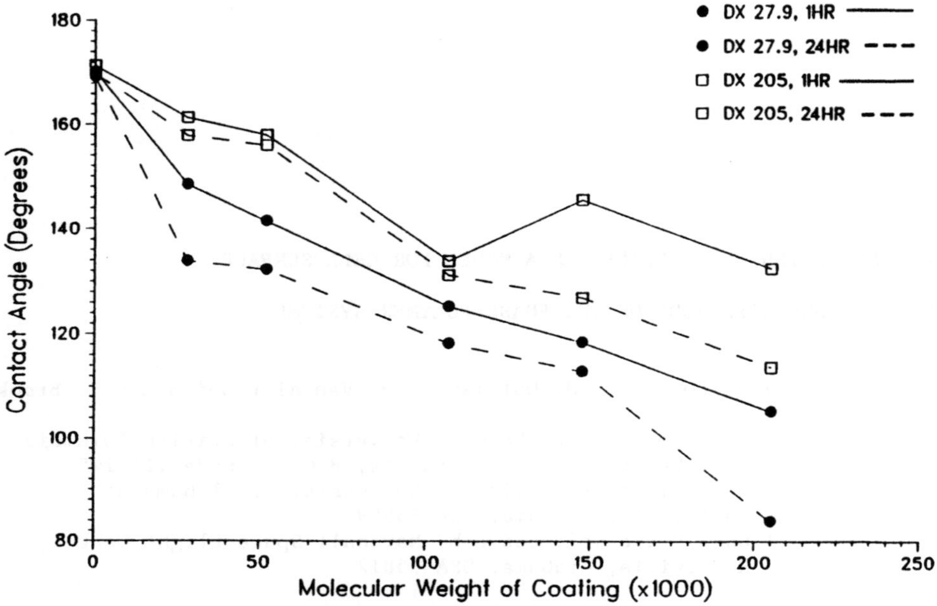

Fig. 1 *The variation of contact angle with the molecular weight of coating*

Our results show a steady decline of the droplet contact angle as the molecular weight of the coating increased. This implies that the interaction between the free dextran in the drop and the bound polymer improves as the size of the bound polymer increases. Such a result might be expected as the ends of the longer chains are probably less constrained in their movement as the distance from the glass surface increases. The strongest interaction is with the low molecular weight free dextran and the largest bound polymer possibly because the smaller 27.9k free dextran can penetrate into the surface coating more completely than the others.

REFERENCES

1 J.M. Harris and M. Yalpani in: "Partitioning in Aqueous Two-Phase Systems," H. Walter, D.E. Brooks and D. Fisher, eds., Academic Press Inc., Orlando (1985)

2 Y. Rotenberg, L. Boruvka and A.W. Neumann, J. Colloid Interface Sci. 93:169 (1983)

A THERMODYNAMIC STUDY OF THE BINDING OF THE *E. coli* F41 ADHESIN TO ITS RECEPTOR, HUMAN GLYCOPHORIN

A.M. Jones and D.E. Brooks

Department of Pathology
University of British Columbia
Vancouver, Canada

Colonisation by many types of bacteria is known to depend upon their ability to adhere to host cells via protein surface appendages known as adhesins. Although the structures of few adhesins are known, they have frequently been thought to be associated with long helical coils of a relatively low molecular weight subunit which form linear appendages known as pili or fimbriae. Pili may reach lengths of a micron or more and are readily visible in EM images of the bacteria which bear them. Not all adhesins are associated with such structures, however, and non-linear, afimbrial adhesins whose geometry has not been studied but which also contain many copies of a subunit polypeptide have been described. In a number of cases, bacteria are known to bind to specific structures on the host surface via their adhesins. Detailed studies on adhesin-receptor interactions are few, however, and no molecular binding studies have been reported to date. Although adhesion to host cells is often associated with the presence of pili or afimbrial aggregates, there is little evidence relevant to the actual location of the binding sites for receptors. Genetic manipulation of some *E. coli* strains has shown, in fact, that the binding sites are not located on the pilin monomer, but occur on a protein present in much lower concentration whose physical relationship to the pili is unknown. In other systems there is evidence that adhesion occurs only at the tip of the pili. Only in one case of gonnocal pili have results been obtained which indicate that the adhesion binding site is located on a pilin monomer, and in this case the receptor was not known.

In the present work we have carried out a study of the binding of a solubilised cell surface receptor, human glycophorin, to the afimbrial adhesin isolated from F41+ *E. coli* under equilibrium conditions utilising a Dextran 500/PEG 8000 two-phase system to separate the free from bound species. The adhesin was isolated by mechanically shearing *E. coli* strain 1593, which is F41+, followed by ammonium sulphate precipitation and gel filtration on Sephacryl S-500, from which the adhesin is excluded. Boiling in SDS produces a dominant monomer of apparent molecular weight 29K on SDS gels as well as small, variable amounts of an 18K band. Human glycophorin was isolated by standard procedures from type MM red cell ghosts as a mixture of glycophorins A, B and C. Glycophorin was radiolabelled with I-125 using Enzymobeads to a level of 0.07 moles/mole. The adhesin, in its native form, was labelled via reductive methylation in the presence of sodium cyanoborohydride with C-14 formaldehyde to a level of 6 moles/mole of 29K monomer. Neither labelling procedure effected the binding reaction nor the behaviour of the proteins in the two-phase system used.

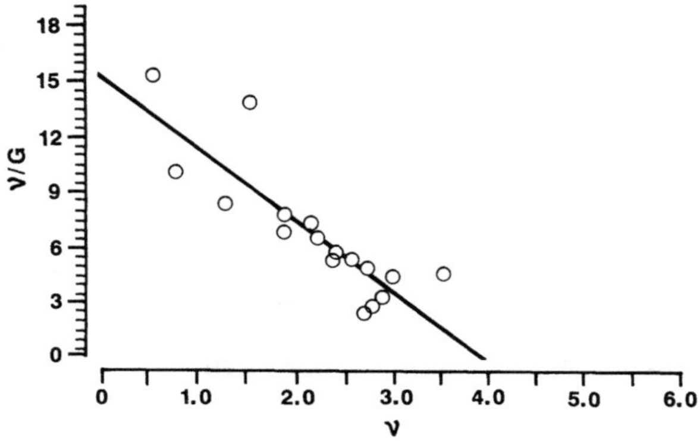

Fig. 1 *A Scatchard plot of* ν/G versus ν

The two-phase system utilised contained 5% w/w Dextran T500, 4% w/w PEG 8000, 150 mM NaCl, 1 mM pH 7.4 phosphate buffer and 0.0416% w/w Tween 80. The detergent was included to prevent aggregation of the adhesin in the phase system; it did not interfere with the adhesin-glycophorin reaction under the conditions used. In this system 21°C, the adhesin partitioned 97% into the bottom phase. Glycophorin partition was concentration dependent, decreasing from 2.6 to 1.6 (top/bottom) over the range 2 to 33 μg/ml in the top phase. Addition of glycophorin had no effect on the partition coefficient of the adhesin (presumably because of its very high molecular weight) but the binding reaction reduced the amount of free glycophorin in the top phase in a concentration dependent manner. Hence, the concentration of free glycophorin in the bottom phase in equilibrium with that bound to the adhesin could be calculated from the concentration of free glycophorin in the top phase and the glycophorin partition coefficient at that concentration. The amount bound was then given by the difference between the total amount of glycophorin in the bottom phase and the amount free in that phase.

Two glycophorin isolations were carried out using blood from different individuals and complete binding isotherms measured for each. The results of one such experiment are shown in Fig. 1. The data is plotted in the form of a Scatchard plot, (ν/G) versus ν, where ν = number of moles of glycophorin bound per 29K adhesin monomer and G = concentration of glycophorin (μmoles/l) in equilibrium with the adhesin. If the binding is "simple" and the polymers have no effect on the reaction, the absolute value of the slope gives the association constant and the intercept on the ν axis is equal to n, the number of sites per monomer occupied at saturation.

In the two experiments performed, the values for the dissociation constant were 3.8×10^{-7} and 2.8×10^{-7} and the corresponding n values were 5.6 and 4.0. The average value of n = 4.8 overestimates the true values because (1) it assumes all the mass of the material bound was glycophorin A or B (which are the only species which bear the M/N blood group and act as receptors) while the preparations used contained all three species and were approximately 47% by weight glycolipid; (2) glycophorin, even in the presence of strong detergents, exists in solution as a dimer. Corrections for these factors lead to a value of n = 1.2. This result indicates clearly that the adhesin from F41+ E. coli bears on binding site per 29K monomer. Hence, the high molecular weight afimbrial structures responsible for adhesion of this strain bear an extremely large number of binding sites, making them very effective attachment structures.

We thank the Medical Research Council of Canada for support.

TEMPERATURE DEPENDENCE OF THE PHASE EQUILIBRIA IN THE PEG-DEXTRAN SYSTEM - A THEORETICAL AND EXPERIMENTAL STUDY

Åke Sjöberg, Gunnar Karlström
Håkan Wennerström and Folke Tjerneld

Chemical Center, P. O. Box 124
S-221 00 Lund, Sweden

In a previous paper we have concluded that polymer incompatibility in aqueous two-polymer systems are due to an unfavourable segment-segment interaction between the two polymers. In order to further study the molecular interactions leading to phase separation, the temperature dependence for the system poly(ethylene glycol)-dextran-water has been investigated.

Our experimental observations show that moderate temperature changes have only a small effect on the phase diagram. The extent of the two-phase area decreases slightly and the slope of the tie lines are changed. When the temperature is increased above 90°C the two-phase region grows dramatically with increasing temperature.

This behaviour with first increasing and then decreasing solubility with temperature cannot be described by classical Flory-Huggins theory for a polymer1-polymer2-solvent system.

Recently one of us presented a statistical mechanical mean field theory for the clouding phenomenon of PEG in water. In this description the emphasis is put on the conformational equilibria of the PEG chain. Since it is possible to have different dihedral angles around the C-C and the C-O bonds in the PEG chain, there is a vast number of different conformations differing in dipole moment and energy. At ambient temperatures, a few low energy polar conformations dominate and at higher temperatures there are several less polar conformations of higher energy populated. As the less polar conformations become more abundant, the polymer becomes less hydrophilic which ultimately leads to a phase separation.

On the basis of these ideas, we have combined the Flory-Huggins treatment of the two polymer system of ref. 1 with the more detailed model of the PEG-water system of ref. 2 to obtain the following expressions for the energy (or enthalpy when the system is assumed incompressible):

$$U=M_o\{w_{1L}*P*\phi_1*\phi_2+w_{1H}*(1-P)*\phi_1*\phi_2+w_{13}*\phi_3+w_{L3}*P*\phi_2*\phi_3$$
$$+w_{H3}*(1-P)*\phi_2*\phi_3+w_{LH}*P*(1-P)*\phi_2*\phi_2$$
$$+w_{HH}*(1-P)*(1-P)*\phi_2*\phi_2/2\}$$

the corresponding expression for the entropy is:

$$S/R=-M_o\{\phi_1*\ln(\phi_1)+(\phi_2/M_2)*\ln(\phi_2)+(\phi_3/M_3)*\ln(\phi_3)+\phi_2*[P*\ln(P)+(1-P)*\ln((1-P)/F)]$$

the different components are 1, water; 2, PEG; and 3, dextran. Index L stands for the polar "low temperature" form of PEG and H for the less polar "high temperature" form of PEG. F which is the statistical weight of the high temperature form relative to the low temperature form has been given the value 8[2] and P is the fraction "low temperature" form of PEG. M_o is the total size of the lattice, and M_i and ϕ_i are the degree of polymerisation and the volume fraction respectively for species i. The interaction-energy parameter characterising the interactions between a segment of type i and a segment of type j is denoted w_{ij} and is related to the ordinary χ parameters through

$$\chi_{ij}=w_{ij}/RT$$

With only one adjustable parameter, a combination of w_{L3} and w_{H3}, it is then possible to calculate the temperature dependence for the system for a fixed set of segment segment interactions.

With this new model the theoretical description of the PEG-dextran-water system has been greatly improved relative to a conventional Flory-Huggins description. The model calculations can predict the decreased solubility above 90°C. In this model phase, separation at low and moderate temperatures is due to PEG-dextran interactions and at temperatures above ~90°C the phase separation is also due to PEG-water interactions.

REFERENCES

1 Å. Gustafsson, H. Wennerström, F. Tjerneld, Polymer 27:1768 (1986)
2 G. Karlström, J. Phys. Chem. 89:4962 (1985)

Chapter 7

Biotechnology: Downstream Processing

LARGE SCALE PROTEIN RECOVERY USING AQUEOUS PHASE SYSTEMS

Maria-Regina Kula

Institut für Enzymtechnologie der Universität Düsseldorf
Postfach 20 50, D-5170 Jülich, FRG

Aqueous phase systems provide not only a gentle environment for biologically active proteins [1,2] but offer unique possibilities for application in downstream processing, especially in large scale. Most important, a very difficult mechanical separation step can be replaced by an extraction process, which is thermodynamically controlled and allows the separation of cells or cell debris from a soluble protein by partition into opposite phases under suitable conditions [3-6]. Poly(ethylene glycol) (PEG)-salt systems, as well as two-polymer systems, may be employed for this purpose, the former appears to give a higher selectivity in protein partition and leads to an enriched product in the first extraction step in high yield [3-5]. PEG is commercially available as a bulk product with several molecular weight fractions in the interesting range 600-8000 dalton. PEG is usually chosen also as one of the polymers if a two-polymer system is investigated. The second component may be either purified dextran fractions (1,2), crude dextran (7), starch derivatives (8), or other polymers like polyvinylalcohol (9). The selection of the basic phase system depends on the partition coefficient of the product of interest and on economic factors, which are closely linked to the concentration required for phase formation. Single step yields and simultaneous purification of product have to be considered also in the design of the primary extraction step. The necessary separation of the two immiscible liquid phases can be carried out by conventional equipment at unit gravity or accelerated by centrifugal forces without major problems [3,7,10-12].

Besides simple physical extraction, a reactive extraction may be applied, utilising complex formation of the desired product in one of the phases for improved separation. This is accomplished by confining a ligand, preferentially a biospecific ligand, into one phase by covalent binding of the ligand to one of the polymers constituting the system, e.g. PEG. Only two-polymer systems, such as PEG-dextran or PEG-derivatised starch, are suitable for an affinity extraction [2,9,13]. PEG-salt phases form only at rather high ionic strength which, besides phase formation, leads to a dissociation of most ligand-protein complexes under these conditions [13,14].

In contrast to affinity chromatography based on similar basic principles, ligand utilisation is higher in partition since there is practically no mass transfer resistance and equilibrium is obtained within seconds [15,17]. Also, particle-loaded phases may be processed without difficulties [13,16].

The extraction technology for proteins can be characterised by:

- short process times, high yield and high productivity
- option for continuous and automated operation
- low investment, energy and labour cost
- comparatively high expenses for chemicals, which rise linearly with scale

The last point becomes very important for processing in scale ≥1 ton of biomass [18]. Phase formation is the basis of the partition and extraction technology. Phase formation occurs only if limit concentration of the phase-forming chemicals in the suspensions or solutions are exceeded as determined by the corresponding phase diagram. To a first approximation the consumption of chemicals is governed by the volume processed and not by the amount of product. Therefore prior reduction of process volumes is important for economic operations. For intracellular products this can be readily accomplished in the harvesting step. The high biomass concentration thus achieved even contributes to the phase formation in aqueous systems designed to isolate intracellular proteins [3,4]. For extracellular products very efficient recycling operations have to be devised if cheap bulk products are considered [19]. For high priced pharmaceuticals produced by recombinant animal cell lines, the cost for phase-forming chemicals is tolerable in view of the high yield, volume reduction and enrichment obtained by the extraction step [20]. Recycling is also desirable to save costs of waste water treatment and diminish environmental loads [18]. PEG and dextran, possibly also the new starch derivatives, can be biologically degraded, aerobically as well as anaerobically, and present no special disposal problems. The technology for the removal of phosphates or sulphates is also available. Recycle or waste treatment has to be taken into account for the analysis of total processing cost. First studies show that a large portion of the chemicals involved in aqueous two-phase extraction can be recycled [16,21,22]. Recycling of polymer bound ligands is also vital for affinity extraction processes. In one process design PEG-dextran systems are alternated with PEG-salt systems. Addition of phosphates to the top phase of the first extraction step dissociates the affinity complex and partitions product and ligand into opposite phases under suitable conditions, enabling ligand concentration and recycling and product separation at the same time [13].

The high standard of technical development and also economic analysis indicate that partition in aqueous phase systems is a viable, new unit operation in protein recovery and enrichment. If highly purified products are requested, further chromatographic steps can be integrated without complications inherent to the extraction technology [20,23].

REFERENCES

1 P.-Å. Albertsson, "Partition of Cell Particles and Macromolecules," 3rd edition, Wiley Interscience, New York (1986)

2 H. Walter, D.E. Brooks and D. Fisher, eds., "Partitioning in Aqueous Two-Phase Systems: Theory, Methods, Uses and Application to Biotechnology," Academic Press, New York (1985)

3 H. Hustedt, K.H. Kroner and M.-R. Kula, Applications of phase partitioning in biotechnology, in: "Partitioning in Aqueous Two-Phase Systems: Theory, Methods, Uses and Application to Biotechnology," p 529, Academic Press, New York (1985)

4 M.-R. Kula, Liquid-liquid extraction of biopolymers, in: "Comprehensive Biotechnology," Vol. 2, p 451, A. Humphrey and C.L. Cooney, eds., Pergamon Press, New York (1985)

5 M.-R. Kula, K.H. Kroner and H. Hustedt, Purification of enzymes by liquid-liquid-extraction, in: "Advances in Biochemical Engineering," Vol. 24, p 73, A. Fiechter, ed., Springer-Verlag, Berlin (1982)
6 M.-R. Kula, K.H. Kroner, H. Hustedt, S. Granda and W. Stach, Verfahren zur Abtrennung von Enzymen, Deutsches Patent Nr. 26 39 129, US-Patent Nr. 41 44 130
7 K.H. Kroner, H. Hustedt and M.-R. Kula, Evaluation of crude dextran as phase forming polymer for the extraction of enzymes in aqueous two-phase systems in large scale, Biotechnol. Bioeng. 24:1015 (1982)
8 F. Tjerneld, S. Berner, A. Cajarville and G. Johansson, New aqueous two-phase systems based on hydroxypropyl starch useful in enzyme purification, Enz. Microb. Technol. 8:417 (1986)
9 F. Tjerneld, New polymers for producing aqueous two-phase systems, in: "Microorganisms in aqueous two-phase systems," IVA Rapport 330, Royal Swedish Academy of Engineering Sciences, Stockholm (1987)
10 K.H. Kroner, H. Hustedt, S. Granda and M.-R. Kula, Technical aspects of separation using aqueous two-phase systems in enzyme isolation processes, Biotechnol. Bioeng. 20:1967 (1978)
11 M.-R. Kula, K.H. Kroner, H. Hustedt and H. Schütte, Technical aspects of extractive enzyme purification, in: "Biochemical Engineering II," A. Constantinides, W.R. Vieth, K. Venkatasubramanian, eds., Ann. N.Y. Acad. Sci., Vol 369:341 (1981)
12 K.H. Kroner and H. Hustedt, Separation of aqueous two-phase systems by gravity settling," in: "ISEC 86, Preprints Intern. Solvent Extraction Conf.," Vol III, p 703, DECHEMA, Frankfurt (1986)
13 A. Cordes and M.-R. Kula, Process design for large-scale purification of formate dehydrogenase from candida boidinii by affinity partition, J. Chromatography, 376:375 (1986)
14 K.H. Kroner, A. Cordes, A. Schelper, M. Morr, A.F. Bückmann and M.-R. Kula, Affinity partition studies with glucose-6-phosphate-dehydrogenase in aqueous two-phase systems in response to triazin dyes, in: "Affinity Chromatography and Related Techniques," p 491, T.C.J. Gribnau, Visser and R.J.V. Nivard, eds., Elsevier, Amsterdam (1982)
15 P.-F. Fauquex, H. Hustedt and M.-R. Kula, Phase equilibration in agitated vessels during extractive enzyme recovery, J. Chem. Techn. Biotechnol. 35B:51 (1985)
16 H. Hustedt, K.H. Kroner, U. Menge and M.-R. Kula, Protein recovery using two-phase systems, Trends in Biotechnology, 3:139 (1985)
17 J.C. Janson, Large scale affinity purification - state of the art and future prospects, Trends in Biotechnology, 2:1 (1984)
18 K.H. Kroner, H. Hustedt and M.-R. Kula, Extractive enzyme recovery: economic considerations, Proc. Biochem. 19:170 (1984)
19 G.H. Schoutens and P.J. Kerkhof, Aqueous two-phase extraction for protein recovery: analysis of cost effective parameters, Proc. 4th Eur. Cong. Biotechnol., 2:625 Elsevier, Amsterdam, (1987)
20 U. Menge, E. Fraune, J. Lehmann and M.-R. Kula, Purification of proteins from cell culture supernatants, Develop. Biol. Standard. 66:391, S. Karger, Basel (1987)
21 N. Papamichael, B. Börner and H. Hustedt, Improved economies of enzyme recovery by aqueous phase extraction by recycling of phase chemicals: a case study, Proc. 4th Eur. Cong. Biotechnol. 2:519 (1987)
22 H. Hustedt, Extractive enzyme recovery with simple recycling of phase forming chemicals, Biotechnol. Lett. 8:791 (1986)
23 H. Schütte, W. Hummel, H. Tsai and M.-R. Kula, L-leucine dehydrogenase from bacillus cereus, Appl. Microb. Biotechnol. 22:306 (1985)

CONTINUOUS CROSSCURRENT EXTRACTION OF PROTEINS IN PROCESS SCALE

Helmut Hustedt, Karl-Heinz Kroner, and Neophytos Papamichael

GBF, Gesellschaft für Biotechnologische Forschung mbH
Mascheroder Weg 1, D-3300 Braunschweig, FRG

Automated continuous cross-current extraction of intracellular microbial enzymes from disrupted cells has been studied using a two-stage plant capable of processing 20-30 kg biomass per hour. The results reveal the feasibility of this advanced method of downstream processing by aqueous two-phase extraction.

INTRODUCTION

During past years extraction in aqueous phase systems [1] has been developed to a new downstream processing line, mainly for intracellular microbial proteins (e.g. 2,3). The technology may be envisaged as an alternative to the conventional primary recovery route: homogenate clarification by filtration/centrifugation ultrafiltration precipitation centrifugation/filtration. But it is also used for extracellular proteins from microorganisms as well as from animal cell cultures, such as for recovery of interferon from human fibroblast cells [2]. A number of extraction processes has already been transferred successfully to industry by our group. Protein extraction is usually carried out batchwise by 2, or in some cases 3, subsequent extraction steps [2,3]. A recent progress in this field is continuous processing exploiting the principle of crosscurrent extraction.

General features of continuous crosscurrent extraction

A flow scheme of continuous 2-stage crosscurrent extraction is shown in Fig. 1. Generally continuous processing may:

- increase space time yield
- allow fully automated operation
- increase the number of extraction parameters
- result in more homogeneous product
- increase the activity yield in the case of labile proteins
- have a potential for efficient recycling of the extraction media
- facilitate fully use of continuous fermentation
- require some efforts for process monitoring and control
- need higher reliability of machinery

Some of these topics are discussed here using, as process examples, the extraction of fumarase from *Saccharomyces cerevisiae*, the extraction of

Table 1 *Basic data of the extraction processes (two stages)*

Enzyme	Organism	Purification factor	Overall yield (%)	Biomass* load (%)
Fumarase	Baker's yeast	13	75	25
Fumarase	*Brevibacterium ammoniagenes*	20	75	20
Penicillin acylase	*E. coli*	10	78	20

* in the primary phase system

fumarase from *Brevibacterium ammoniagenes* and of penicillin acylase from a highly producing recombinant E. coli. Their basic data are collected in Table 1. The experiments were conducted in a newly developed two-stage plant operated under computerised process control with flow rates in the range of 100 l primary phase system per hour corresponding to biomass throughputs of 20-30 kg per hour. This plant was designed by us after some successful preliminary experiments with the fumarase/*B. ammoniagenes* process [4,5] and investigations concerning the continuous recovery of aspartase from *E. coli* [4,5]. Construction of the plant was conducted by the Westfalia Separator AG (Oelde/FRG).

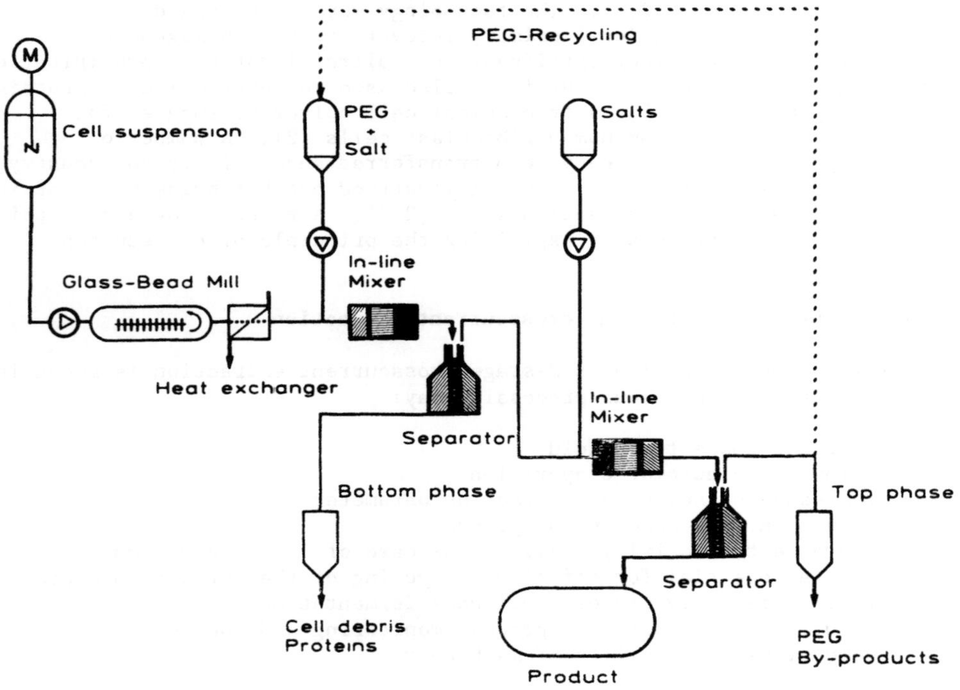

Fig. 1 *Flow scheme (including cell disruption by wet milling in a glass bead mill) of continuous crosscurrent extraction of proteins*

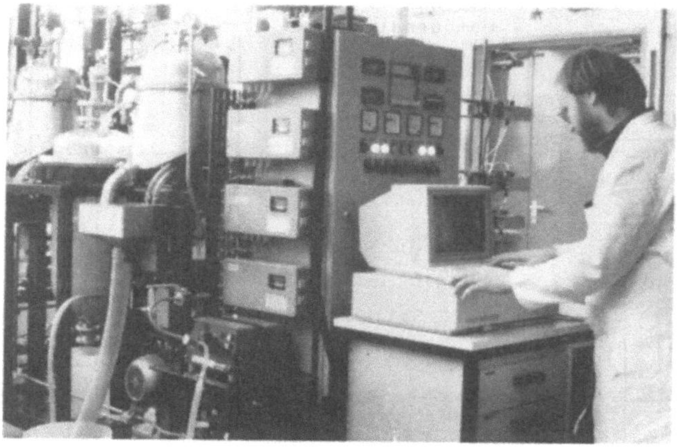

Fig. 2 *Photograph of the crosscurrent extraction plant / capacity ~600 kg wet biomass per day* (24 h)

Extraction Plant

A photograph of the plant, in which the scheme shown in Fig. 1 is realised, is given in Fig. 2. It contains two pilot size separators of type SA 1 (Σ-factor 1460 m^2, Westfalia Separator AG). The corresponding process control scheme is outlined in Fig. 3. The pressures serve as alarm functions. In addition to the instrumentation shown in Fig. 3 a time-based automated sludge discharge control unit was installed in the plant.

Continuous Crosscurrent Extraction of Fumarase from Baker's Yeast

To meet the high demand for biomass of continuous extraction (~20 kg per h) a two-step extraction procedure for fumarase from baker's yeast has been developed as a model process [6]. This material is available in any quantity at comparatively low cost (but has the drawbacks of limited reproducibility and contamination by varying amounts of filter aids leading to problems during phase separation, see below). The physico-chemical parameters of the process liquids of importance for operation are reported elsewhere [7].

Mixing and phase equilibration have been studied in both extraction steps using Kenics static in-line mixers (Chemineer Ltd., Derby, Great Britain) with internal diameters of 4.9 and 8.0 mm showing efficiencies close to 100% in adequate flow rate ranges [7,8]. This confirms that attainment of phase and partition equilibrium is not a limiting or complicating factor in continuous processing (see ref. 9 for comparison).

The second requirement for processing with a stage efficiency of about 100% is the complete separation of the phases. Broad experience has already been accumulated by our group in this field showing that this is generally possible with high performance by using common disk stack separators operated either in a liquid-liquid mode or, when the bottom phase exhibits high viscosities (η >500 m Pas), solid-liquid made with nozzle separators [2,3]. When using the SA-1 separator with the normal disk set (disk gap 0.4 mm, Σ-factor ~1460 m^2) for the separation of the primary phase system in the fumarase/baker's yeast process, nearly complete phase separation was obtained

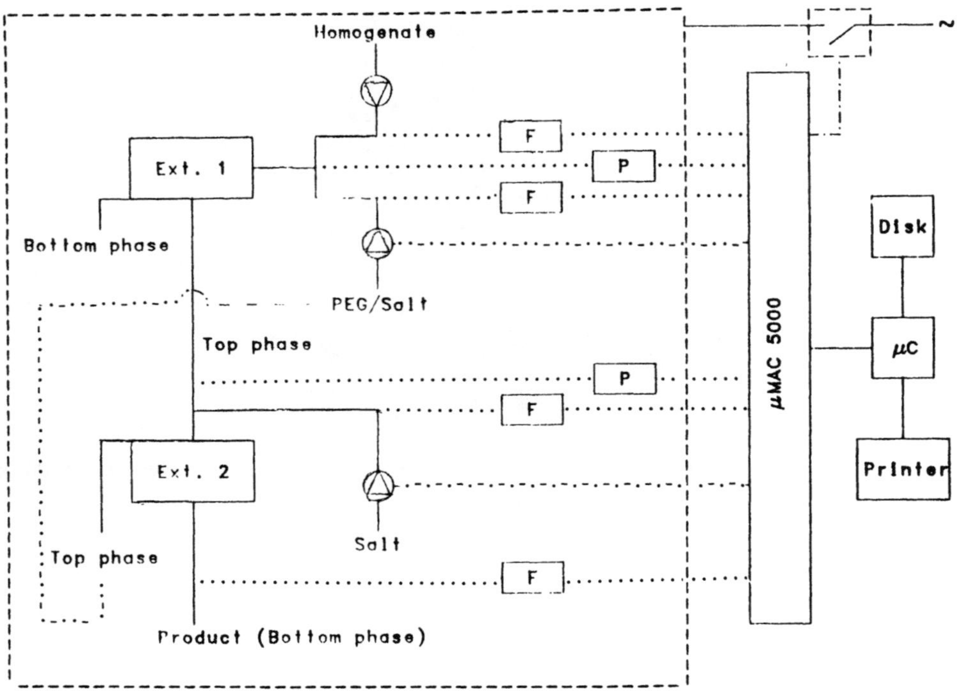

Fig. 3 *Flow scheme of continuous two-stage crosscurrent extraction with computerised process control (F = flow meter, P = pressure transducer ·········· measured signal,—·—·— control line, —··—··— option for PEG-recycling)*

with flow rates up to about 90 l/h. This is relatively low and is due to the small density difference (55 kg/m^3) between the phases. However, the separation efficiency decreased after a short processing time due to the irreversible deposition of contaminating filter aids (Kieselgur) onto the disks. This problem was overcome by using a disk set with a larger gap spacing of the disks (0.8 mm); but this reduced the Σ-factor to ~900 m^2 and correspondingly the purity of the bottom phase decreased when the feed rate exceeded about 60 l/h, which means that product was lost with increased feed rates.

Figs. 4-6 reflect a fully automated two-step run of the baker's yeast process conducted over about 6 h according to the scheme outlined in Fig. 3. The feed rate to the first separator was about 100 l/h (corresponding to a biomass throughput of 28 kg/h). Under these conditions about 15% loss of top phase containing the product occurred with the bottom phase; but this flow rate was used to simulate an average process. Automated sludge discharge was carried out in 30 min. intervals, which helped to maintain high and constant phase purities during the whole process. The data presented show a high reliability and stability over the whole process time. The enzyme and total protein concentration in the product enriched phase corresponded to those in the batch process. Small periodical changes are observed in the flow rates of the separator effluents, which are probably due to the formation of deposits in the separator bowl. (The feeds to the 1st separator were constant). After sludge discharge the original flow conditions were always restored. The negative peaks in the pressure and the flow rate curves are due to the sludge discharge.

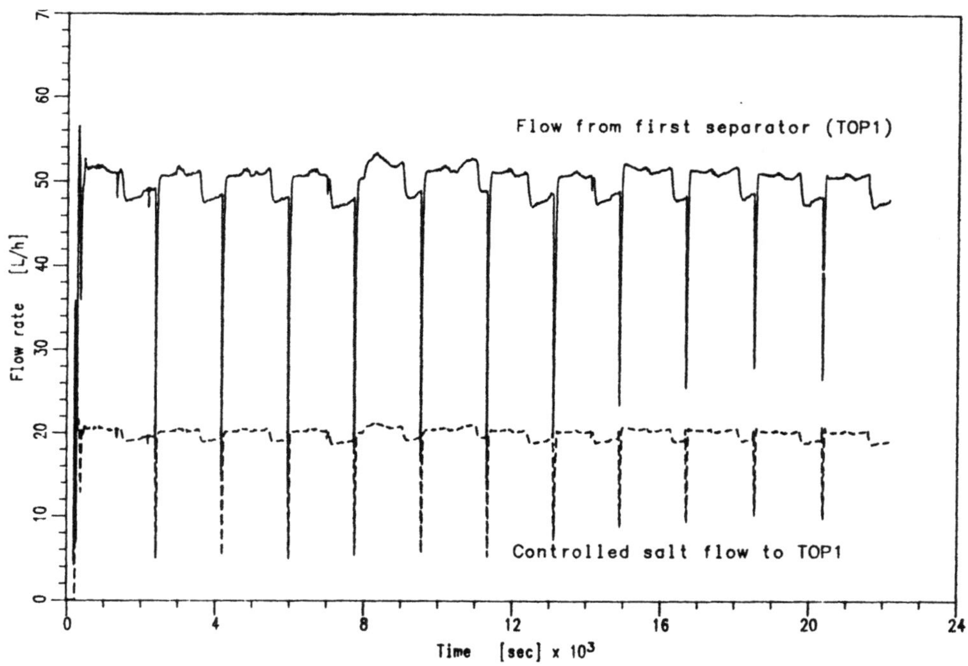

Fig. 4 *Time course of the flow of the primary top phase and of the flows to the 2nd separator (reproduced from [7])*

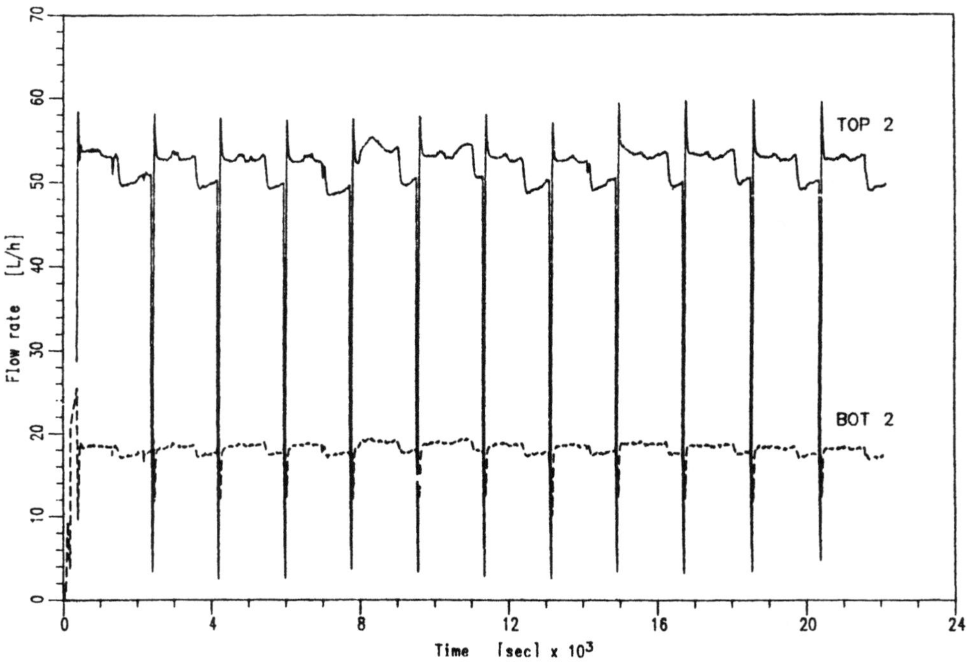

Fig. 5 *Course of the flows from the 2nd separator (top 2 = 2nd top phase, flow measured; bot 2 = 2nd bottom phase, flow calculated via a flow balance) (reproduced from [7])*

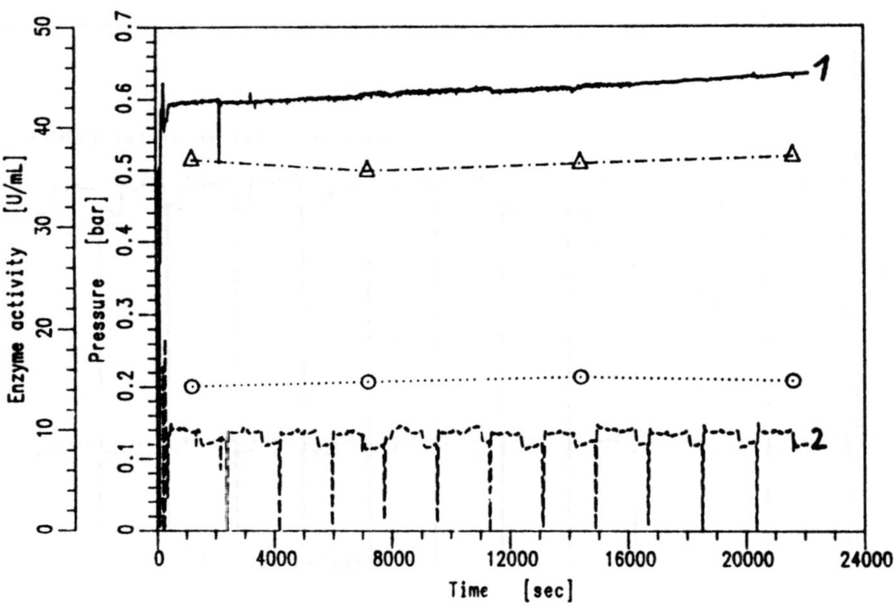

Fig. 6 *Course of enzyme concentration in the phases and pressures before the static mixers (Kenics, i.d. 8.1 mm); 1.2: pressure before 1st and 2nd mixer; symbols: enzyme concentration in 1st top phase (o·······o) and 2nd bottom phase* (Δ-·-·-Δ) (reproduced from [8])

Generally, the temperatures of the phases from the separators are slightly higher than in the feeds, but quite constant. Usually the temperature rise will not present problems. Otherwise the temperature could be adjusted by the inclusion of a heat exchanger before the respective separator.

Recovery of two biocatalysts for technical application

The concept described was also tested by running the fumarase/ *Brevibacterium ammoniagenes* process in the plant. Here the small gap disk stack could also be used for the separation of the primary phase system enabling throughputs up to about 110 l/h with acceptable phase purities, e.g. about 140 kg of biomass (wet weight) corresponding to ~700 l primary phase system) was processed within 7.5 h with the same final yield and purity as obtained by the comparable batch process. Data from this run are given in Fig. 7 showing again the feasibility of the approach.

The extraction plant has also been applied successfully to the recovery of penicillin acylase from a genetically engineered *E. coli*; here 100 kg biomass were processed within 4.5h, again with identical yield and product purification as in the batch process conducted in parallel. The use of a static mixer with a larger diameter (~12 mm) at the 2nd stage to reduce the back pressure to the 1st separator would increase its capacity for this application to 25-30 kg biomass per hour.

Recycling of phase components

It is reasonable to assume that much longer process times than reported here can be realised without stopping for cleaning. (In an early preliminary

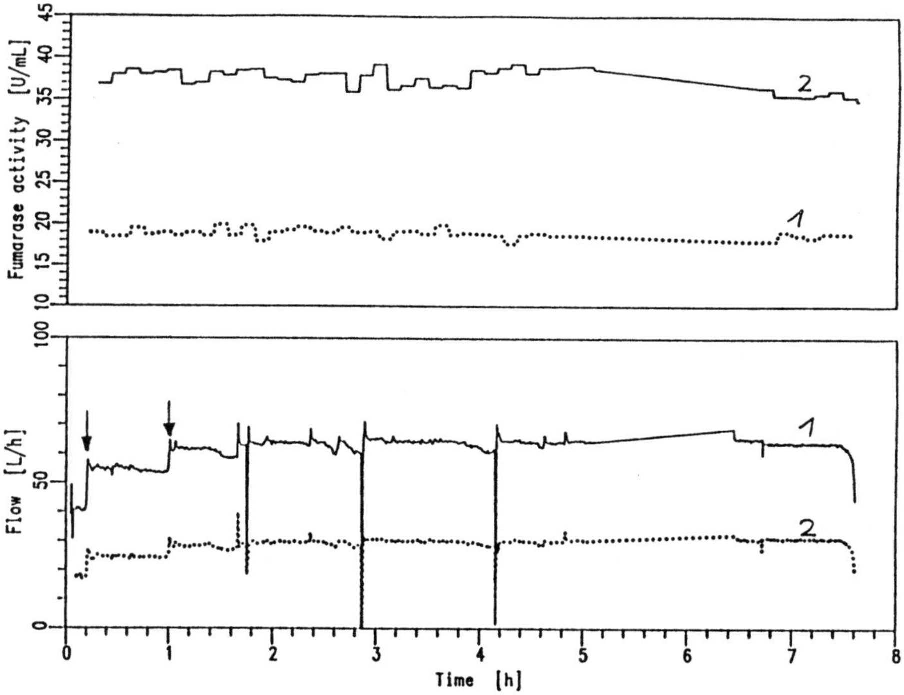

Fig. 7 *Continuous crosscurrent extraction of fumarase from B. ammoniagenes-time course of the flows of primary top (1) and secondary bottom phase (2) and enzyme concentration in these phases processing ~22 kg biomass per hour (~100 l primary phase system per h). (The arrows (↓) mark points at which the feed rate of the cell homogenate was increased; the enzyme concentrations were determined by automated on-line analysis)*

run with manual process control, the machine was, e.g. operated for 14 hours without stopping). The processing of large volumes of phase systems means a relatively high consumption of chemicals for generating the phase systems. Therefore, the biomass load of the primary phase system should be as high as possible, at least 20% (wet weight). To improve the economy of continuous cross-current extraction in commercial scale further, therefore, recycling of the extraction media has been attempted in a simple way: The PEG (and salt in this phase) is recycled by reuse of the secondary top phase in the primary extraction (see Fig. 1). The main portion of the salt, which is in the secondary bottom phase (i.e. the salt enriched phase) is recycled as the permeate obtained by concentrating the phase by ultrafiltration. The principal feasibility of this approach has been demonstrated recently for the fumarase/baker's yeast process; fourfold recycling of PEG- and salt phase in batch has been found possible without reductions in product purity or activity yield [6]. The recycling of the PEG-phase in a semicontinuous way has also been proven for continuous crosscurrent operation. The considerable savings in material and effluent treatment cost (the latter reflects the situation in the F.R.G.) can be seen from the PEG mass balance presented in Table 2. First experiments to realise continuous "in-line" PEG-recycling

Table 2 *PEG-balance and cost reduction with regard to recycling of the 2nd top phase/two-step extractive purification of fumarase from baker's yeast (data based on processing of 1 ton biomass, wet weight)*

process step/ liquid	PEG pro-cessed	COSTS for:		SAVINGS for:		SAVINGS in total
		PEG	effluent treatment[c]	PEG	effluent treatment[c]	
	kg	DM[d]	DM[d]	DM[d]	DM[d]	DM
total processing without recycling	680	2040	1160			
cell debris containing waste phase (→losses)	47	141				
recycling of the 2nd top phase (→reuse)	a 634 b 507			a 1902 b 1521	a 1080 b 864	a 2982 b 2385

[a] extrapolation for infinitesimal continuous recycling
[b] fourfold semicontinuous recycling (within the process)
[c] according to the German law for waste water treatment based on COD
[d] German Marks

with online monitoring of product concentration and purity and process control by the computer were also successful [10]. Inline recycling has the advantage that no storage of the process liquids to be reused is necessary; the whole process could be carried out with a third to a quarter of the PEG (and half of the salt if the salt phase is recycled, too) needed normally. A further reduction of the PEG consumption, e.g. by including continuous regeneration of the secondary top phase, should be possible.

Consequences from the short residence time

The residence times in continuous crosscurrent extraction are very short: the time from mixing until phases are separated is in the range of ~10-30 sec for each stage. This has several benefits in comparison to batch processing with residence times of several hours.

1 Operation can be carried out at comparatively extreme pH-values without activity losses (the pH-value is an important factor for steering the partition of proteins).

2 Operation at higher temperatures without significant product denaturation is facilitated offering an additional factor for the control of the protein partition.

3 The product yield will obviously be enhanced when processing labile products.

4 The homogeneity of the product will be improved.

For special purposes, for example a pH-treatment in a phase system, the residence time can be increased without undue disturbance of the residence time distribution.

Concluding remarks/outlook

The feasibility of two-stage continuous crosscurrent extraction of proteins in aqueous phase systems has been approved and its performance and potential discussed in comparison with batchwise production. Continuous processing has been shown to have several benefits; in addition, it should facilitate establishing integrated downstream processing lines spanning cell harvesting to final ultrafiltration [11,12]. Further advances might be expected, beside others, from continuous reactive extraction with carrier recycling and from direct coupling of extraction with chromatographic systems to enable automated continuous or semi-continuous high purification.

REFERENCES

1 P.-Å. Albertsson, "Partition of Cell Particles and Macromolecules," 3rd edition, Wiley-Interscience, New York (1986)
2 H. Hustedt, K.-H. Kroner, U. Menge and M.-R. Kula, Protein recovery using two-phase systems, Trends Biotechnol. 3:139, (1985)
3 H. Hustedt, K.-H. Kroner and M.-R. Kula, Applications of phase partitioning in biotechnology, in: "Partitioning in Aqueous Two-Phase Systems. Theory, Methods, Uses and Applications to Biotechnology," H. Walter, D.E. Brooks and D. Fisher, eds., Adademic Press, New York (1986)
4 H. Hustedt, K.-H. Kroner, H. Schütte and M.-R. Kula, Extractive purification of enzymes, in: "Enzyme Technology," R.M. Lafferty, ed., Springer-Verlag, Berlin (1983)
5 H. Hustedt, K.-H. Kroner and M.-R. Kula, Continuous enzyme purification by crosscurrent extraction, Proc. 3rd Eur. Congr. Biotechnol. vol. II, Verlag Chemie, Weinheim (1984)
6 H. Hustedt, Extractive enzyme recovery with simple recycling of phase forming chemicals, Biotechnol. Lett. 8:791 (1986)
7 H. Hustedt, B. Börner, K.-H. Kroner and P. Papamichael, Fully automated continuous crosscurrent extraction of enzymes in a two-stage plant, Biotechn. Techn. 1:49 (1987)
8 H. Hustedt, K.-H. Kroner and N. Papamichael, Recovery of microbial proteins by continuous crosscurrent extraction, ISEC '86, vol. 3:597, Dechema, Frankfurt 1986
9 P.F. Fauquex, H. Hustedt and M.-R. Kula, Phase equilibration in agitated vessels during extractive enzyme recovery, J. Chem. Technol. Biotechnol, 35B:51 (1985)
10 N. Papamichael and H. Hustedt, Improved economics of extractive enzyme recovery by recycling: a case study, in: "Separations Using Aqueous Phase Systems: Applications in Cell Biology and Biotechnology," D. Fisher and I.A. Sutherland, eds., Plenum, New York (1989)
11 H. Hustedt, B. Börner, N. Papamichael and R. Krützfeldt, Efficient downstream processing of proteins, Proc. 4th Eur. Congr. Biotechnol., vol. 2:531, Elsevier, Amsterdam (1987)
12 H. Hustedt, R. Krützfeldt and N. Papamichael, Fast downstream processing of proteins by flocculation and liquid extraction, in: "Separations Using Aqueous Phase Systems: Applications in Cell Biology and Biotechnology," D. Fisher and I.A. Sutherland, eds., Plenum, New York (1989)

AQUEOUS TWO-PHASE PARTITION IN BIOCHEMICAL RECOVERY FROM MAMMALIAN CELL CULTURE

Jon G. Huddleston and Andrew Lyddiatt

Biochemical Recovery Laboratory
Department of Chemical Engineering
University of Birmingham
P. O. Box 363, Birmingham B15 2TT, UK

A study of the fractionation of monoclonal antibodies and other solutes in spent mammalian cell culture media has been undertaken with aqueous PEG-phosphate partition systems. Manipulation of pH and average molecular weight of polymers yielded systems capable of separating transferrin from other components. Introduction into dispersed phases of a mixture of anti-BSA Sepharose and human-IgG silica promoted selective recovery of BSA, anti-huIgG monoclonal antibody and transferrin in a "four phase", batch system distinguished by the discrete separation of antibody product and deleterious proteolytic activity into separate aqueous phases.

INTRODUCTION

In direct response to the interest generated by the application of aqueous two-phase partition systems in biochemical downstream processing [1], and the recoveries and efficiencies reported for various systems [2], a study has been established to critically compare and/or integrate the new methods with unit operations currently exploited in laboratory and production plant. Such include microfiltration, ultrafiltration and selective adsorption which are under study in our laboratory using protein recovery from mammalian cell culture, microbial fermentation and waste streams as research vehicles. Preliminary work has concentrated upon the application of economic PEG-phosphate systems to the purification of monoclonal antibodies (MCAB) specific for the $C\gamma 2$ domain of human immunoglobulin G constant chain (huIgG-Fc). The antibodies are produced in batch or continuous suspension cultures (0.5 to 5 litres) of murine hybridomas in serum-based media containing albumin, transferrin, acidic protease and MCAB product as the major solutes at culture harvest [3]. Concomitant with recoveries of MCAB, the potential for the fractionation of albumin and transferrin for culture replenishment, and the removal of deleterious protease contamination has been studied.

MATERIALS AND METHODS

Murine hybridomas were grown in continuous and batch cultures as previously described [3]. Where appropriate, cells were harvested by gravity sedimentation in continuous culture, or centrifugation or cross flow microfiltration in Millipore apparatus equipped with HVLP 0.45 μm membranes. Anti-human IgG monoclonal antibody in clarified media was concentrated 2 to 5 fold by ultrafiltration in similar apparatus equipped with PTGC 10K MWCO membranes (0.024 m^2).

Phase systems (PEG 400-, PEG 1000-phosphate; supplied by Sigma UK) were prepared double strength in aqueous solution (w/w) and final concentrations were achieved for batch experiments (0.02 to 2 litres) by addition of aqueous samples (e.g. culture media). Adjustments to pH were made in dispersed phases by addition of orthophophosphoric acid, and phase separation achieved by centrifugation or gravity settling. Distribution of protein components was evaluated by SDS-PAGE analysis of dialysed or diluted samples [4], and protein quantified by the method of Bradford [5]. Immunological activity of anti-huIgG MCAB was determined by affinity HPLC or conventional plate ELISA using purified human IgG as antigen and sheep-anti-mouse IgG-HRP conugate as second antibody [6]. Where appropriate, dialysed samples or diluted samples were analysed for carbohydrate [7], nucleic acid [8] and acidic protease [9], whilst PEG was quantified in samples by the method of Skoog [10].

Ultrafiltration was studied in recoveries from top (22% PEG) and bottom (2% PEG) phases containing cell culture components using crossflow ultrafiltration as described above. Flux was measured in steady state conditions at recirculation rates of 800 ml/minute and transmembrane pressures of 10 psi, and samples taken for PEG quantitation and the determination of retention coefficients (RCO). The procedure was repeated at successively reduced PEG concentrations following the establishment of original clean water flux.

Anti-BSA Sepharose was prepared by coupling specific monoclonal antibody (supplied by Binding Site Ltd., University of Birmingham) to CNBr-activated Sepharose CL-4B (supplied by Pharmacia) following a modified method of Desai et al [11] to yield ligand concentrations of 2 to 5 mg/ml settled agarose. Hu-IgG silicas (Spherisorb VLS 280 Å, 20 nm; supplied by Phase Separations, UK) were prepared with ligand concentrations of 5 to 30 mg/g dry silica by coupling pure human IgG to aldehyde-activated diol silicas as previously described [6]. Adsorbents were collected into mini-columns, washed with phosphate buffered saline (PBS pH 7.5) and eluted in 3M KSCN in PBS. Molecular purity of BSA and anti-huIgG was determined by SDS-PAGE, and immunological activity by ELISA.

RESULTS AND DISCUSSION

In this preliminary study, PEG-phosphate systems were selected for reasons of economy and their potential for maximal fractionation, subsequent recycling of the major components, e.g. cell culture media [12], and the separation of MCAB product from deleterious proteases.

Polymer concentrations in excess of 25% PEG 1000 promoted near quantitative precipitation of impure MCAB. Successful partition of MCAB and the majority of media proteins was achieved at pH 9 in a 16% PEG 1000 phosphate system with increased recoveries of MCAB relative to BSA at pH 7.0 (Fig. 1). An estimated six theoretical stages would be required of a countercurrent separator to produce product purities greater than 90% under these conditions [13]. In batch systems, yields of immunologically active MCAB rarely exceeded 30% since material was lost in aggregates and precipitates formed on the PEG side of the phase interface. Preliminary work indicates that such insoluble accumulations would interfere with the establishment of countercurrent operations.

Reduction of the average molecular weight of PEG polymers increased immunological yields to 70% of feedstream concentrations in 14% PEG 400: PEG 1000 (1:1)-phosphate at pH 8.8 (Fig. 2). In this system, the partition coefficients of albumin and transferrin are 9 and 0.7 respectively, and thus selective fractionation is possible particularly in low protein media (albumin:transferrin <5:1 compared to 10:1 herein). Carbohydrates and nucleic acids routinely partition to phosphate-rich, bottom phases in these systems, whilst the successful partition of whole cell culture broths has been constrained in all studies by precipitation at phase interfaces.

The recovery of product and media components from partitioned phases was undertaken exploiting ultrafiltration of separated top and bottom phases. A decline in transmembrane flux was observed with increasing solute concentrations (Fig. 3) although the efficient use of 10K MWCO membranes was improved by dilution of feedstreams and increase of available membrane area. However, the dilution necessary to effect diafiltration of PEG and/or phosphate (2 to 10 fold), the resulting dilution of polymer permeate and the extended process times were considered inappropriate to the study.

A major obstacle to product recovery from batch systems was the contamination of MCAB in PEG-rich phases with albumin and acidic protease. Solutions to this problem were sought in the introduction of adsorbents to dispersed systems. Hydrophobic adsorbents (phenyl-, octyl-Sepharose) were insufficiently selective to contribute to efficient separation. The strength of hydrophobic binding forces involved (IgG>BSA>transferrin) reflects the partition coefficients observed in PEG-phosphate systems in the absence of adsorbents and the system is incompatible with the high specification of purity demanded.

Protein-A Sepharose, as reported by Hedman [14], partitions to the top phase of PEG-phosphate systems and was found in this study to selectively bind the majority of human IgG introduced into a dispersed 14% PEG-phospate system (data not shown). However, the binding of anti-huIgG of murine origin to protein-A is unacceptably weak under the conditions of pH and ionic strength in this system. Sepharose particles derivatised with human IgG or anti-BSA murine IgG partition preferentially to PEG-rich top phases where they selectively bind anti-IgG MCAB or albumin respectively. Either of these components may be individually recovered in high yield and purity by dosing dispersed phases with a single adsorbent and conventional elution of solid phase collected into mini-columns following partition. Increased advantage can be gained from the finding that anti-BSA Sepharose and huIgG silicas partition to top and bottom phases respectively in 14% PEG-phosphate systems when separated by controlled centrifugation (500 x g for 10 minutes in batch experiments). Introduction of a mixture of selective silica and agarose into dispersed phases in the presence of clarified cell culture media promoted the partition of MCAB product to phosphate-rich bottom phase as a complex with huIgG silica (Fig. 4). Acidic proteases remained in the top phase together with albumin specifically bound to anti-BSA Sepharose. The application of two adsorbents characterised by unique binding properties and different partition behaviour permits the selective recovery of the major components of serum-based, animal cell culture. Pure MCAB as product, and BSA for media renewal may be eluted from selective adsrobents whilst transferrin may be concentrated from bottom phase by ultrafiltration. Deleterious proteases remain in solution in the top phase of these systems.

Batch systems (20 g) were sampled and analysed by ELISA and dye binding assay as described in Materials and Methods.

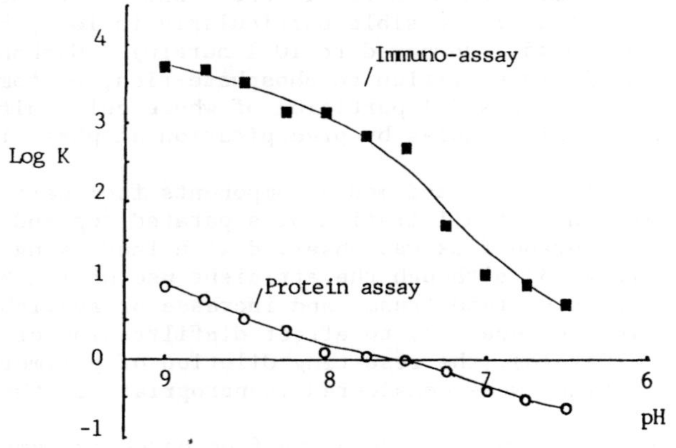

Fig. 1 *Partition of MCAB from serum-based culture fluid in 16% PEG/phostate: variation of partition coefficient with pH*

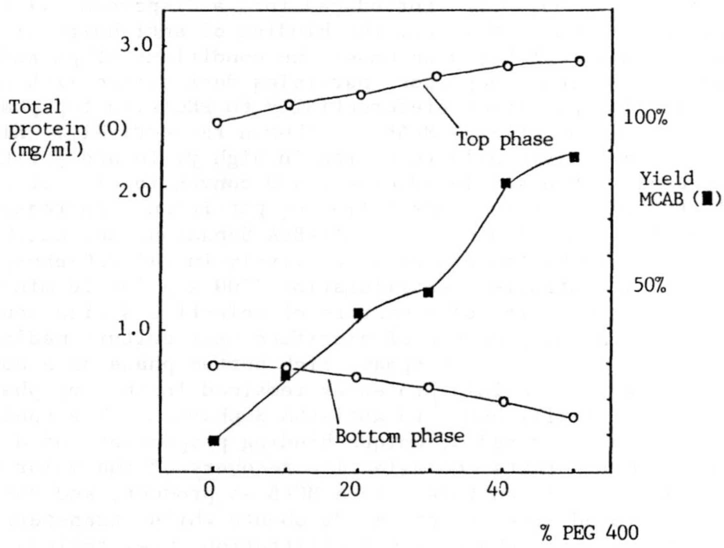

Fig. 2 *Recovery of MCAB from 14% PEG/phosphate systems: variation of PEG 400:PEG 1000 (Experimental system described in Fig. 1)*

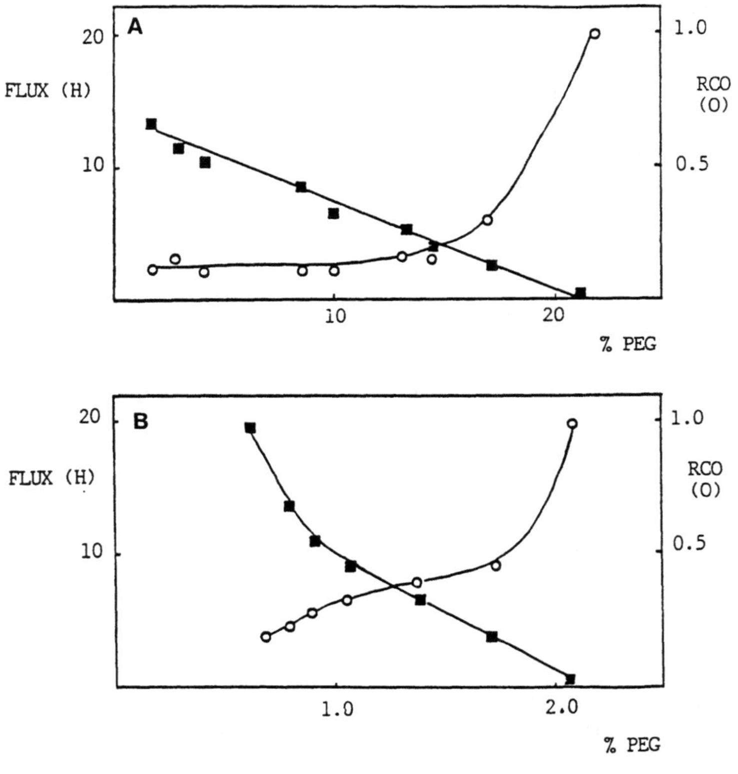

Fig. 3 *Product recovery from PEG/phosphate partition: ultrafiltration (10K MWCO) of top and bottom phases*

FEEDSTREAM :	Anti-huIgG-Fc	(30-150 μg/ml)
	BSA	(2.5 mg/ml)
	Transferrin	(300 μg/ml)
	Acidic proteases	(10-50 μg/ml)

DISPERSED PHASES :	14% PEG 400:PEG 1000 (1:1)-PHOSPHATE	
ADSORBENTS :	Not included	I Anti-BSA Sepharose II HuIgG-HPLC silica
PARTITION :	<u>Top phase</u>	<u>Top phase</u>
	*Anti-huIgG BSA Protease	BSA-adsorbent I Protease
	<u>Bottom phase</u>	<u>Bottom phase</u>
	BSA Transferrin	*Anti-huIgG-adsorbent II Transferrin

Fig. 4 *PEG/phosphate partition of spent culture media: application of affinity adsorbents. Batch experiments conducted on a 20 g PEG-phosphate system. Biological activity determined by ELISA, total protein by dye binding assay, and protein purity determined by SDS-PAGE as described in Materials and Methods*

Experiment conducted in a 2 litre batch system containing culture medium (10 g total protein). Transmembrane flux measured in steady state conditions at each concentration of PEG determined for estimation of retention coefficient (RCO) as described in Materials and Methods. Original clean water flux re-established before each determination.

CONCLUSION

We conclude that operational pH and average polymer molecular weight require careful optimisation for the efficient recovery of immunologically active MCAB from clarified animal cell cultures. Similar attention to conditions diminishes interference from interface precipitation and promotes successful separation of albumin and transferrin as useful components of spent media. Product and component recovery may be effected from phases by ultrafiltration providing a 2 to 10 fold dilution of permeate and prolonged process times are acceptable. The introduction of a mixture of immunoaffinity agarose and silica enables efficient fractionation of the major components in clarified cell culture media in a "four phase" partition system.

ACKNOWLEDGEMENT

We acknowledge the gift of murine hybridomas from the Department of Immunology, University of Birmingham.

REFERENCES

1 H. Schutte, K.H. Kroner, W. Hummel and M.-R. Kula, Recent developments in separation and purification of biomolecules, Ann. N.Y. Acad. Sci. 413:270 (1983)
2 M.-R. Kula, K.H. Kroner and H. Hustedt, Purification of enzymes by liquid-liquid extraction, Advan. Biochem. Eng. 24:73 (1982)
3 J. Rudge, M.A. Desai, S.A. Shojaosadaty and A. Lyddiatt, Continuous culture of murine hybridomas with integrated recovery of monoclonal antibodies, in: "Modern Approaches to Animal Cell Technology," R.E. Spier and J.B. Griffiths, eds., Butterworth, London
4 U.K. Laemmli, Cleavage of structural protein during the assembly of the head of bacteriophage T4, Nature, 227:680 (1970)
5 M.M. Bradford, A rapid and sensitive method for the quantitation of microgram quantities of protein utilising the principle of protein-dye binding, Anal. Biochem. 72:249 (1976)
6 S.A. Shojaosadaty, Application of affinity HPLC to the monitoring and quantitation of biological process operations, Conference Proceedings EFB4, Amsterdam 3:136 (1987)
7 D.T. Plummer, Introduction to practical biochemistry (2nd edition), McGraw-Hill, London (1978)
8 D. Herbert, P.J. Phipps and R.E. Strange, in: "Methods in Microbiology", J.R. Norris and D.W. Ribbons, eds., Vol. 5B:285, Academic Press, London (1971)
9 B. Kassell and P.A. Meitner, Bovine pepsinogen and pepsin, in: "Methods in Enzymol.," Academic Press, London (1970)
10 B. Skoog, Determination of poly(ethylene glycols) 4000 and 6000 in plasma preparations, Vox. Sng. 37:345 (1979)
11 M.A. Desai, A. Lyddiatt, J. Rudge, A.W. Sansome-Smith and A.B. Stevens, Biochemical and physical characterisation of a composite solid phase developed for large scale biochemical adsorption, Conference Proceedings EFB4 Amsterdam 2:517 (1987)
12 K.H. Kroner, H. Hustedt and M.-R. Kula, Extractive enzyme recovery-economic considerations, Process Biochem. 19:67 (1984)
13 C.J.O.R. Morris and P. Morris, Countercurrent distribution, in: "Separation Methods in Biochemistry," Pitman, London (1976)
14 P.O. Hedman, Protein adsorbents intended for use in aqueous two-phase systems, Anal. Biochem. 138:411 (1984)

PRODUCTION OF ß-GALACTOSIDASE FUSED PROTEIN A AND THE PURIFICATION OF IT IN AN AQUEOUS TWO-PHASE SYSTEM

Andres Veide, Lars Strandberg, Halldis Hellebust and Sven-Olof Enfors

Department of Biochemistry and Biotechnology
The Royale Institute of Technology
S-100 44 Stockholm, Sweden

We have studied the production of the genetically fused protein Staphylococcal protein *A/Escherichia coli* ß-galactosidase (SpA-ßgal). The cultivation conditions were found to have a profound effect on the overall productivity of SpA-ßgal. Biologically active SpA-ßgal was produced corresponding to 12% of the cell dry weight. The purification of SpA-ßgal was based on the partitioning properties of ß-galactosidase during extraction in an aqueous two-phase system consisting of poly(ethylene glycol) 4000 and potassium phosphate. Stability studies of SpA-ßgal revealed that the ß-galactosidase moiety remained intact both with respect to size and enzyme activity, while the SpA moiety was completely degraded by *E. coli* proteases.

INTRODUCTION

There is a growing interest in the development of processes for the isolation and purification of proteins. A reason for this is the fast growth of the recombinant DNA (rDNA) technology allowing bacteria to overproduce native, as well as foreign proteins. The rDNA technology can also be used to facilitate the purification of proteins. For example, the fusion of human Insulin-Like Growth Factor I (IGF-I) to the IgG-binding protein A was utilised to design a large scale process for the production of IGF-I with *E. coli* [1].

We have studied the production of the genetically fused protein Staphylococcal protein *A/E. coli* ß-galactosidase (SpA-ßgal). There are few data concerning large-scale cultivation of rDNA products under controlled conditions [2]. Therefore the influence of different cultivation conditions on the production of SpA-ßgal with an *E. coli* strain was investigated [4].

A critical step in the purification of intracellular proteins is the primary purification, which includes removal of bacterial cell debris. A continuous process for large-scale isolation of ß-galactosidase from a suspension of disintegrated *E. coli* cells has been developed. In poly(ethylene glycol) (PEG) 4000/potassium phosphate aqueous two-phase systems (pH 7), all cell debris, the major part of the unwanted proteins, and nucleic acids were partitioned to the denser salt-rich phase. Since ß-galactosidase had an extreme partition coefficient, compared to the average *E. coli* proteins, ß-galactosidase could be collected in a small PEG-rich top phase without sacrificing too much of the recovery [3]. Based on the partitioning properties of ß-galactosidase in the PEG 4000/potassium phosphate aqueous system, a similar purification procedure was investigated for SpA-ßgal.

CULTIVATION

We have studied the influence of different cultivation conditions on the production of SpA-ßgal with *E. coli* RR1 lacZ ΔM15 (pRIT1, pNF2690) [4]. The strain was developed by Uhlén and co-workers [5]. The fused protein has the enzymatic capacity of ß-galactosidase as well as the IgG-binding capacity of protein A [5,6]. The production has been studied both under the uncontrolled conditions in shake flask cultures and under controlled fermenter conditions. The amount of produced, biologically active SpA-ßgal was strongly dependent on the cultivation conditions.

The expression of SpA-ßgal is under the control of the lambda phage promoter P_R and its temperature sensitive repressor cI 857 [5]. The construction allows for the possibility to control the expression by the culture temperature and thereby to separate the growth phase from the production phase. The production is so intensive that no cell growth occurs. At 30°C the repressor is active and the cells can grow and at 42°C the repressor is inactive and SpA-ßgal is produced. The experimental protocol for the examinations of the influence of the cultivation conditions has been described in detail elsehwere [4].

We have found that when cultivated in shake flasks the level of SpA-ßgal raised during the first few hours after temperature induction, but thereafter declined to a zero level. Electrophoresis showed that after two hours in the production phase, approximately 50% of the produced SpA-ßgal existed in an insoluble, inactive form in the cells, and, after 7 hours, all of the product existed in this form, as so-called inclusion bodies. This inactivation could have been caused by the falling pH in the shake flask culture or by some action of the acetic acid that was produced. Undissociated acetic acid has the ability to penetrate the cell wall and cell membrane and dissociate within the cells and thereby lowering the intracellular pH [7]. The inactivation could also have been caused by some other conditions, for example, by a lowered intracellular redox potential caused by oxygen starvation.

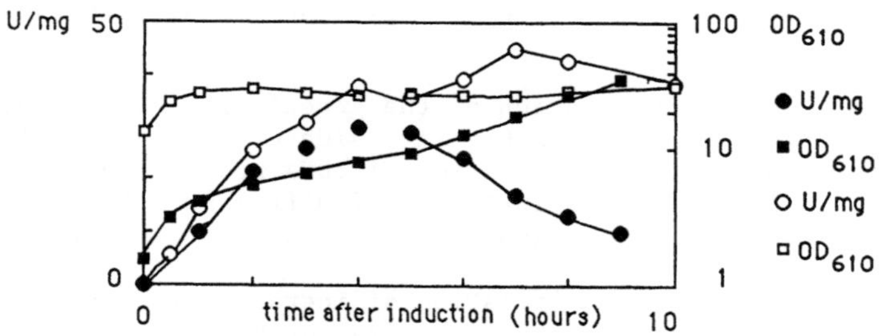

Fig. 1 *Specific activity of SpA-ßgal (U/mg cell dry weight) and cell growth (OD610) as a function of time after induction, determined as described elsewhere [4]. Shaded symbols indicate fermenter culture induced at a low cell density and open symbols induction at a high cell density*

When production was performed under the controlled conditions in a fermenter, most of the product was found in a soluble and active form. The time of temperature induction was found to be important (Fig. 1). When induced at a low cell density (0.4 g cell dry weight/litre), the culture segregated into a non-growing, SpA-ßgal producing population and a growing, non-productive population, as determined by enzyme assays and viable count on different agar medium. The specific activity decreased after 4 hours in the production phase, although the total activity increased. This decrease of specific activity was caused by dilution of the culture with growing, non-productive cells. When induced at a higher cell density (4 g cdw/l), no segregation occurred. The absence of segregation, when the production was induced at a high cell density, could have been caused by some cultivation condition, for example, high levels of accumulated acetic acid which inhibits growth [8]. The maximal amount of produced SpA-ßgal was 50 U/mg cell dry weight (with a cell density of 8 g cdw/l) which corresponds approximately to 12% of the cell dry weight. This was obtained in the fermenter when induced at a high cell density. Thus, by late induction it was possible to circumvent the problem with plasmid instability and to achieve an overall high productivity.

LIQUID/LIQUID EXTRACTION OF CELL DISINTEGRATES FROM THE SPA-ßGAL AND THE ß-GALACTOSIDASE STRAINS

The partitioning of SpA-ßgal in PEG 4000/potassium phosphate systems (pH 7) resulted in partition coefficients in the range of 5-250 depending on system composition and type of ions included in the systems. The partition coefficient for protein A in the same kind of systems were about 0.2. Thus the partitioning properties of ß-galactosidase in PEG 4000/potassium phosphate systems could be utilised for the purification of SpA-ßgal, but could also be of interest for separation of other ß-galactosidase fused proteins.

An increased adsorption to the interface of cell debris was observed when cell disintegrate from the SpA-ßgal producing *E. coli* strain (SpA-ßgal strain) was partitioned in phase systems (pH 7), where the PEG 4000 and potassium phosphate concentrations were increased [9]. In similar experiments with cell disintegrate from the ß-galactosidase producing *E. coli* strain (ßgal strain), the cell debris moved from the bottom phosphate phase to the top PEG phase with increasing concentration of phase forming constituents [3]. Our findings that the biopolymers included in the cell disintegrate contributed to the phase formation and thus altered the position of the binodial curve towards lower concentrations of PEG 4000 and potassium phosphate were in agreement with observations made by Kula et al [10]. By the addition of broken cells, phase systems were formed in the one-phase area below the binodial curve, which was given for the inert PEG 4000/potassium phosphate system (pH 7) [9].

The formation of two phases below the binodial has also been reported for a PEG 1550/potassium phosphate system. In this system it was found that cell debris generally partitioned into the bottom phase in a wedge-shaped area along the upper part or the bionodial [11]. However, when the partitioning of cell debris from the ßgal strain and from the SpA-ßgal strain were compared in PEG 4000/salt systems, the cell debris from the SpA-ßgal strain were located at the interface in all systems, while the cell debris from the ßgal strain preferred the bottom phase to a greater extent [9]. Thus, the position in a phase system at which cell debris were caught by the interface seemed to vary with the type of *E. coli* strain. The strain dependent partitioning of cell debris was also seen for three other *E. coli* strains in PEG 4000 or PEG 6000/potassium phosphate systems. The distribution of the cell debris from one of the strains changed from the interface to the

Table 1 *Partitioning of cell debris from the SpA-ßgal producing E. coli strain in PEG/potassium phosphate (PP) aqueous two-phase systems (pH 7) where the molecular weights of PEG were changed* *

Phase system composition			Top phase	Interface	Bottom phase
PEG		PP			
molecular weight	(% (w/w))	(% (w/w))			
400	10.8	20.6	+	(+)	-
1500	7.6	15.5	(+)	+	(+)
4000	5.3	12.5	-	+	(+)
6000	4.2	12.0	-	+	(+)
10000	5.6	10.0	-	-	+
20000	5.6	10.0	-	-	+

* Thawed cells, 50% wet weight, were diluted in 50 mM potassium phosphate, pH 7, and disintegrated in a French press. In each experiment, 1g of disintegrated cell suspension was added, giving a final concentration of 6.2% (w/w)

+ = turbid phase (+) = slightly turbid phase - = clear phase

potassium-rich bottom phase when the molecular weight of PEG was increased [9]. The cell debris from the SpA-ßgal strain was removed from the PEG-rich top phase to the interface and finally to the bottom phase when the molecular weight of PEG was increased from 400 to 20 000 in PEG/potassium phosphate system (pH 7), as illustrated in Table 1. The compositions of the phase systems were chosen in such a way that approximately the same length of tie line was obtained in all systems.

All these results indicated differences in the surface properties of the cell debris from the various strains. Individual extraction systems must be designed for each strain. Differences in the surface structure have been used for the separation of various *E. coli* strains [12,13].

The partitioning of ß-galactosidase and SpA-ßgal to the PEG phase was influenced by the addition of biological material to the phase systems. The increased load of cell disintegrate in a given phase system reduced the partition coefficient for ß-galactosidase, the recovery of ß-galactosidase, and the purification factor [3]. However, by decreasing the concentrations of PEG 4000 and potassium phosphate in a phase system containing a given amount of cell disintegrate, i.e. decreasing the length of the tie line and thereby decreasing the interfacial tension, the reduction of the partition coefficient and the recovery of ß-galactosidase could be counteracted [3]. In a similar partitioning experiment using disintegrate from the SpA-ßgal strain, also the recovery of SpA-ßgal decreased with increasing length of the tie line (Fig. 2). However, at the same time, an increase of the partition coefficient was observed for SpA-ßgal (Fig. 2). This was probably a consequence of increased partitioning to the interface of SpA-ßgal when the interfacial tension in the partitioning system was increased. Thus, while ß-galactosidase moved from the top phase to the bottom phase with increasing length of tie line, SpA-ßgal moved from both phases to the interface. More data from this experiment are shown in Table 2.

Table 2 *Partitioning of cell disintegrate from the SpA-ßgal strain in PEG 4000/potassium phosphate (PP) aqueous two-phase systems of different compositions (see Fig. 2 for a comparison)* [a)]

Phase system composition		Cell debris partitioning			Phase volumes			Partition coefficients for total protein [b)]
PEG 4000 (%(w/w))	PP (%(w/w))	Top	Inter	Bottom	Top (ml)	Inter (ml)	Bottom (ml)	
7.7	11.6	-	+	-	0.1	4.0	5.0	1.6 [c)]
7.3	11.0	-	+	(+)	1.5	2.3	5.2	0.7
7.0	10.5	-	+	(+)	2.0	0.9	6.0	0.7
6.9	10.4	-	(+)	+	2.3	0.5	6.5	0.7
6.8	10.2	-	(+)	+	2.2	0.4	6.5	0.6
6.7	10.1	-	-	+	2.2	0	6.8	0.7
6.6	9.9	Single phase				9.2		Single phase

a) The cell disintegrate was obtained by diluting a 50% (w/w) SpA-ßgal cell suspension (in 50 mM potassium phosphate, pH 7, 10 mM $MgCl_2$) two times in 50 mM potassium phosphate, pH 7, 10 mM $MgCl_2$ and disintegrating it in a French press. In each experiment, 5 g of disintegrated cell suspension (17 430 units of SpA-ßgal) was added, giving a final concentration of 12.5% (w/w). The total weight of each system was 10 g. The temperature in the phases during this experiment was 24-26°C.

b) Protein concentration was determined at 595 nm after treatment with CBB G 250 [14]

c) The analysis was performed on the interface including the top phase

\+ = turbid phase (+) = slightly turbid phase - = clear phase

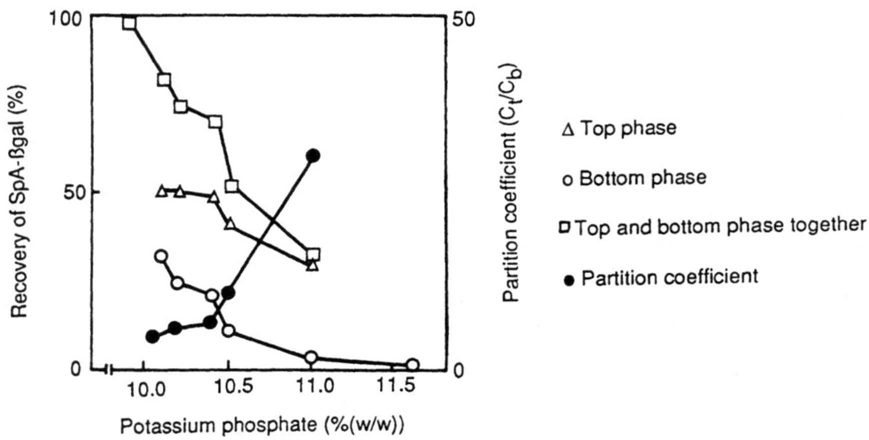

Fig. 2 *Recoveries of SpA-ßgal in top and bottom phases, and partition coefficients, after partitioning of cell disintegrate in PEG 4000/potassium phosphate aqueous two-phase systems (pH 7) of different compositions. The ß-galactosidase activity of SpA-ßgal was determined as the rate of o-nitrophenyl-ß-D-galactopyraniside hydrolysis at 25°C, as described elsehwere [15]. (See Table 2 for a comparison)*

The partitioning of SpA-ßgal to the PEG phase was also affected by the additional salts present in PEG 4000/potassium phosphate system during the extraction. When a cell debris free crude extract from the SpA-ßgal strain suspended in potassium phosphate buffer was added to a PEG 4000/potassium phosphate system (pH 7), almost all of the SpA-ßgal was partitioned to the PEG top phase. However, if crude extract suspended in Tris-HCl buffer, including $MgCl_2$ and NaCl was partitioned in the same kind of system, most of the SpA-ßgal was confined to the interface. A very dense interface was observed in both cases [9]. It has been shown that the addition of Tris-HCl to a pure PEG 4000/potassium phosphate system (pH 7) alters the position of the binodial towards lower concentrations of PEG 4000 and potassium phosphate [3], and thereby increases the length of the tie line for a given phase composition. Thus, the partitioning behaviour of SpA-ßgal could be due to the change of interfacial tension. It has been reported that the conformation of ß-galactosidase is affected by Mg^{2+} [16,17]. Hence the altered partitioning of Spa-ßgal could be a reflection of this.

The partitioning experiments with ß-galactosidase and SpA-ßgal showed that the composition of the phase systems used in the cell debris removal step had to be chosen carefully. Especially in the case of SpA-ßgal, a favourable position was found only very close to the binodial, where the cell debris was partitioned totally to the bottom phase, without getting caught by the interface, and only a minor amount of SpA-ßgal was adsorbed at the interface.

STABILITY

Stability studies of SpA-ßgal in cell disintegrates and in a purified preparation revealed that the ß-galactosidase moiety of the molecule remained intact both with respect to size and enzyme activity, while the SpA moiety was completely degraded by *E. coli* protease (Fig. 3). In a purified extract, obtained after two-phase partitioning in a PEG 4000/potassium phosphate system (pH 7) and subsequent ultrafiltration, the degradation of the SpA moiety proceeded but at a much lower rate, showing that the main part of the proteolytic activity against SpA was removed by this purification procedure. The degradation of SpA-ßgal seemed to proceed via more stable intermediate compounds, represented by ß-galactosidase molecules having three, two or one complete SpA moiety, revealed by the gel filtration. The degradation could be partly inhibited by o-phenanthroline (10 mM), but not by PMSF (10 mM) or EDTA (10 mM). The part of SpA-ßgal found in inclusion bodies was not subjected to proteolytic degradation.

E. coli has several proteases, some of which are used for protein modification while others are responsible for the complete degradation of intracellular proteins. Ten soluble proteases have been purified from *E. coli* [18,19,20]. The best characterised E. coli protease is the La protease. It is an ATP dependent serine protease and cleaves peptide bonds in hydrophobic sequences of the protein substrates and catalyses the rate limiting step in the degradation of abnormal proteins in the cell [21]. The fact that the reaction was not inhibited by PMSF and that protein A is considered as a hydrophilic protein [6], suggested that the La protein was not involved in the degradation of SpA-ßgal.

The proteolytic degradation of a protein might not be recognised as a problem during small scale preparations. However, when the process scale is increased, the time needed for cell separation and further purification increases. Thus an increase of time-dependent losses of product, such as proteolytic degradation, should be expected.

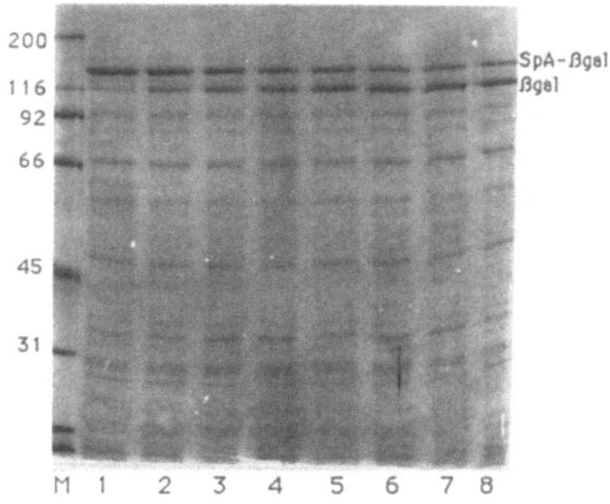

Fig. 3 *SDS-PAGE of disintegrated cells, 25% wet weight, incubated at 37°C. Each sample contains approximately 1 unit ß-galactosidase. Lane M, marker proteins shown as molecular weight in kD.Lane 1-8, stored 0, 0.5 1, 1.5, 2, 3, 4.5 and 5 hours, respectively*

CONCLUSIONS

Liquid/liquid extraction in aqueous two-phase systems could be included in the development of fast and efficient primary purification procedures for microbial intracellular proteins. However, it was shown that the phase system must be designed very carefully to obtain favourable partitioning conditions with respect to both cell debris and product. The importance of a fast treatment during the recovery of the target protein becomes obvious if the protein is exposed to time-dependent loss, as, for instance, proteolysis. Moreover, the need for a fast treatment should be expected to increase with increasing scale of operation.

If the partitioning of the product protein to the cell debris free phase is high, the extraction step can be utilised as a protein fractionation and concentration step at the same time as it is used for primary recovery. As a way to increase the partitioning of proteins to the PEG phase, the bounding of target proteins to ß-galactosidase on a gentic level is suggested.

REFERENCES

1 T. Moks, L. Abrahmsén, B. Österlöf, S. Josephson, M. Östling, S.-O. Enfors, I.-L. Persson, B. Nilsson and M. Uhlén, Large-scale affinity purification of human insuline-like growth factor I from culture medium of E. coli, Bio/technology 5:379 (1987)

2 D.W. Zabriskie and E.J. Arcuri, Factors influencing productivity of fermentations employing recombinant microorganisms, Enzyme Microbiol. Technol. 8:706 (1986)

3 A. Veide, T. Lindbäck and S.-O. Enfors, Continuous extraction of ß-galactosidase from E. coli in an aqeuous two-phase system: effects of biomass concentration on partitioning and mass transfer, Enzyme Microb. Technol. 6:325 (1984)

4 L. Strandberg, A. Veide and S.-O. Enfors, Production of the hybrid protein Staphylococcal protein A/Escherichia coli ß-galactosidase with E. coli, J. Biotechnol., in press
5 B. Nilsson, L. Abrahamsén and M. Uhlén, Immobilisation and purification of enzymes with Staphylococcal protein A gene fusion vectors, EMBO 4:1075 (1985)
6 M. Uhlén, The gene for Staphylococcal protein A and its use in biotechnology, Thesis, Royal Institute of Technology, Stockholm (1984)
7 C.V. Salmond, R.G. Kroll and I. Booth, The effect of food preservatives on pH homeostasis in Escherchia coli, J. Gen. Microbiol. 130:2845 (1984)
8 T. Yano, H. Mori, T. Kobayashi and S. Shimizu, Reusability of broth supernatant as medium, J. Ferment. Technol. 58:259 (1980)
9 A. Veide, L. Strandberg and S.-O. Enfors, Extraction of ß-galactosidase fused protein A in aqueous two-phase systems, Enzyme Microb. Technol., in press
10 M.-R. Kula, K.H. Kroner and H. Hustedt, Purification of enzymes by liquid/liquid extraction, in: "Advances in Biotechnology," A. Fiechter, ed., Springer-Verlag, Berlin (1972)
11 H. Hustedt, K.H. Kroner, U. Menge and M.-R. Kula, Aqueous two-phase systems for large-scale enzyme isolation processes, in: "1st European Congress on Biotechnology-Preprints (Part 1)" (1978)
12 P.-Å. Albertsson and G.D. Baird, Countercurrent distribution of cells, Exp. Cell Res. 28:296 (1962)
13 K.-E. Magnusson and O. Stendahl, Partitioning of bacteria, virus and phage, in: "Partitioning in Aqueous Two-Phase Systems," H. Walter, D.E. Brooks and D. Fisher, eds., Academic Press, London (1985)
14 M.M. Bradford, A rapid and sensitive method for the quantitation of microgram quantities of protein utilising the principle of protein-dye binding, Anal. Biochem. 72:248 (1976)
15 A. Veide, A.-L. Smeds and S.-O. Enfors, A process for large-scale isolation of ß-galactosidase from E. coli in an aqueous two-phase system, Biotechnol. Bioeng. 25:1789 (1983)
16 J.-P. Tenu, O.M. Viratelle and J. Yon, Kinetic study of the activation process of ß-galactosidase from E. coli by Mg^{2+}, Eur. J. Biochem. 26:112 (1972)
17 G.S. Case, M.L. Sinott and J.-P. Tenu, The role of magnesium ions in ß-galactosidase-catalysed hydrolyses, Biochem. J., 133:99 (1973)
18 A.L. Goldberg, K.H.S. Swamy, C.H. Chung and F.S. Larimore, Proteases in Escherichia coli, Methods in Enzymology 80:680 (1981)
19 J.E. Roseman and R.L. Levine, Purification of a protease from Escherichia coli with specificity for oxydised glutamine synthetase, J. Biol. Chem. 262:2110 (1987)
20 Y. Katayama-Fujimura, S. Gottesman and M.R. Maurizi, A multiple-component, ATP-dependent protease from Escherichia coli, J. Biol. Chem. 262:4477 (1987)
21 L. Waxman and A.L. Goldberg, Selectivity of intracellular proteolysis: protein substrates activate the ATP-dependent protease (La), Science, 232:500 (1987)

CONTINUOUS AFFINITY PARTITIONING IN A POLY(ETHYLENE GLYCOL)-REPPAL PES AQUEOUS TWO-PHASE SYSTEM

Shaun P. Foaden [1], Howard A. Chase [1] and Christopher R. Lowe [2]

[1] Department of Chemical Engineering
University of Cambridge
Pembroke Street, Cambridge CB2 3RA, UK

[2] Institute of Biotechnology
University of Cambridge
Downing Street, Cambridge CB2 3EF, UK

The starch-based polymer, Reppal Pes, has been used to form two-phase affinity systems with dyed-poly(ethylene glycol). Lactate dehydrogenase has been purified from a crude rabbit muscle extract in these systems using the triazine dyes Cibacron blue F3G-A and Procion Orange H-ER bound to poly(ethylene glycol). Processing parameters have been optimised in small-scale batch experiments and the results used to formulate systems for continuous affinity partitioning of rabbit muscle lactate dehydrogenase in a simple two-stage mixer-settler unit. Polymer-bound ligand from the eluting stage has been recycled into the affinity partitioning stage, thereby reducing the amount of immobilised ligand required.

INTRODUCTION

Affinity partitioning is capable of processing large amounts of biomaterial, producing endproducts of high purity [1,2] without the limitations such as matrix fouling, difficulties of scale-up and long equilibration times associated with solid-phase separation techniques [3,4]. Despite these apparent attractions, affinity partition has failed to become a widely accepted unit operation in commercial separations. This has been due mainly to the high cost of the fractionated dextran which, together with poly(ethylene glycol), is used to form the most popular aqueous two-phase system. Most large-scale applications exploit the less expensive poly(ethylene glycol)-salt systems which, however, lack the resolving power of affinity systems and are thus of limited use in enzyme purification [5].

Alternatives to fractionated dextran, such as crude dextran, Aquaphase PPT [7] and Reppal Pes [8,9], have been investigated. In this report the performance of Reppal Pes in an affinity partitioning system for the purification of lactate dehydrogenase has been systematically evaluated.

Despite the existence of a considerable body of liquid-liquid extraction experience in the chemical processing industry, no real attempt has been made to use affinity partitioning in a large-scale, continuous process. Most large-scale applications to date have been performed as batch processes, and continuous applications have relied on centrifugation to separate the phases thereby raising capital and operating costs [10,11].

In this work, a simple mixer-settler has been used for the continuous processing of two-phase systems as a first step towards the development of a simple unit operation based on standard chemical engineering practices.

EXPERIMENTAL

Materials

Poly(ethylene glycols) (Mw = 3350 and Mw = 8000) were obtained from Sigma Chemical Company (Poole, UK). Reppal Pes (Mw = 114 000) was purchased from Reppe Glykos AB (Växjö, Sweden). Cibacron blue F3G-A was purchased from Pierce (UK) Ltd (Cambridge, UK) and Procion Orange H-ER was obtained from ICI (Blackley, UK). The dye - poly(ethylene glycol) conjugates were synthesised using the method of Johansson and Joelsson [12]. DEAE-cellulose anion exchanger (capacity 0.91 $meq.g^{-1}$, medium mesh) was purchased from the Sigma Chemical Company (Poole, UK). Rabbit muscle lactate dehydrogenase (L-Lactate:NAD oxidoreductase, EC 1.1.1.27; 550 $units.mg^{-1}$ at 25°C) was purchased from The Boehringer Corporation (Lewes, UK) as were all auxiliary enzymes, substrates and co-enzymes. A crude rabbit muscle extract (57 $units.mg^{-1}$ lactate dehydrogenase and 63 $units.mg^{-1}$ pyruvate kinase at 25°C), bovine albumin (fraction V powder, 96-99% albumin) and Coomassie Brilliant Blue G (90% dye content) were obtained from Sigma Chemical Company (Poole, UK). All other chemicals were of analytical grade.

Analytical Two-Phase Affinity Systems

Two-phase systems were prepared from stock aqueous solutions of the polymers, 40% (w/w) poly(ethylene glycol) and 30% (w/w) Reppal Pes (Mw = 114 000). Appropriate amounts of these stock solutions were weighed out into test tubes and mixed with sodium phosphate buffer (0.5M pH = 7.3), dye-poly(ethylene glycol) solution, water and enzyme sample (10 - 800 $units.ml^{-1}$) to give a final system mass of 3.0 g. The systems were mixed by inverting 20 times and then centrifuged for 5 minutes at 20°C and 200 rpm in a Sorvall RT600B refrigerated centrifuge. Samples of top and bottom phases were taken and assayed for lactate dehydrogenase and protein content in order to determine the partitioning between the phases. The partition coefficient is defined as the ratio of the concentrations of enzyme in the top and bottom phases respectively. The purification ratio is defined as the ratio of enzyme specific activity in the product phase to that in the enzyme feed solution. All work, unless otherwise stated, was performed at room temperature (~20°C).

Continuous Affinity Partitioning

A two-stage mixer-settler was constructed of Perspex with each stage containing a mixing chamber of 55 ml and a settling chamber of 600 ml separated by a simple weir. Belt-driven, shaft-mounted barrel impellers, powered by a 240V a.c. 50/60 Hz shaded pole motor were used in the mixing chambers. Masterflex peristaltic pumps (Cole-Palmer Instrument Company, Chicago, USA) were used to pump the various streams. The system layout is shown in Fig. 1.

Using the stock polymer solutions and appropriate amounts of sodium phosphate buffer (1.0M, pH = 7.3), water and poly(ethylene glycol)-bound dye, two-phase polymer systems were made up and pumped into the affinity extraction stage (Stage 1) mixing chamber at 400-760 $ml.h^{-1}$. The enzyme feed solution (lactate dehydrogenase = 38 $units.mg^{-1}$ protein, 1.10 mg $protein.ml^{-1}$ in 50 mM sodium phosphate buffer, pH 7.3) was also pumped into this chamber at 80-152 $ml.h^{-1}$.

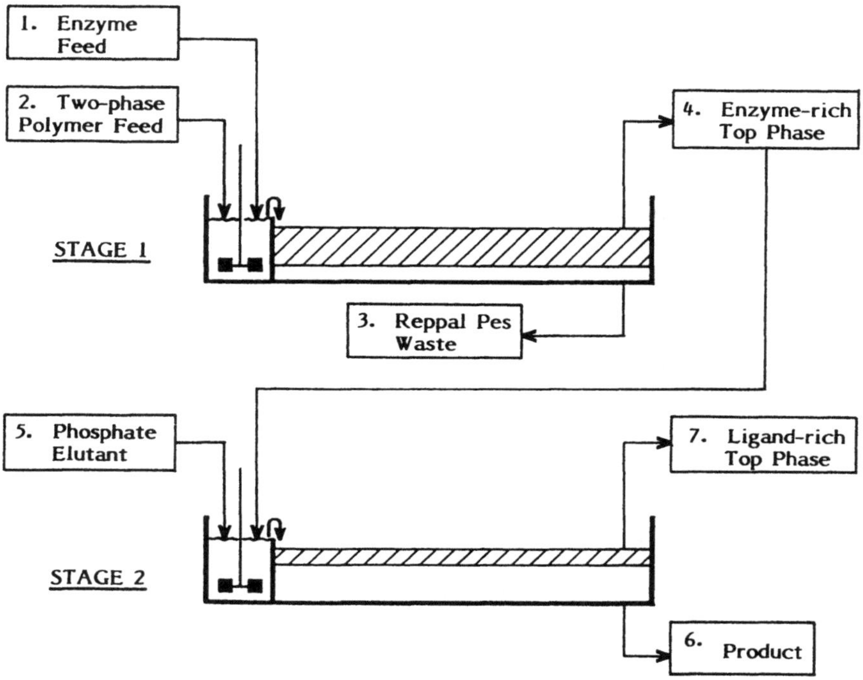

Fig. 1 *Schematic diagram of two-stage mixer-settler.*
Stage 1 = Poly(ethylene glycol) - Reppal Pes affinity partition stage
Stage 2 = Poly(ethylene glycol) - Potassium phosphate elution stage

Once the settling chamber had been filled to its operating level, separated bottom phase was pumped to waste and enzyme-rich top phase was pumped to the salt eluting stage (Stage 2). In the mixing chamber of the second stage, this top phase was mixed with the eluting potassium phosphate solution (2.3-2.8M, pH 7.3). After reaching operating level in the settling chamber the bottom phase product stream was removed at 200-635 $ml.h^{-1}$. Top phase from Stage 2 was either pumped to waste or recycled directly to the Stage 1 mixing chamber. When the system was operated with this recycle the composition of the two-phase feed to the first stage was adjusted to ensure that the overall polymer composition in Stage 1 did not deviate from the pre-recycle value. Stream flow-rates and compositions are shown in Table 1.

Table 1 *Flow-rate and Phase Composition Data for Continuous Affinity Partitioning*

System[a] No.	Stream[b] No.	Flow rate ml.h^{-1}	PEG[c] %(w/w)	RP 100 %(w/w)	Sodium Phosphate mM	Dye-PEG %(w/w)	Dye mM	KPhosphate %(w/w)
1.	1	152	--	--	50	--	--	--
	2	760	11.9	14.3	50	0.12	0.36	--
	3	244	5.5	34.3	50	0.05	0.18	--
	4	662	12.0	2.2	50	0.11	0.35	--
	5	331	--	--	--	--	--	35.0
	6	636	3.4	--	--	0.03	0.10	14.7
	7	366	13.0	--	--	0.12	0.39	9.5
2.	1	80	--	--	50	--	--	--
	2	238	--	20.1	--	0.05	0.05	--
	3	146	0.8	30.3	124	0.04	0.04	--
	4	334	11.4	0.5	124	0.51	0.51	--
	5	148	1.7	--	--	--	--	31.5
	6	320	0.5	--	--	--	--	15.9
	7	162	23.3	--	--	0.95	1.01	5.0

[a] Refer to Table 2
[b] Refer to Fig. 1
[c] Abbreviations: PEG = poly(ethylene glycol); RP 100 = Reppal Pes (Mw = 114 000); KPhosphate = Potassium Phosphate

Samples of the final product stream were taken every 5-10 min until a steady-state enzyme concentration was reached and every 25 min thereafter. The samples were assayed for lactate dehydrogenase activity and protein content in order to determine the specific activity and, hence the purification ratio. Final product and enzyme feed solution were also assayed for pyruvate kinase and malate dehydrogenase activities. When recycling top phase from the second stage the conductivity of the mixed solution in the first stage was monitored to ensure that there was no significant change in the ionic strength.

Assays

A Perkin-Elmer Lambda 5 Thermostatted Spectrophotometer was used to determine the activities of lactate dehydrogenase [13], pyruvate kinase [14] and malate dehydrogenase [14]. General protein content was determined according to Bradford [15] with bovine serum albumin as standard.

RESULTS

Analytical Two-Phase Affinity Systems

Partitioning of lactate dehydrogenase in poly(ethylene glycol) - Reppal Pes systems was found to produce similar results to those reported for poly(ethylene glycol) - dextran systems [16]. Equivalent concentrations of Reppal Pes and dextran produce two-phase systems with poly(ethylene glycol); however, the binodials for the Reppal Pes systems are flatter than those with dextrans of similar molecular weight and the bottom phases consequently contain a higher percentage of poly(ethylene glycol). Increasing the sodium phosphate buffer concentration in the systems from 5 to 150 mM increased thepurification ratio from 2.8 to 5.5 with a concomitant drop in yield of about 30%. The effect of the ligand concentration was found to agree closely with the theory of Flanagan and Barondes [17]. Increasing the general protein concentration up to 4.0 $mg.ml^{-1}$ had no significant effect on the purification ratio.

Continuous Affinity Partitioning

In order to determine the residence time distribution of the mixer-settler unit, a lactate dehydrogenase solution of 18.3 $units.ml^{-1}$ was fed into the mixer-settler at 152 $ml.h^{-1}$, together with a two-phase poly(ethylene glycol) (Mw = 3350) and Reppal Pes (Mw = 114 000) system at 760 $ml.h^{-1}$. The overall composition of the system in the mixer was 10% (w/w) poly(ethylene glycol), 12% (w/w) Reppal Pes and 0.04% (w/w) Cibacron blue F3G-A - poly(ethylene glycol). Product stream was removed from the bottom of the second stage at 588 $ml.h^{-1}$ and the lactate dehydrogenase concentration measured at 3 min intervals. Using the resultant enzyme activity profile in the product stream, the mean residence time, for both stages combined, was calculated according to the method of Hill [18], to be 68 min. A steady-state enzyme concentration was achieved in the product stream 120 min after commencing the enzyme feed.

The results of two purification runs, with and without recycle of the polymer-bound ligand, are displayed in Table 2. In System 1 no recycle of Stream 7 (see Fig. 1) was employed and Cibacron blue F3G-A - poly(ethylene glycol) (Mw - 3350) was consumed at 1.01 $g.h^{-1}$. System 2, however, was run with total recycling of Stream 7 directly into the Stage 1 mixing chamber. This reduced the consumption of Procion Orange H-ER - poly(ethylene glycol) (Mw = 8000) to 0.12 $g.h^{-1}$ from 1.79 $g.h^{-1}$ for the same system without recycle.

The activities of pyruvate kinase and malate dehydrogenase were not measured in System 1; however, the results from System 2 indicate a 24.3% recovery of pyruvate kinase activity and a 77.2% recovery of malate dehydrogenase activity in the product stream.

The settling tank levels were controlled by manually adjusting the flow-rates of the Stage 1 bottom phase (Stream 3) or the product stream (Stream 6). Stage 1 top phase and, when recycled, Stage 2 top phase, could not be used to control levels because any change in their flow-rates would have altered the composition in the two stages.

DISCUSSION

Reppal Pes has been successfully used as a substitute for dextran in extractive bioconversions [12] and in affinity partition systems using biospecific chromatographic particles [13]. In this work, Reppal Pes has been exploited as a polymer for formulating two-phase systems for the large-scale continuous purification of a typical diagnostic enzyme, lactate dehydrogenase. Reppal Pes is ideally suited to such an application since it produces two-phase systems with poly(ethylene glycol) at similar concentrations to dextran, but is considerably less expensive.

The results of the small-scale batch experiments, although not fully reported here, demonstrated the suitability of using Reppal Pes in affinity partitioning systems. Process parameters, such as the polymer concentration, ionic strength, ligand concentration and protein concentration, were all found to influence the partition of lactate dehydrogenase in poly(ethylene glycol)-Reppal Pes systems in a comparable manner to that in poly(ethylene glycol)-dextran systems, and the general trends applicable to partition in poly(ethylene glycol)-dextran systems can also be applied in this system.

Continuous affinity partitioning, using gravity settling to separate the phases, has been shown to be a workable alternative to processes based on centrifugal separation. In processing runs of up to 24 hours, lactate dehydrogenase was recovered in high yield (>90%) and with more than two fold purification. With recycling of the eluting stage top phase, consumption of dyed-poly(ethylene glycol) was reduced by some 93.3% without any drop in enzyme recovery or purification. Pyruvate kinase activity was reduced by 76%, however, as reported elsewhere [11], malate dehydrogenase was co-extracted with the lactate dehydrogenase.

As a result of limitations on the amount of enzyme starting material available, the maximum general protein content used in the continuous processes was 0.18 $mg.ml^{-1}$. This is considerably lower than the maximum tested in the small-scale experiments (4.0 $mg.ml^{-1}$) and indicates that a 22 fold increase in enzyme throughput is possible.

The results reported here indicate that inexpensive Reppal Pes-poly(ethylene glycol) aqueous two-phase systems perform equally as well as the traditional dextran-poly(ethylene glycol) systems in affinity partitioning processes. When combined with the advantages of a simple, economical continuous liquid - liquid extraction process based on standard chemical engineering technology, the new system makes affinity partitioning an attractive alternative to other purification techniques.

REFERENCES

1 A. Cordes and M.-R. Kula, Process design for large-scale purification of formate dehydrogenase from Candida boidinii by affinity partition, J. Chromatography 376:375 (1986)

2 A. Veide, A.-L. Smeds and S.-O. Enfors, A process for large-scale isolation of ß-galactosidase from E. coli in an aqueous two-phase system, Biotechnol. Bioeng. 25:1789 (1983)

3 C.R. Lowe, New developments in downstream processing, J. Biotechnol. 1:3 (1984)

4 P.-F. Fauquex, H. Hustedt and M.-R. Kula, Phase equilibration in agitated vessels during extractive enzyme recovery, J. Chem. Tech. Biotechnol. 35B:51 (1985)

5 G. Johansson, Aqueous two-phase systems in protein purification, J. Biotechnol. 3:11 (1985)

6 K.H. Kroner, H. Hustedt and M.-R. Kula, Evaluation of crude dextran as phase-forming polymer for the extraction of enzymes in aqueous two-phase systems in large-scale, Biotech. Bioeng. 24:1015 (1982)

7 F. Tjerneld, S. Berner, A. Cajarville and G. Johansson, New aqueous two-phase system based on hydroxypropyl starch useful in enzyme purification, Enzyme Microb. Technol. 8:417 (1986)

8 R. Kaul and B. Mattiasson, Extractive bioconversion in aqueous two-phase systems. Production of prednisolone from hydrocortisone using Arthrobacter simplex as catalyst, Appl. Microbiol. Biotechnol. 24:259 (1986)

9 B. Mattiasson and T.G.I. Ling, Efforts to integrate affinity interactions with conventional separation technologies. Affinity partition using biospecific chromatographic particles in aqueous two-phase systems, J. Chromatography 376:235 (1986)

10 K.H. Kroner, H. Hustedt, S. Granda and M.-R. Kula, Technical aspects of separation using aqueous two-phase systems in enzyme isolation processes, Biotech. Bioeng. 20:1967 (1978)

11 M. Joelsson and G. Johansson, Sequential liquid-liquid extraction of some enzymes from porcine muscle using polymer-bound triazine dyes, Enzyme Microb. Technol. 9:233 (1987)
12 G. Johansson and M. Joelsson, Preparation of Cibacron Blue F3G-A poly(ethylene glycol) in large-scale for use in affinity partitioning, Biotech. Bioeng. 27:621 (1985)
13 L.A. Decker, ed., "Worthington Enzyme Manual," Worthington Biochemical Corp., Freehold NJ (1977)
14 "Biochemica Information I," Boehringer Mannheim GmbH, Mannheim (1973)
15 M.M. Bradford, A rapid and sensitive method for the quantitation of microgram quantities of protein utilising the principle of protein-dye binding, Anal. Biochem. 72:248 (1976)
16 G. Johansson and M. Joelsson, Liquid-liquid extraction of lactate dehydrogenase from muscle using polymer-bound triazine dyes, App. Biochem. Biotech. 13:15 (1986)
17 S.D. Flanagan and S.H. Barondes, Affinity partitioning - a method for purification of proteins using specific polymer-ligands in aqueous polymer two-phase systems, J. Biol. Chem. 250:1484 (1975)
18 C.G. Hill, Jr., "An Introduction to Chemical Engineering Kinetics and Reactor Design," Wiley, New York (1977)

FAST DOWNSTREAM PROCESSING OF PROTEINS BY FLOCCULATION AND LIQUID EXTRACTION

H. Hustedt, R. Krützfeldt and N. Papamichael

GBF, Gesellschaft für Biotechnologische Forschung mbH
Mascheroder Weg 1
D-3300 Braunschweig, FRG

The initial downstream processing of intracellular microbial enzymes is usually conducted in the sequence: cell harvesting → cell disruption → clarification → initial enrichment. The efficiency of the conventional recovery route - most importantly the space-time yield, but also the activity yield - can often be increased considerably by using flocculation, mechanical cell disruption and liquid extraction. A further enhancement of the space-time yield or reduction in process time, respectively, may be obtained by the integration of single steps to semi or fully continuous processes. These topics are discussed in this paper using the downstream processing of fumarase from Brevibacterium ammoniagenes as a case example based on experiments conducted in pilot/commercial scale.

The performance of cell harvesting by centrifugation was found to be very low. Much better results were obtained with flocculation by charged polymers in conjunction with centrifugation using disk stack separators or a decanter with flow rates up to ~2.5 m^3/h [1]. For cell disruption only wet milling worked satisfactorily. The problem of product clarification could not be solved by filtration or centrifugation, but by application of aqueous two-phase extraction. Besides separation of cell debris, the two-step extraction process developed results in a ~20 fold purification (on the basis of specific activity) with an overall yield of 75%. The procedure was transformed into a continuous process exploiting the principle of crosscurrent extraction; a newly developed plant operated under computerised process control and processing ~25 kg biomass per hour was used during this work [2]. After desalting by ultrafiltration and polishing filtration, the enzyme is sufficiently pure to be used as biocatalyst for L-malate production. Semi-continuous operation of the whole procedure enables the processing of ~200 kg biomass (wet weight) in two 8 h shifts using a SA-7 separator (Σ-factor 7000 m^2, Westfalia Separator AG, Oelde/FRG), 2 LME-20 bead mills (grinding chamber 22 l, Netzsch Ltd., Selb/FRG), the crosscurrent extraction plant (with two separators, type SA-1, Σ-factor 1460 m^2, Westfalia Separator AG, Oelde/FRG), and an ultrafiltration plant (see flow scheme). Fully continuous processing would reduce the net operation time to 8 h. A further scale-up of the process line appears to be easy. The majority of intracellular enzymes could be recovered according to the scheme discussed; bead milling may be replaced in many cases by high pressure homogenisation. Of course, high purification can be carried out after the ultrafiltration step or sometimes applying a phase directly to chromatography, too.

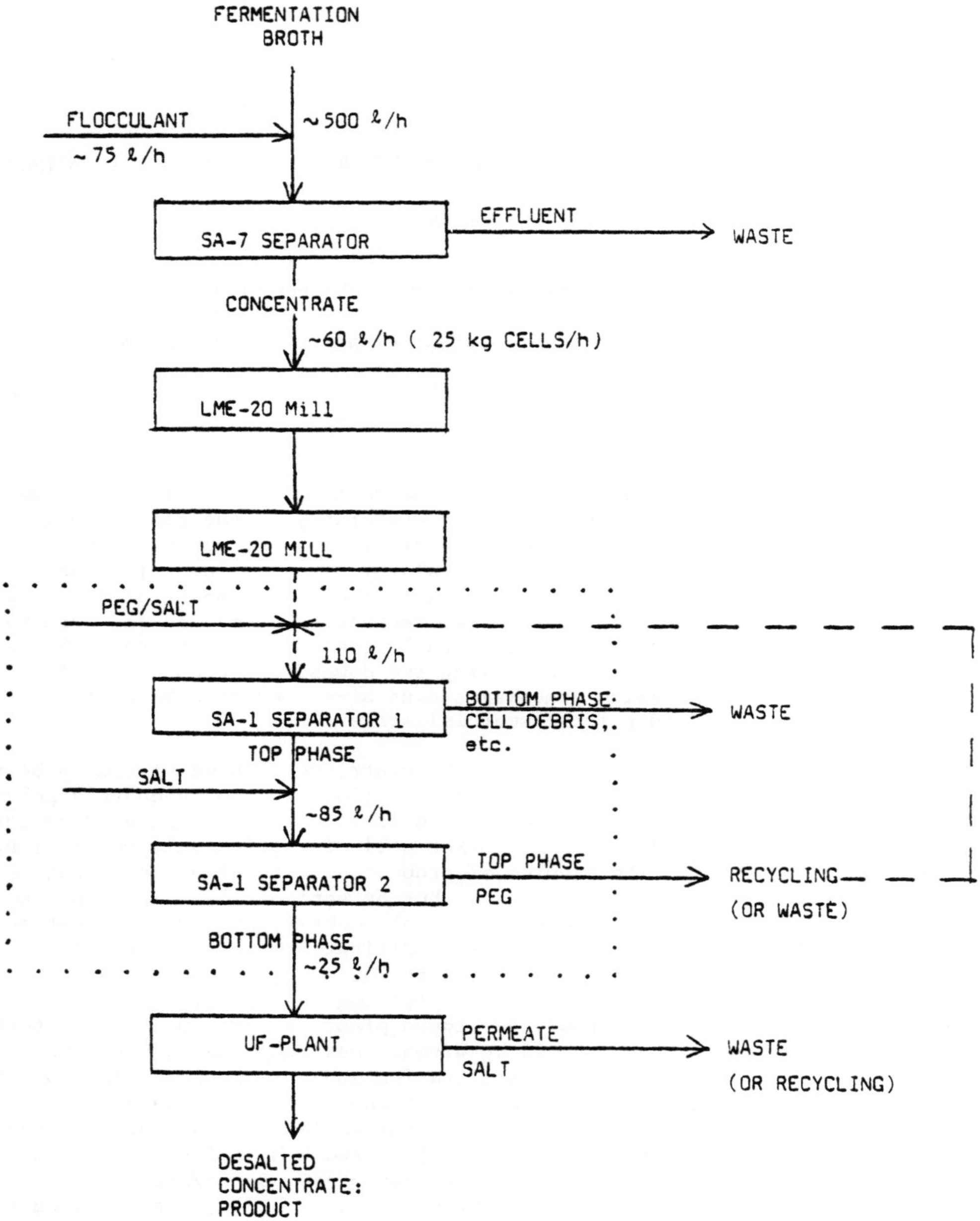

Fig. 1 *Flow schemes of (semi) continuous downstream processing of intracellular microbial proteins based on flocculation, wet milling and crosscurrent extraction (⋮⋮ crosscurrent extraction plant; data given are specific for the fumarase/Brevibacterium ammoniagenes process)*

REFERENCES

1 H. Hustedt, N. Papamichael, K.H. Kroner, R. Krützfeldt, Synthesis of efficient downstream processing lines for proteins," in: "Biotechnology in Germany 1987," Dechema, Frankfurt (1987)

2 H. Hustedt, K.H. Kroner and N. Papamichael, in: "Separations Using Aqueous Phase Systems: Applications in Cell Biology and Biotechnology," D. Fisher and I.A. Sutherland, eds., Plenum, New York (1989)

GRAVITY SETTLING OF APS - CHARACTERISTICS AND PROCESS APPLICATIONS

K.H. Kroner and H. Hustedt

GBF Gesellschaft für
Biotechnologische Forschung mbH
D-3300 Braunschweig, FRG

Aqueous two-phase systems (APS) have been described as useful for the extractive purification of enzymes in large scale [1] and also for applications in bioconversion processes [2]. Albertsson [3] has described these systems and their use in biochemical research basically. At large-scale processing phase separation is conducted either by using common centrifugal separators or by settling under unit gravity. This paper analyses the settling behaviour of model systems with regard to their physico-chemical properties; further the application of settling in extractive enzyme recovery processes is considered.

For the basic studies a set of PEG 1550/potassium phosphate as well as a set of PEG 4000/Dextran 500 systems, each with increasing tie-line length, were selected; these systems are representative for most of the APS used thus far in preparative scale. Tables 1 and 2 show their physico-chemical data measured. All parameters examined changes with increasing tie-line length. In the case of the PEG-dextran systems the data show quite low interfacial tensions in the range from $3.1 \cdot 10^{-4}$ - $4.1 \cdot 10^{-2}$ mN/m and density differences between the phases of 18 - 79 kg/m^3. The viscosity of the continuous PEG-rich phase changes only slightly (~4 mPa.s). In the case of the PEG/salt systems the interfacial tension is 1-2 orders of magnitude higher (0.01 - 0.5 mN/m); the density differences are slightly higher (38 - 108 kg/m^3); the viscosities of the top phases are somewhat higher, those of the bottom phases are much lower than for PEG/dextran systems. Thus, a different settling behaviour for both types of system can be expected and is generally observed.

The basic settling behaviour was studied with both sets of 'model-systems' described in simple glass cylinders in laboratory scale (0.1 - 1 L), with a height to diameter ratio of around 2 - 3. The bottom phase was the disperse one in all cases; the volume of settled disperse phase was read over time. Generally, the settling behaviour depends on the length of the tie-line of the systems. Settling times were in the range of 5 - 10 minutes for the PEG/salt systems and 15 - 40 minutes for PEG/dextran systems, respectively (for comparison see [3]). The change of the settling velocity is nearly linear in the range of 10 - 30% for the tie-line length and with similar steepness for both systems. In this range the separation velocity of the PEG/dextran system is about a factor of 2 - 3 lower comparing to the PEG-salt system, although the density difference/viscosity quotient is higher. This reflects the strong influence of the interfacial tension.

Table 1 *Physico-chemical data of PEG 1550/potassium phosphate two-phase systems* *

PEG (%)	potassium phosphate (%)	tie-line length (%)	volume ratio (V_t/V_b)	$\Delta\rho$ (kg/m^3)	ηtop (mPa·s)	ηbott (mPa·s)	σ (mN/M)
11.0	12.4	5.6	0.9	38	5.7	2.0	$0.1 x 10^{-1}$
14.0	12.0	21.9	1.3	70	7.8	1.9	$1.0 x 10^{-1}$
16.0	12.0	27.8	1.4	82	9.7	1.9	$2.2 x 10^{-1}$
19.0	12.0	35.7	1.4	108	12.9	2.4	$5.1 x 10^{-1}$

* pH = 7.0

The use of gravity settling in large-scale enzyme preparations has been described already in the literature, mostly in subsequent partition steps without solid cell material present. A few data are published also with phase systems including cell homogenates (collected data see [4]). The practical applications are restricted mainly to PEG/salt systems. Commonly simple cylindrical vessels are used in 100 L scale or m^3-scale, too, operating in a batchwise mode. For systems containing high amounts of solids separation values between 85 - 93% have been reported with time needed between 10 and 20 h [4]. These times are one order of magnitude higher as for systems being free of solids (30 - 90 minutes).

Continuous operation with gravity settlers is generally possible, too. The separation performance of a horizontal tube settler (glass-tube settler, Quickfit, Wiesbaden, FRG) with a length of 1 m and a diameter of 10 cm using a PEG/phosphate system was 40 - 80 l/h. These values are around a factor of two lower as can be assumed from static settling experiments. Some influence of the flow through the apparatus must be considered such as disturbance of the settling area, redispersion due to the low interfacial tension and entrapping of very small droplets of the disperse phase in the more viscous continuous phase. This problem could be reduced by introducing wears or baffles and some of such improvements have been already reported [4].

Table 2 *Physico-chemical data of PEG 4000/Dextran 500* * *two phase systems*

PEG (%)	dextran (%)	tie-line length (%)	volume ratio (V_t/V_b)	$\Delta\rho$ (kg/m^3)	ηtop (mPa·s)	ηbott (mPa·s)	σ (mN/M)
6.0	5.8	6.2	1.5	18	4.1	43	$0.03 x 10^{-2}$
6.5	6.1	10.5	1.4	32	3.6	100	$0.12 x 10^{-2}$
7.0	6.6	14.4	1.5	48	3.7	145	$0.81 x 10^{-2}$
8.0	7.6	19.8	1.7	64	4.3	303	$2.0 x 10^{-2}$
9.0	8.5	24.8	1.9	79	4.4	364	$4.1 x 10^{-2}$

* systems include 1 mM potassium phosphate buffer, pH 7.0;
Dextran 500 was purchased from Pfeiffer & Langen, Dormagen, FRG

REFERENCES

1 M.-R. Kula, K.H. Kroner and H. Hustedt, Purification of enzymes by liquid-liquid extraction, in: "Advances in Biochemical Engineering," A. Fiechter, ed., Springer, Berlin (1982)

2 B. Mattiasson, M. Larsson, Bioconversion in aqueous two-phase systems, in: "Third Europ. Congress on Biotechnol.," Verlag Chemie, Weinheim (1984)

3 P.-A. Ålbertsson, Partition of cell particles and macromolecules, Wiley Interscience, New York (1971)

4 H. Hustedt, K.H. Kroner, M.-R. Kula, Applications of phase partitioning in biotechnology, in: "Partitioning in Aqueous Two-Phase Systems - Theory, Methods, Uses and Applications to Biotechnology," H. Walter, D.E. Brooks, D. Fisher, eds., Academic Press, New York (1985)

IMPROVED ECONOMICS OF EXTRACTIVE ENZYME RECOVERY BY RECYCLING: A CASE STUDY

N. Papamichael and H. Hustedt

Gesellschaft für Biotechnologische Forschung mbH (GBF)
Mascheroder Weg 1
D-3300 Braunschweig, FRG

A wide variety of enzymes and other biologically active proteins have been isolated by aqueous phase extraction [1] and the technology is already applied in industry. The technique has several advantages; a drawback is the relatively high consumption of chemicals to form the aqueous phases which becomes significant in large scale production. The economics of recovery, however, may be improved by reusing the phase-forming chemicals as has already been demonstrated in batch [2] and also in continuous operation with batchwise recycling [3]. We have also been investigating the feasibility of continuous recycling and some results are presented here. The isolation of the enzyme fumarase from *S. cerevisiae* was adopted as a model. The separation was carried out continuously in a phase system comprising PEG (MW 1500) and potassium phosphate in 2 consecutive stages, a first stage which removes cell debris, DNA and other proteins, and a second stage for the

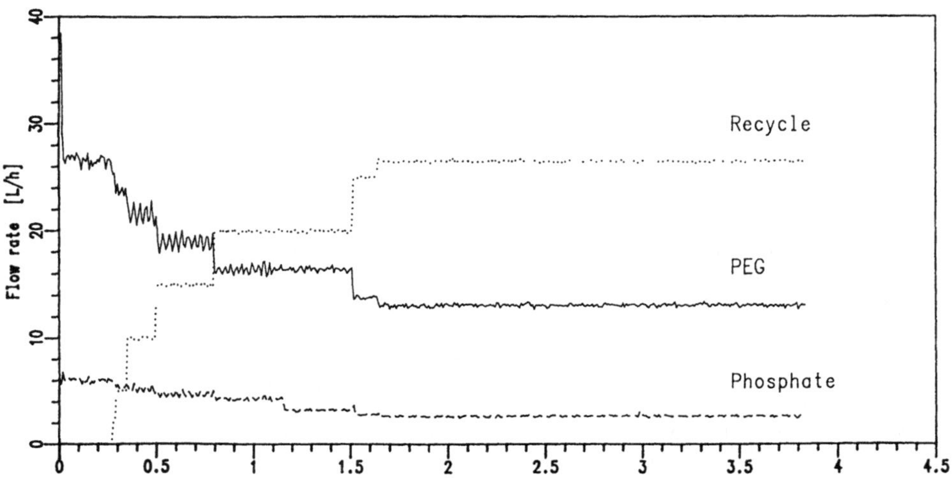

Fig. 1 *Flow rates with recycle*

further purification of the enzyme and the recovery of PEG [2]. The isolation was carried out in a computerised pilot-scale plant with a capacity of approximately 100 l/h total liquid throughput or 150 tonnes per annum wet cell mass. The phases were separated by centrifugal separators of type SA-1 (Σ-factor of 1460 m^2) (Westfalia Separator AG, Oelde, FRG) see [4]. The top (PEG-rich) phase from the second stage was recycled continuously to the first stage [4] while monitoring the protein and enzyme levels automatically in 3 streams.

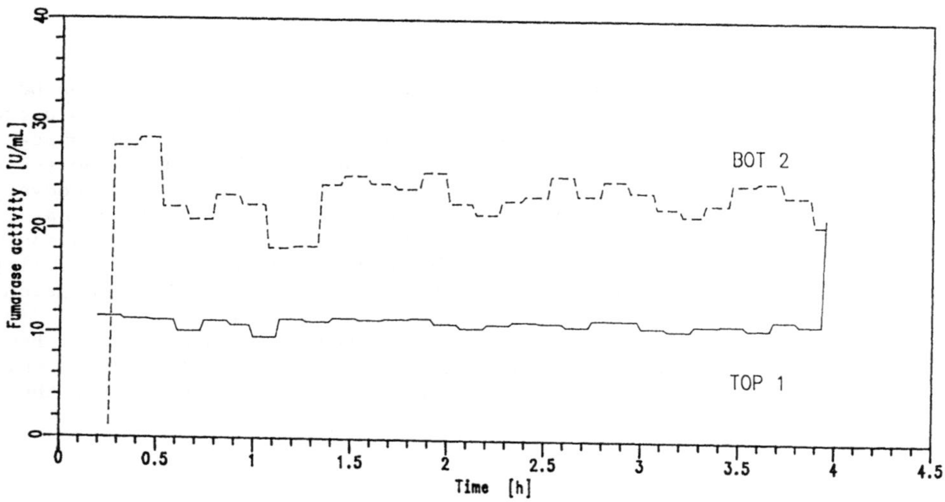

Fig. 2 *Fumarase activities during recycling*

Results of a recycle are presented in Figs. 1 and 2. The recycling of the TOP 2 phase presents no serious problems. The preset flow rate increase in recycle is reflected also by a drop in PEG and phosphate flow adjusted by the computer. The recycle was taken to approximately 60% of the TOP 2 flow rate, achieving a saving of over 50% in PEG and fresh phosphate flow (corresponds to 29% total phosphate reduction since approximately 11% is also present in the cell homogenate) (see Fig. 1). The fumarase activity [2] and total protein concentrations (data not shown) remain relatively constant throughout the run. These data were used to calculate the running costs at this scale of operation (ca. 90t wet biomass/annum) with the following assumptions: 6000 h operation, installed investment $160,000, raw material and utility costs as FRG 1987, 2 DM/$ conversion rate. Table 1 presents this analysis for the case of a non-recycled and recycled system. The savings made are significant for the complete system and represent a major proportion of the raw material costs. The savings would be still more significant using

Table 1 *Cost analysis for fumarase from S. cerevisiae*

Point	Non-Recycled		Recycled		Savings
	Batch	Cont.	Batch	Cont.	B/C (%)
M	1.98	2.48	1.13	1.51	43 / 39
L	1.07	0.80	1.33	0.73	-24 / 8
I (total)	100	125	100	160	0 /-28
E	0.012	0.014	0.013	0.016	- 8 /-19
T	5.21	5.12	4.93	3.92	5 / 23

T = 1.13 M + 2.6 L + 0.16 I + 1.13 E ($ 000's/t wet cells) (see [5] for comparison), T = total production cost, M = raw materials (PEG, phosphate, water, KOH), L = labour (6000 h operation, 3 shifts), I = installed investment, E = utilities (power)

a larger extraction plant more adequate with regard to the processing of tons of biomass. A plant with two separators of type SA-7 (Σ-factor 7000 m^2) would process 75-150 kg biomass per h.

CONCLUSIONS

This paper shows that i) recycling in continuous mode is feasible with no deleterious effects on enzyme recovery; ii) savings are significant even though investment requirements increase; and iii) the extraction conditions determining the recycle concentrations are important for the recycle flow rate. However, in general before developing recycling processes the biomass load in the primary phase system should be optimised.

REFERENCES

1 H. Hustedt, K.H. Kroner, U. Menge and M.-R. Kula, Trends in Biotechnology 3:139 (1985)
2 H. Hustedt, Biotechnol. Lett. 8:191 (1986)
3 H. Hustedt, unpublished results
4 H. Hustedt, K.H. Kroner and N. Papamichael, in: "Separations Using Aqueous Phase Systems: Applications in Cell Biology and Biotechnology," D. Fisher and I.A. Sutherland, eds., Plenum, New York (1989)
5 K.H. Kroner, H. Hustedt and M.-R. Kula, Proc. Biochem. 19:170 (1984)

THE USE OF PEG-NADH/Na_2CO_3 AQUEOUS TWO-PHASE SYSTEMS FOR THE LARGE-SCALE PREPARATION OF TECHNICAL GRADE PEG-(M_r 20,000)-N^6-(2-AMINOETHYL)-NADH

Andreas F. Bückmann, Michael Morr and Maria-Regina Kula [1]

Gesellschaft für Biotechnologische Forschung (GBF)
Mascheroder Weg 1, D-3300 Braunschweig, West Germany

[1] Present address: Institut für Enzymtechnologie der Universität Düsseldorf in der Kernforschungsanlage Jülich, P.O.Box 2050, D-5170 Jülich, West Germany

A procedure for the preparation of technical grade poly(ethylene glycol) (PEG) (M_r 20,000)-N^6-(2-aminoethyl)-NADH was developed. Using the reaction mixture after the alkylation of NAD by ethyleneimine, containing 25% unreacted NAD, 67% (N(1)-(2-aminoethyl)-NAD, and 8% (N(1)-NAD by-products, technical grade PEG (M_r 20,000)-N(1)-(2-aminoethyl)-NAD was synthesised by the carbodiimide coupling method. After chemical reduction to PEG (M_r 20,000)-N(1)-(2-aminoethyl)-NADH and Dimroth rearrangement, technical grade PEG (Mr 20,000)-N^6-(2-aminoethyl)-NADH was obtained with an overall yield of 33.5% [Table 1, Fig. 1]. By insertion of a cation-exchange purification step in batch mode after the coupling step and partitioning in PEG-NADH/Na_2CO_3 aqueous two-phase systems before and after the Dimroth-rearrangement step >98% of the uncoupled NAD(H), N(1)-(2-aminoethyl)-NAD(H), and alkylated NAD(H) by-products could be removed. Because large volumes may be handled easily in these operations, the procedure developed for the preparation of technical grade PEG(M_r 20,000)-N^6-(2-aminoethyl)-NADH can now be adapted to large-scale synthesis [1].

Table 1 *Summary of the synthesis of technical grade PEG (M_r 20,000)-N^6- (2 AE)-NADH*

Step		mmol	Presence of uncoupled NAD(H) and alkylated NAD(H) (mmol)
I	$PEG(M_r 20,000)-[NH-CO-CH_2-CH_2-COOH]_2$ ↓ Reaction mixture containing NAD(35 mmol), N(1)-(2AE)-NAD(93.8 mmol), N(1)-NAD by-products (11.2 mmol) EDC, pH 4.6-4.7, room temperature ↓ PEG (M_r 20,000)-N(1)-(2AE)-NAD	76.2 [a]	63.8 [a]
II	Batch treatment with AGW50-X4, pH 1.4, 4°C ↓ PEG(M_r 20,000)-N(1)-(2AE)-NAD	52.2 [a]	9.3 [a]
III	Chemical reduction $Na_2S_2O_4$, pH 8.0, 45°C ↓ PEG (M_r 20,000)-N(1)-(2AE)-NADH	52.3 [b]	9.3 [b]
IV	1st PEG(M_r 20,000)-N(1)-(2AE)-NADH/Na_2CO_3 aqueous two-phase system pH 10, 4°C ↓ PEG(M_r 20,000-N(1)-(2AE)-NADH	51.2 [b]	5.0 [b]
V	Dimroth rearrangement pH 11, 65°C ↓ PEG(M_r 20,000)-N^6-(2AE)-NADH	50.0 [b]	6.2 [b]
VI	2nd PEG(M_r 20,000)-N^6-(2AE)-NADH/Na_2CO_3 aqueous two-phase system pH 10.5, 4°C ↓ PEG(M_r 20,000)-N^6-(2AE)-NADH	47.0 [b]	0.9 [b]

Note:
Overall yield 33.5% (I→VI); yield 62% (II→VI);yield 90% (III→VI); PEG = poly(ethylene glycol); 2AE = 2-aminoethyl.
[a] Based on ϵ_{259} = 18.000 $M^{-1}cm^{-1}$; [b] Based on ϵ_{340} = 6220 $M^{-1}cm^{-1}$

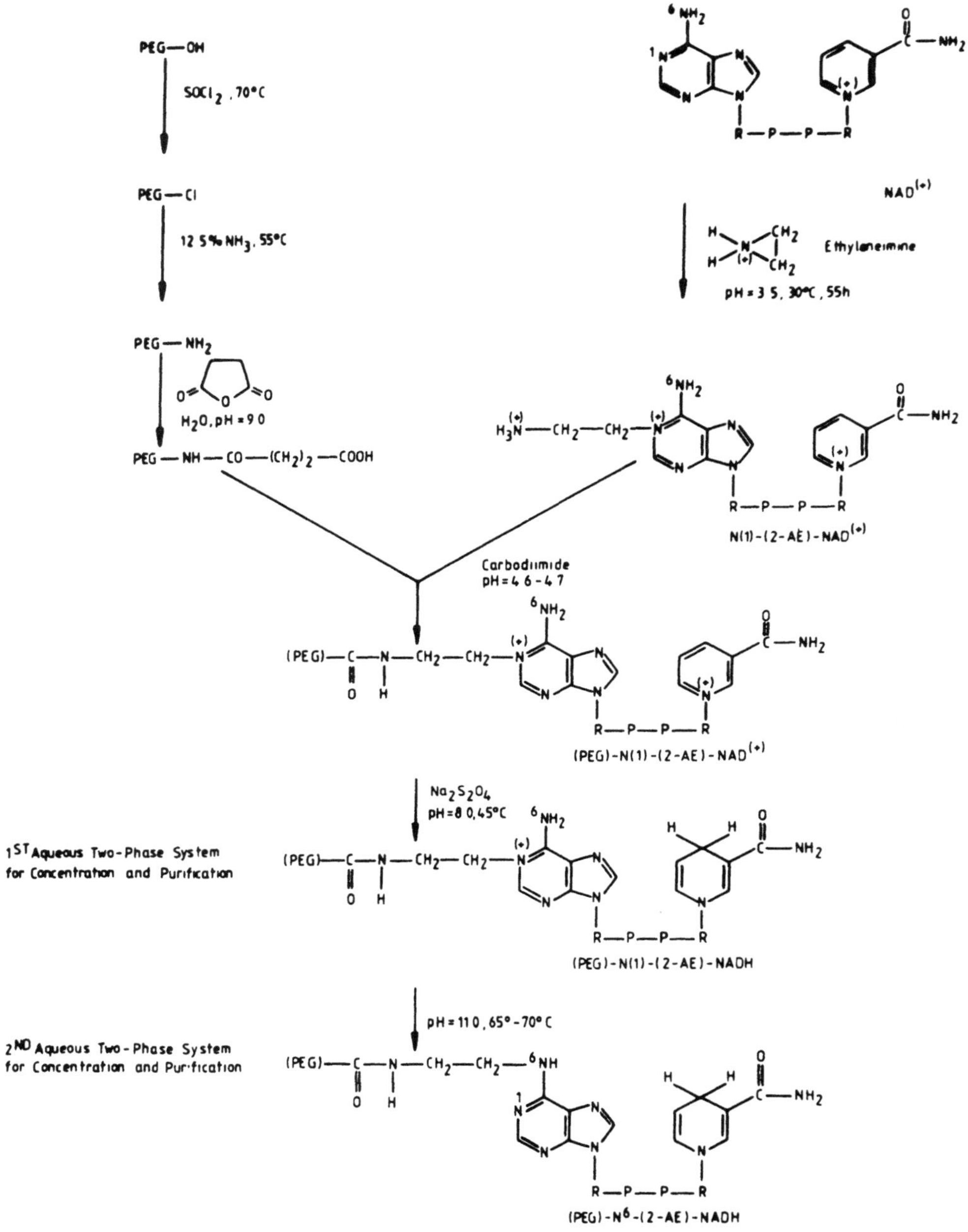

Fig. 1 *Reaction pathway of the simplified synthesis of PEG (M_r=20,000)-N^6-(2-aminoethyl)-NAD(H)*

REFERENCE

1 A.F. Bückmann, Biotechnol. Appl. Biochem. 9:258 (1987)

Chapter 8

Biotechnology: Extractive Bioconversions

EXTRACTIVE BIOCONVERSIONS IN AQUEOUS PHASE SYSTEMS

Bo Mattiasson and Torbjörn G.I. Ling

Department of Biotechnology, Chemical Center
University of Lund, P. O. Box 124
S-221 00 Lund, Sweden

INTRODUCTION

The use of biocatalysts to perform conversions in processes is very appealing in many aspects. The reaction conditions - atmospheric pressure, ambient temperature, water environment and steroselectivity, all favour for biocatalysis. The drawbacks have, however, not been listed frequently, but should, of course, be considered as important. Bioconversions are often slow in relation to chemical processes, they yield a dilute product stream in water, and are often product inhibited. When designing biocatalytic processes, it is thus important to find conditions that fully exploit the advantages offered by the biocatalyst without neglecting the challenges raised by the mentioned limitations.

One way to improve productivity has been to operate at a substantially increased substrate concentration. This has been accomplished by, e.g. dissolving the substrate in an organic solvent and then running the bioconversion in an organic medium. This approach has in the last few years attracted much attention and we are now beginning to understand how to design such unnatural systems [1,2,3]. If the bioorganic synthesis approach is unphysiological, there is another main approach that is more gentle. By continuous extraction of the product during the conversion, the equilibria are changed and more substrate is converted. This approach can be called extractive bioconversion. Several strategies are gathered under this headline, Table 1, and among these is the use of aqueous phase system as a reaction medium. The principle is to keep the phase systems in suspension, thereby having the biocatalyst (enzyme and/or cell) confined in one phase while the product partitions more or less to the other phase.

Table 1 *Examples on methods applied in extractive bioconversion*

Extraction with organic solvents
Extraction with aqueous phase systems
Extraction with supercritical fluids
Use of solid sorbents
Integration with membrane filtration
Pervaporation
Perstraction
Vacuum fermentation
Cell recycling using flocculation and centrifugation

WHAT CAN AQUEOUS PHASE SYSTEMS OFFER?

* The high content of water (80 - 95%) in both phases has been regarded as an advantage. This is especially so for processes involving cell growth where the cells must stay viable and also proliferate.

* The low surface tension between the phases makes partitioning of enzymes possible, without loss of activity. In the case of oil/water or water/air interfaces, there is a high probability of denaturation of the protein taking place. At the interface in aqueous phase systems no such problems appear.

* The content of polyols present in most aqueous phase media may help to stabilise enzymes.

* Due to the low surface tension it is quite easy to maintain a suspension by using low energy input. This reduces the risks of harmful shear forces.

* The small droplets generated in such a phase system give short distances and large surface areas, facilitating mass transfer.

WHAT SPECIFIC CRITERIA SHOULD BE SET WHEN AQUEOUS TWO-PHASE SYSTEMS ARE TO BE USED IN BIOCONVERSIONS?

In practical applications of aqueous two-phase systems most of the discussion centres around the costs of the two polymers (salt/polymer systems will in most cases involve ionic strengths, which are too high rendering them unusable. So far there are no reports on the use of halophilic organisms in such systems). PEG is regarded as being acceptable in price, but this consideration is taken with the assumption that the PEG-rich phase can be recirculated. The bottom phase polymer is more critical. Dextran, which has been used in many of the more fundamental studies on aqueous two-phase systems, and also in the initial studies on bioconversion systems, is far too expensive. Reppal PES - a starch derived polymer seems to fulfill the role of dextran very well, at a fraction of the cost.

Biodegradability may be a problem when long term processes are to be run. PEG is stable under the conditions normally applied, whereas the polysaccharides constituting the bottom phase may give problems. Dextran is, together with several other polysaccharides, suitable as a phase constituent since, under normal conditions, there is a very low risk that an organism cultivated will produce suitable enzymes (or an enzyme preparation shall contain impurities) that will start to degrade the polymer. However, in the case of starch and cellulose-based phase components, there are definitely complications. The same situation is also valid for proteinaceous phase components.

Starch is successfully degraded in phase systems (see further below under the examples). When amylases might be present, it is not recommended to use native starch. Reppal PES, which is a starch derivative developed for use in aqueous two-phase systems [4], consists of modified starch molecules. Higher degrees of modification have been found to have better resistance to enzymatic degradation, Fig. 1.

When the conversion is finished, it is desirable to separate the product and also biocatalyst in a simple and convenient way. As is seen below there are different ways to carry out these separations.

Contamination. If large scale processes are to be set up based on conversions in aqueous two-phase systems, then one can also predict pollution

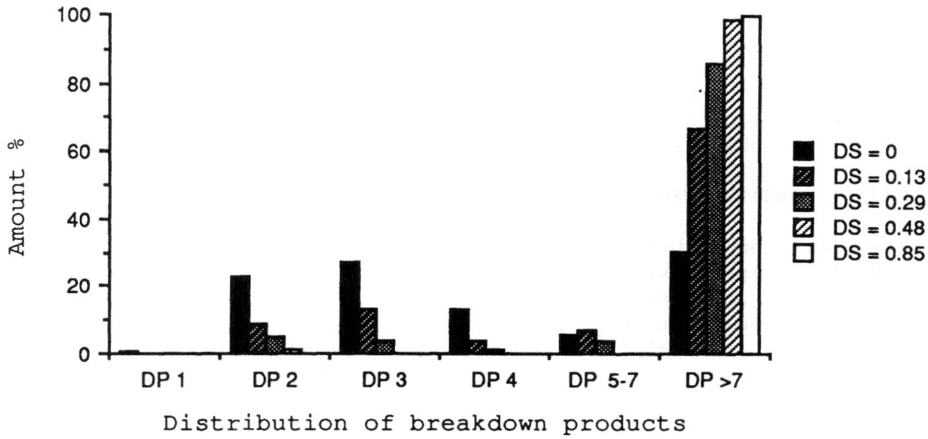

Fig. 1 *Enzymatic degradation of starch substituted with hydroxypropyl groups. The degree of substitution, DS, is defined as mol hydroxypropyl groups per mol saccharide units. The length of the polymer chain, DP, counts as saccharide units*

by phase components. By working with biodegradable phase constituents, these problems may be reduced substantially.

Membrane influence. It has been reported that high concentrations of PEG are a good medium for causing cell fusions and other alterations of cell membranes. One has therefore to be extremely careful when operating with cells in concentrated polymer solutions.

METHODS OF PRODUCT AND/OR PHASE CONSTITUENT RECOVERING

Since the biocatalyst may be confined to one of the phases, the other phase should be regarded as the extractive phase. By varying the relative volumes between the phases, it is then possible to remove a formed product quite efficiently from the site of synthesis, even if the partition cosntant is not favourable.

In the early experiments on bioconversions in aqueous two-phase systems, ethanol production from sugar was studied [5]. Here the product was removed by distillation from the top phase while the yeast cells stayed in the bottom phase. The remaining phase system was used in repetitive fermentations.

Other examples are separation based on differences in size, in charge and in hydrophobicity. Size exclusion is used to recover low molecular weight compounds from biodegradations of macromolecules. In these cases it is possible to keep the catalyst and the substrate preferentially in one of the phases, while the product will partition fairly evenly between the phases. By passing the phase without the catalyst over a membrane unit with pores that will let the product pass but retain the polymers, a simple and efficient separation may be achieved [6,7]. Furthermore, the use of membranes of varying hydrophobicities, counters the added benefit of a slight additive effect due to partition besides the size exclusion effects. Charged molecules may be recovered by using ion exchange systems and hydrophobic compounds by applying hydrophobic sorbents [8].

Recovery of the biocatalyst may be a bit more difficult since the size of the polymer and the biocatalyst may, in some cases, be in the same range. Differences in charges and in hydrophobicity may however be used.

ECONOMICAL CONSIDERATIONS

The operation of bioconversions in aqueous two-phase systems must be economically competitive with other ways to run the processes. It has been found that when phase systems have been applied to protein purification, the cost of the phase constituents is the dominating element of the total process cost. Thus, it is crucial to reduce these costs as far as possible. A typical phase system (7% w/w PEG and 12% w/w dextran or Reppal PES) has a cost per litre of \$25 and \$1.5, respectively.

An evaluation of the economy must take all factors into account, so that the cost of the polymers is balanced, for example, by a higher productivity or reduced equipment cost. At present no such study has been presented, so it is difficult to state the competitiveness of the technology.

EXAMPLE: MACROMOLECULAR SUBSTRATES → SMALL MOLECULES

The applications of immobilised enzymes have dealt mainly with conversion of low molecular weight compounds. Macromolecular substrates cause problems due to diffusion restrictions which cause very inefficient mass transfer. Particulate substrates are even more unattractive for the conventional technology with immobilised biocatalysts. It was thus quite natural to first focus the attention on this area when designing systems for bioconversions by utilising aqueous two-phase systems. As is stated above, mass transfer is facilitated due to the large surface area. Furthermore, since there are no solid supports and thus no mechanical hindrances, one can also expect that macromolecular and particulate substrates can be converted.

Starch Degradation to Glucose

The most well-established system for degradation of macromolecular substrate is that of starch degradation. In the normal way this is first done batchwise with the enzyme in free solution to achieve a partially degraded dextran which, in a subsequent step, may be further degraded to glucose using an immobilised step. By applying aqueous two-phase technology, one can foresee a suitable process if the product can be removed from the site of production. A general complication when working with soluble macromolecules is that, at higher concentrations, they contribute to the properties of the two-phase system, leading to a gradually changing phase system as the degradation proceeds. In a recent report, it was shown that the enzymatic degradation of starch led to a shrinkage of the relative phase volume of the starch-rich bottom phase in an ordered way; furthermore, one can use this diminution as a measure of the status of the reaction [9]. This observation obviously leads to the development of a new way to follow and control such a complex process, simply by following the volume of one of the phases. Fig. 2 shows how the phase volume of the bottom phase is related to the amount of glucose formed.

Partition studies showed that the particulate starch was all recovered from the bottom phase. Most of the soluble starch was found there as well as the alpha-amylase, whereas the glucoamylase was also recovered from the top

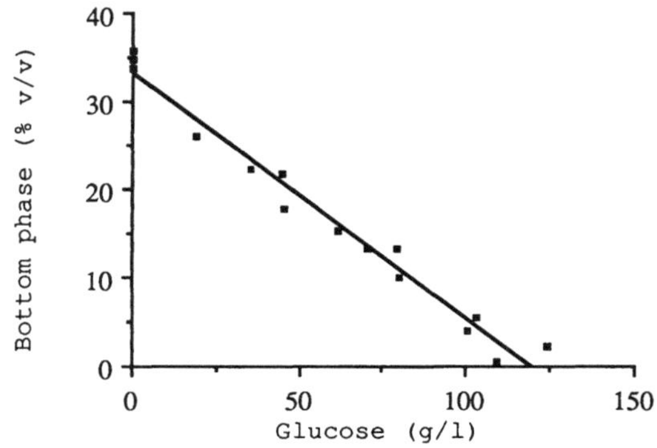

Fig. 2 *Volume reduction of the bottom phase (in% of total volume) as a function of enzymatic starch degradation by glucoamylase. The phase system consisted of PEG 20M (10% w/w) and waxy maize starch, partially hydrolysed (19.5% w/w initially) [9]*

phase. Glucose, the product, was evenly distributed between the phases. Since the enzyme was also present in the top phase, it was not possible to simply remove the top phase and replace it with fresh polymer - the loss in enzyme would make such an approach totally uneconomical. Instead the top phase was passed over an ultrafiltration unit where glucose was separated off and the macromolecules were recycled back to the process.

This approach offers certain advantages: since the enzymes are recycled they can be reused several times, allowing the process to be run with a higher starting concentration, which in turn leads to a faster process. Furthermore, we have observed that the enzymes are stabilised by the polymers which constitute the phase system, a fact which further improves the performance of the two-phase concept [10].

Cellulose to Glucose

Cellulose is an even more complicated substrate to degrade than starch. Because of its crystallinity, there is a tendency for the enzymes to adsorb very strongly to the particulate cellulose. The enzymes involved are endo- and exocellulases and ß-glucosidase, which hydrolyse the cellobiose molecules formed.

After the initial work in this area [11] several improvements have been reported [12] but still the basic problems of crystallinity of the substrate and the low accessibility to the enzymes remain. Various ways of pretreatment of the substrate have been tried, but more processing introduces the higher costs.

A problem that was encountered in these studies was that the bottom phase polymer that was used to stabilise the cellulose-rich phase, raw dextran, was slowly degraded by impurities in the enzyme preparations. Thus, it would be advantageous to use systems of only cellulose and PEG or to add a nondegradable polymer to stabilise the bottom phase. However, in the latter case environmetal problems may appear.

EXAMPLES: FERMENTATIVE CONVERSION OF SMALL MOLECULES

There are many examples described of product inhibited fermentations. Production of solvents and small organic acids are carried out by organisms that are inhibited by the elevating product concentrations generated during the fermentation. Here, it is vital to efficiently remove the products from the site of formation and thereby reduce the negative influence on the productivity exerted by the product. These systems have been quite natural to use when increasing productivity by extractive bioconversion.

Production of Ethanol from Glucose

Ethanol production by *Saccharomyces cerevisiae* ceases when the ethanol concentration reaches 10%, but long before that a dramatic drop in productivity is observed. It would thus, from the biological point of view, be advantageous to ferment in very dilute solutions that will not exert any appreciable inhibition of the process. However, the costs of the further upgrading are too high to allow this. A break even point has been calculated to be in the region of 6% ethanol. The two major costs for ethanol production have been identified as raw material cost and energy cost for distillation.

By applying extractive fermentation, it might be possible to efficiently remove the product from the site of the cells and enrich the product further in a later step. If this process is carried out with cell recycling in any form, the need for new cell mass decreases, making it possible to convert a higher percentage of the carbon source into ethanol.

Acetone-Butanol Fermentation Using *Clostridium acetobutylicum*

When producing solvents by *Clostridium acetobutylicum*, a content of only 2% is inhibitory to further production. This system is therefore a good model when studying the effects of extractive removal of the product from the fermentation medium.

Since the partition constant for the small molecules is quite close to unity it is important to search for the greatest difference in partitioning. A large number of phase systems have been investigated (Table 2). The best results were obtained in systems of 6% Dextran T-40 and 25% PEG-8000. A typical experiment is illustrated in Fig. 3. Here it is seen that the onset of butanol production is quicker in the phase system and that the peak value is higher [13]. When setting up continuous processes this observation might be valuable. In the experiments reported here the butanol content decreased in later investigations; this has been reported as being a consequence of a changed metabolic behaviour of the cells caused by the high polymer content in the medium [14].

ONE STEP BIOCONVERSIONS

Production of Prednisolone from Hydrocortisone Using *Arthrobacter simplex* as Catalyst

The bacterium *Arthrobacter simplex* was able to grow in a two-phase system and the cells could be revitalised after a period of use. The use of an aqueous two-phase medium made it possible to operate at a higher substrate concentration than in pure buffer solution [8]. Furthermore, it was shown that, when semicrystalline fermentations were carried out, complex conversion was achieved in contrast to the situation in pure buffer. The cells were

Table 2 *Partition of ethanol, acetone and butanol in different phase systems*

Polymer	Conc % w/w	Polymer	Conc % w/w	V_T/V_B	Ethanol K_{part}	Acetone K_{part}	Butanol K_{part}
PEG	15	Dextran	3	8:1	1.2	1.3	1.6
"	17	"	3	9:1	1.3	1.5	1.9
"	20	"	3	12:1	1.2	1.5	1.9
"	15	"	4	7:1	1.2	1.3	1.6
"	10	"	6	3:1	1.5	1.0	1.3
"	15	"	6	4:1	1.3	1.5	1.8
"	25	"	6	6:1	1.9	1.9	2.0
PVA	8	"	2	8:1	1.0	1.2	1.2
"	8	"	2.25	4:1	1.1	1.1	1.3
"	8	"	2.5	3:1	1.1	1.3	1.3
Pluronic	14	"	4	5:1	1.4	1.4	1.5
"	14	"	6	4:1	1.3	1.4	1.4
Ucon	14	"	4	4:1	1.3	1.1	1.4
"	15	"	4	5:1	1.1	1.3	1.5
"	12	"	5	3:1	1.2	1.3	1.3

PEG = poly(ethylene glycol), PVA = polyvinylalcohol. Pluronic and Ucon = copolymers of PEG and poly(propyleneglycol). V_T and V_B = volume of the top and bottom phases

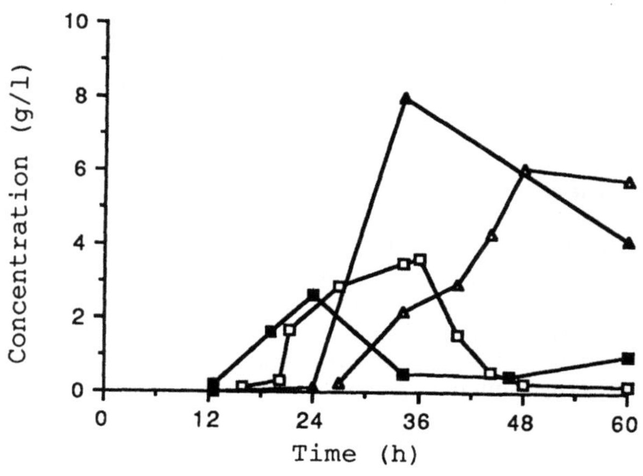

Fig. 3 *Production of butanol (triangles) and butyric acid (squares) in batch (open symbols) and aqeuous two-phase systems (closed symbols). The batch experiment was run in a medium consisting of glucose, peptone, yeast extract and salts. The two-phase system consisted of Dextran T-40, 6% (w/w) and PEG 8000, 25% (w/w) dissolved in the medium. The ratio of the phase volumes was 6 (top/bottom) [13]*

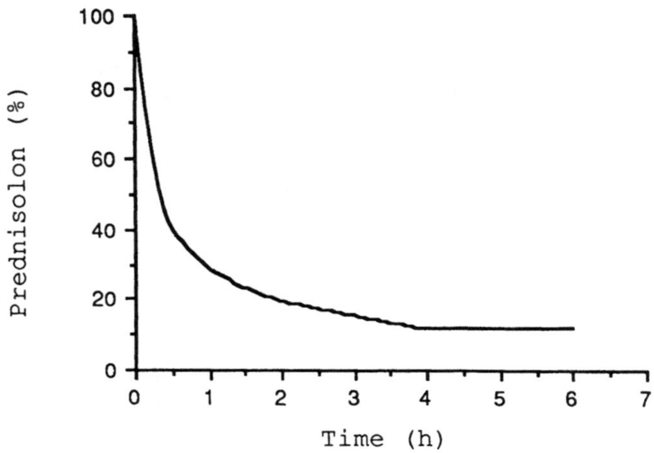

Fig. 4 *Time dependence for adsorption of prednisolon to Amberlite XAD-4 from the top phase in a PEG/dextran system [8]*

recovered from the bottom phase, whereas the steroid was partitioned to the top phase. The further isolation of product from the phase system was achieved by adsorbing the steroid to a hydrophobic adsorbent, Amberlite XAD-4 (Fig. 4).

The advantages in performing a process like this in phase systems, as compared to buffer medium are: the complex bioconversion, the ease of separating off the cells and isolating the product in a subsequent step. A further advantage is that no organic solvent is added, allowing the cells to stay active for a longer time. A draw-back with the set up was the cost of the dextran-based system. Therefore, similar experiments were carried out also in Reppal-PES and comparable results were achieved in those systems.

Other cases where it may be advantageous to perform an extractive one-step bioconversion is when a product is formed that severely influences the reaction. For example, by splitting off the side chain of penicillin G for production of 6-APA, a proton is formed which lowers the pH, this, in turn, produces the risk of hydrolysation of the APA formed and deactivation of the enzyme. In order to achieve short processing time, very short columns are used for operation of these immobilised systems. A careful study on the potential of applying aqueous phase systems in this case was recently presented [15].

Cell-culture

In cultivation of cells that produce self-toxic products, it may be necessary to efficiently remove the products formed. In their by now classical work, Hedén and Puziss demonstrated that cultivation of *Clostridium tetanii* could be much more efficient in a two-phase system as compared to conventional fermentation medium [16]. The reason for this is that a toxin produced which is a potent protease that will attack the cells themselves. By extractive removal of the product an increase in productivity of 50 times was achieved.

CONCLUDING REMARKS

There are still only a few examples of reactions carried out successfully in aqueous two-phase systems. One limiting factor for a more universal exploitation has been the cost of the phase components. Another, even worse, is the conservatism amongst the potential users. It took several decades before protein purification by extraction in aqueous two-phase systems became an accepted technology, and it may also be a long lag-phase in this case before people are convinced.

ACKNOWLEDGEMENT

Part of this project was supported by the National Swedish Board for Technical Development.

REFERENCES

1 L.E.S. Brink and J. Tramper, Optimisation of organic solvents in multiphase biocatalysis, Biotechnol. Bioeng. 27:1258 (1985)

2 C. Laane, S. Boeren and K. Vos, On optimising organic solvents in multi-liquid-phase biocatalysis, Trends Biotechnol. 3:251 (1985)

3 M. Reslow, P. Adlercreutz and B. Mattiasson, Organic solvents for inorganic synthesis. 2. Influence of log P and water solubility in solvents on enzymatic activity, in: "Biocatalysis in Organic Media," C. Laane, J. Tramper and M.D. Lilly, eds., Elsevier, Amsterdam (1987)

4 H. Nilsson, G. Kjellén, T.G.I. Ling and B. Mattiasson, Reppal-PEX - A new starch derived polymer for applications in aqueous two-phase systems, in: "Proceedings 4th European Congress on Biotechnology," Vol. 2, O.M. Neijssel, R.R. van der Meer and K.Ch.A.M. Luyben, eds., Elsevier, Amsterdam (1987)

5 I. Kühn, Alcoholic fermentation in an aqueous two-phase system, Biotechnol. Bioeng. 22:2393 (1980)

6 M. Larsson and B. Mattiasson, Novel process technology for biotechnological solvent production, Chem. Ind. 12:428 (1984)

7 B. Mattiasson and M. Larsson, Extractive bioconversions with emphasis on solvent production, in: "Biotechnology and Genetic Engineering Reviews," Vol. 3, G.E. Russel, ed., Intercept Ltd, Newcastle upon Tyne, UK (1985)

8 R. Kaul and B. Mattiasson, Extractive bioconversion in aqueous two-phase systems. Production of prednisolone from hydrocortisone using Arthrobacter simplex as catalyst, Appl. Microbiol. Biotechnol. 24:259 (1986)

9 M. Larsson and B. Mattiasson, Characterisation of aqueous two-phase systems based on polydisperse phase forming polymers. Enzymatic hydrolysis of starch in a PEG-starch aqueous two-phase system, Biotechnol. Bioeng. 31:979 (1988)

10 M. Larsson, V. Arasaratnam and B. Mattiasson, poster presented at 5th Int. Conf. on Partition in Aqueous Two-Phase Systems, Oxford, UK, 23-28 August 1987

11 B. Hahn-Hägerdal, B. Mattiasson and P.-Å. Albertsson, Extractive bioconversion in aqueous two-phase system. A model study on the conversion of cellulose to ethanol, Biotechnol. Letters 3:53 (1981)

12 I. Persson, F. Tjerneld and B. Hahn-Hägerdal, Semicontinous cellulase production in an aqueous two-phase system with *Trichoderma reesei* Rutgers C30, Enzyme Microb. Technol. 6:415 (1984)

13 B. Mattiasson, Application of aqueous two-phase systems in biotechnology, Trends in Biotechnol. 1:16 (1983)

14 B. Mattiasson and B. Hahn-Hägerdal, Microenvironmental effects on metabolic behaviour of immobilised cells. A hypothesis, Eur. Appl. Microbial. Biotechnol. 16:52 (1982)

15 E. Andersson, B. Mattiasson and B. Hahn-Hägerdal, Enzymatic conversion in aqueous two-phase systems: Deacylation of benzylpenicillin to 6-aminopenicillanic acid with penicillin acylase, Enzyme Microb. Technol. 6:301 (1984)
16 M. Puziss and C.G. Hedén, Toxin production by *Clostridium tetani* in biphasic liquid cultures, Biotechnol. Bioeng. 7:355 (1965)

PROCESS SIMULATION OF CONTINUOUS BIOCONVERSIONS IN AQUEOUS TWO-PHASE SYSTEMS

Christian Reisinger [1], Michael Narodoslawsky [2] and Anton Moser [1]

[1] Institute of Biotechnology
Microbiology and Waste Technology
Schlögelgasse 9, A-8010 Graz, Austria

[2] Institute of Chemical Engineering
Inffeldgasse 25, A-8010 Graz, Austria

Progress in biotechnology largely depends on fermentation optimisation and improvement in downstream processing. These improvements get strongly supported if an integration of fermentation and product extraction can be verified in one processing unit. Continuous bioconversions in aqueous two-phase systems show a great potential of combining these two unit operations. General requirements are the capability of cells and substrates to enrich in one of the phases and the tendency of products to diffuse in the opposite phase. This integrated process is most promising for reaction systems, which exhibit kinetics characterised by product inhibition and, in addition, the process facilitates cell recycling and supports an increase in cell concentration. However, the industrial realisation of this technique is still not fully developed mainly due to difficulties in measurement and control, and to a lack of profitable products. In this work the conversion of glucose to ethanol is used as a model study to develop mathematical optimisation criteria and to create a "feasibility window" for the ongoing search for products.

INTRODUCTION

Aqueous two-phase systems are formed by a mixture of two immiscible polymers with water and tend to distribute molecules, cells or particles unequally between the two phases. A separation occurs within a short period due to a thermodynamic equilibrium usually simply represented by a phase diagram. The volume ratio of the phases and the partition of certain substances build up an extraction system, which can be helpful in a variety of separation problems [1,2].

The broadening use of aqueous two-phases extends its applications to integrated bioprocessing techniques, which benefit from certain properties of these systems, e.g. biocompatibility, lower interfacial tension. Continuous bioconversions represent the most promising area, where an integration, in this case a combination of fermentation and extraction in one processing

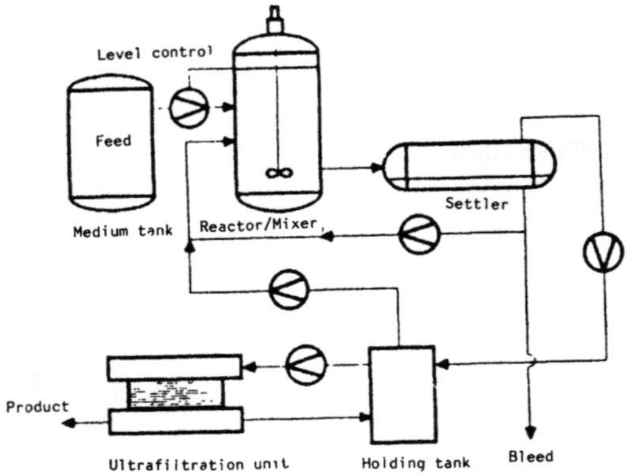

Fig. 1 *Flow chart of a continuous bioconversion unit with cell recycle*

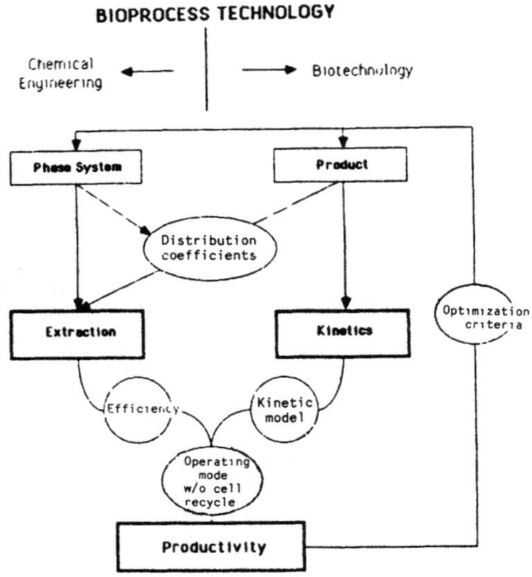

Fig. 2 *Principle of model development for continuous bioconversion*

unit, leads to a significant increase in the productivity of the entire reactor system. Additionally, cell recycling is facilitated and higher cell concentrations are possible.

General requirements are the capability of cells and substrates to enrich in one of the phases and for the products to diffuse into the opposite phase.

An experimental set-up for this process has been developing at the University of Lund [3-5]. A general flow chart of a continuous bioconversion plant with cell recycling is shown in Fig. 1 and serves as a basis for mass balances. Cells are recycled with the bottom phase, the product is recovered from the top phase. Polymers in the product stream are rejected in the ultrafiltration unit, which reduces the polymer loss to the bleed stream. The quantity of the bleed stream is generally governed by the dilution rate and the elimination of otherwise enriching by-products and dead cells [6].

An industrial realisation of a developing technique requires a structured process development, including the mathematical modelling for an estimation of theoretical limitations and optimisation [7]. This simulation serves as a tool for the development of mathematical optimisation criteria and creates a "feasibility window" for the ongoing search for profitable products.

MODEL DEVELOPMENT

The combination of chemical engineering and biotechnology provides the rationale for the model structure and results in a simple model development (Fig. 2). One has to be aware that the mathematical model reflects only part of reality and that the manipulation of the model gives the information we need about that part of reality leading to a specified goal [8].

Our goal is threefold:

a) Sensitivity analysis of kinetic and extraction parameters.
b) Estimation of theoretical productivity increase.
c) Estimation of kinetic criteria for further development ("Feasibility Window")

Goals a) and b) require the choice of a proper model system, preferably a system which exhibits kinetics characterised by product inhibition. For the continuous bioconversion of ethanol to glucose by yeast most information was found including economic evaluation of alternative processes [9]. The PEG 20M - crude dextran phase system reduces polymer cost to one tenth of the price of a fractionated dextran system [10] and was therefore chosen for a first analysis. The following assumptions reduce the reliability of the process simulation and require an investigation of their impact on the output:

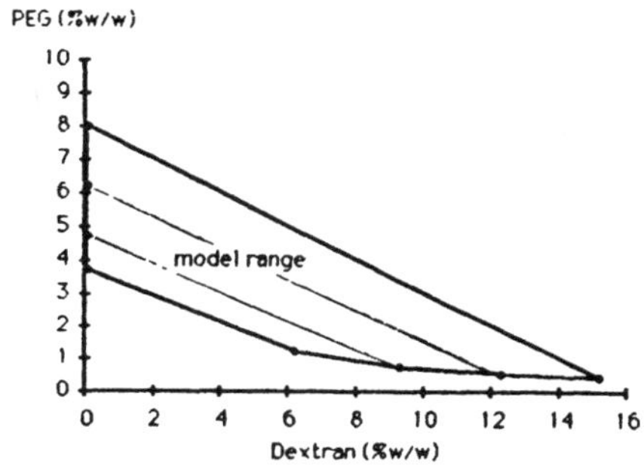

Fig. 3 *Calculated phase diagram for PEG 20M - crude dextran*

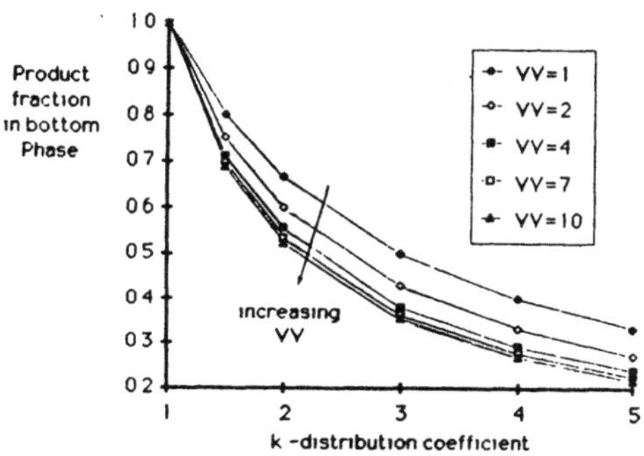

Fig. 4 *Product concentration at various extraction conditions*

a) The fermentation is restricted to the bottom phase.
b) The different kinetic models employed hold in the entire range of analysis.
c) The extraction efficiency is 100%.
d) Transport limitations are neglected.
e) The extracellular ethanol concentration controls the product inhibition.

The mathematical phase diagram for the PEG 20M - crude dextran system is based on a numerical input and excludes the area close to the critical point, where the separation phenomenon disappears (Fig. 3). The advantageous reduction of the product concentration by extraction depends on the volume ratio and the distribution coefficient. The fraction of the formed product, which remains in the fermentation phase is given by equation 1 and Fig. 4.

$$\lambda = \frac{P_B}{P} = \frac{1 + VV}{1 + VV*k} \quad (1)$$

Different kinetic models are investigated to determine the importance of kinetic parameters (equations 2-5).

a) Two-parameter model [9]

$$\mu = \mu_{max} \left[\frac{S}{S + K_S}\right] \left[1 - \frac{P}{P_{max}}\right]^n \quad (2)$$

b) One-parameter model combined with substrate inhibition [11]

$$\mu = \mu_{max} \left[1 - \frac{P}{P_{max}}\right] \frac{S}{\left[S + K_S + \frac{S^2}{K_S \omega}\right]} \quad (3)$$

c) Two-parameter model based on average product concentrations [12]

$$\mu = \left[\mu_0 - aP(b - P)\right] \frac{S}{S + K_s} \quad (4)$$

d) One-parameter model [12,14-17]

$$\mu = \mu_{max} \frac{S}{S + K_s} \frac{K_P}{P + K_P} \quad (5)$$

$$P = Y_{P/S} (S_F - S) \quad (6)$$

Mass balance lead to the result that the specific growth rate equals the dilution rate (equation 7) for continuous operation with and without cell recycle as well. The productivity term, however, has to be defined differently for the two cases. The combination of equations 1, 5, 6 and 7 (using model d) as an example leads to a quadratic equation for the product concentration (equation 8).

$$\mu = D \quad (7)$$

Table 1 *Parameters used in this simulation (unless varied)*

μ_{max}	0.5 h^{-1}	K_S	3.3 g/l
K_P	5.2 g/l	$Y_{P/S}$	0.43 g P/g S
$Y_{X/S}$	0.1 g cells/g S	VV	4.0
k	1.5	S_F	100.0 g/l

$$P^2 + P \left[\frac{K_P}{\lambda} - Y_{P/S} (K_S + S_F) - \frac{\mu_{max}}{D} \frac{K_P}{\lambda} \right]$$

$$+ \frac{Y_{P/S} K_P}{\lambda} \left[\frac{\mu_{max}}{D} S_F - (K_S + S_F) \right] = 0 \qquad (8)$$

The productivity term for operation without cell recycle is given by equation 9 and uses the product concentration in the top phase and the dilution rate D as the reciprocal value of the mean hydraulic residence time to represent the size of the reactor.

$$Pr_{w/o} = P_T D = k.\lambda.P.D \qquad (9)$$

At operation modes with cell recycle the dilution rate does not represent the hydraulic residence time, so the following definition comes into account (equation 10). The specific growth rate μ can be calculated for a given substrate conversion. Note - that the productivity is inverse proportional to the conversion and depends on the cell concentration.

$$Pr_w = P_T \; D_H = k\lambda\mu X_{max} \frac{Y_{P,S}}{Y_{X,S}} \qquad (10)$$

The parameters used for the following analysis are given in Table 1 and represent typical data for ethanol production.

RESULTS AND DISCUSSION

Polymer losses modify the extraction system, so we first focus on the impact of changes in polymer concentration. Losses in the bleed stream are of major concern, since a good performance of the ultrafiltration unit can reduce the loss in the product stream to a negligible value. The inverse effects of changes in polymer concentration are shown in Fig. 5. As the bleed stream is richer in dextran, we expect an increase in the volume ratio. On the one hand, this has only little effect on the resulting product concentration (Fig. 4), on the other hand, the cell concentration in the bottom phase becomes smaller and reduces the productivity proportionally, according to equation 10.

A graphical presentation of equations 8 and 9 is given in Fig. 6. As shown clearly, the use of aqueous phases couples an increase in productivity with a better conversion and, in addition, the point of maximum productivity refers to a smaller residence time (= higher dilution rate). Questioning the kinetic model employed a comparison has to be made. The major difference between the models is the curve shape which borders the kinetic area we are aiming at (as indicated in Fig. 7 for models a and d).

From an economical point of view, the use of aqueous phases requires a certain productivity increase to compensate the additional operating costs. A required minimum increase refers to a particular kinetic parameter of the model chosen (K_P value of model d in Fig. 9). Regardless of the model, a kinetic area can be defined, where economic operation is feasible (Fig. 8). This means that fermentations whose kinetic data are out of the shaded area do not fulfill our requirements (in this case 40% increase) and are not the subject of further investigations.

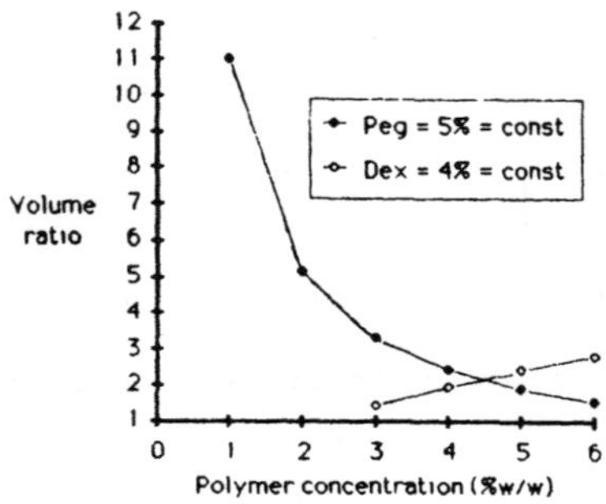

Fig. 5 *Change of volume ratio with polymer loss*

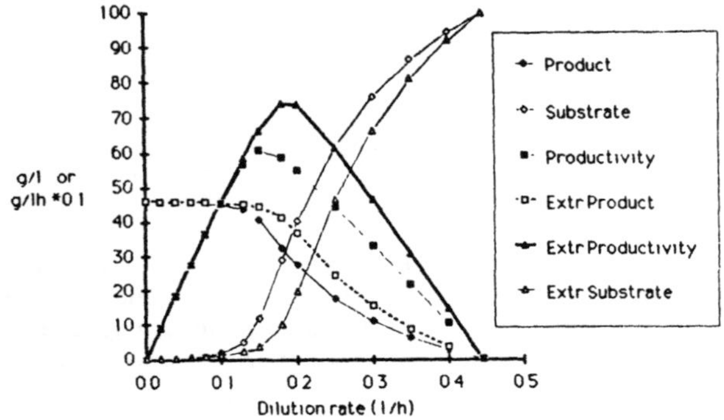

Fig. 6 *Possible productivity increase at chemostat operation*

Fig. 9 shows the relative productivity increase for both operation modes. The line without cell recycle refers to the maximum productivity, as indicated in Fig. 6; for cell recycle a 95% conversion is considered. Since the productivity term for cell recycle is proportional to the maximum feasible cell concentration, the decision on which mode to operate is based on this concentration.

As one would expect, the strongest influence on the productivity is controlled by the distribution coefficient (Fig. 10). We observe an almost linear proportional productivity increase, especially for small values. Unfortunately, small molecules, like ethanol, tend to distribute rather

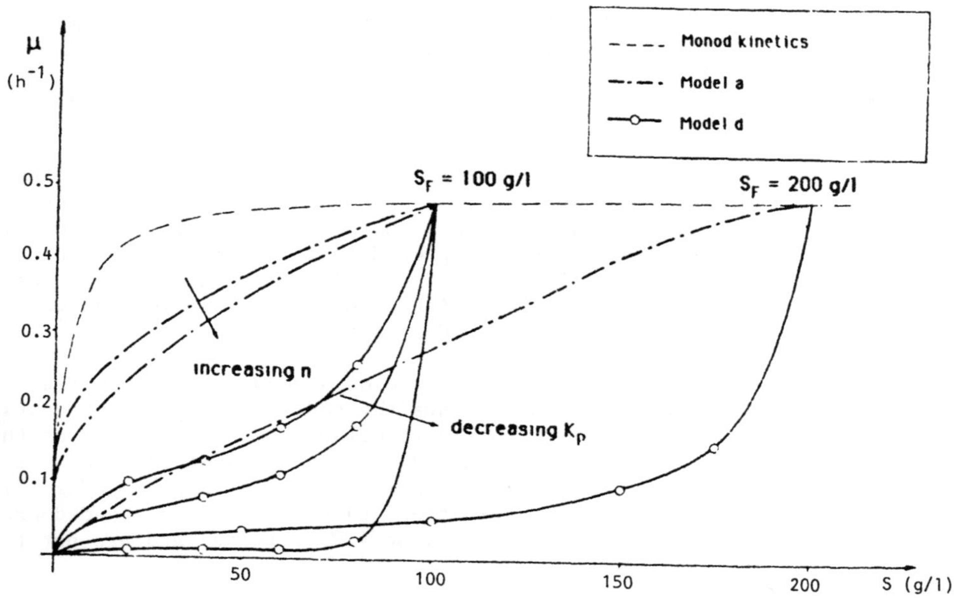

Fig. 7 *Influence of kinetic parameters on the curve shape*

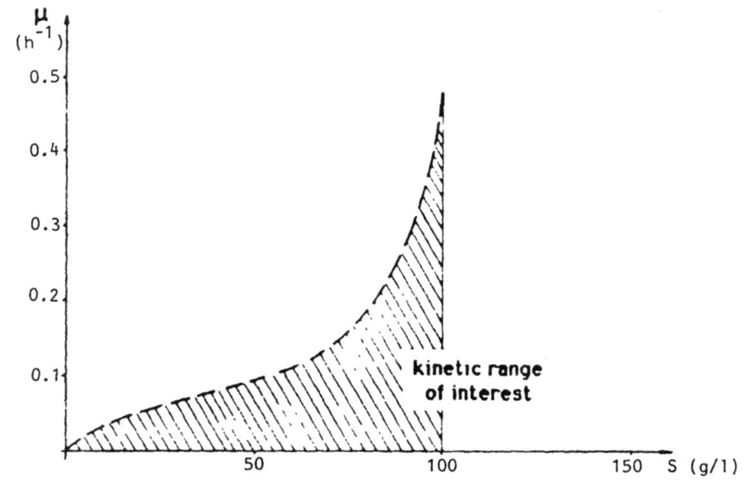

Fig. 8 *Kinetic area of a certain productivity increase*

equally between the two phases, so that the above mentioned criteria are of equal importance. Nevertheless, successful investigations, aiming at a shift in the distribution, can have a tremendous effect on the entire development.

In conclusion, continuous bioconversions in aqueous two-phase systems prove their economical relevance. Based on the simulation, one can eliminate certain reactions and can concentrate the biotechnological development on fermentations which meet the discussed conditions.

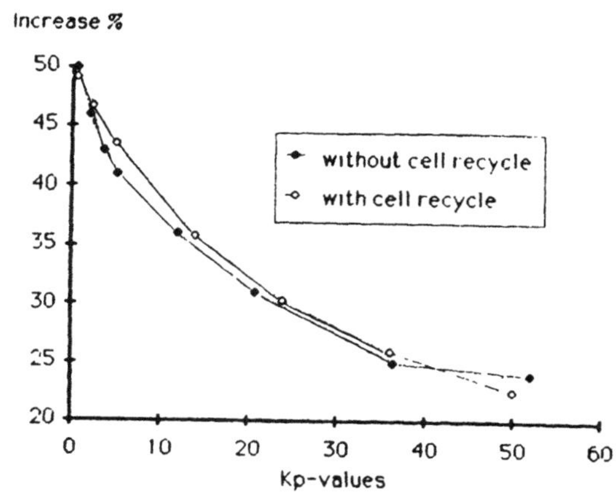

Fig. 9 *Increase in productivity at different kinetic conditions*

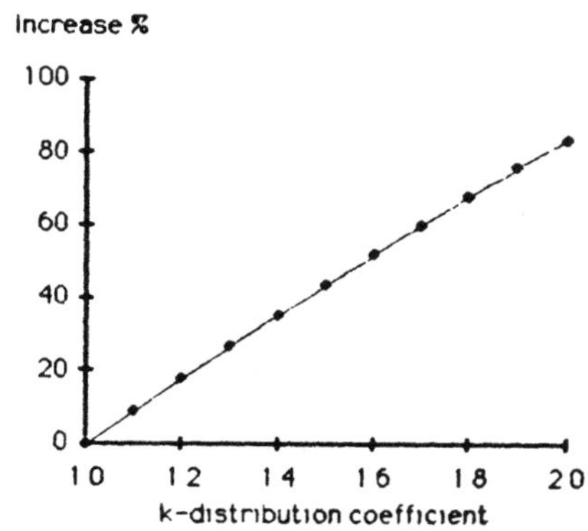

Fig. 10 *Increase in productivity with changing k*

NOMENCLATURE

P	product conc. formed [g/l]	VV	ratio top/bottom phase
P_T	product conc. in top phase [g/l]	$Y_{P/S}$	yield coefficient [g product/g substrate]
P_B	product conc. in bottom phase [g/l]	$Y_{X/S}$	yield coefficient [g cells/g substrate]
k	distribution coeff. = P_T/P_B	X_{max}	maximum feasible cell concentration [g/l]
λ	product fraction in bottom phase = P_B/P	a,b	product inhibition coeff.
μ	specific growth rate [h^{-1}]	D_H	hydraulic dilution rate [h^{-1}]
μ_{max}	max. specific growth rate [h^{-1}]	D	dilution rate [h^{-1}]
μ_o	initial specific growth rate [h^{-1}]	$Pr_{w/o}$	productivity without cell recycle [g/lh]
S	substrate conc. [g/l]	P_w	productivity with cell recycle [g/lh]
S_F	feed substrate conc. [g/l]	ω	substrate inhibition coeff.
K_P	product inhibition const. [g/l]	K_S	Monod constant [g/l]

REFERENCES

1 P.-Å. Albertsson, Partition of Cell Particles and Macromolecules, 3rd ed., John Wiley & Sons (1985)

2 M.-R. Kula, Liquid-liquid extraction of biopolymers, in: "Comprehensive Biotechnology," Vol. 2, Pergamon Press (1985)

3 B. Mattiasson, Applications of aqueous two-phase systems in biotechnology, Trends in Biotech. 1:16 (1983)

4 B. Mattiasson and M. Larsson, Extractive bioconversion with emphasis on solvent production, Biotech. and Gen. Eng. Reviews 3:137 (1985)

5 M. Larsson and B. Mattiasson, Novel process technology for biotechnological solvent production, Chem. and Ind. June (1984)

6 C.W. Lee and H.N. Chang, Kinetics of ethanol fermentations in membrane cell recycle fermentors, Biotech. Bioeng. 23:1105 (1987)

7 A. Moser, Bioprozesstechnik, Springer-Verlag, Berlin (1981)

8 N.W.F. Kossen and N.M.G. Oosterhuis, Modelling and scaling-up of bioreactors, in: "Biotechnology," Vol. 2, Verlang Chemie, (1985)

9 B.L. Maiorella et al, Economic evaluation of alternative ethanol fermentation processes, Biotech. Bioeng. 26:1003 (1984)

10 B. Hahn-Hägerdal, Fermentation and product formation in aqueous two-phase systems with yeast and bacillus Sp., Lectures given at a Workshop of the WP Microbial Physiology of the EFB, University of Lund (1986)

11 T.K. Ghose and R.D. Tyagi, Rapid ethanol fermentation of cellulose hydrolysate II. Product and substrate inhibition and optimisation fermentor design, Biotech. Bioeng. 21:1401 (1979)

12 C.D. Bazua and C.R. Wilke, Ethanol effects on the kinetics of a continuous fermentation with Sacch. cer., Biotech. Bioeng. Symp. 7:105 (1977)
13 G.K. Hoppe and G.S. Hansford, Ethanol inhibition of continuous anaerobic yeast growth, Biotech. Letters 4:39 (1982)
14 S. Aiba et al, Kinetics of product inhibition in alcohol fermentation, Biotech. Bioeng. 10:845 (1968)
15 G.R. Cysewski and C.R. Wilke, Rapid ethanol fermentations using vacuum and cell recycle, Biotech. Bioeng. 19:1125 (1977)
16 F.F. Pironti, Kinetics of alcohol fermentations, Ph.D. Thesis, Cornell University
17 F. Kollerup and A.J. Daugulis, A mathematical model for ethanol production by extractive fermentation in CST fermentor, Biotech. Bioeng. 27:583 (1985)

BASIC TRIALS ON EXTRACTIVE BIOCONVERSION OF GLUCOSE TO LACTIC ACID

W. Hartmeier, C. Bücker, J. Wallrath, T. Hembach and T. Döppner

Institute of Microbiology, Aachen University of Technology
D-5100 Aachen, Federal Republic of Germany

Some basic data on extractive bioconversion of glucose to lactic acid are presented. Using *Lactobacillus delbrückii*, lactic acid fermentation is carried out in a two-phase system consisting of an organic (paraffin) and an aqueous phase. Bacteria and nutrient components are kept nearly quantitatively in the aqueous phase, whereas the lactic acid is considerably transferred into the paraffin phase. Influences of different surfactants and anion carriers on the partition coefficients are evaluated. A new concept of continuous fermentation with simultaneous extraction of the inhibiting product by means of an external extraction loop is discussed.

INTRODUCTION

Lactic acid fermentation is one of the classical batch processes of biotechnology competing with chemical synthesis of lactic acid nowadays. Several approaches to use immobilised bacteria for the continuous production of lactic acid are known from literature [1,2]. The systems used in these published investigations mostly comprise matrix-entrapment of the bacterial

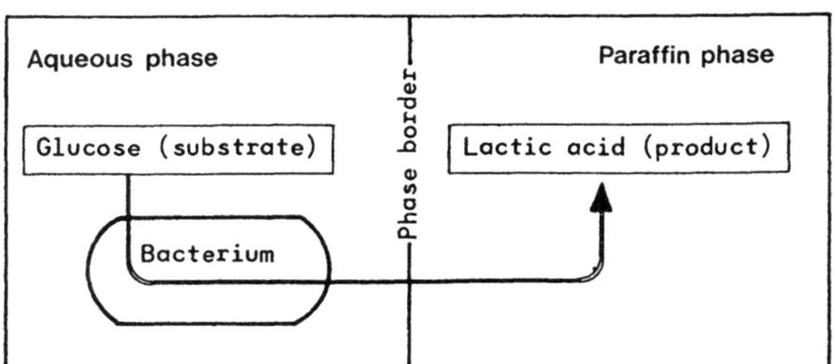

Fig. 1 *Ideal distribution of the components in the two-phase system*

cells. Problems arising from such systems are diffusion limitation by the polymeric matrix and instability of the matrix material in prolonged continuous processing.

As an alternative to common immobilisation techniques, we wanted to introduce a two-phase system as a special form of immobilisation. In such systems a physical confinement, and therewith immobilisation, of the lactic acid bacteria results, that could permit the commercial long-time use of these biocatalysts. Contrary to the present trend to use aqueous two-phase systems, we preferred solvent/aqueous two-phase emulsions because the paraffin was very much cheaper than the polymers (dextran, poly(ethylene glycol) normally used in the aqueous two-phase systems.

In the present paper some basic trials are described using bacteria, immobilised by confinement in the aqueous medium, while the lactic acid was partly transferred to the paraffin phase, which is not water miscible. Fig. 1 shows the ideal distribution of the components of such a two-phase system for lactic acid production. It depends on the proportion of aqueous to organic phase, whether a water-in-oil or an oil-in-water emulsion results, when the mixture is stirred.

MATERIALS AND METHODS

Lactobacillus delbrückii DSM 20074, being identical with *L. delbrückii* ATCC 9649, was used in the present study. This strain showed homofermentative conversion of glucose to D(-)-lactic acid.

The MRS-medium used for cultivation of the lactic acid bacteria had the following composition: 2% glucose, 1% casein-peptone, 1% meat extract, 0.5% yeast extract, 0.5% sodium acetate, 0.2% K_2HPO_4, 0.2% di-ammonium hydrogen citrate, 0.1% Tween 80, 0.02% $MgSO_4 \cdot 7H_2O$, 0.005% $MnSO_4 \cdot 7H_2O$.

The bioconversions were carried out at 45°C in 7 l laboratory fermenters with a stirrer speed of 150 rpm. The conditions were kept anaerobic by slight gasification with nitrogen. The pH-value was kept constant at pH 6.2 by automatic titration with 2 M NaOH.

Trioctylmethylammoniumchlorid (Adogen 464^R, article no. 37076 from Serva, D-6900 Heidelberg) was used as a liquid anion exchanger, soluble in paraffin.

Paraffin (liquid, low viscosity grade, article no. 7174 from E. Merck, D-6100 Darmstadt) was used as organic (not water-miscible) phase in the two-phase systems.

Table 1 *Partition coefficients for lactic acid and cell viability in two-phase systems*

Components of the phases Organic	Aqueous	Partition coefficient	Cell viability
Paraffin	Buffer	0.02	very good
Paraffin	Buffer + Tween 80	0.05	not determined
Paraffin	Buffer + Adogen	0.08	no
Paraffin	Buffer + Span 85	not determined	very good
Paraffin	Buffer + Adogen + Span 85	0.20	no
Paraffin	MSR-medium	0.41	very good

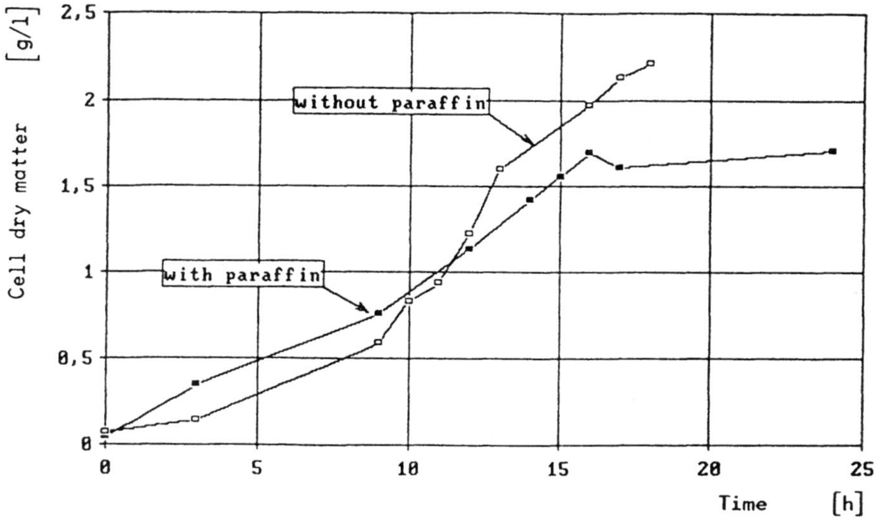

Fig. 2 *Biomass formation in batches with and without paraffin*

In the determinations of the partition coefficient 0.1 M phosphate buffer of pH 6.2 was used.

RESULTS

The partition coefficient K_p in our case is defined as the concentration of lactic acid or glucose in the organic top phase (paraffin phase) divided by the concentration in the bottom phase (aqueous phase).

$$K_p = C_{org}/C_{aqu}$$

In Table 1 partition coefficients and cell viabilities in systems with different aqueous phase are listed. It can be seen that MRS-medium leads to the best partition coefficient of 0.41 and that the cells keep their full viability. Under the same conditions, the partition coefficient for the substrate glucose is less than 0.02, saying that glucose is kept back nearly quantitatively in the aqueous phase. Also the cells almost completely remained in the aqueous phase in all trials.

Theoretically, the addition of the anion carrier Adogen, being soluble in paraffin, could improve the transfer of lactate to the organic phase. This could be observed indeed, but the liquid anion exchanger was rather poisonous for the cells (see Table 1).

Surfactants (like Tween 80 and Span 85, led to an increased partition coefficient too. The viability of the bacteria remained good under addition of Span 85 and Tween 80. MRS-medium contains 0.1% of Tween 80. This surfactant is essential, but not the only reason for the shift of lactic acid distribution towards the organic phase in MRS-medium.

Figs. 2 and 3 give some data on growth and fermentation properties of the bacteria in MRS-medium and in a two-phase system with MRS-medium and paraffin. There is no inhibition of biological activity in the two-phase system as compared with the conventional mono-phase medium. Fig. 2 giving a

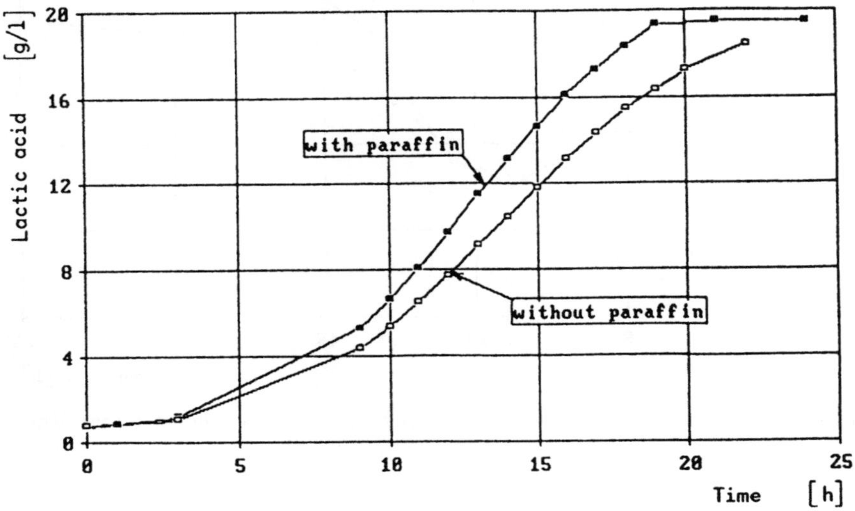

Fig. 3 *Lactate formation in batches with and without paraffin*

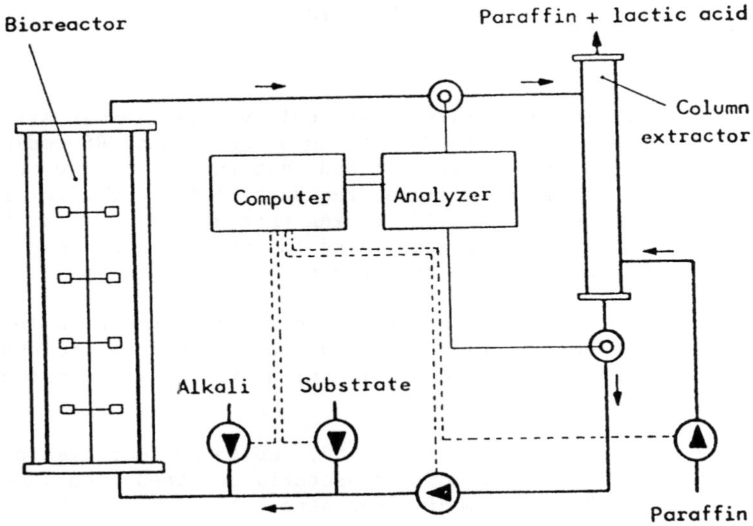

Fig. 4 *Set-up for extractive production of lactic acid*

typical parallel run in two-phase and mono-phase culture shows even better initial growth of the bacteria in the two-phase system. Consequently, the lactate formation in the two-phase system proceeds quicker than in the mono-phase batch (Fig. 3).

The better product formation in the two-phase system can possibly be explained by the partial removal of inhibiting end product from the aqueous phase. Thus, further improvement can be expected by continuous removal of the end product and by increasing the density of the bacteria in the aqueous phase.

CONCLUSIONS

With the two-phase system, consisting of paraffin and MRS-medium, a sufficient confinement (immobilisation) of the biocatalysts (bacteria) in the aqueous phase can be achieved. In such a system, the procedures of lactic acid fermentation and of down-stream processing can be combined, due to transfer of the lactic acid into the paraffin phase. Continuous operation in a stirred two-phase system could be a promising alternative to the classical batch processes. Basing on these results, a continuous extractive bioconversion, as shown in Fig. 4, is under further investigation.

REFERENCES

1 H. Büyükgüngör and H. Büschelberger, Production of lactic acid from lactose by free and immobilised *Lactobacillus bulgaricus*, Biotech. Forum 3:80 (1986)

2 S.L. Stenroos, Y.Y. Linko and P. Linko, Production of L-lactic acid with immobilised *Lactobacillus delbrückii*, Biotechnol. Lett. 4:159 (1982)

3 A. Tuli, R.P. Sethi, P.K. Khanna, S.S. Marwaha and J.F. Kennedy, Lactic acid production from whey permeate by immobilised *Lactobacillus casei*, Enz. Microb. Technol. 7:164 (1985)

INTEGRATION OF BIOCONVERSION AND DOWN-STREAM PROCESSING - STARCH HYDROLYSIS IN AN AQUEOUS TWO-PHASE SYSTEM

Mats Larsson, Vasanthy Arasaratnam and Bo Mattiasson

Department of Biotechnology, Chemical Center
University of Lund
P. O. Box 124, S-221 00 Lund, Sweden

Starch hydrolysis is one of the major processes catalysed by enzymes in industrial operation today. Since it deals with the enzymatic degradation of a macromolecular or even particulate substrate, the use of soluble enzymes has been preferred. Since the enzymes used are rather cheap, it has been possible to operate the process under economically acceptable conditions, in spite of the consumption of enzymes. However, the reaction time is long because the biocatalytic density is kept rather low, since the soluble enzymes are not reused.

Consumption of enzymes and long reaction time are the characteristic features of the processes when dealing with macromolecular or particulate substrates.

In earlier reports we have demonstrated the feasibility of carrying out enzymatic conversions in aqueous two-phase systems, performing the process in one of the phases and extracting the product to the other phase continuously [1,2]. This would give an opportunity of carrying out enzymatic degradation of macromolecular substrates without losing the enzyme used.

Integration of the bioconversion step and the first step of down-stream processing can be used as a means of increasing the productivity of bioprocesses. This integration also gives the possibility to run the bioconversion in a continuous mode. We demonstrate the use of an aqueous two-phase system in combination with ultrafiltration to accomplish this.

Conversion of native starch to glucose by α-amylase (Termamyl, Novo) and glucoamylase (SAN 150, Novo) was carried out in a mixer-settler reactor containing an aqueous two-phase system composed of 5% PEG 20M and 3% crude dextran. The top-phase in the settling part of the reactor was continuously ultrafiltered. In this way, a continuous stream of glucose was obtained; the phase forming polymers, as well as the starch degrading enzymes, were recycled to the reactor, and clogging of the ultrafiltration membranes was avoided.

This process was carried out continuously for 8 days using a 10% starch suspension (waxy maize) as substrate. The productivity during the experiment was about 13 g/l*h. The enzyme activities in the top and bottom phases and in the mixing chamber were also monitored throughout the experiment.

The settling time for phase systems containing solid starch and varying amounts of PEG and crude dextran was studied.

The ratio between α-amylase and glucoamylase, to obtain optimal activity, as well as the activity and stability of enzyme mixtures, was studied both in buffer media and in media containing the polymers to be used in the enzyme reactor. The enzymes were found to be more active and stable in media containing polymers than in the buffer solutions used.

The optimum pH, temperature and ionic strength were determined.

REFERENCES

1 B. Mattiasson, Trends in Biotechnology, Vol. 1 (1983)
2 B. Mattiasson and M. Larsson, Biotech. Genetic Eng. Reviews, Vol. 3 (1985)

Chapter 9

Emerging Technology: New Methodology Involving Liquid–Liquid Chromatography and Countercurrent Distribution

LIQUID-LIQUID PARTITION CHROMATOGRAPHY OF BIOPOLYMERS IN AQUEOUS TWO-PHASE POLYMER SYSTEMS

Werner Müller

R & D Chromatography
E. Merck
D-6100 Darmstadt 1, FRG

The classical concept of liquid-liquid partition chromatography (LLPC) presented by Martin & Synge in 1941 may be successfully applied for chromatographic separations of biopolymers when aqueous-aqueous two-phase polymer systems, introduced for the partition of biopolymers and cells by P.-Å. Albertsson in 1958 are used. The problem of immobilising one of the two aqueous phases selectively to a support could be solved by exploiting the incompatibility between a polymer grafted on the support surface and one of the polymers forming the two-phase system. For Albertsson's two-phase system, based on poly(ethylene glycol) ("PEG") and dextran ("Dx"), a support coated with polyacrylamide repels the PEG-rich phase since polyacrylamide and PEG turned out to be incompatible. This repulsion causes, on the other hand, a binding of the Dx-rich phase to the support.

The preparation of LLPC columns with Albertsson's system is briefly described. The potential of such columns for separating proteins and nucleic acids is documented by various examples.

INTRODUCTION

Liquid-liquid partition chromatography (LLPC), as introduced by Martin & Synge in 1941 [1] and successfully applied over several decades for the isolation of natural products, is based on the principles of countercurrent distribution (CCD), which separates the constituents of a sample by exploiting differences in their partition coefficients in the two-phase system used. The transfer of the stepwise CCD process into the dynamic LLPC process is effected by the continuous movement of one of the two phases along the second one immobilised in the form of droplets or a continuous layer on a suitable support material, thus forming the stationary phase of the LLPC column.

In the classical version of LLPC organic-aqueous or organic-organic two-phase systems were used of which the more polar phase was bound to silica, cellulose or others, mostly polysaccharide-based support materials. The application of this technique only required that the material to be fractioned was sufficiently soluble in the two-phase system used in the column. This trivial condition prohibited its application for separating biopolymers like proteins, nucleic acids, or polysaccharides. This restriction was overcome - at least in principle - when P.-Å. Albertsson discovered an aqueous-aqueous two-phased system based on poly(ethylen glycol) ("PEG") and dextran ("Dx") in water which he introduced for the separation of macromolecules and particles like cells in 1958 [2]. It took a further twenty years, however, until the first LLPC columns, working with Albertsson's system, were described in the literature [3,4]. Their development became possible when it was detected that cellulose binds with sufficient preference the Dx-rich phase of Albertsson's system thus allowing an immobilisation of this phase in a chromatographic column. Such columns were well suited for the separation of nucleic acids, but useless for proteins due to insufficient phase-binding capacity and irreversible adsorption of some proteins to the cellulose. The support problem was finally solved by grafting linear polyacrylamide chains on the surface of hydrophilic chromatographic support materials carrying primary or secondary aliphalic hydroxyl groups on their surface [5].

In this article a brief insight to the functioning of the poly-acryl-amide layers on support surfaces is given followed by a short description of a few basic properties of the PEG/Dx system of Albertsson, as far as their knowledge is essential for successful handling of LLPC columns prepared with this system. In the last section some examples, documenting the capabilities of these columns in separating proteins and nucleic acids, are presented.

POLYACRYLAMIDE - A POLYMER INCOMPATIBLE WITH POLY(ETHYLENE GLYCOL)

When aqueous-organic two-phase systems are used for LLPC, many hydrophilic support materials may be found which bind the aqueous phase by preferential solvation of their surface, thus forming the stationary liquid phase in a column. If an aqueous-aqueous two-phase system is to be used, a selective phase binding can only be achieved by making use of other principles. One of these manifests itself by the formation of two phases, when two polymers are codissolved in a given solvent exceeding certain critical concentrations. According to the theories on polymers in solution of Flory [6] and Huggins [7] from 1941, such a phase separation takes place when the segment interaction of the polymers is thermodynamically unfavourable. This "incompatibility" effect is responsible, for instance, for the formation of two phases in the case of Albertsson's system consisting of PEG and Dx in water.

More recently it was found in our laboratory that linear polyacrylamide and PEG form another pair of incompatible polymers. This observation provided a simple recipe of how to produce support matrices able to bind the Dx-rich phase of Albertsson's system. By coating the support surface with polyacrylamide, the PEG-rich phase should be repelled due to the polymer incompatibility, thus forcing the Dx-rich phase to wet such a surface. Since a coating with polyacrylamide may be easily achieved by grafting polymerisations on materials carrying primary or secondary aliphatic hydroxyl groups on their surface, this concept could be successfully realised [5] using Fractogels^{R} HW and LiChrospher^{R} Diols as base materials.

Beside the incompatibility effect mentioned, one might consider an intrinsic affinity between Dx and polyacrylamide as an additional source for the selective phase binding observed. However, all attempts to detect such an affinity by chromatographic retardation measurements of dextran on polyacrylamide columns failed.

Under the aspect presented, it is not too surprising that cellulose binds the Dx-rich phase of Albertsson's PEG/Dx system. The chemical similarity between cellulose and dextran infers an incompatibility between PEG and the cellulose causing again a selective binding of the Dx-rich phase by the latter. This interpretation is also supported by the fact that starch and PEG are incompatible as reported recently [8]. Further experimental evidence for the validity of the basic concept given has been communicated in a previous article [9].

APPLICATION OF THE PEG/DX TWO-PHASE SYSTEM FOR THE LLPC OF BIOPOLYMERS

General rules for the application

The basic protocol describing how to prepare an LLPC column is given in the paper of Martin & Synge [1]. When the PEG/Dx system of Albertsson is to be used as phase pair, the binding of the Dx-rich phase is effected most easily by suspending polyacrylamide-coated support particles* previously washed with Dx-rich phase, in a 1:1 mixture of both phases, washing away the excess of Dx-rich phase by PEG-rich phase, suspending the material in PEG-rich phase and pouring the slurry into a chromatographic tube** The settled packing is equilibrated with PEG-rich phase containing all the salts and buffer materials necessary for performing the planned separation. The sample dissolved in PEG-rich phase, or - if present as a liquid - adjusted to the same PEG concentration as present in the PEG-rich phase may then be loaded on the column and the constituents eluted with mobile phase. In many cases the separation of a series of similar compounds, or the complete removal of all sample constituents from the column require the application of an elution gradient acting by a salt-exchange at constant ionic strength.

Properties of the aqueous poly(ethylene glycol)-dextran two-phase system relevant for its use in LLPC of biopolymers

In LLPC the migration rate of a solute passing through a column of a given ratio of stationary to mobile phase is exclusively governed by its partition coefficient in the two-phase system used. Therefore, only those properties of the PEG/Dx system will be treated here which are related to the partitioning of biopolymers in this system. (For more comprehensive information the reader is referred to references 10 and 11). In this respect it is important to know the following few facts:

i) The partition coefficients K of proteins and nucleic acids are size dependent. The K values (defined as c_{top}/C_{bottom}) decrease exponentially with increasing size. This rule, not too strictly followed by proteins even at their isoelectric points [12], agrees with the semi-empirical law of Bronstedt [13] and has been successfully applied to the fractionation of DNA restriction fragments by size [3-5].

---

* Polyacrylamide-coated support materials, based on hydrophilic vinyl polymers, are available from E. Merck, Darmstadt, under the name of Li-Par-Gel^R. The support materials are offered in two Dx-phase-coated versions and in an uncoated form. The corresponding PEG/Dx phases are also available (LiParSolv^R moph and staph).

** Suitable chromatographic columns in a transparent inert version, suitable for HPLC and FPLC in dimensions from 10 to 26 mm i.d. and 0-600 mm in length are available from E. Merck, Darmstadt ("Superformance^R" series, Merck).

ii) Electrolytes added to PEG-Dx system in concentrations keeping the total ionic strength below 0.5 create electrostatic potential differences between the phases in the order of a few millivolts, if the partition coefficients of the anions and the cations differ appreciably. Charged biopolymers dissolved in such systems undergo shifts in their partitioning so as to reduce the potential differences between the phases. Since these shifts are exponential with the number of charges of the solute and the phase potential difference, salt exchange gradients applied in the mobile phase provide a powerful tool for removing very large biopolymers from a column. Detailed information about suitable combinations of cations and anions are given in [4,10,11,14].

    At higher salt concentrations, i.e. at ionic strengths above 1, nucleic acids move completely into the Dx-rich phase, whilst proteins partition more and more towards the PEG-rich one [10].

iii) The PEG-rich (upper) phase of Albertsson's system is slightly more hydrophobic than the Dx-rich bottom phase, thus increasing the partition coefficients of hydrophobic solutes relative to its size.

iv) The molecular weight of the phase-forming polymers influence the partitioning of macromolecules insofar as the latter is expelled from a phase if the polymer forming that phase is increased in size, and vice versa. This effect is especially pronounced if the molecular weight of the macromolecule dissolved is in the same range as the phase-forming polymer.

Making use of the properties and effects listed above for LLPC mainly consists in balancing the partition coefficient of sample constituents of interest into a range optimal for its separation. At a ratio of stationary to mobile phase in the column of about 0.5-1, the partition coefficient of the components should approach values between 0.4 and 0.1. This range may be reached in most cases by adjusting the parameters listed above. In some cases, however, this adjustment is difficult to achieve due to strong intrinsic affinities of sample constituents for one of the phase-forming polymers. In such cases extreme pH-values in the system may be helpful in order to mobilise as many electrostatic charges on the solute as possible. This renders its partition more sensitive to electrostatic potential differences between the phases which may be controlled by salts as mentioned above. If this also fails, the addition of organic solvents as recently described [15], provides another means of adjusting the partition coefficients of the solutes. The latter manipulation proved also to be useful when the solubility of the material to be separated is limited in the PEG/Dx system. This holds for example for many immunoglobulins (W. Müller, unpublished). In cases in which all the manipulations listed fail, other polymer systems may be taken into consideration, provided one of the phase-forming polymers is incompatible with linear polyacrylamide which is easy to test. Such alternatives are under investigation in our laboratory.

Applications of PEG/Dx systems for the separation of proteins and nucleic acids

Proteins. Before applying a freshly packed column for a specific separation problem, it is recommended that a test run be performed using two readily available coloured test proteins of which the partition coefficients may be easily determined. The combination of myoglobin and peroxidase (from horseradish) has been used routinely for this purpose in our laboratory. From the entering of the two proteins into the column packing any imperfections concerning the fit of the plunger or the packing at the column head may be detected.

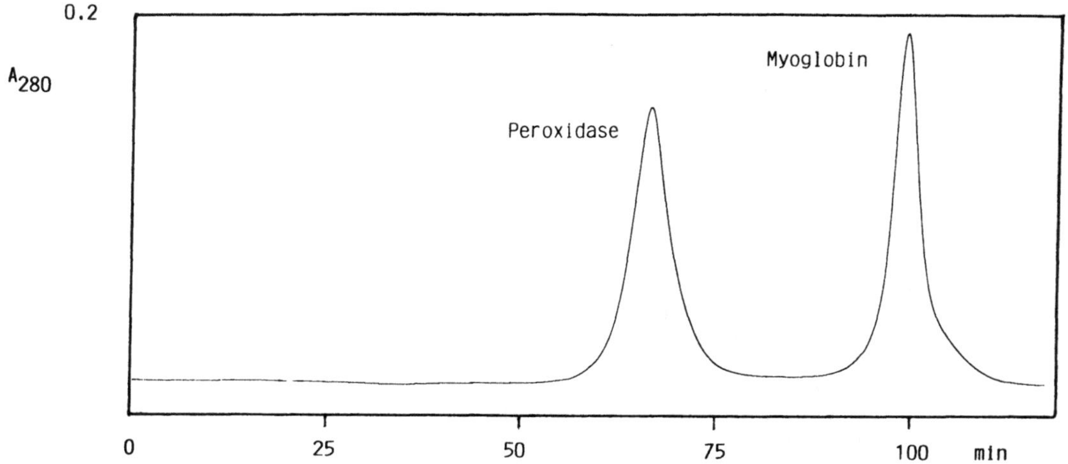

Fig. 1 *Test run for the determination of the amount of stationary and mobile phase and the performance of an LLPC column. Support: polyacrylated LiChrospher^R-Diol (1000A/5µm); system: PEG/Dx 6000/40, 10 mM Na-phosphate, 75 mM KBr, 1 mM Na-azide, pH 7.2; flow rate: 0.3 ml/min; column: 28 x 1 cm Superformance^R; 23°C; test compounds: peroxidase from horseradish and myoglobin*

From the difference between the elution volumes of the two proteins, the amount of stationary and mobile phase in the column may be determined. For further information the reader is referred to a previous article [14] or to a pamphlet on LLPC, edited by E. Merck, Darmstadt, which is available on request. A typical elution pattern of the two proteins is shown in Fig. 1 of which the legend contains all information on the conditions of the run.

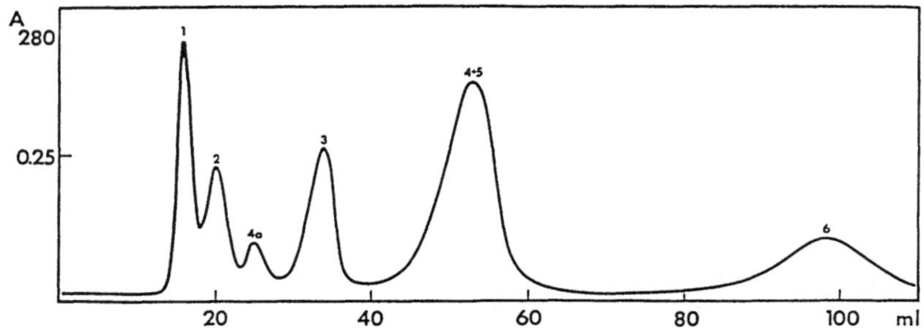

Fig. 2 *Separation of some smaller proteins by LLPC on a LiParge-750 column (32x1 cm) in the same system as given for Fig. 1; flow: 0.5 ml/min; 23°C; 1: chymotrypsinogen A, 2: peroxidase, 3: carbonanhydrase, 4: ß-lactoglobulin, (4a an impurity in that protein), 5: ovotransferrin, 6: ovalbumin*

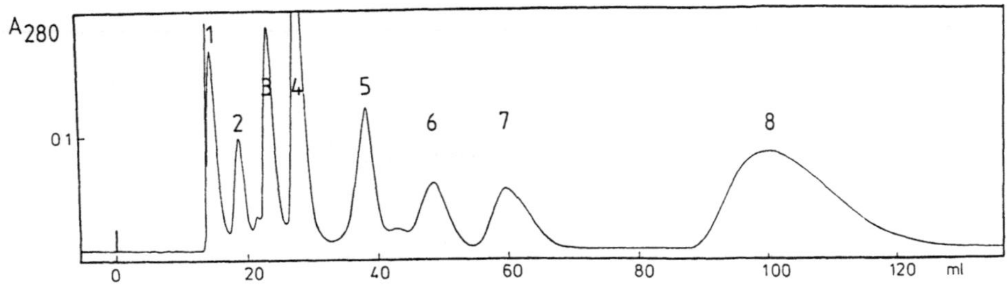

Fig. 3 *Separation of eight smaller proteins (13-68 kd) by LLPC on polyacrylated LiChrospher^R-Diol (1000A/10μm). System: PEG/Dx 6000/40, 10 mM cacodylate, 75 mM Na-perchlorate, 25 mM Li-acetate, pH 7.2; flow: 0.3 ml/min; 23°C; column: 30x1 cm Superformance^R; 1: lysosyme, 2: peroxidase, 3: cytochrome C, 4: myoglobin, 5: ß-lactoglobulin, 6: conalbumin, 7: ovalbumin, 8: human serum albumin*

In Fig. 2 the isocratic separation of a set of smaller proteins on a 20-40 μm support (LiParGel^R) is shown. ß-Lactoglobulin and ovotransferrin coelute in one peak. Changing slightly the salt composition in the mobile phase and choosing a microparticular support material (e.g. of 10 um particle diameter) based on LiChrospher^R Diol (1000 A mean pore size) not only separates the two proteins just mentioned, but allows one to separate 8 different proteins ranging from 13 to 68 kd in an isocratic run. The elution pattern is shown in Fig. 3.

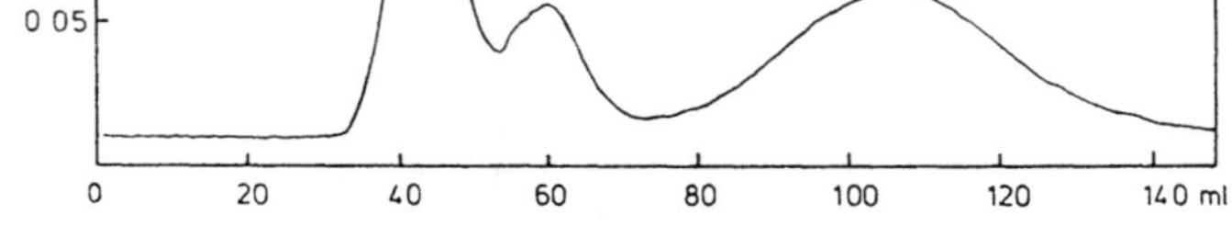

Fig. 4 *Separation of three dehydrogenases by LLPC on LiPargel 750; column: 30x1.6 cm Superformance^R); system: PEG/Dx 6000/40; gradient elution from 15 mM PO_4 (A) to 5 mM PO_4, (B) pH 8.4, 100 ml grad. followed by 50 ml B. Flow: 1 ml/min, 23°C; ADH = alcohol dehydrogenase, FDH = formate dehydrogenase, LDH = lactate dehydrogenase (about 3 mg of each)*

A synthetic mixture of several dehydrogenases ranging from 110 to 145 kd may be separated under isocratic conditions at pH 8.5 and low total ionic strength. The elution pattern is shown in Fig. 4. It may be of interest that the addition of NAD or NADH in concentrations above their Michaelis constants to the system barely influences the behaviour of these enzymes.

The first application of the LLPC for isolating some proteins from a natural mixture was presented by Heubner et al [16] who were able to isolate, for the first time, human sex hormone and corticosteroid binding globulins from serum in a homogeneous form. The isolation procedure, starting by an ammonium sulfate precipitation followed by affinity chromatography and ion-exchange chromatography on Mono Q, was completed by the LLPC step, the only method able to separate the globulins from residual serum albumin and a few other non-identified components.

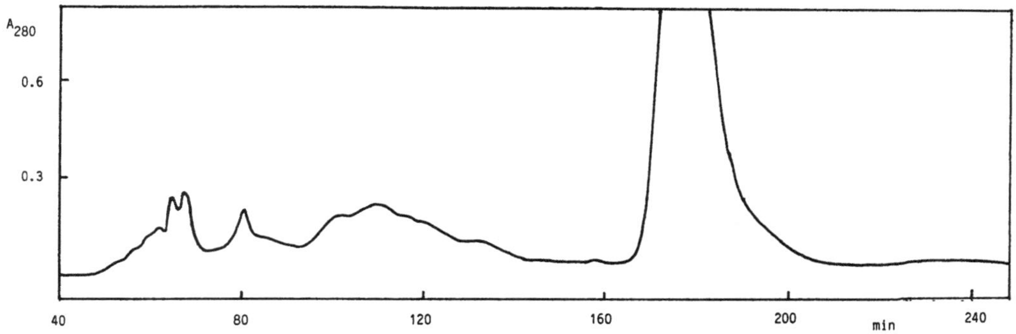

Fig. 5 *Fractionation of goat serum (0.5 ml) on polyacrylated LiChrospher^R-Diol (1000A/5 μm) by LLPC in PEG/Dx 6000/40 in a gradient from (A) 75 mM KBr to (B) 33 mM Li_3citrate with 10 mM Na-PO_4, pH 7.2, serving as basic buffer in both gradient solutions. Column: 30x1 cm Superformance^R, flow: 0.3 ml/min, 23°C; (the immunoglobulins precipitated by the PEG in the system were removed from the sample by centrifugation)*

The fractionation of a crude serum on a microparticulate support column is shown in Fig. 5. The late elution of the albumin which is typical for all sera tested [19] permits its complete removal from all the other proteins which is not easy by other methods.

The fractionation of immunoglobulins in standard PEG/Dx-systems suffers from a rather low solubility of these proteins in PEG-rich media. However, adding betaine or N-methylpyrrolidone to the phase pair improves the reliability to some extent without enabling one to fractionate this class of component on a preparative scale [Fig. 6a and b]. The isolation of monoclonal antibodies from mouse ascites fluid is shown in Fig. 7.

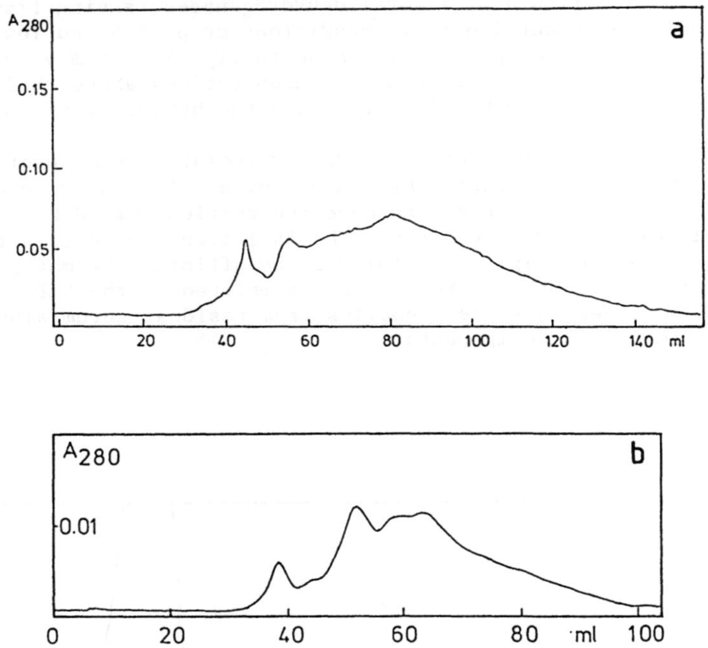

Fig. 6 *Fractionation of humn IgG by LLPC: System: PEG/Dx 6000/40; column: 30x1.6 cm Superformance*[R]*; support: LiPargel 750, 23°C*
a) in 10 mM Na-PO_4, 25 mM KBr, 100 mM betain, pH 8.3; flow: 1.0 ml/min
b) in 10 mM $Na_2B_4O_7$, 10 mM KBr, 100 mM NaCl, pH 9.2; flow: 0.5 ml/min

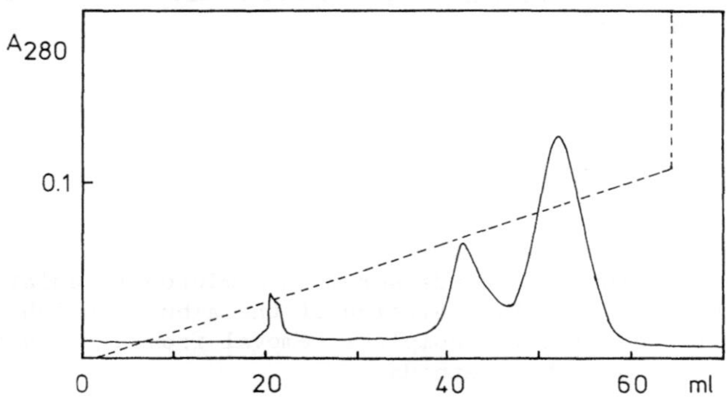

Fig. 7 *Isolation of monoclonal antibodies (mab) from mouse ascites fluid. 1.1 A_{280} units of the mab fraction obtained after prefractionation on an anion-exchanger (Fractogel*[R] *EMD-TMAE 650 (s)) were loaded on a 31x1 cm Superformance*[R] *column of polyacrylated LiChrospher*[R]*-Diol equilibrated with PEG/Dx 6000/40 containing 60 mM $NaClO_4$, 10 mM Tris.HCl, pH 8.3 (=A) and eluted by a gradient of 100 ml going from A to 67 mM Li_3citrate in 10 mM Tris.HCl, pH 8.3 (=B). The broken line indicates the shape of the gradient*

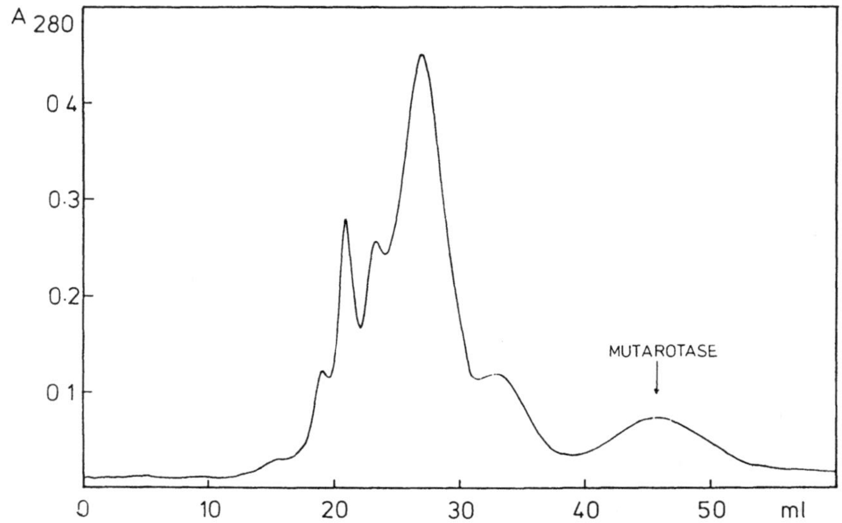

Fig. 8 *Isolation of mutarotase froma crude bacterial extract (Acinetobacter) by LLPC. 0.25 ml of the crude extract (15 mg protein) were adjusted to a final concentration of 9% (w/w) PEG and 50 mM Li_2SO_4, 10 mM PO_4, 10 mM NaOAc, pH 5.0, and eluted with mobile phase containing the same salts at the same pH, LiParGel column 30 x 1 cm, flow: 0.3 ml/min*

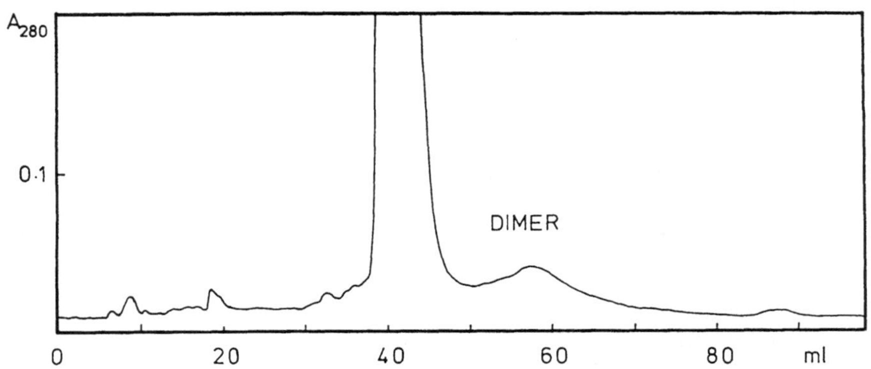

Fig. 9 *Separation of dimers and higher aggregates from human serum albumin - monomers by LLPC. 10 mg of commercial HSA were run through a 30x1 cm Superformance^R column of polyacrylated LiChrospher^R-Diol (1000A/10µm) in PEG/Dx 6000/40. Elution by a gradient from 20 mM Tris. OAc, 10 mM $NaClO_4$, pH 8.4 (A) to 30 mM Li_3citrate, 20 mM Tris.citrate, pH 8.4 (B), 50 ml gradient, further elution by B. Flow: 0.4 ml/min, 23°C*

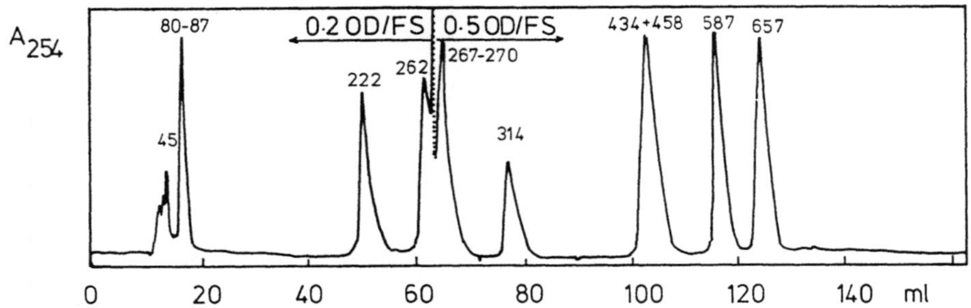

Fig. 10 *Separation of DNA restriction fragments (pDS1 plasmid x HaEIII) by LLPC: 10 A_{260}-units were loaded on a 25x1 cm Superformance[R] column of polyacrylated LiChrospher[R]-Diol (4000A/10μm) in PEG/Dx 8000/500. The elution was effected by a gradient from (A=) 240 mM NaOAc to (B=) 240 mM LiOAc both in 10 mM Na-cacodylate, pH 6.5, at 37°C at a flow rate of 0.3 ml/min. Shape of the gradient: 0-31 ml:0% B, 31-62 ml: 0-20% B, 62-235 ml: 20-85%. The numbers above the peaks indicate the size of the fragments in base pairs*

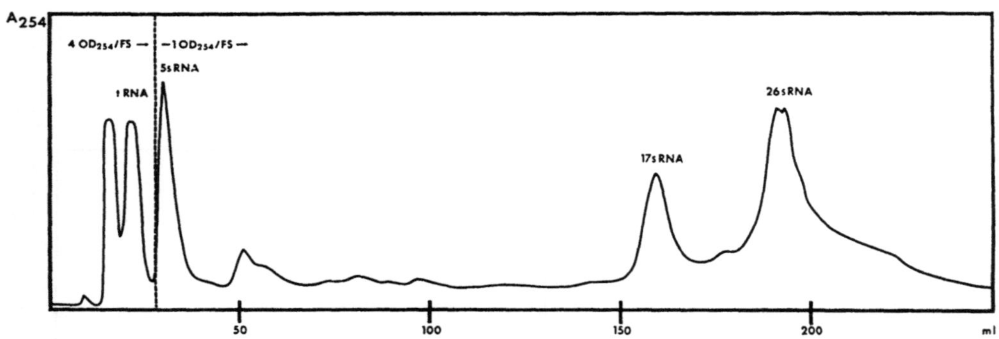

Fig. 11 *Fractionation of total RNA from yeast by LLPC: 105 A_{260}-units of RNA were fractionated on a 25x1 cm Superformance[R] column of polyacrylated LiChrospher[R]-Diol (1000A/10μm) in the system PEG/Dx 8000/500 at 50°C. The elution was effected by a gradient from (A=) 240 mM NaOAc, to (B=) 67 mm Li_3citrate, both in 10 mM Na-cacodylate, pH 6.2. Gradient volume: 250 ml, flow rate: 0.3 ml/min*

In Fig. 8 the purification of mutarotase from a crude bacterial extract is shown. The basic enzyme migrating in the last peak is well separated from the acid by products by salts added to the system which positivate the mobile phase with respect to the stationary one.

In Fig. 9 the separation of oligomers of human serum albumin from the (dominating) monomer is shown. This separation, normally achieved on large gel permeation columns, may be obtained by partition chromatography within two to three hours.

Nucleic acids. The fractionation of nucleic acids by LLPC in PEG-Dx two-phase systems has been reported in the past in several publications [3-5]. It could be shown that DNA fragments may be separated by size in a range from about 10 bp to about 30 kb. The chromatographic separation of DNA fragments by base composition [17,18] exceeds by far the resolution obtained by any centrifugation methods without being restricted by a minimum DNA size. RNAs are fractionated by size over the whole range of the natural occurring species [17]. The fact that the system works at constant ionic strength avoids precipitations of the larger species which necessarily takes place when ribosomal RNAs are to be fractionated on ionexchangers. In Figs. 10 and 11 examples for resolution of DNA restriction fragments and RNAs from yeast are given.

REMOVAL OF THE PHASE FORMING POLYMERS FROM PEAK FRACTIONS

For preparative work the removal of the phase-forming polymers from the peak fraction of the material of interest is required. This is easily achieved by passing the collected fractions through a small bed of an ion-exchanger, or - after adding sufficient ammonium sulfate - through a hydrophobic interaction gel, or through hydroxylapatite. The neutral, hydrophilic polymers pass through these adsorbents from which the bound compounds may be recovered in concentrated form by a step elution with a suitable elution medium.

CONCLUSION

The broad application range of LLPC in aqueous two-phase polymer systems, based on the versatility of these method, provides an efficient new tool for the fractionation of biopolymers in their native state. To some extent the procedure combines the separation criteria of most of the classical separation techniques:

- size fractionations are basically possible using systems without potential differences between the phases
- separations, according to charge, are possible by the use of "charged" systems at pH values which maximise the number of positive or negative net charges on the analyte
- separations, according to hydrophobicity, always take place to some extent due to the relative hydrophobicity of the PEG-rich phase. The effect may be emphasised by adding surfactants as Brij, Triton or fatty acid derivatives of the phase-forming polymers to the system
- affinity separations may be performed by applying suitable affinity labels bound to PEG or Dx to the system.

It is obvious that a clear distinction by which of the first three modes listed above a separation was achieved can barely be made because the partition coefficient which determines the elution position of a given compound in the two-phase system depends largely on the surface properties of the solute. The mode combination often, however, renders separations possible which normally would require several consecutive runs in conventional systems.

REFERENCES

1 A.I.P. Martin, and R.L.M. Synge, A new form of chromatogram employing two liquid phases Biochemical J. 35:1358 (1941)
2 P.-Å. Albertsson, Particle fractionation in liquid two-phase systems, Biochim. Biophys. Acta 27:378 (1958)
3 W. Müller, H-J. Schutetz, C. Guerrier-Takada, P.E. Cole and R. Potts, Size fractionation of DNA fragments by liquid-liquid chromatography, Nucleic Acids Res. 7:2483 (1979)
4 W. Müller and G. Kütemeier, Size fractionations of DNA fragments ranging from 20-30,000 base pairs by liquid-liquid chromatography, Eur. J. Biochem. 128:231 (1982)
5 W. Müller, New phase supports for liquid-liquid partition chromatography of biopolymers in aqueous poly(ethylene glycol)-dextran systems, Eur. J. Biochem. 155:213 (1986)
6 P.J. Flory, Thermodynamics of high polymer solutions, J. Chem. Phys. 9:660 (1941)
7 M.L. Huggins, Solutions of long-chain compounds, J. Chem. Phys. 9:440 (1941)
8 F. Tjerneld, S. Berner, A. Cajarville and G. Johansson, New aqueous two-phase system based on hydroxypropyl starch useful in enzyme purification, Enzyme Microb Technol. 8:417 (1986)
9 W. Müller, New phase supports for partition chromatography of biopolymers, in: "Macromolecular Symposia Series," Hütlig und Wepf Verlag, Basel, Heidelberg, New York, in press
10 P.-Å. Albertsson, in: "Partition of Cell Particles and Macromolecules," 2nd ed., Almquist Wicksell, Stockholm, Wiley (Futuresciences), New York (1971)
11 G. Johansson, in: "Partitioning in Aqueous Two-Phase Systems," H. Walter, D.E. Brooks and D. Fisher, eds., Academic Press (1985)
12 S. Sasakawa and H. Walter, Partition behaviour of native proteins in aqueous dextran-poly(ethylene glycol) phase systems, Biochemistry 11:2760 (1972)
13 J.N. Bronstedt, Molekülgrösse under Phasenverteilung, J. Phys. Chem. 157a:257 (1931)
14 W. Müller, Liquid-Liquid Partition Chromatography (LLPC) (Part 1): History and Principles, Kontakte, Darmstadt, 3:3 (1986)
15 G. Johansson and G. Kopperschläger, Effects of organic solvents on the partitioning of enzymes in aqueous two-phase systems, J. Chromatogr. 388:295 (1987)
16 A. Heubner, O. Belovsky, W. Müller, H.-J. Grill, B. Manz, M. Juchem and K. Pollow, Application of liquid-liquid partition chromatography in the simultaneous purification of sex-hormone-binding globulin and corticosteroid-binding globulin, J. Chromatogr. 397:419 (1987)
17 W. Müller, Liquid-Liquid Partition Chromatography (LLPC) (Part 2): Nucleic Acids, Kontakte, Darmstadt, 1:45 (1987)
18 W. Müller, in: "Partitioning in Aqueous Two-Phase Systems," H. Walter, D.E. Brooks and D. Fisher, eds., Academic Press (1985)
19 W. Müller, A. Heubner and J. Kaniecki, Liquid-Liquid Partition Chromatography (LLPC) (Part 3): Proteins, Kontakte, Darmstadt, 1:54 (1988)

APPLICATION OF LIQUID-LIQUID PARTITION CHROMATOGRAPHY (LLPC) IN THE PREPARATION OF STEROID BINDING PROTEINS

Arnulf Heubner *, Michael Juchem *, Werner Müller + and Kunhard Pollow *

* Department of Experimental Endocrinology
Johannes Gutenberg University Mainz
Langenbeckstrasse 1, 6500 Mainz, FRG

+ E. Merck, Forschung Reagenzien
Frankfurter Strasse 250, 6100 Darmstadt, FRG

Two human serum proteins, i.e. sex hormone binding globulin (h-SHBG) and corticosteroid binding globulin (h-CBG), rat corticosteroid binding globulin (r-CBG), and progesterone binding globulin (PBG) from new guinea pig were purified by the application of three different modes of chromatography. The proteins were purified by affinity chromatography and anion exchange chromatography. Fractions containing the steroid binding proteins were finally purified by liquid-liquid partition chromatography on LiParGel 750 (Merck, Darmstadt, FRG). This chromatographic sequence clearly separated the steroid binding proteins from other proteins, mainly from serum albumin without a loss of protein and completely retaining the binding affinity towards steroids.

INTRODUCTION

Isolating defined proteins from serum is of interest in understanding the influence of serum proteins on metabolism and how the serum proteins are regulated. Therefore the purification and determination of the concentration of various serum proteins becomes more and more important for research and diagnostic reasons. The induction of antibodies as well as the biochemical characterisation often requires a purified protein which is, on the one hand, immunogenic and, on the other hand, purified in a way which does not or only slightly effects the structure and biological function of the protein. The binding characteristics and specificity of a polyclonal antiserum are dependent on the quality and to a certain extent on the purity of the antigen. Moreover, for the preparation of monospecific antibodies, a highly pure antigen is needed. Corticosteroid binding globulin (h-CBG, r-CBG), sex hormone binding globulin (h-SHBG) and progesterone binding globulin (PBG) are circulating plasma glycoproteins with molecular weights of about 70000. The proteins are thought to be synthesised in the liver. CBG, a protein which binds corticosteroids and progesterone, was first described in 1956 [1]. SHBG, a protein that binds testosterone and estradiol, was first described in

1966 [2]. Progesterone binding globulin (PBG) is a polydisperse glycoprotein [3] which appears only in pregnant hystricomorph rodents, such as guinea pigs [4]. PBG binds progestins, androgens and corticosteroids with high affinity [5-7].

EXPERIMENTAL

Steroids

3H-Cortisol (specific activity 85 Ci/nmol), 3H-dihydrotestosterone (DHT, specific activity 118 Ci/mmol) and 3H-progesterone (specific activity 130 Ci/mmol) were obtained from New England Nuclear (Boston, MA, USA). The radiolabelled steroids were kept in benzene-ethanol (85:15, v/v) at 4°C and checked regularly for purity by thin-layer chromatography. Unlabelled steroids were purchased from Serva (Heidelberg, FRG) or Merck (Darmstadt, FRG).

Synthesis of Affinity Gels

The affinity gels were synthesised as previously described [8]. In brief: for the CBG-binding affinity gel 1,12-diaminododecane was bound as a spacer to tresyl-activated Sepharose (Pharmacia, Uppsala, Sweden). The ligand for CBG (17α-hydroxy-4-androsten-3-one-17ß-carboxylic acid N-hydroxysuccinimide ester) was then bound to the gel. For SHBG-binding affinity gel 15 Aminopentanoic acid was bound as a spacer to tresyl-activated Sepharose. Then 17 ε - aminomethyl-5α-androstane-3ß, 17-diol was combined via the spacer with the Sepharose.

Sera

Rat serum and serum from pregnant new guinea pigs were a generous gift from Schering (Berlin, FRG). Human pregnancy serum was pooled. All sera were stripped of endogenous steroids with charcoal (2 g per 100 ml for 2 h at 23°C) and kept at -30°C.

Quantification of specific steroid binding sites

The binding capacities were determined by multi-point titration and Scatchard analysis of the data [9]. Sera or aliquots of the purified proteins were diluted (1:50 for human pregnancy serum and rat serum, 1:200 for new guinea pig serum, and 1:500 for the purified proteins) in PENG buffer (10 mM KH_2PO_4 - 10 mM K_2HPO_4 - 1.5 mM EDTA-3 mM NaN_3 - 10% glycerol, pH 7.5). For the dilution of the purified proteins 0.1% bovine serum albumin was added to the PENG buffer. The diluted samples were incubated for 16 h at 4°C with various concentrations of 3H-cortisol (for the measurement of CBG-binding capacity), 3H-dihydrotestosterone (for the measurement of SHBG-binding capacity) or 3H-progesterone (for the measurement of PBG-binding capacity) over a range calculated to saturate the binding capacity (1-16 nmol/l). Additional samples were incubated with tritiated ligands in the presence of a 200-fold excess of unlabelled ligands in order to determine the extent of non-specific binding. Unbound ligand was removed from protein associated by incubation (5 min 4°C) with 500 ul of a suspension of dextran-coated charcoal (DCC). Then the samples were centrifuged at 1500 x g for 10 min, 500 ul of the supernatant were withdrawn and counted for radioactivity. For the preparation of DCC, 5 g Norit A (Serva) were suspended in doubly distilled water. The charcoal particles were repeatedly allowed to settle and the fines were removed. The pelleted charcoal was finally suspended in 1 litre PENG buffer and 0.5 g of dextran T500 (Pharmacia) was added.

High Performance Liquid Chromatography

Chromatography was carried out with the Pharmacia FPLC System, connected to an LB 510 two channel UV Radioactivity Chromatography System (Berthold, Wildbad, FRG) by the following method.

A prepacked Mono Q anion exchange column (HR 5/5, Pharmacia) was equilibrated in T buffer (20 mM Tris;HCL, pH 7.5) at room temperature and, after sample injection, eluted at a flow rate of 2 ml/min with a sodium chloride gradient formed by T buffer and TN buffer (20 mM Tris-HCL-1M NaCl, pH 7.5); gradient, 0-4 ml 0% TN buffer, 4-30 ml 0-30% TN buffer, 30-36 ml 30-100% TN buffer, 36-40 ml 100% TN buffer. Proteins were monitored at 280 nm. Fractions of 500 μl were collected. The radioactivity was monitored in 30 μl of each fraction by liquid scintillation counting (Ready-Solv HP scintillation cocktail and LS 6800 scintillation counter; Beckman, Munich, FRG). For the characterisxation of the proteins in crude serum, the radioactivity was measured with the Berthold LB 510 system.

Liquid-Liquid Partition Chromatography [8,10,11]

The support particles (LiParGel 750 or polyacrylated LiChrospher 1000, Merck, Darmstadt, FRG) were suspended in about three volumes of lower phase (LiParSolv 40/6000 staph, Merck) containing 10 mM phosphate (ph 7.4) and packed at 20°C and a pressure of 0.05 MPa into a column 30 x1 cm, equipped with a heating mantle. The column was equilibrated with the upper phase A (LiParSolv 40/6000 moph) at a flow rate of 0.2 ml/min until the eluate was clear. This indicated that excess of lower phase had been removed from the column. Upper phase A contained 10 mM phosphate-75 mM KBr, pH 7.4. Upper phase B contained 10 mM phosphate-25 mM Li-citrate, pH 7.4.

In the sample solutions of serum or steroid binding proteins to be fractionated, sufficient solid PEG 6000 was dissolved to reach the same concentration as was present in the mobile phase. A 0.5 - 1 ml volume of sample was injected into the column, which was installed in the Pharmacia FPLC system. The column was eluted with upper phases (200 ml) at a flow rate of 0.2 ml/min. Fractions were collected with a FRAC 100 fraction collector (Pharmacia) in the peak collection mode and were assayed for concentration of steroid binding proteins.

Purification of Steroid Binding Proteins

a) h-CBG and h-SHBG: Volumes of 200 ml human pregnancy serum were fractionated by precipitation with ammonium sulphate. The 50% pellet and the 80% pellet were dissolved in 1 l of PENG buffer and incubated with 5 g of affinity gel (wet weight) (50% pellet for SHBG and 80% pellet for CBG purification) for 2 h at 4°C. The gel was then washed on a scintered glass funnel with 200 ml of PENG buffer and eluted with a high molar excess of DHT (for SHBG, 10 ml of $3x10^{-4}$ solution) and cortisol (for CBG, 10 ml of $3x10^{-4}$ solution). Aliquots (100-200 mg) of the gels were eluted with 2 ml of tritiated hormones (64 nM). The mixture of both eluates was chromatographed on a Mono Q column. Fractions containing bound radioactivity were pooled, concentrated with Centricon 10 (Amicon, Witten, FRG) and further purified by LLPC. Separation of SHBG and CBG from dextran and PEG was achieved by anion exchange chromatography (Mono Q).

RESULTS AND DISCUSSION

Many methods have been described for the purification of SHBG and CBG [12-20] on the basis of affinity chromatography with different types of columns, used as a second purification step, e.g. hydroxylapatite, anion

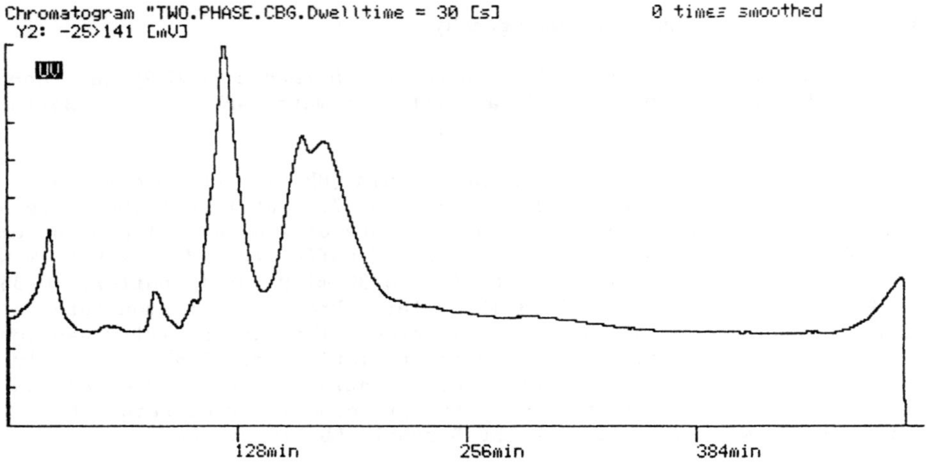

Fig. 1 *Fractions eluted from the Mono Q column with bound 3H-Cortisol were pooled, concentrated and further purified by LLPC (LiParGel 750). Sample was eluted with LiParSolv 40/6000 moph containing 10 mM* NaH_2PO_4 *(pH 7.4) and 75 mM KBr at a flow rate of 0.2 ml/min*

exchange, Cibacron Blue Sepharose, gel filtration and preparative isoelectric focusing as second or third purification step. Most purification procedures led to apparent homogeneity with molecular weight, determined by SDS-PAGE, of 60000-70000 for CBG (double band) and 40000-52000 for SHBG (either a single or double band).

We purified h-CBG, r-CBG and SHBG by a combination of affinity chromatography, FPLC and anion exchange chromatography and LLPC. Ammonium sulphate precipitation and affinity chromatography achieved a 100-fold (h-CBG, SHBG) and 150-fold purification of the proteins. Subsequent anion exchange chromatography, the second purification step, resulted in a further 10-fold purification (15-fold for r-CBG). We could obtain apparent homogeneity of the purified proteins using only those one or two fractions of the Mono Q eluate with maximum bound radioactivity, but, as the purified molecules are so heterogenous with respect to their charge and carbohydrate content, a loss of protein had to be accepted. Since LLPC, the third purification step, allows a clear separation of the steroid binding proteins from other proteins, mainly albumin, we could use all fractions with specifically bound tritiated hormones for further purification. After the last purification step the recovery was 48% for SHBG, 69% for h-CBG and 55% for r-CBG. We obtained a 2272-fold purification for h-CBG, 3220-fold for r-CBG and 2648-fold for SHBG. The results obtained by SDS-PAGE are in good agreement with previous estimates (66000 and 70000 for h-CBG and r-CBG, and 43000 and 45000 for SHBG). We observed a microheterogeneity of h-CBG and SHBG which had not been described before and which is probably due to the purification procedure. Isoelectric focusing of the purified proteins shows a pattern of protein bands over a wide pH-range (pH 5.2-5.7 for SHBG and 3.5-4.4 for h-CBG) [8]. Moreover, we could separate the two bands of human CBG, which are characteristic for this protein in PAGE, by means of column chromatography for the first time (Fig. 1). The purified proteins were

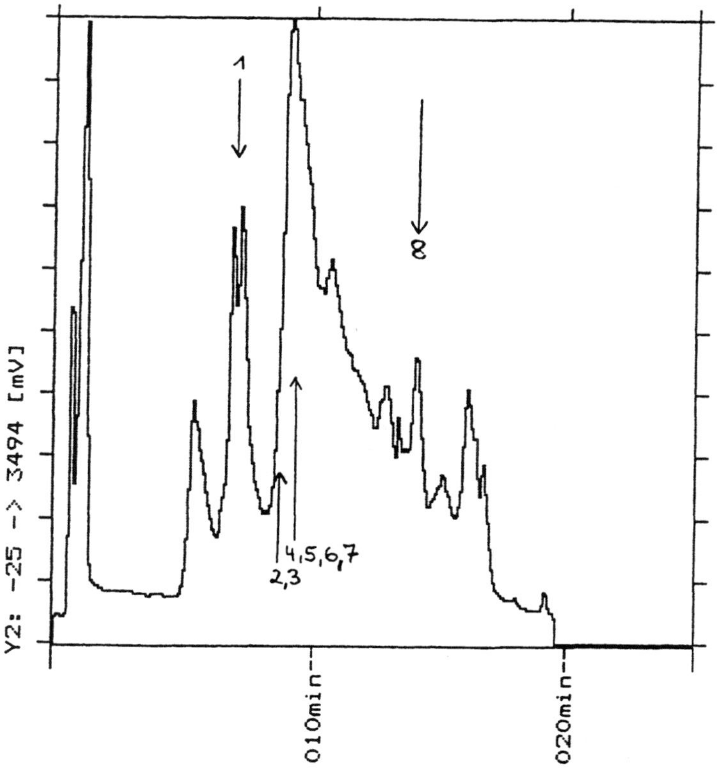

Fig. 2 *Anion exchange chromatography (Mono Q) of human pregnancy serum (10 mg total protein in 500 µl). The column was eluted as described under Experimental. The numbers represent the elution of the following proteins: 1: transferrin; 2: C3-activating protein; 3: SHBG;* 4: α1-acid glycoprotein; 5: α1-antitrypsin; 6: CBG; 7: albumin; 8: ceruloplasmin

compared with those in crude serum by anion exchange and LLPC. Figs. 2 and 3 show the elution of crude serum on a Mono Q anion exchange column and on a column with polyacrylated LiChrospher as support using a Dextran T40/PEG 6000 system. The comparison of the two elution pattern shows that the classical liquid-liquid chromatography in its new version, using aqueous two-phase polymer systems as the poly (ethylene glycol)-dextran-water system of P.-Å. Albertsson [11], should provide a versatile tool because it operates on the basis of the protein partitioning between two aqueous phases.

PBG from new guinea pig was purified by ammonium sulphate precipitation anion exchange chromatography and LLPC (data not shown). Using anion exchange chromatography, we could demonstrate that PBG eluted as a broad peak, overlapping with other proteins, mainly serum albumin. In LLPC, as the final purification step, PBG eluted as a double peak approximately 50 ml before albumin, having molecular weights of 115000 and 80000 in SDS-PAGE.

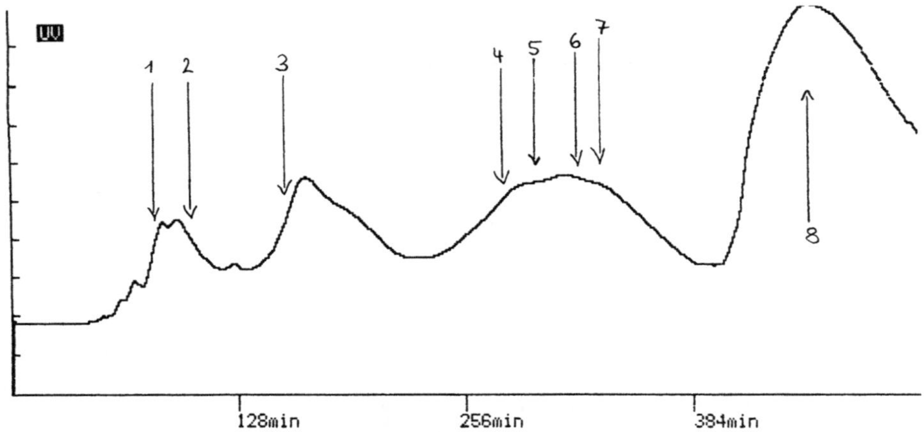

Fig. 3 *LLPC of human serum (30 mg total protein in 1 ml) using polyacrylated LiChrospher 1000 as support. Proteins were eluted with LiParSolv 40/6000 moph containing 10 mM NaH_2PO_4 (pH 7.0), 75 mM KBr at a flow rate of 0.2 ml/min. The numbers represent the following proteins: 1: α1-acid glycoprotein; 2: SHBG; 3: CBG; 4: α1-antitrypsin; 5: C3-activating protein; 6: ceruloplasmin; 7: transferrin; 8: albumin*

The results demonstrate that LLPC presented the only possibility for clearly separating SHBG, h-CBG, r-CBG and PBG from other proteins, mainly albumin, without a loss of binding activity towards steroids and heterogeneity, which is due to the carbohydrate content of the proteins.

ACKNOWLEDGEMENT

This paper contains work that forms part of the PhD Thesis of Michael Juchem at the Johannes Gutenberg University Mainz.

REFERENCES

1 W.H. Daughaday, Evidence for two corticosteroid binding systems in human plasma J. Lab. Clin. Med. 48:799 (1956)
2 C. Mercier-Bodard, A. Alfsen and E.E. Baulieu, in: "Proceedings of the Second Symposium on Steroid Hormones," International Congress Series 101, Exerpta Medica Foundation, Amsterdam (1966)
3 R.M. Burton, G.B. Harding, W.R. Aboul-Hasn, D.T. MacLaughlin and U. Westphal, Progesterone-binding globulin from the serum of pregnant guinea pigs, a polydisperse glycoprotein Biochemistry 13:3554 (1974)
4 R.B. Heap and D.V. Illingworth, The maintenance of gestation in the guinea pig and other hystricomorph rodents: changes in the dynamics of progesterone metabolism and the occurrence progesterone-binding globulin (P.B.G.), Symp. Zool. Soc. Lond. 35:385 (1974)
5 O.H. Lea, Isolation and characterisation of a progesterone- and testosterone-binding globulin from pregnant guinea pig serum, Biochim. Biophys. Acta 317:351 (1973)

6 E. Milgrom, P. Allouch, M. Atger and E.E. Beaulieu, Progesterone-binding plasma protein of pregnant guinea pigs, J. Biol. Chem. 248:1106 (1973)
7 W.J. Raymoure and R.W. Kuhn, Steroid-binding proteins in guinea pig milk and plasma, Endocrinology 106:1747 (1980)
8 A. Heubner, O. Belovsky, W. Müller, H.J. Grill, B. Manz, M. Juchem and K. Pollow, Application of liquid-liquid partition chromatography in the simultaneous purification of sex-hormone-binding globulin and corticosteroid-binding globulin, J. Chromatogr. 397:419 (1987)
9 G. Scatchard, The attractions of proteins for small molecules and iones, Ann. N.Y. Acad. Sci. 51:660 (1949)
10 W. Müller, Eur. J. Biochem. 155:213 (1986)
11 P.-Å. Albertsson, in: "Partition of Cell Particles and Macromolecules," 2nd ed., Almquist and Wicksell, Stockholm (1971)
12 P.H. Petra, Serum sex steroid-binding protein - purification, characterisation and immunological properties of the human and rabbit proteins, J. Steroid Biochem. 11:253 (1979)
13 W. Rosner and R.N. SMith, Isolation and characterisation of the testosterone-estradiol-binding globulin from human plasma. Use of a novel affinity column, Biochemistry 14:4813 (1975)
14 P. Fernlund and C.-B. Laurell, A simple two step procedure for the simultaneous isolation of corticosteroid binding globulin and sex hormone binding globulin from human serum by chromatography on cortisol-sepharose and phenyl-sepharose, J. Steroid Biochem. 14:545 (1981)
15 G.L. Hammon, P.A. Robinson, H. Sugino, D.N. Ward and J. Finne, Physicochemical characteristics of human sex hormone binding globulin: evidence for two identical subunits, J. Steroid Biochem 4:815 (1986)
16 P.H. Petra and J. Lewis, Modification in the purification of the sex steroid-binding protein of human serum by affinity chromatography, Anal. Biochem. 105:165 (1980)
17 C.Y. Cheng, N.A. Musto, G.L. Gunsalus and C.W. Bardin, Demonstration of heavy and light protomers of human testosterone-estradiol binding globulin, J. Steroid Biochem. 4:1379 (1983)
18 I.R. Lee, L.C. Greed, J.D. Wetherall and R. Hähnel, A rapid procedure for the isolation of human sex steroid binding protein, Clin. Chim. Acta 125:255 (1982)
19 M.J. Iqbal and M.J. Johnson, Purification and characterisation of human sex hormone binding globulin, J. Steroid Biochem. 10:535 (1979)
20 K.E. Mickelson, G.B. Harding, M. Forsthoefel and U. Westphal, Steroid-protein interactions. Human corticosteroid-binding globulin: characterisation of dimer and electrophoretic variants, Biochemistry 21:654 (1982)

COLUMN BASED LIQUID/LIQUID SEPARATION OF CELLS USING AQUEOUS POLYMERIC TWO-PHASE SYSTEMS

D.R. Skuse and D.E. Brooks

Departments of Chemistry and Pathology
University of British Columbia
2211 Wesbrook Mall, Vancouver, B.C., Canada V6T 1W5

Column-based aqueous polymeric two-phase systems were used to effect separation of dog and human erythrocytes on the basis of their interactions with a PEG-oleate affinity ligand. The resolution obtained was substantially higher than would be expected in a corresponding single tube experiment. Cell recoveries were also very high. However, elution volumes were independent of PEG-oleate concentration indicating that multiple partition steps, analagous to CCD, did not occur.

INTRODUCTION

The suitability of aqueous polymeric two-phase systems for the separation of cells is well established [1,2]. When it is desirable to separate cell types having similar partition coefficients, a single partition step does not give an efficient separation. In these cases it is necessary to use a multi-stage procedure to improve the resolution. This has most usually been done using a countercurrent distribution (CCD) technique, in which a large number of extraction steps are carried out in sequence [1]. As an alternative, for macromolecular separations, Müller developed a column-based technique in which the dextran-rich phase of a dextran/poly(ethylene glycol) (PEG) phase system is immobilised onto chromatography beads derivatised with polyacrylamide. The PEG-rich phase is used as the eluent [3]. Müller showed that this technique gave a considerable increase in resolution and saving in time over CCD.

We have adapted Müller's technique and investigated its applicability to the separation of cells. As a model we have investigated the separation of erythrocytes from humans and dogs, this being a well documented [4] and easily performed single tube separation. The separation is based on the differing binding affinities of erythrocytes for a ligand consisting of a PEG head group esterified to a fatty acid. The species-dependent ability of PEG-fatty acids to bind to erythrocytes has been attributed to hydrophobic interactions between the fatty acid of the affinity ligand and membrane lipid modified by interactions with sphingomyelin and phosphatidylcholine head groups [4]. Dog erythrocytes have high affinity for the ligands and human erythrocytes low. In phase systems formed by mixing dextran T40 and PEG 8000 and suitable buffers, in the absence of affinity ligand both dog and human erythrocytes accumulate at the interface between the phases [5]. The addition of PEG-oleate, which partitions strongly into the upper PEG-rich phase, causes the erythrocytes to re-partition into the upper PEG-rich phase. By controlling the concentration of affinity ligand it is possible to separate dog and human erythrocytes between the interface and the upper phase to varying degrees.

In a previous communication [5] we have shown that a mixture of fresh human and dog erythrocytes could be separated using a column-based technique. Columns consisted of agarose beads (diameter ~250 μm) with a bound layer of dextran-rich phase. Cell mixtures were loaded and incubated to allow partition onto the column. Non-partitioned cells were eluted with PEG-rich phase. Dog erythrocytes were eluted by including PEG-oleate into the eluent. Human erythrocytes were then recovered by washing the beads with phosphate buffered saline (PBS). Purities >75% were obtained. Cell recoveries were ~85% of the cells loaded.

In the present paper we report in more detail on the partition behaviour of fixed dog and human erythrocytes on columns consisting of polyacrylamide derivatised silicic chromatography beads (diameter ~65 M) with bound layers of the dextran-rich phase of the same phase system. PEG-rich phase was used as the eluent. Cells were pre-equilibrated with the appropriate concentration of PEG-oleate in PEG-rich phase and eluted isocratically with no incubation period. Elution profiles and cell recovery data for dog and human erythrocytes were obtained independently as a function of PEG-oleate concentration. This information was then used to effect separation of a dog/human erythrocyte mixture.

MATERIALS AND METHODS

Preparation of Two-Phase Systems

A phase system containing 3% dextran T40 (Pharmacia, Dorval, Quebec, Canada), 7.7% PEG 8000 (Union Carbide, Century Grade, Piscataway, New Jersey, USA), 0.079 M dibasic sodium phosphate, 0.025 M monobasic sodium phosphate and 0.006 M sodium azide (pH 7.2) was used. The phase system was mixed well and allowed to settle for 72 hr before the phases were separated. The system was prepared and all experiments carried out at 22°C.

Preparation of Chromatography Beads

Fractogel TSK HW-65F (BDH, Vancouver, British Columbia, Canada) silicic chromatography beads were used. The beads were prepared for reaction by the following wash cycle: three washes with 50 cm^3 aliquots of distilled water; one wash with a 50 cm^3 aliquot of a solution containing 0.2 M sodium acetate and 0.001 M trisodium ethylene diaminetetraaacetic acid monohydrate (EDTA); four washes with 50 cm^3 aliquots of distilled water.

The reaction vessel was a 500 cm^3 three-neck round-bottom flask. Neck one held a nitrogen inlet, neck two a reflux condenser topped with an N_2 outlet via a paraffin bubbler, and neck three a stoppered separating funnel with a pressure equalising arm.

Washed beads (50 cm^3) and an acrylamide solution (40 g in 200 g distilled water) were placed in the reaction vessel and 0.2 M cerium IV ammonium nitrate ($CeIV(NH_4)_2(NO_3)_6$) in 1 M nitric acid (11.6 cm^3) was placed in the separating funnel. The apparatus was degassed, by bubbling nitrogen with magnetic stirring, for one hour. The reaction was initiated by adding the cerium IV solution to the acrylamide and beads, and allowed to proceed with magnetic stirring under nitrogen flow. After three hours the reaction was stopped by exposing the reaction mixture to air. The beads were washed with twenty, 50 cm^3 aliquots of distilled water and stored at 4°C in 0.006 M sodium azide.

Preparation of Erythrocytes

Fresh blood was collected into citrate or heparin from human and dog donors, respectively. The erythrocytes were washed three times in 10 volumes of PBS.

Packed dog erythrocytes (approx. 0.2 cm_3) were incubated with 0.1 cm^3 of [^{51}Cr] sodium chromate (3.7 x 10^7 Bq cm^{-3}) (Amersham, Montreal, Quebec, Canada) and PBS (0.2 cm^3) at 37°C for 1 hr. The cells were washed with five 1 cm^3 aliquots of 1% glutaraldehyde in PBS and fixed by rotating end-over-end at 30 rpm for 20 hr in 1% glutaraldehyde in PBS (5 cm^3).

Human erythrocytes were labelled in similar fashion: packed erythrocytes (approx. 0.2 cm^3) were rotated end-over-end at 10 rpm for 1 hr with 0.001 cm^3 of [^{125}I] potassium iodide (3.7 x 10^9 Bq cm^{-3}) (Amersham), four Iodobeads (Pierce, Rockford, Illinois, USA) and PBS (0.5 cm^3). The cells were washed with five, 1 cm^3 aliquots of 1% glutaraldehyde in PBS (5 cm ^{3}) for 20 hr.

Fixed but not radio-labelled dog and human erythrocytes were also prepared: packed cells (1 cm^3) in 1% glutaraldehyde in PBS (10 cm^3) were rotated end-over-end at 30 rpm for 20 hr.

Radio-labelled and unlabelled cells were blended to yield stock cell suspensions containing 1 x 10^8cells cm^{-3} and 2 x 10^5 Bq cm^{-3} in 1% glutaraldehyde in PBS.

Cell concentrations were determined by impedance cell counting (Electrozone Celloscope, Particle Data Incorporated, Elmhurst, Illinois, USA). Activities were measured using a gamma counter (1282 Computagama, LKB, Bromma, Sweden).

Load mixes for column experiments (suspensions of cells in PEG-rich phase) were prepared from the above stocks by mixing dog and/or human erythrocytes as required, and PEG-rich phase containing the appropriate concentration of PEG-oleate, such that the final cell concentration was 1 x 10^8 cm^{-3} (5 x 10^7 cm^{-3} in each cell type for separation experiments).

Preparation of columns

Derivatised beads (20-25 cm^3) were washed with three 50 cm^3 aliquots of a solution containing 0.79 M dibasic sodium phosphate, 0.025 M monobasic sodium phosphate and 0.006 M sodium azide (pH 7.2) and one 25 cm^3 aliquot of dextran-rich phase. The beads were then resuspended in dextran-rich phase (50 cm^3) and rotated end-over-end at 30 rpm for 48 hr. Columns of the resultant dextran-rich phase covered beads were packed at flow rates of 6 cm^3hr^{-1} to give beads of 1.0 cm x 0.7 cm^2. Each column was used only once.

Chromatographic procedure

Aliquots of the above described erythrocyte load mixes were loaded onto the column and eluted isocratically at 6 cm^3hr^{-1} (the optimum flow rate for maximum cell recovery) for 2 hr. The eluent was collected in 1.2 cm^3 (12 min) aliquots using an automatic fraction collector. Cells remaining on the column at the end of this period were removed by washing with PBS. Erythrocyte concentrations were determined using the gamma counter. The percentage of the total of each type of cell loaded present in each aliquot was calculated.

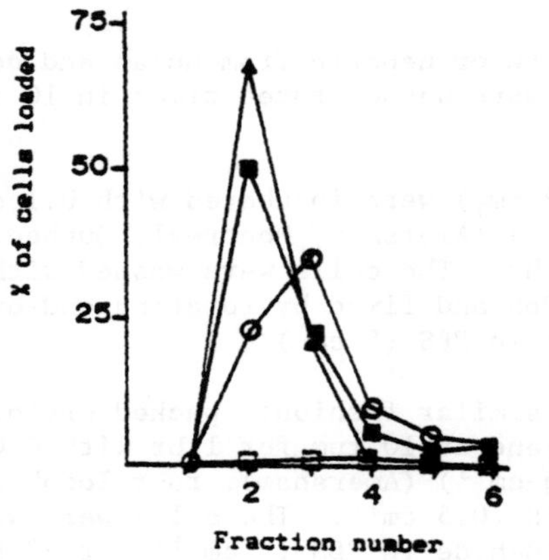

Fig. 1 *Elution profile of dog erythrocytes from an aqueous liquid/liquid phase partitioning column at various PEG-oleate concentrations:* ▲ *2.5×10^{-6} M,* ■ *1.2×10^{-6} M,* ○ *0.6×10^{-6} M,* □ *0.3×10^{-6} M*

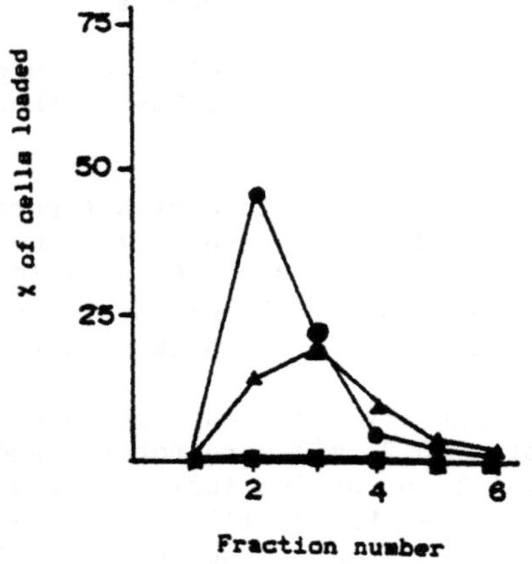

Fig. 2 *Elution profile of human erythrocytes from an aqueous liquid/liquid phase partitioning column at various PEG-oleate concentrations:* ● *5×10^{-6} M,* ▲ *2.5×10^{-6} M,* ■ *1.2×10^{-6} M*

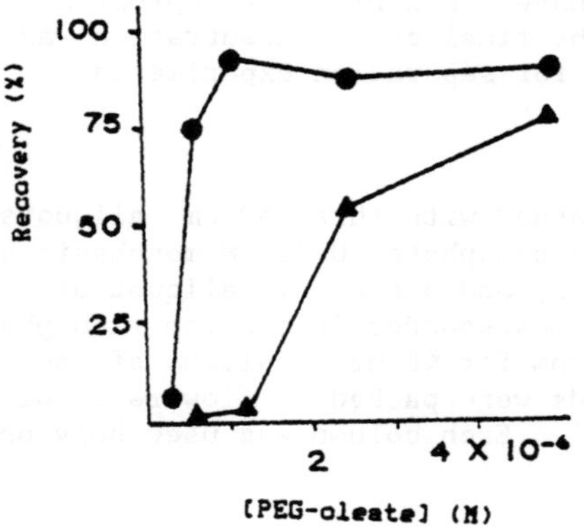

Fig. 3 *Recovery of dog,* ● *, and human,* ▲ *, erythrocytes from an aqueous liquid/liquid phase partitioning column at various PEG-oleate concentrations*

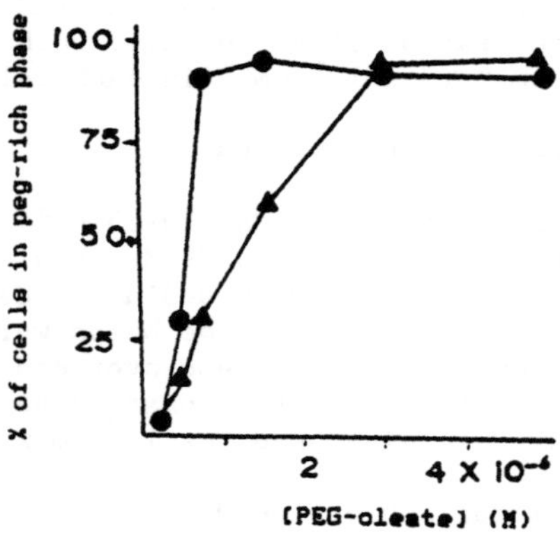

Fig. 4 *Single tube partitioning behaviour of dog,* ● *, and human,* ▲ *, erythrocytes at various PEG-oleate concentrations*

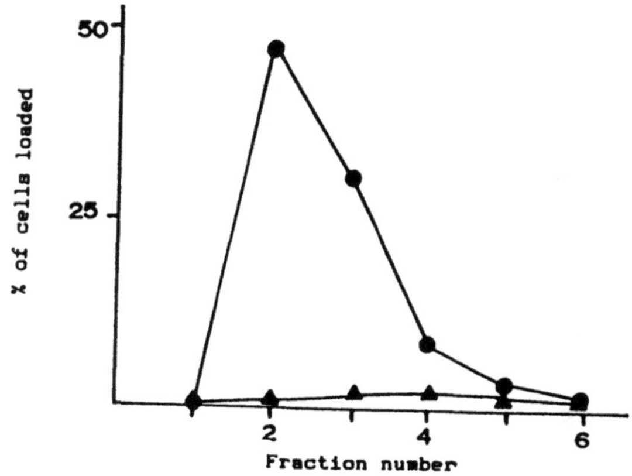

Fig. 5 *Elution profile of dog,* ● *, and human,* ▲ *, erythrocytes from an aqueous liquid/liquid phase partitioning column at PEG-oleate concentrations* 1.25×10^{-6} *M*

Single Tube Experiments

The experimental procedure has been described in detail elsewhere [5,6].

RESULTS AND DISCUSSION

Elution profiles for dog (Fig. 1) and human (Fig. 2) erythrocytes and cell recovery data (Fig. 3) were obtained independently as a function of PEG-oleate concentration. These figures suggest the following:

i) There are critical PEG-oleate concentrations below which no cell elution is observed. The critical concentrations are 0.3×10^{-6} *M* for dog erythrocytes and 1.25×10^{-6} *M* for human erythrocytes.
ii) Above these critical concentrations, cell recovery increases with PEG-oleate concentration. Recoveries can be as high as ~90% of the cells loaded for dog and ~80% for human erythrocytes.
iii) Elution volume is only proportional to PEG-oleate concentration at concentrations which give poor recoveries. Thus it seems that the behaviour of the cells in practically useful column-based phase systems varies from that described by Müller for macromolecules in that multistep partitioning does not occur. This suggests that, for cells, an analogy with CCD is inappropriate.

However, comparison of the human and dog erythrocyte recovery versus PEG-oleate concentration profiles (Fig. 3) with the percentage PEG-rich phase partition of human and dog erythrocytes versus PEG-oleate concentration profile (Fig. 4) suggests that column-based separation of dog and human erythrocytes should have a higher resolution than corresponding single tube experiments.

Column-based separations in the PEG-oleate concentration range 1.2 - 3.5 $\times 10^{-6}$ *M* were carried out. As expected the best separation occurred at 1.2×10^{-6} *M* PEG-oleate (Fig. 5), where 92% of the dog erythrocytes were eluted with only 7% of the human cells. The purest fraction contained 49% of the dog cells loaded and only 0.4% of the human. This separation is considerably better than would be expected from a corresponding single tube separation.

CONCLUSIONS

A column-based technique can be used to separate human and dog erythrocytes from a mixture on the basis of their differential interactions with PEG-oleate affinity ligand. The resolution obtained in such a separation is substantially higher than would be obtained in a corresponding single tube experiment. Total recoveries of cells are also very high.

However, column-based cell partitioning is not analagous to CCD. There is no change in elution volume with change in PEG-oleate concentration which suggests that multistep partitioning does not occur.

ACKNOWLEDGEMENTS

The financial support of the Medical Research Council of Canada is gratefully acknowledged.

REFERENCES

1 P.-Å. Albertsson, in: "Partition of Cell Particles and Macromolecules," Wiley Interscience, New York (1971)
2 H. Walter, D.E. Brooks and D. Fisher, eds., "Partitioning in Aqueous Two-Phase Systems," Academic Press, Florida (1986)
3 W. Müller, New phase supports for liquid-liquid partition chromatography of biopolymers in aqueous poly(ethylene glycol)-dextran systems. Synthesis and application for the fractionation of DNA restriction fragments, Eur. J. Biochem. 155:213 (1986)
4 E. Ericksson, P.-Å. Albertsson and G. Johansson, Hydrophobic surface partition of erythrocytes studied by affinity partition in aqueous two-phase systems, Molecular and Cellular Biochemistry 10:123 (1976)
5 D.R. Skuse, W. Müller and D.E. Brooks, Column chromatographic separation of cells using aqueous polymer two-phase systems, Analytical Biochemistry, submitted for publication
6 K.A. Sharp, M. Yalpani, S.J. Howard and D.E. Brooks, Analytical Biochemistry, 154:110 (1986)

A NEW APPROACH TO COUNTERCURRENT DISTRIBUTION COMBINING SEPARATION WITH ANALYSIS IN THE BIOMEK AUTOMATED LABORATORY WORKSTATION

Ian Sutherland [1], Somaiya Siddiqi [1], Richard Keightley [2] and Derek Fisher [3]

[1] Engineering Department
National Institute for Medical Research
Mill Hill, London NW7 1AA, UK

[2] Beckman (UK) Ltd
High Wycombe, Bucks HP12 4JL, UK

[3] Department of Biochemistry
Royal Free Hospital Medical School
Rowland Hill Street, London NW3 2PF, UK

A method of programming the Biomek Automated Laboratory Workstation has been developed to perform countercurrent distribution separations of both solutes and cells using aqueous two-phase systems. Studies involving an 11-transfer CCD show good correlation between experimental and theoretical distributions. The separation of high and low partition coefficient cells have shown that the Biomek's mixing and transfer regimes work well for both particulates and solubles. These results could form the basis of a new approach to the separation of viable biological material and extend the range of an already versatile piece of laboratory equipment.

INTRODUCTION

The Beckman Biomek 1000 Automated Laboratory Workstation (Fig. 1) is designed for applications involving rapid, accurate, and reproducible liquid handling operations [1]. The Biomek has a range of pipetting and dispensing tools, fluid reservoirs, dispensing trays and tube racks, as well as means for taking multiple absorbance measurements with a special optical density accessory. It is widely used for the enzyme linked immunosorbent assay (ELISA) for qualitative and quantitative determinations of antibody and antigen and the production of monoclonal antibodies through hybridoma techniques [2]. Both of these techniques have a series of labour intensive, pipetting, dispensing and transfer operations which are most suited to automation.

The adaptability and ease of programming of the Biomek make it a suitable candidate for replacing other labour intensive laboratory procedures. One such procedure which is widely used in the separation technology field is

Fig. 1 *The Biomek Automated Laboratory Workstation*

countercurrent extraction or distribution (CCD). Widely used in the 1950's for the purification of natural products using aqueous/organic phase systems [3], it has largely given way today to more continuous processes, such as HPLC and CCC [4]. However, one of its major applications today is in the separation and purification of viable biological material by partitioning in the aqueous two-phase polymer systems developed by Albertsson [5]. This technology is becoming increasingly used for cell and organelle separations on an analytical scale [6,7] and has recently been promoted for the large scale extraction of proteins and enzymes [8]. Separations using polymer phase systems can be based on surface charge, hydrophobicity [7] or ligand affinity [9]. Often adequate separation can be achieved in a single extraction step, but more frequently multiple-extraction or countercurrent distribution is necessary.

While there has been some success in transferring this technology to liquid-liquid chromatography systems [10] and column chromatography [11], there is no doubt that the simplicity and theoretical predictability of countercurrent distribution make it a prime candidate for future separation and analysis if suitable methodology can be found that will appeal to modern laboratory practices. Currently the methodology is limited to 100 transfer thin layer apparati [12] which are quite labour intensive and require full-time technical support to operate. Furthermore, such systems are not readily available commercially, which is limiting the more widespread use of the technique.

In contrast, the Biomek is already available commercially and has a range of applications for the modern laboratory. Its versatility would be further enhanced if it could be used as a sensitive separation and analysis system as well.

This paper examines the feasibility of using the Biomek to perform CCD using aqueous two-phase systems and identifies potential areas of further development which could turn the Biomek into an extremely versatile separation facility.

ADAPT-CCD

We have chosen to adopt the term ADAPT-CCD for three reasons. Firstly, it draws attention to the "adaptability" of the Biomek systems; secondly, depending on the particular separation protocol required, the system "ADAPTS" to suit it and, thirdly, the mnemonic "ADAPTS" stands for "Automated Dispensing And Pipetting Transfer System" which incorporates all the essential ingredients of this approach to CCD.

The process is fully "Automated" so that once the phase systems and samples have been placed in the appropriate reservoirs, it can be left to do the separation and analysis unattended. It uses its "dispensing" mode to meter out the appropriate quantities of the phase systems in each tube; its "Dispensing and Pipetting" mode to achieve the CCD mixing operation; its "Pipetting and Transfer" mode to transfer one of the phases from one row of tubes to the next and its time-out facility to achieve the appropriate settling times. In addition, its compact size will allow CCD to be performed in a laminar flow cabinet (i.e. a sterile environment), all components being either autoclavable or sterilised disposables.

METHODOLOGY

All feasibility studies were performed in a single 8 x 12 tube rack, the tube dimensions being 40mm x 8mm with a total capacity of 1 ml.

Preparation of Phase Systems

The phase systems used consisted of Dextran T500 (4.75% w/w), PEG 6000 (4.75% w/w) in 0.11 M sodium phosphate buffer, pH 6.8. The phase system was mixed and allowed to settle at 25°C for 24 hours in a separating funnel. Upper PEG-rich phase and lower dextran-rich phase were drawn off and 35 ml of each phase placed in two of the Biomek reservoirs. Eight samples, made up according to the experimental protocol outlined below, were pipetted into column 12. Once these simple operations were performed the Biomek was ready to start its ADAPT-CCD programme which is outlined schematically in Fig. 2.

Preparation of Samples

Soluble Dyes :- two dyes were used, congo-red (1 mg/ml) and neutral-red (1.8 mg/ml) with partition coefficients of 5.46 and 1.46 respectively. Samples for CCD were made up in the upper PEG-rich phase, with four samples of congo-red and four of neutral-red in each run.

Mammalian erythrocytes :- rat, human and rabbit erythrocytes were used, fixed with 1% v/v glutaraldehyde (Ladd Chemical Co., Burlington, Vermont). Samples were washed three times in 0.11 M sodium phosphate buffer (NaPB), spun down in a Beckman microfuge and resuspended. Samples for CCD were prepared by diluting 20 μl of packed cells with 700 μl of PEG-rich upper phase. Eight separate samples were prepared for each CCD run comprising:

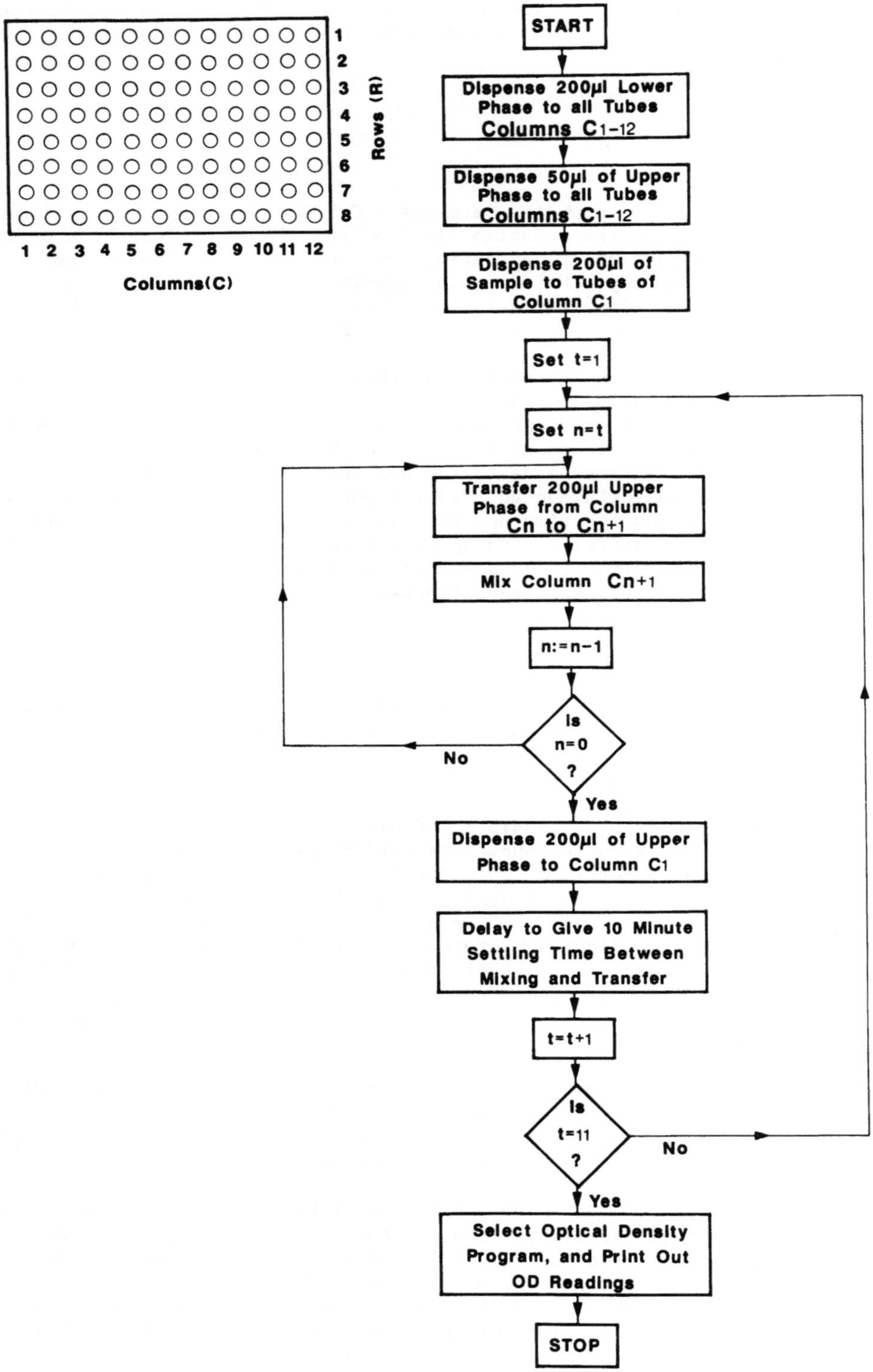

Fig. 2 *Schematic outline of the Biomek ADAPT-CCD programme*

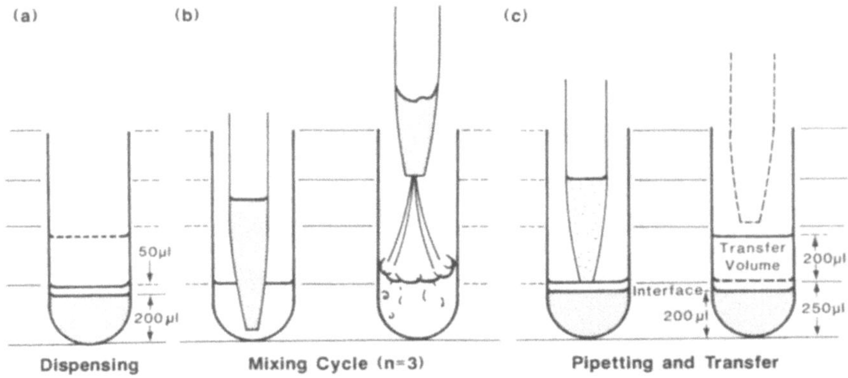

Fig. 3 *The dispensing, mixing and transfer protocol for the Biomek ADAPT-CCD system*

2 rat, 2 human, 2 rabbit, and mixtures of rat and human and rabbit and human erythrocytes.

Countercurrent Distribution using the Biomek

Lower (dextran-rich) phase (200 µl) is initially dispensed into all the tubes using the multiple pipetting head (eight channels) followed by 50 µl of upper phase (Fig. 3a). Eight separate 200 µl samples are then transferred from column 12 to the first column of tubes, where they are mixed using three successive pipetting and dispensing operations (Fig. 3b).

Following a 10 minute settling period, 200 µl of the upper phase is transferred from the first to the second column of tubes (Fig. 3c) where the mixing cycle is repeated. Fresh PEG-rich upper phase (200 µl) is then dispensed into the first column, and the mixing process repeated. This cycle of mixing, settling and transfer, while not simultaneous is arranged to leave identical settling times following the mixing of each column of tubes. After 11 such mixing/transfer cycles, an optical density programme is selected which successively mixes and samples from each tube, transferring respective samples to a microtitre plate where they are sequentially diluted and mixed before being automatically scanned for optical density at selected wavelengths (in this case 540 nm).

Analysis of Fractions

Dye fractions were analysed automatically using the automated optical density facilities of the Biomek. Cell fractions were analysed using a Coulter Counter (model 2B1) by first breaking the phases by mixing 0.46 ml of Isoton with the contents of each partition tube. Aliquots of 0.2 ml were then taken from each tube, and diluted with a further 10 ml of Isoton for each Coulter count. Mixtures of erythrocytes were analysed using a channelyser (Coulter Electronics Ltd) to discriminate between the two cell populations.

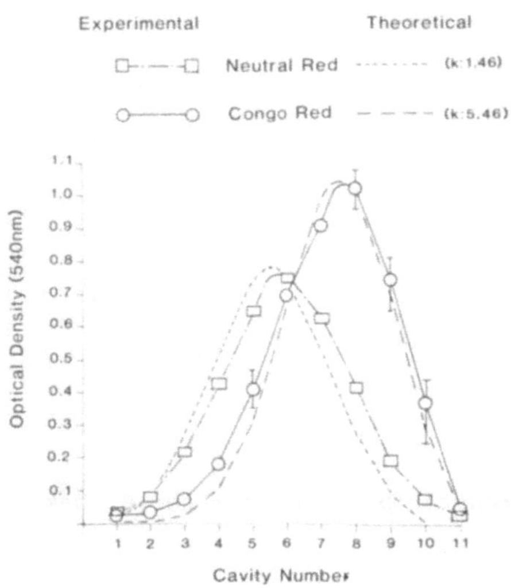

Fig. 4 *Theoretical and experimental distribution profiles for two soluble dyes*

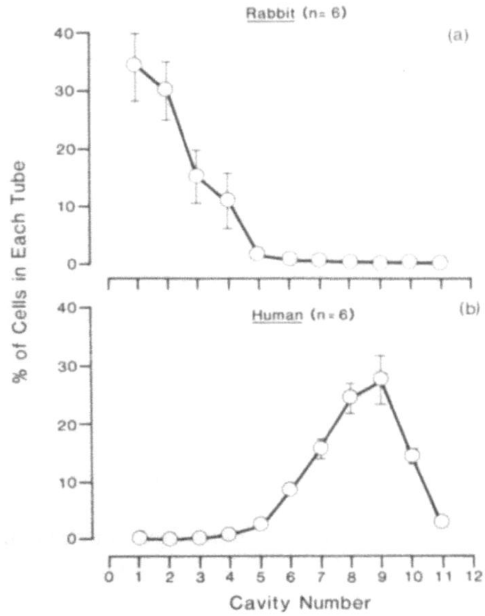

Fig. 5 *Distribution profiles for a) rabbit and b) human erythrocytes run separately*

RESULTS

Partitioning of Soluble Dyes

The distribution profiles at an optical density of 540 mm are shown in Fig. 4 for two dyes of differing partition coefficient following an 11-transfer CCD. The experimental distributions show the mean and standard deviation for congo-red (n=4) and neutral-red (n=4) and are compared with theoretical distributions based on independently measured (n=5) partition coefficients (k=5.5 and 1.5 respectively).

Partitioning of Fixed Erythrocytes

Distribution profiles from 11-transfer CCD runs on the Biomek Automated Laboratory Workstation are shown for rabbit and human cells run separately in Fig. 5a and 5b, and together in Fig. 6. The results in Fig. 5 are the mean and standard deviation of six separate distribution profiles for these separate runs. The distribution profiles of the mixture of rabbit and human erythrocytes for the three runs are shown separately in Fig. 6 as it was difficult to maintain the same ratio of human/rabbit cells for each separate run. Distribution profiles for rat erythrocytes and mixtures of rat and human erythrocytes were also obtained but are not shown as the rat profiles were too similar to the human ones in this phase system to portray any significantly different information.

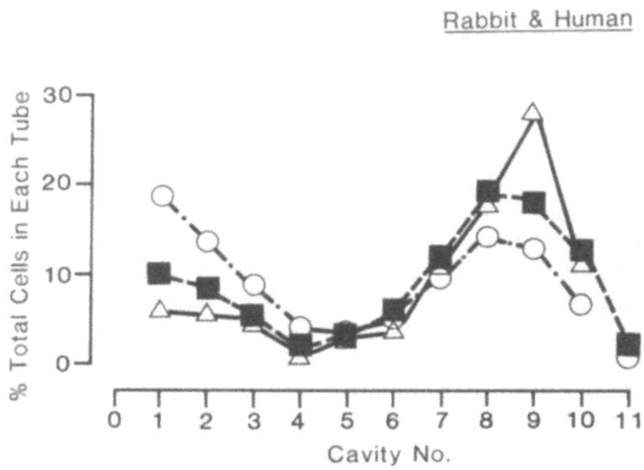

Fig. 6 *Distribution profiles of a mixture of rabbit and human erythrocytes*

DISCUSSION

The experimental results using dyes show good Gaussian distributions when compared to the theoretical profiles and the low standard deviations indicate excellent reproducibility. The experimental curves tend to "lead" the theoretical ones, suggesting that some lower phase transfer may be occurring along the chain. A volume analysis of the amount of upper and lower phase in each tube along the transfer chain suggested a successive loss from the first tube of approximately 5 μl of lower phase per transfer. As transfers progressed a successive loss of upper phase was also noted in the leading transfer tube. These accumulative transfer errors can however be compensated for by programme modification.

One concern when first programming the Biomek was that interface material would be sucked up during the transfer process. The distribution profile in Fig. 5a for the rabbit cells, that partition predominantly toward the interface, suggests that this is not occurring. It should be noted that a 50 μl volume (2 mm vertical height) of upper phase is left between the bottom of the pipette and the interface during transfer to safeguard against this effect. We have shown that if this distance is less than 2mm interface material can be transferred.

Another concern was whether adequate mixing could be achieved using the standard dispensing and pipetting actions of the Biomek. Tests using dyes showed that taking up 200 μl from the bottom of each tube (often containing the more viscous lower phase) and then re-dispensing from the top of the tube, would create a homogeneous mix within three cycles.

The successful programming of the Biomek to perform 11-transfer CCD for both soluble and particulate material is encouraging for many reasons. The inherent adaptability of the process make it suitable for a range of other related operations as outlined schematically in Fig. 7. We have just outlined the feasibility of using the system for a limited number of transfers using a single 8 x 12 tube rack. It should now be possible to extend the versatility of the process by doing multiple transfers of fewer samples, namely 4 x 24T or 1 x 96T. The major limitation of the process is not so much the number of transfers, but the number of successive mixing, settling and transfer steps that can be fitted into the settling period.

We have estimated that between 15-24 stages are possible with settling times in the order of 10-15 mins. This would mean that 8 x 24T CCD would be possible using two tube racks and settling times in the order of 15 mins. More transfers would require less samples and an intermediate single head transfer stage as outlined in Fig. 8. Here a 4 x 24T CCD system is illustrated, using a single 12x8 tube rack, where the four upper phase transfers from column 12 row 1-4 are transferred using the single-tip pipetting tool to column 1 rows 5-8, while all other transfers are performed using the multi-tip (n=8) pipetting tool. Interestingly, shorter settling times with relatively incomplete phase separation, may provide improved separations [13].

Perhaps one of the most exciting prospects is an automated screening programme for unknown material that would set up a matrix varying polymer concentration in one direction and surface charge, hydrophobicity or ligand concentration in the other. Such a programme could dispense from stock solutions, perform one step mixing, settling and transfers before sampling both upper and lower phase to obtain partition coefficients. A typical scheme varying polymer and salt concentrations is shown in Fig. 9. Stock solutions of 20% dextran T500, 40% PEG 6000, 0.5 M NaCl, 0.5 M NaPB, distilled water and sample would be held in the reservoir system and a program selected to dispense the appropriate phase systems in a 6 x 8 array of tubes such that polymer concentration varied by row and salt concentration by column. Each row of tubes would be successively mixed and the upper phases transferred to columns 7-12 once the phases had settled. The phase system would then be successively diluted to break the phases and then scanned at the appropriate wavelength to obtain optical density readings of upper and lower phase, and a subsequent 2-dimensional partition map.

The availability of such "easy to use" protocols for screening new products, identifying the appropriate phase systems and then performing the appropriate fractionations could well provide the framework for this new technology to develop. By encouraging the use of stock solutions of polymers, a market environment will be created leading to their standardisation and routine production. The recent development of highly specific separation methodology such as immunoaffinity partitioning [14,15] highlights the need for fast limited transfer CCD techniques. The use of the Biomek could therefore take two-phase aqueous extraction into routine laboratory and industrial use in the future.

REFERENCES

1 R. Keightley, Principles of automated diluting and pipetting procedures, Medical Technologist & Scientist (November 1986)
2 E.W. Stewart and N.C. Appleby, A flexible integrated system for automation of hybridoma and ELISA procedures, American Biotechnology Laboratory (September/October 1986)
3 L.C. Craig and H.O. Post, Apparatus for countercurrent distribution, Analytical Chemistry, 21:500 (1949)
4 I.A. Sutherland, Countercurrent chromatography, Laboratory Practice (February 1987)
5 P.-Å. Albertsson, Partition of cell particles and macromolecules, 2nd Edition, Wiley Interscience, New York (1971)
6 P.-Å. Albertsson, Separation of particles and macromolecules by phase partition, Endeavour 1:69 (1977)
7 C. Larsson, in: "Isolation of membranes and organelles from plant cells," J.L. Hall and A.L. Moore, eds., Academic Press, London (1983)

8 H. Hustedt, K.H. Kroner and M.-R. Kula, Applications of phase partitioning in biotechnology, in: "Partitioning in aqueous two-phase systems," H. Walter, D.E. Brooks and D. Fisher, eds., Academic Press, London New York (1985)
9 S.D. Flanagan, P. Taylor and S.H. Barondes, Affinity partitioning of acetylcholine-receptor enriched membranes and their purification, Nature (London) 254:441 (1975)
10 I.A. Sutherland, D. Heywood-Waddington and Y. Ito, Countercurrent chromatography, J. Chromatog. 384:197 (1987)
11 W. Müller, Application of PEG-Dx systems to liquid-liquid partition chromatography of proteins and nucleic acids, ibid : ?? (1988)
12 P.-Å. Albertsson, Thin layer countercurrent distribution, Anal. Biochem. 11:121 (1965)
13 L.J. Karr, S.G. Schafer, J.M. Harris, J.M. Van Alstine and R.S. Snyder, Immuno-affinity partition of cells in aqueous polymer two-phase systems, J. Chromatog. 354:269 (1986)
14 K.A. Sharp, M. Yalpani, S.J. Howard and D.E. Brooks, Synthesis and application of a poly(ethylene glycol)-antibody affinity ligand for cell separations in aqueous polymer two-phase sysems, Anal. Biochem. 154:110 (1986)
15 D. Fisher and H. Walter, Cell separations and subfractionations by countercurrent distribution in two-polymer aqueous phase systems depend on non-equilibrium conditions, Biochim. Biophys. Acta 801:106 (1984)

ELUTION COUNTERCURRENT DISTRIBUTION

Eva Hultin and Per-Åke Albertsson

Department of Biochemistry, Chemical Center
University of Lund
P. O. Box 124, Lund S-221 00, Sweden

SUMMARY

A thin-layer countercurrent distribution apparatus was modified so that it can be utilised for the single withdrawal technique [1], here called elution countercurrent distribution.

Thylakoid membrane vesicles from spinach chloroplasts were subjected to the elution countercurrent distribution. A gradient of NaCl was used in the eluting upper phase to elute the different fractions in the order of their affinity for the moving upper phase.

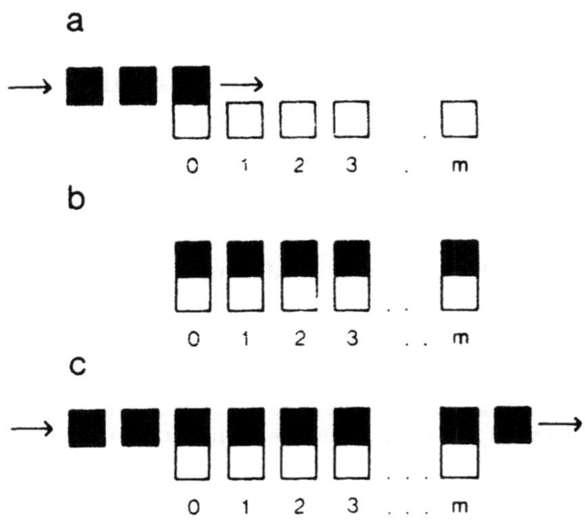

Fig. 1 *The fundamental process in comparison with elution countercurrent distribution, schematically.*
a) Start of both the fundamental process and the elution countercurrent distribution.
b) Completion of the fundamental process.
c) Elution countercurrent distribution after m+1 transfers.

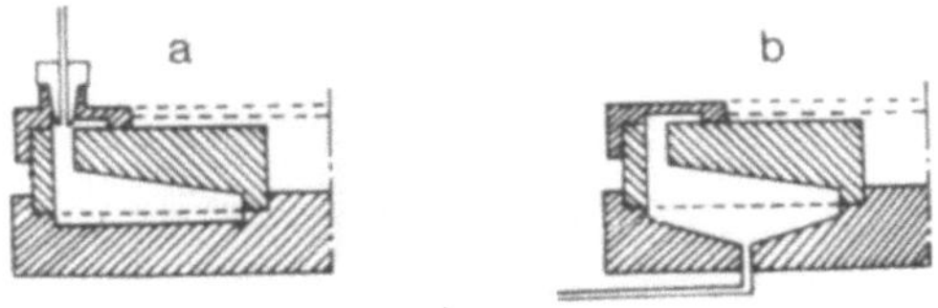

Fig. 2 *The modifications of a thin-layer countercurrent distribution apparatus necessary for elution countercurrent distribution. For explanation see text*

INTRODUCTION

In the conventional countercurrent distribution (the fundamental process) the number of transfers is limited to the number of partition cavities. Thus, if the number of partition cavities is m+1 (numbered 0,1,...,m), the number of transfers is m (Fig. 1). In the elution countercurrent distribution the number of transfers is unlimited; after m transfers the moving upper phase is withdrawn from the extraction train and fresh phase is introduced in cavity number 0 (Fig. 1). Substances will leave cavity number m in the order of their partition coefficients, those having higher partition coefficients eluting first. The fact that fresh upper phase is introduced in cavity number 0 after each transfer makes the elution countercurrent distribution technique ideal for gradient elution.

MATERIALS AND METHODS

Apparatus. A thin-layer countercurrent distribution apparatus [2] with 60 cavities was modified for the purpose of the elution countercurrent distribution technique.

1 A hole was made in the cover above cavity 0 to allow addition of an aliquot of upper phase into this cavity [Fig. 2a].
2 Lower cavity 56 was modified as shown by the vertical cross section in Fig. 2b to allow removal of each upper phase transferred to this cavity.
3 A hole was made in the cover above cavity 57 to allow addition of an aliquot of pure upper phase into this cavity to wash away remaining phase.
4 A hole was made in lower cavity 58 to allow withdrawal of wash phase.

Phase system. The phase system used contained 5.9% (w/w) Dextran T500, 5.9% (w/w) poly(ethylene glycol), 10 mM sodium phosphate buffer (pH 7.4), 5 mM NaCl and 20 mM sucrose.

Elution countercurrent distribution of sulfuric acid. In order to test the apparatus, sulfuric acid was used as a model substance. Sulfuric acid was included in cavity 3. The procedure continued until all the sulfuric acid had emerged with the effluent. After the run aliquots from each withdrawn fraction were titrated with NaOH. A comparison of the experimental curve with the theoretical one showed good agreement.

Preparation of inside-out vesicles. Inside-out vesicles from spinach chloroplasts (B3) were obtained by press treatment and phase partition as described previously [3].

Sonication procedure. A phase system (10 ml), including the B3 vesicles and also containing 1 mM $MgCl_2$ was sonicated in a Branson B30 sonifier equipped with a 3/4 in. tip for 2x30 s under continuous cooling. The ultrasonic exposure had an intensity output of 5 with 20% duty pulses.

Elution countercurrent distribution of membrane vesicles. Cavities numbered 0-2 and 8-59 were loaded with phase system. The sample system containing the membrane vesicles was loaded in cavities 3-7. The settling time was 8 min and the shaking time was 30 s. To create a salt gradient, the aliquot added to cavity 0 was altered after 20 transfers to a phase with a reduced concentration of NaCl (2.5 mM). 139 transfers were carried out and after the run the fractions were diluted with buffer. The fractions were analysed for absorbance at 680 nm (relative thylakoid vesicle content) and chlorophyll a/b ratio according to Arnon [4].

RESULTS AND DISCUSSION

Inside-out thylakoid vesicles were resolved into two major and several minor fractions, all having different chlorophyll a/b ratios. Sonicated inside-out thlakoid vesicles were resolved into at least five different fractions with different chlorophyll a/b ratios. The fact that distinct peaks are obtained indicates that the photosynthetic membrane is organised in discrete domains. Such good resolutions were not obtained earlier when the fundamental process was used. The advantages compared to the fundamental process are that:

- Many more transfers can be carried out with a given number of partition cavities.
- The procedure is very suitable for gradient elution.

In this work a gradient in salt concentration was used but also other types of gradients can be used, for example, a gradient in polymer concentration or a biospecific ligand bound to poly(ethylene glycol). Elution countercurrent distribution can be performed in any thin-layer countercurrent distribution apparatus after only a few rather simple modifications.

REFERENCES

1 T.P. King and L.C. Craig, Meth. Biochem. Anal. 10:201 (1962)
2 A. Melis, P. Svensson and P.-Å. Albertsson, Biochim. Biophys. Acta 85:402 (1986)
3 P.-Å. Albertsson, Anal. Biochem. 11:121 (1965)
4 D.I. Arnon, Plant Physiol. 24:1 (1949)

GRAVIMETRIC ANALYSIS OF PHASE SYSTEMS USING A MICROWAVE OVEN AND THE RAPID DETERMINATION OF PHASE DIAGRAMS

Timothy Treffry and Bhavwanti Sheth

Biochemistry Department
University of Sheffield
Sheffield S10 2TN, UK

The first step in the preparation of a phase diagram is the determination of the polymer content of the phases. In the method described here, the dextran content of a dextran:PEG phase system is determined using a polarimeter with a cell with a 5.025 cm path length. Under these conditions the rotation recorded has the same value as the dextran concentration in grams per decilitre. The cell is also designed to be used with viscous solutions so that stock solutions of dextran (at 20%) can be standardised directly, avoiding dilution errors and the sometimes inadvertent confounding of w/w and w/v measures. The w/v value is converted to w/w using the formula of Brooks et al [1]. At this point it should perhaps be noted that when using a polarimeter to determine the dextran content of a phase from a two-phase system, the temperature of the polarimeter cell must not be lower than that at which the phases were equilibrated. At lower temperatures the phases will become turbid as a new equilibrium is established.

Total solute is determined gravimetrically using a glass fibre filter, a quick-acting sensitive balance (Mettler AE100) and a domestic microwave oven. The method requires that the dry weight of the filter is recorded. It is again weighed after the addition of a convenient amount of the test solution and after drying in the microwave oven.

The method is rapid and at least as accurate and reproducible as other methods provided certain difficulties are avoided. Filter papers are not satisfactory, except possibly in a very dry climate, as when dried they absorb water from the atmosphere when being weighed. We use 2.5 cm glass-fibre filters (Whatman GFC). It is important that these are only supported at the edge. We use inverted disposable plastic weighing boats in which a hole slightly smaller than the filter has been punched but any lightweight ring placed on the balance and tared would do. The filter, which can be labelled in pencil if required, is placed on the support and weighed. We find they do not need pre-drying. A few drops of the test solution are added. A greater amount improves accuracy but it must not be enough to exceed the absorptive capacity of the filter and thus spread to wet the support. Wet weight is recorded. Each filter is then transferred to a watch glass in the microwave oven using fine forceps. Again, the principle is that the filter is only supported at its edge. In our oven, drying is done in two stages: three minutes at low power followed by 3 minutes at high power. If the initial drying is too rapid some solute is lost. In establishing a methodology, use is made of solutions of known concentrations. When the dry

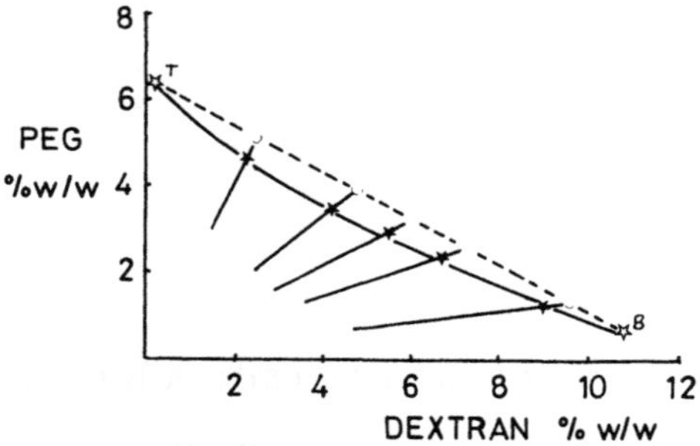

Fig. 1 *Phase Diagram for 4% w/w aqueous solutions of PEG6000 and 4.6% w/w Dextran T500 at 25°C. Composition of top-phase T, and bottom-phase B. Tie-line *----*. Binodal *——* is established where dilution of phase mixtures o causes loss of turbidity on mixing*

weight has been determined, total solute w/w can be calculated. To obtain the PEG content the dextran content determined by polarimetry is subtracted. Where other solutes are present in significant amounts and their bulk concentration is known, they are assumed to be distributed equally between the phases and subtracted accordingly.

When the compositions of the top and bottom phases have been determined in this way, they can be plotted on a phase diagram and a tie-line drawn. Top and bottom phases are mixed together in convenient proportions by weight to give, say, 5 g systems, and their bulk compositions represented on the tie-line. If, for example, top and bottom phases are mixed in the ratio 2:1, the composition of the mixture is represented by a point on the tie-line one third of the way between the top and bottom phase compositions. Small (recorded) aliquots of water are added to each minisystem until it is found to remain optically clear on vortexing and has thus changed from a two-phase to a homogeneous system. Dilution will change the composition of the system from that represented on the tie-line directly towards the origin of the phase diagram. The point of transition for each minisystem can be plotted and these points connected to give the binodal. This method of determining the binodal by dilution has also been reported by Albertsson [1].

REFERENCE

1 P.-Å. Albertsson, "Partition of Cell Particles and Macromolecules," John Wiley, New York (1986)

RAPID PARTITIONING OF HUMAN NEUTROPHILS BY CCD

Paul Eggleton *, Ian A. Sutherland + and Derek Fisher *

* Biochemistry Department, Royal Free Hospital School of Medicine University of London, London NW3 2PF, UK and
\+ Engineering Department, National Institute for Medical Research Mill Hill, London NW7 1AA, UK

The ability of countercurrent distribution (CCD) to fractionate cells on the basis of differences in their surface charge and hydrophobic characteristics makes CCD potentially a powerful technique to investigate correlation between surface properties of neutrophils and their functional activities.

Standard thin-layer CCD techniques for cells involves exposure of neutrophils to dextran and PEG for up to 6 hours. Neutrophils are short lived cells with a half life of 6-7 hours in the circulation. They are also very sensitive to excessive manipulation and environmental change [1,2]. In our experience, exposure of isolated neutrophils to phase reagents for 4-5 hours led to gradual morphological deterioration of the cells with evidence of degranulation in many of the cells. We have therefore developed a simple manual CCD method which allows neutrophils to be partitioned within 2 hrs. This reduces the time that the cells are exposed to the phase reagents, allowing neutrophils to be extracted with preservation of morphology and function.

Neutrophils (1-2 x 10^8) were isolated from 20 ml aliquots of EDTA anti-coagulated blood by a differential centrifugation procedure following NH_4Cl lysis [3] washed in Hank's balanced salt solution (HBSS) and suspended in 1.4 ml top phase. The phase system was composed of 5% w/w Dextran T5000 and 6.5% w/w PEG 6000 in 270 mosM sodium phosphate buffer, pH 7.2. A hand-held, hand-operated modified 30 chamber version of the Pritchard device [4,5], with a bottom chamber volume of 0.65 ml was used at room temperature. The bottom plate was filled with 0.6 ml bottom phase, and the top plate placed in position. 0.6 ml of top phase was added to all chambers except the first two, which received a load mix (0.6 ml) containing approximately 1 x 10^8 neutrophils. The phases were mixed by manual inversion of the apparatus 15 times. The unit was then placed on its side in a horizontal position so that the phases formed thin layers. After 2.5 minute settling time, the CCD unit was returned to its vertical position and the top plate moved over to the next bottom chamber. After 20 operations of this mixing, settling and transfer cycle, the contents were emptied into plastic tubes and 0.6 ml HBSS added to break the phase system.

Cells were concentrated by centrifugation at 400 x g for 10 minutes, washed and resuspended in HBSS. The cell containing-fractions were united to form 4 equal pools (P1-P4), each containing 25% of the recovered cells. Each pool was tested for their ability to phagocytose *Staphylococcus aureus*.

Fig. 1 shows a typical CCD profile of neutrophils and the phagocytic capacity of the four pools. With increasing partition ratio, the number of bacteria engulfed by the neutrophils decreased.

We have found that this manual CCD method allows neutrophils to be subjected to CCD with speed and ease, providing cells in good morphological and functional order. It is being used for more detailed studies of neutrophil function related to surface charge properties.

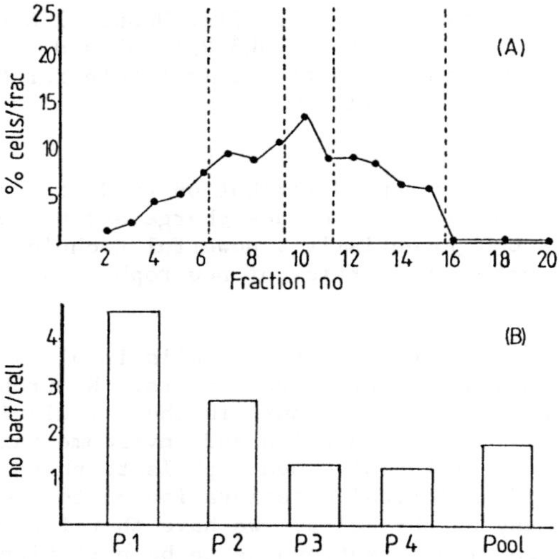

Fig. 1 *CCD profile of human neutrophils (A) and phagocytic capacity of the neutrophil pools (B)*

REFERENCES

1 R.L. Berkow, S.J. Weisman, D. Tzeng, R.A. Hoak, F.W. Kleinhaus, S. Barefoot and R.L. Baehnner, J. Lab. Clin. Med. 104:698 (1984)

2 L.A. Guthrie, L.C. McPhail, P.M. Henson and R.B. Johnston, J. Ex. Med. 160:1656 (1984)

3 P. Eggleton, N. Crawford and D. Fisher, in: "Separations Using Aqueous Phase Systems: Applications in Cell Biology and Biotechnology," D. Fisher and I.A. Sutherland, eds., Plenum, New York (1989)

4 R.N. Pritchard, J.A. Halpern, B.C. Halpern and R.A. Smith, Biochim. Biophys. Acta 404:289 (1975)

5 I.A. Sutherland, in: "Partitioning in Aqueous Two-Phase Systems. Theory, Methods, Uses & Applications to Biotechnology," H. Walter, D.E. Brooks and D. Fisher, eds., Academic Press, Orlando (1985)

QUANTITATION OF THE RESOLVING POWER OF CCD USING HUMAN ERYTHROCYTES

O.M. Smith and D. Fisher

Department of Biochemistry
Royal Free Hospital School of Medicine
Rowland Hill Street, London NW3 2PF, UK

Although CCD has been applied to the separation and analysis of a number of different cell types, there has been little attempt to quantitate the resolving power of this technique. In this study, the resolving power of CCD has been examined using neuraminidase-treated and untreated human erythrocytes as model cells. By treating erythrocytes with different concentrations of neuraminidase, cell populations differing in their sialic acid content, and hence their surface negative charge, were obtained. The ability of CCD to resolve neuraminidase-treated and control erythrocytes, exhibiting increasingly similar amounts of surface sialic acid, was determined using 30-transfer CCD and a phase system of high electrostatic potential difference between the phases ($\Delta\Psi$). The limit of resolution of CCD was expressed in terms of the μg difference in sialic acid content between the two red cell populations which were just resolved.

1 ml of washed, packed erythrocytes was incubated with 0.2 ml of neuraminidase (ranging in concentration from 1.0 to 0.11 U/ml) for 10 minutes at 37°C. The amount of sialic acid released was determined by the thiobarbituric acid assay of Warren [1]. Cells treated with 1.0 or 0.33 U/ml enzyme were fractionated separately but simultaneously by 20-transfer CCD in a high $\Delta\Psi$ phase system composed of: 5% Dextran T500 (Pharmacia), 3.7% PEG 8000 (Union Carbide), 0.03 M NaCl, 0.09 M sodium phosphate (pH 6.8) and 5% FCS. Red cells treated with 0.17 or 0.11 U/ml enzyme and control cells were compared using 30-transfer CCD and the high $\Delta\Psi$ phase system described above. To render CCD more sensitive regarding the separation of cell populations differing only slightly in their sialic acid content, cells were subjected to the [^{51}Cr]-chromate radiolabelling procedure of Walter and Krob [2-3] prior to CCD.

The CCD profiles of erythrocytes, treated with 1.0 or 0.33 U/ml neuraminidase, were shifted leftward relative to that for control cells by 7 and 4 chambers, respectively. An obvious shift in CCD profile location for cells, treated with 0.17 U/ml enzyme compared to control cells, was not seen when using 20 transfer CCD. This was attributable to the small difference in sialic acid content (23 μg/10^{10} cells) between the control and treated cell populations. To determine whether resolution of these two cell populations by CCD was possible, a mixture of radiolabelled, treated and unlabelled, control cells were fractionated using 30-transfer CCD. The resulting peak of radioactivity (indicating the location of treated cells) was situated to the left of the overall cell distribution profile. This demonstrates that CCD is capable of resolving two red cell populations differing in their sialic acid content by 23 $\mu g/10^{10}$ cells. When the same procedure was applied to the fractionation of cells treated with 0.11 U/ml enzyme and control cells, the radioactivity peak coincided with the cell distribution profile indicating that CCD is unable to distinguish between these two cell populations which differed in their sialic acid content by 9 $\mu g/10^{10}$ cells.

In conclusion, CCD is able to discriminate between two red cell populations differing in their sialic acid content by as little as 23 $\mu g/10^{10}$ cells. The latter represents the limit of resolution of CCD in a high $\Delta\Psi$ phase system.

REFERENCES

1 L. Warren, J. Biol. Chem. 234:1971 (1959)
2 H. Walter and E.J. Krob, Cell Biophys. 5:205 (1983)
3 H. Walter and E.J. Krob, Cell Biophys. 5:301 (1983)

Chapter 10

Emerging Technology: New Polymers and Phase Systems

NEW POLYMERS FOR AQUEOUS TWO-PHASE SYSTEMS

Folke Tjerneld

Department of Biochemistry, Chemical Center
University of Lund
P. O. Box 124, S-221 00 Lund, Sweden

New aqueous polymer two-phase systems have been studied. The aim has been to develop phase systems suitable for large scale applications. Derivatives of starch, cellulose and poly(vinyl alcohol) were investigated. The phase separation in aqueous solutions of the polymers was studied, as well as the partitioning in the new phase systems of macromolecules, cell particles and cells. With cellulose derivatives it has been found that phase systems can be formed at very low polymer concentrations. In phase systems based on poly(vinyl alcohol) new partitioning properties have been obtained.

INTRODUCTION

There is a need to develop new aqueous two-phase systems suitable for large scale biotechnical processes. For separations on a laboratory scale in biochemistry and cell biology, the two-phase systems composed of dextran and poly(ethylene glycol) (PEG) have been almost exclusively used [1,2]. For large-scale applications this system is too expensive due to the high cost of fractionated dextran [3]. Crude, unfractionated dextran has been used as a substitute for dextran in enzyme extractions [4] and in bioconversions [5,6]. A drawback with crude dextran is, however, the high molecular weight and the high viscosity. For large-scale enzyme extractions the PEG-salt systems are extensively used [7,8]. This system is inexpensive but has the drawback that the salt concentration is high in both phases [1,7]. The use of affinity ligands is thus restricted in the PEG/salt systems, and these systems may denature sensitive biological structures.

Our first objective was to develop an inexpensive substitute for dextran with similar partitioning properties. It was then possible to use the large amount of experimental experience with the PEG/dextran systems [1,2] when performing partitions with the new system. At a later stage we have been studying polymers with new properties, such as phase separations at very low polymer concentrations or the formation of phase systems based only on synthetic polymers.

We have studied derivatives of starch, cellulose and poly(vinyl alcohol) for use in aqueous two-phase systems).

HYDROXYPROPYL STARCH AS A SUBSTITUTE FOR DEXTRAN

Starch is an inexpensive polymer available in large quantities. The main problem with starch as a substitute for dextran, however, is the strong gel forming tendencies of starch solutions (retrogradation). The gel formation can be overcome by derivatisation of the starch molecule with hydroxyalkyl ether groups [9]. We have studied hydroxyethyl and hydroxypropyl derivatives of starch for use in aqueous phase systems. The best stability in solution was obtained with hydroxypropyl derivatives. Solutions of this polymer are stable at 20% concentration at +20°C and +4°C more than 14 days [10].

Hydroxypropyl starch forms two-phase systems with PEG [10-12]. We have demonstrated how HP-starch can be used as a substitute for dextran in aqueous phase systems [10-12]. The partitioning of proteins and affinity partitioning of enzymes have been studied in HP-starch/PEG systems [10]. Proteins of human serum have been fractionated by countercurrent distribution in HP-starch/PEG phase systems and comparisons were made with fractionations in dextran/PEG systems [13]. The HP-starch polymer is now marketed under the trade-name Aquaphase PPT by Perstrop Biolytica AB, Lund, Sweden.

We have tested the new polymer in large-scale (up to 50 L) affinity extractions of lactate dehydrogenase from swine muscle [14]. The dye Procion yellow HE-3G, bound to PEG in the top phase, was used for the extraction. To speed up the separation of the phases, a centrifugal separator (Alfa-Laval LAPX 202) was used. Both PEG and ligand-PEG could be effectively recycled [14]. Two process alternatives were compared (Table 1), direct homogenisation of muscle in the phase system and the preparation of a muscle extract by removal of debris by centrifugation. Direct homogenisation in the presence of ligand-PEG simplifies the process, but more ligand-PEG is consumed and the capacity of muscle per litre phase system is lower.

Table 1 *Large-scale affinity extraction of lactate dehydrogenase from swine muscle. Phase system: Aquaphase PPT 20%, PEG 8000 7.6%, 40 mM sodium phosphate pH 7.9. Volumes: 10 L (homogenisation in system), 50 L (muscle extract). Separator: Alfa-Laval LAPX 202. Speed of separation: 0.5 - 2.0 L/min*

	Muscle homogenised in system	Muscle extract
Concentration of Procion yellow PEG (% of total PEG)	2	1.5
Muscle (g) used per L system	100	160
Initial LDH concentration (kU/L system)	66	70
Recovery of LDH in top phase:		
(kU/L system)	32	58
(kU/kg muscle)	312	364
(kU/g ligand-PEG)	21	51
Specific activity (U/mg protein)	114	126
Degree of purification (relative to protein)	10x	11x

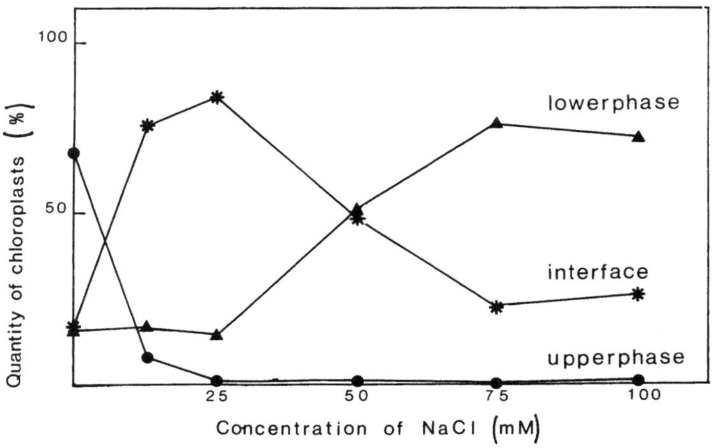

Fig. 1 *Influence of salt composition on the partition of class II chloroplasts. Phase system: 13% w/w Aquaphase PPT, 4.5% w/w PEG 8000, 10 mM sodium phosphate pH 7.4*

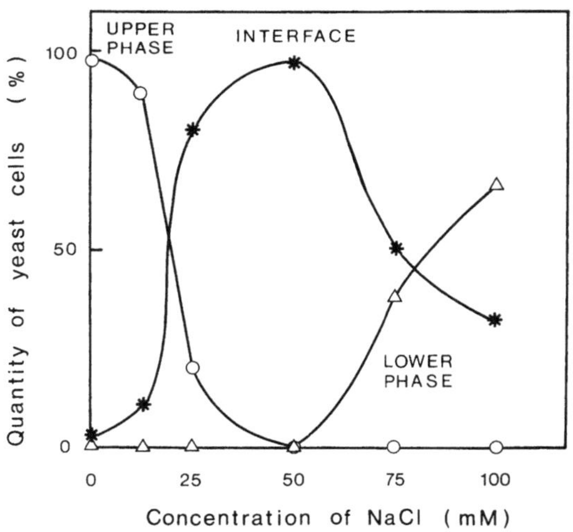

Fig. 2 *Influence of salt composition on the partition of yeast (Saccharomyces cerevisiae) cells. Phase system: 13% w/w Aquaphase PPT, 5% w/w PEG 8000, 10 mM sodium phosphate pH 7.0*

Accordingly a better recovery of LDH is obtained with the muscle extract, which gives significantly lower total costs when calculated per kU of enzyme compared with direct homogenisation [14].

The effect of salt composition on the partitioning of cell particles and cells in HP-starch/PEG systems has been studied. In Fig. 1 is shown how the partitioning of chloroplasts is influenced by the salt composition. A similar investigation has been made on the partitioning of yeast cells (Fig. 2). At 10 mM sodium phosphate the chloroplasts and the yeast cells are partitioned to the top phase. With increasing NaCl concentration the chlorplasts and the yeast cells are distributed almost totally to the interface, and at 100 mM the distribution is between the interface and the lower phase. The partitioning can thus be influenced by the salt composition in a similar way as in dextran/PEG systems.

AQUEOUS TWO-PHASE SYSTEMS AT LOW POLYMER CONCENTRATIONS

The aqueous two-phase system composed of dextran and methylcellulose is known [1]. This system is interesting in that very low polymer concentrations (1-2% total polymer) are sufficient to obtain phase separation. We have studied the polymer ethyl hydroxyethyl cellulose (EHEC) in order to obtain a polymer which is both inexpensive and can be used at low concentrations in aqueous phase systems. For large scale work there are important economical advantages with aqueous phase systems formed at low concentrations of polymers [3].

EHEC is a nonionic, watersoluble cellulose derivative, which is synthesised from cellulose by substitution with ethyl and ethylene oxide (EO) groups. The EO groups can form short polyethylene oxide side chains on the cellulose molecule [15]. EHEC can be obtained in a range of molecular weights and with different degrees of substitution of ethyl and EO groups [16].

EHEC forms aqueous phase systems with PEG only at unpractical high concentrations, probably due to the ethylene oxide chains on EHEC which makes the two polymers too similar chemically. On the other hand EHEC phase separates at low concentrations with dextran and with hydroxypropyl starch. In Fig. 3 the phase diagrams are shown for EHEC of different molecular weights together with Dextran 500. The phase separation is obtained at significantly lower polymer concentrations compared to the PEG-dextdran system [1].

The phase separation in the system EHEC/HP-starch is shown in Fig. 4. Due to the lower molecular weight of the HP-starch molecule compared to Dextran 500, the phase separation is obtained at slightly higher polymer concentration. But compared to the phase system with PEG/HP-starch [10] there is a strong reduction in polymer concentration when EHEC is used instead of PEG.

EHEC forms stable solutions both at room temperature and at lower temperatures (+4°C). It is thus a suitable phase-forming polymer. The partitioning of proteins in phase systems, based on EHEC and HP-starch, can be influenced in the same way as in PEG-dextran systems. In Fig. 5 we have studied how the partitioning of two enzymes can be influenced by the polymer concentration and the ionic composition. As in PEG/dextran systems, the enzymes are partitioned more to the bottom phase with increasing polymer concentration. The effect of salt composition on the partition is qualitatively similar to what is observed in PEG/dextran systems. The negatively charged enzyme, glucose-6-phosphate dehydrogenase, is partitioned

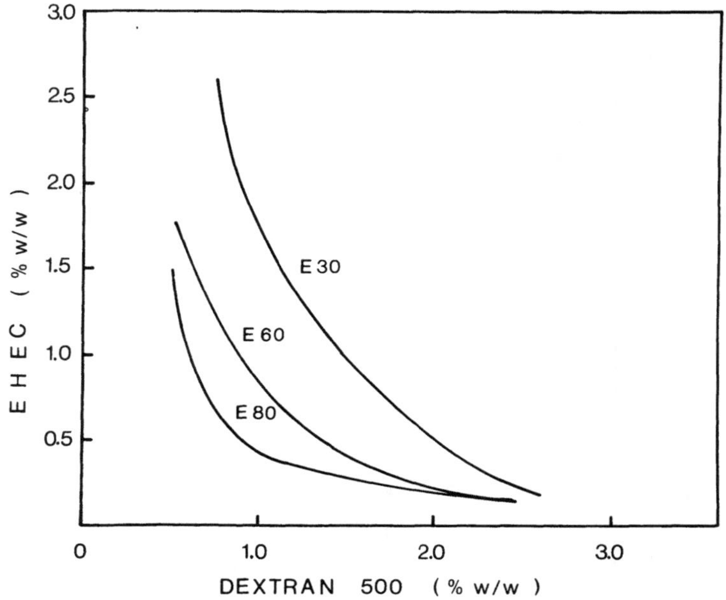

Fig. 3 *Phase diagrams for the system EHEC/Dextran T-500/water at 20°C. The molecular weights (M_n) of EHEC 30, 60 and 80 are 30000, 60000 and 80000, respectively*

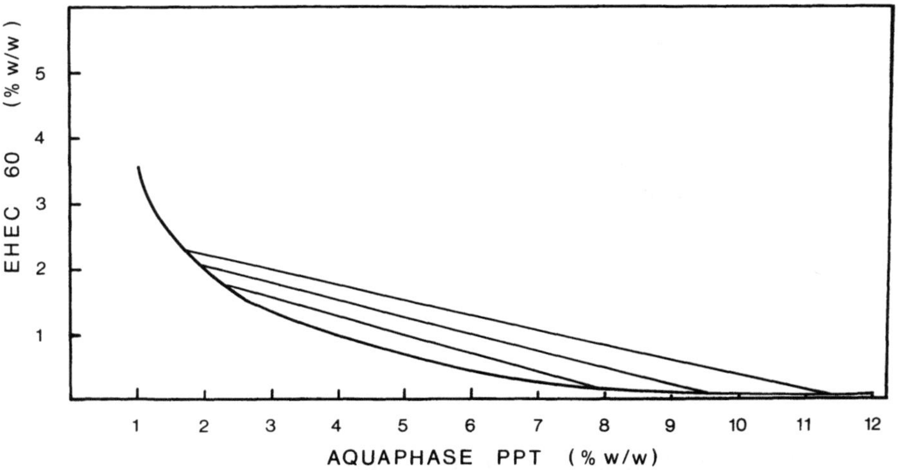

Fig. 4 *Phase diagram for the system EHEC 60/HP-starch (Aquaphase PPT)/water at 20°C*

more to the bottom phase when sodium chloride is in excess over the phosphate buffer [1].

The advantage of EHEC as phase forming polymer is the low concentrations of both polymers that are needed in phase systems based on EHEC. The drawback is that phase systems with EHEC have longer settling times than both dextran/PEG and HP-starch/PEG systems. However, in technical processes centrifugal separators can be used for the reduction of settling times.

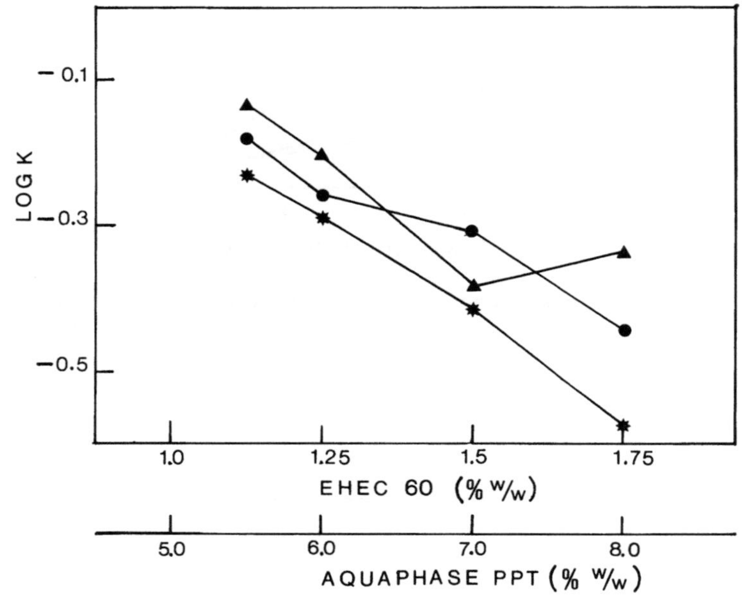

Fig. 5 *Effect of polymer concentration and ionic composition on the partition coefficient (K) in the system EHEC 60/HP-starch (Aquaphase PPT). Lactate dehydrogenase in 25 mM Na-phosphate pH 6.8 (▲), glucose-6-phosphate dehydrogenase in 25 mM Na-phosphate pH 6.8 (●) and in 25 mM Na-phosphate plus 0.2 M NaCl (✱)*

AQUEOUS TWO-PHASE SYSTEMS BASED ON POLY(VINYL ALCOHOL)

Previously used aqueous phase systems with two polymers have been based either on one carbohydrate polymer (dextran, HP-starch) and one synthetic polymer (PEG), or on two carbohydrate polymers (EHEC/HP-starch, dextran/Ficoll). We have now developed a functioning two-phase system based on two synthetic polymers: poly(vinyl alcohol) (PVA) and PEG. This system has partitioning properties that are different from the PEG/dextran systems.

PVA is obtained by hydrolysis of poly(vinyl acetate) and is available in a wide range of molecular weights and with different degrees of hydrolysis, i.e. different amounts of residual acetate groups. PVA polymers that are fully hydrolysed (99-100%) give stable solutions at 20°C but are gel-forming at +4°C. With a degree of hydrolysis of 90% the PVA polymer gives stable solutions both at +4°C and at +20°C.

Fig. 6 shows a phase diagram for a system with PVA of 99% hydrolysis: PVA 22000/PEG 20000/water. This phase system can be used at room temperature, but at +4°C the PVA-rich bottom phase forms a gel.

The partitioning of enzymes in PVA-PEG systems is demonstrated in Fig. 7. Enzymes are partitioned strongly to the top phase in these systems and the partition coefficient increases with increasing polymer concentration. This is the opposite effect compared with PEG/dextran and EHEC/HP-starch systems (see Fig. 5).

We have used the system PVA 14000/PEG 8000 at 28°C for the cultivation of the fungus *Aspergillus phoenicis* QM 329 for the production of ß-glucosidase [17]. This fungus also produces amylases which degrade a starch-based two-phase system. In the PVA/PEG system the fungus is partitioned to the interface and the ß-glucosidase to the top phase.

In order to obtain a PVA/PEG phase system which can be used also at low temperatures, we have tested PVA polymers with degrees of hydrolysis between 75 and 98%. The best results have been obtained with 90% hydrolysed PVA with

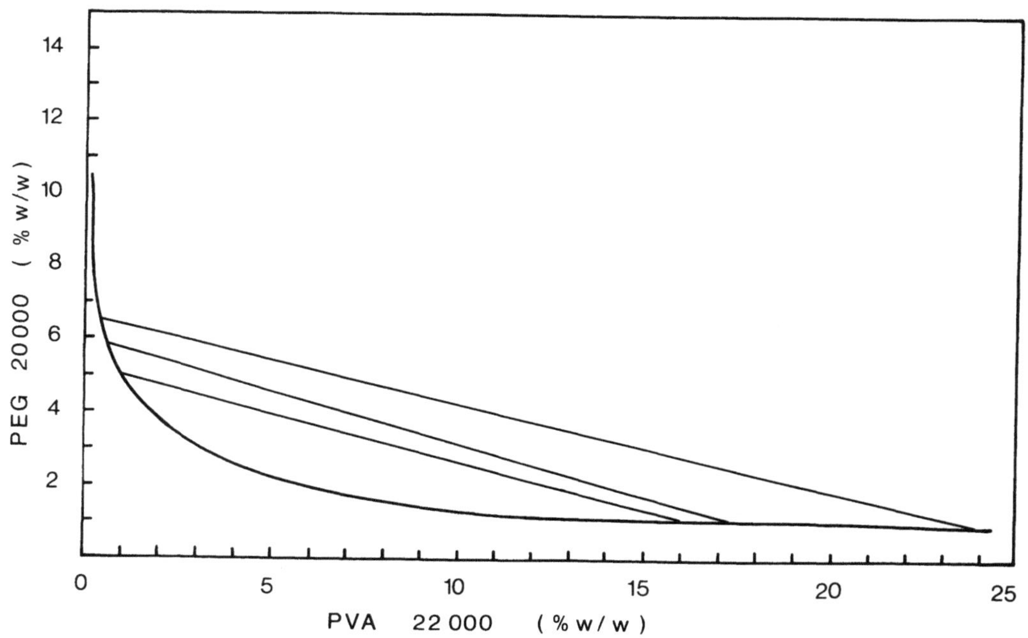

Fig. 6 *Phase diagram for the system PEG 20000/PVA 22000/water at 20°C*

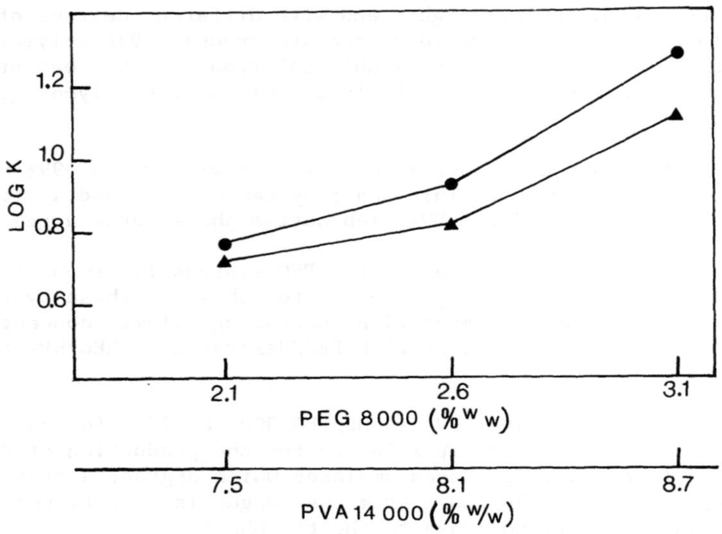

Fig. 7 *Log K as a function of polymer concentration in the phase system PEG 8000/PVA 14000. Buffer: 25 mM sodium phosphate, pH 6.8. Lactate dehydrogenase (▲), glucose-6-phosphate dehydrogenase (●)*

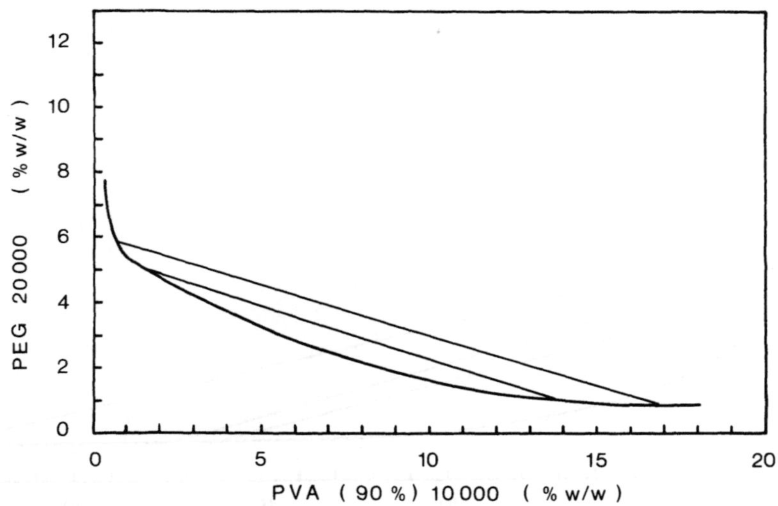

Fig. 8 *Phase diagram for the system PEG 20000/PVA 10000, 90% hydrolysed/water at 20°C*

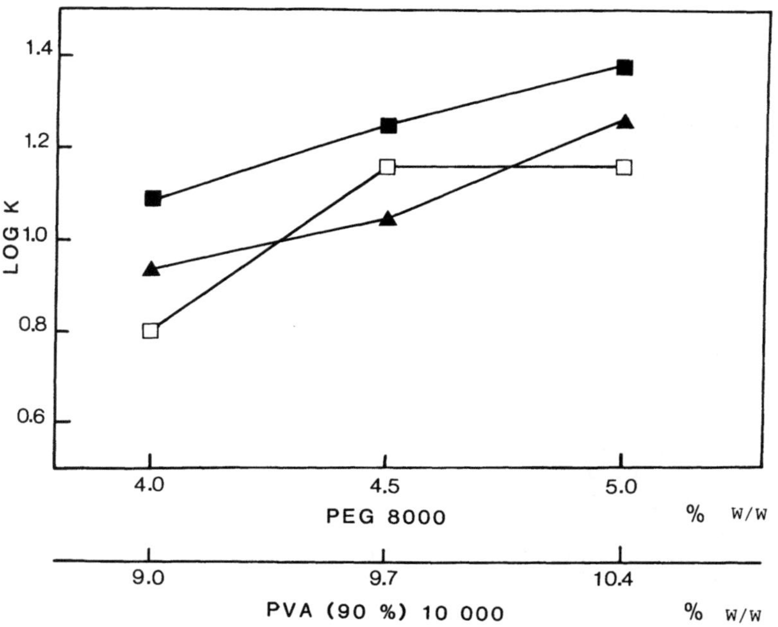

Fig. 9 *Effect of polymer concentration and ionic composition on the partition coefficient (K) in the system PEG 8000/PVA 10000, 90% hydrolysed. Lactate dehydrogenase in 25 mM Na-phosphate pH 6.8 (▲), glucose-6-phosphate dehydrogenase in 25 mM Na-phosphate pH 6.8 (■) and in 25 mM Na-phosphate plus 0.2 M NaCl (□)*

a molecular weight of 10000. A phase diagram for this polymer together with PEG 20000 is shown in Fig. 8. The PVA-rich bottom phase is in this system fluid both at room temperature and at +4°C.

The partition of enzymes in PEG-PVA (90%) systems has been studied (Fig. 9). The partitioning is strongly towards the top phase, just as in systems with 99% hydrolysed PVA (see Fig. 7), and log K increases with increasing polymer concentration. The effect of a change in the salt composition is qualitatively similar to what is observed in PEG/dextran systems. By using an excess of sodium chloride over the phosphate buffer it is possible to decrease the partition coefficient for negatively charged proteins and obtain the opposite effect with positively charged proteins [1].

CONCLUSIONS

New polymer phase systems can be developed with properties that are important for biotechnical applications, such as phase systems based on inexpensive polymers and at low polymer concentrations. At the same time we will obtain phase systems with new partitioning properties for macromolecules, cell particles and cells, which can be used for separation purposes in biochemistry and cell biology. The study of new phase systems also increases the knowledge of the basic physical-chemical mechanisms for phase separation and partitioning [18].

ACKNOWLEDGEMENTS

This research was supported by the National Swedish Board for Technical Development (STU). For the work on ethyl hydroxyethyl cellulose the author is grateful for the cooperation of Dr Olle Stenberg, Berol Kemi AB, Stenungsund, Sweden.

REFERENCES

1 P.-Å. Albertsson, "Partition of Cell Particles and Macromolecules," 3rd ed., Wiley, New York (1986)
2 H. Walter, D.E. Brooks and D. Fisher, "Partitioning in Aqueous Two-Phase Systems: Theory, Methods, Uses, and Applications to Biotechnology," Academic Press, New York (1985)
3 K.H. Kroner, H. Hustedt and M.-R. Kula, Extractive enzyme recovery: economic considerations, Process Biochem. 19:170 (1984)
4 K.H. Kroner, H. Hustedt and M.-R. Kula, Evaluation of crude dextran as phase-forming polymer for the extraction of enzymes in aqueous two-phase systems in large scale, Biotechnol. Bioeng. 24:1015 (1982)
5 F. Tjerneld, I. Persson, P.-Å. Albertsson and B. Hahn-Hägerdal, Enzymatic hydrolysis of cellulose in aqueous two-phase systems. I. Partition of cellulases from *Trichoderma reesei*, Biotechnol. Bioeng. 27:1036 (1985)
6 F. Tjerneld, I. Persson, P.-Å. Albertsson and B. Hahn-Hägerdal, Enzymatic hydrolysis of cellulose in aqueous two-phase systems. II. Semicontinuous conversion of a model substrate, Solka Floc BW200, Biotechnol. Bioeng. 27:1044 (1985)
7 M.-R. Kula, K.H. Kroner and H. Hustedt, Purification of enzymes by liquid-liquid extraction, Adv. Biochem. Bioeng. 24:73 (1982)
8 A. Veide, A.-L. Smeds and S.-O. Enfors, A process for large-scale isolation of ß-galactosidase from *E. coli* in an aqueous two-phase system, Biotechnol. Bioeng. 25:1789 (1983)
9 E.T. Hjermstad, Starch hydroxyethyl ethers and other starch ethers, in: "Industrial Gums - Polysaccharides and Their Derivatives," R.L. Whistler and J.N. BeMiller, eds., Academic Press, New York (1973)
10 F. Tjerneld, S. Berner, A. Cajarville and G. Johansson, New aqueous two-phase system based on hydroxypropyl starch useful in enzyme purification, Enzyme Microb. Technol. 8:417 (1986)
11 Å. Gustafsson, H. Wennerström and F. Tjerneld, Aqueous polymer two-phase systems in biotechnology, Fluid Phase Equilibria 29:365 (1986)
12 F. Tjerneld and G. Johansson, U.S. Patent 4,740,304
13 G. Birkenmeier, G. Kopperschläger, P.-Å. Albertsson, G. Johansson, F. Tjerneld, H.E. Åkerlund, S. Berner and H. Wickström, Fractionation of proteins from human serum by countercurrent distribution, J. Biotechnol. 5:115 (1987)
14 F. Tjerneld, G. Johansson and M. Joelsson, Affinity liquid-liquid extraction of lactate dehydrogenase on a large scale, Biotechnol. Bioeng. 30:809 (1987)
15 S. Lindenfors and I. Jullander, Ethylhydroxyethyl-cellulose, in: "Industrial Gums - Polysaccharides and Their Derivatives," R.L. Whistler and J.N. BeMiller, eds., Academic Press, New York (1973)
16 O. Stenberg, Berol Kemi AB, Stenungsund, Sweden, personal communication
17 I. Persson, F. Tjerneld and B. Hagn-Hägerdal, Cultivation and enzyme production in aqueous two-phase systems with *Trichoderma reesei* and with *Aspergillus phoenicis*, Lecture at IVA and EFB Workshop, 10-11 November 1986 in IVA-report 330, Stockholm (1987)
18 Å. Gustafsson, H. Wennerström and F. Tjerneld, The nature of phase separation in aqueous two-polymer systems, Polymer 27:1768 (1986)

PROTEIN PARTITIONING BETWEEN MICROEMULSIONS AND CONJUGATE AQUEOUS PHASES

P.D.I. Fletcher and D. Parrott

School of Chemistry
The University of Hull
Hull, North Humberside HU6 7RX

Water-in-oil (W/O) microemulsions consist of thermodynamically-stable dispersions of tiny water droplets (radii 1-20 nm) in oil stabilised by a surfactant. These phases can co-exist with aqueous phases at equilibrium. Added proteins partition between the two phases to a degree which is sensitive to droplet size, aqueous phase salt concentration and protein charge. Hence these systems offer potential for the selective extraction of proteins. Furthermore, partitioning results for enzymes and substrates are useful in interpreting changes in enzyme catalytic activity in microemulsions as compared with bulk water.

INTRODUCTION

Microemulsions have been the subject of much recent investigation with a view to determining their phase behaviour, structure and properties as novel solvents for chemical (and enzyme) reactions. Facets of all these areas are relevant to the use of microemulsion phases as protein extraction media. In this article we describe some important properties of microemulsions, the partitioning behaviour of proteins in two-phase microemulsions systems and the relationship between partitioning behaviour and enzyme catalytic behaviour in microemulsions. Examples illustrating some general principles are drawn from our recent work.

One-phase W/O microemulsions using the anionic, double-tailed surfactant sodium bis 2-ethylhexyl sulphosuccinate (AOT) may be simply prepared by adding water to AOT solutions in alkane and gently shaking. The transparent, thermodynamically stable microemulsions form in a few seconds. They consist of discrete spherical water droplets surrounded by a monolayer of AOT. In agreement with simple geometrical arguments the droplet size is proportional to the molar ratio of water to AOT (=R). Altering the AOT concentration but maintaining R constant alters the droplet concentration in proportion but does not affect the droplet size [1-9]. Although several microemulsion systems have been studied, the discussion here is limited to one employing AOT as surfactant since this system is particularly well characterised and does not have the added complexity of requiring a co-surfactant.

When n-heptane solutions of AOT are placed in contact with aqueous phases containing NaCl, a two phase system is formed. The upper is a W/O microemulsion containing virtually all the AOT and an amount of dispersed water which is dependent upon the aqueous phase salinity. The aqueous phase contains, in addition to the NaCl, a concentration of monomeric AOT equal to the value of the critical micelle concentration (cmc) in the particular system. As in the one-phase systems, the droplet size is determined by the ratio of dispersed water to AOT and may be calculated according to eqn. 1.

$$\text{water droplet radius } r/\text{nm} = 0.175\ R \qquad \text{eqn. 1}$$

The amount of water dispersed in the microemulsion phase depends on both the AOT concentration and the aqueous phase salt concentration. However, the molar ratio, R, is independent of the AOT concentration and depends only on the aqueous phase [NaCl]. As shown in Fig. 1, the R values are low at high [NaCl] but rise asymptotically as the [NaCl] is reduced and approaches 0.05 mol dm^{-3}. (Below 0.05 mol dm^{-3} NaCl the AOT transfers to the aqueous phase forming an oil-in-water microemulsion in equilibrium with excess oil - so-called "phase inversion"). Hence both the droplet size (through altering [NaCl] and droplet concentration (through altering [AOT]), may be independently varied in the two phase system. The properties of this system are described in detail in references 10-12.

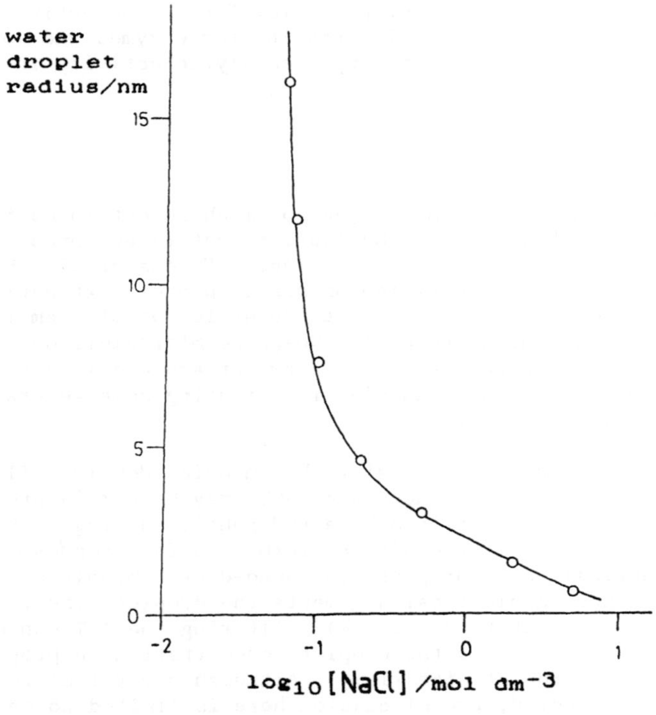

Fig. 1 *Microemulsion droplet size versus aqueous phase NaCl concentration for the AOT/n-heptane system at 25.0°C*

A large variety of enzymes are capable of dispersion within W/O microemulsions with retention of their catalytic activity, although, in general, the kinetic parameters are altered in the microemulsion as compared with bulk water [13,14]. For example, in the case of the alpha chymotrypsin catalysed hydrolysis of N-glutaryl-l-phenylalanine-p-nitroanalide (GPNA), the enzyme turnover number k_{cat} in the AOT microemulsion dispersed water droplets is not much affected relative to its value in bulk water, whereas the Michaelis constant K_M (equal to the dissociation constant of the enzyme/substrate complex) is increased approximately one hundredfold [15]. The use of microemulsions as solvents for enzyme reactions has potential synthetic applications which have been recently reviewed [13,14].

Since enzyme sizes are comparable with microemulsion droplet sizes, proteins in microemulsions are necessarily in close proximity to, and hence interact with the surfactant in monolayer surrounding the droplets. When a protein is allowed to partition between an aqueous phase and the W/O microemulsion phase, this interaction leads to a sensitivity of the degree of partitioning to droplet size (varied by altering the aqueous phase [NaCl] and protein charge (altered by varying the aqueous pH). Hence these systems have potential for the liquid-liquid extraction of proteins as has been demonstrated by the studies of Luis [16] and Hatton [17].

PARTITIONING OF PROTEINS

The partitioning of low molecular weight solutes between AOT stabilised W/O microemulsions and conjugate aqueous phases has been reported [12]. As discussed in reference 12, the partition coefficient in these systems may be expressed in a number of ways. For the case of proteins, it is assumed the protein species are located in the W/O microemulsion phases exclusively within the dispersed water droplets. The microemulsion water droplets are assumed to behave as a separate phase (a "pseudo-phase") within the microemulsion phase. The partition coefficient of the protein $P_{dw/aq}$ is then defined as:

$$P_{dw/aq} = [\text{solute}]_{dw}/[\text{solute}]_{aq} \qquad \text{eqn. 2}$$

where $[\text{solute}]_{dw}$ is the concentration of the partitioning solute expressed per unit volume of the dispersed water within the microemulsion phase (i.e. the solute concentration in the dispersed water "pseudo-phase") and $[\text{solute}]_{aq}$ is the aqueous phase concentration of solute. The measured solute concentrations (per unit volume of total microemulsion) are related to the dispersed water "pseudo-phase" concentrations through the volume fraction of water in the microemulsion phases. This "pseudo-phase" approximation is the same as used in the analysis of kinetics in micellar and microemulsion phases.

As with a conventional partition coefficient, $P_{dw/aq}$ is expected to be constant when the relative volumes of each phase are altered. In the case of the microemulsion phases the volume of dispersed water "pseudo-phase" can be altered by changing the droplet concentration through altering the AOT concentration at constant $[NaCl]_{aq}$. Obviously, changing the salinity (which alters R and hence droplet size) or the pH is expected to affect $P_{dw/aq}$. Fig. 2 shows the measured equilibrium percentages of total chymotrypsin in both phases as a function of AOT concentration for $[NaCl]_{aq} = 0.5$ mol dm^{-3} and a pH of 6.1. Under these conditions the R value is 16.7 and the droplet

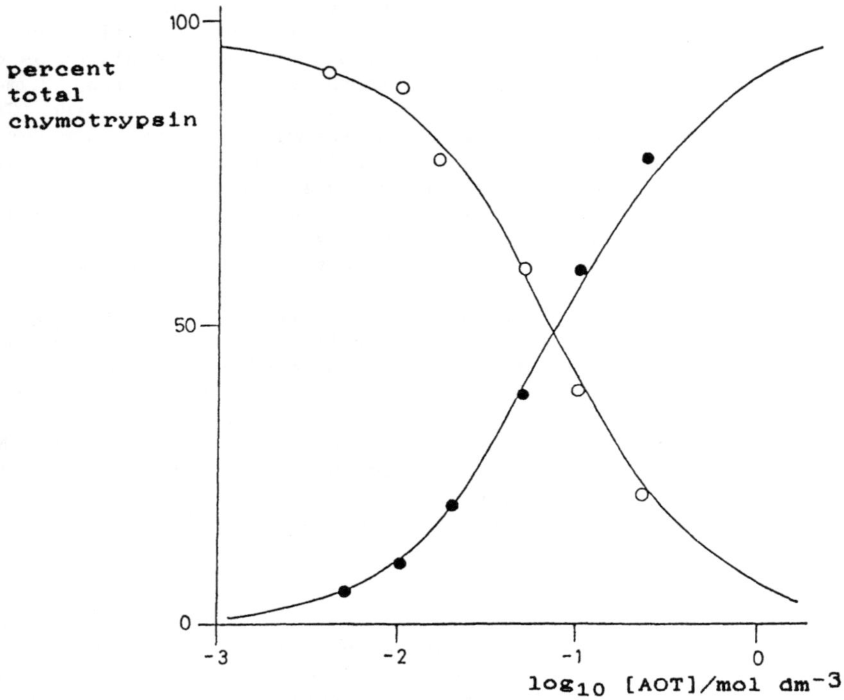

Fig. 2 *Percentage of total chymotrypsin in the microemulsion phase (filled circles) and aqueous phase (unfilled circles) versus [AOT] in the oil phase [18]*

radius is 2.9 nm. The solid lines are calculated using a "best-fit" value of $P_{dw/aq}$ equal to 44. It can be seen that the single value of $P_{dw/aq}$ adequately describes the variation of partitioning with [AOT] indicating that the "pseudo-phase" approximation is valid in this particular case.

As discussed above, the partitioning of proteins is expected to be sensitive to protein charge and droplet size. It must be emphasised, however, that the droplet size can only be varied through altering the salt concentration and observed effects are due to the effects of changing both parameters together. Fig. 3 shows the variation of $P_{dw/aq}$ for chymotrypsin with droplet size and [NaCl] (both shown) at pH values above and below the isoelectric point of the protein (pI = 8.2). Under all conditions the protein shows a greater affinity for the dispersed water as expressed by the values of $P_{dw/aq}$ being greater than unity. Charge effects are large for the larger droplets corresponding to lower electrolyte concentrations. However, even when the net protein charge is negative (the same as the AOT monolayer) $P_{dw/aq}$ is greater than unity. This could be due to a hydrophobic interaction of the protein with the oil/water interface which is sufficient to overcome a repulsive electrostatic interaction. Similar results are obtained for pepsin and cytochrome c when the differences in isoelectric point are taken into account [18]. It can be seen from Fig. 3 that the partition coefficient can be modulated over 2-3 orders of magnitude through changing ionic strength and pH. Further control in the overall degree of partitioning between the two phases can be achieved through changing the AOT concentration.

Lysozyme (pI. 11.1) is highly positively charged at pH values near neutral. Different behaviour is observed for this protein in the AOT system at moderate ionic strengths [18]. When sufficient AOT is present for

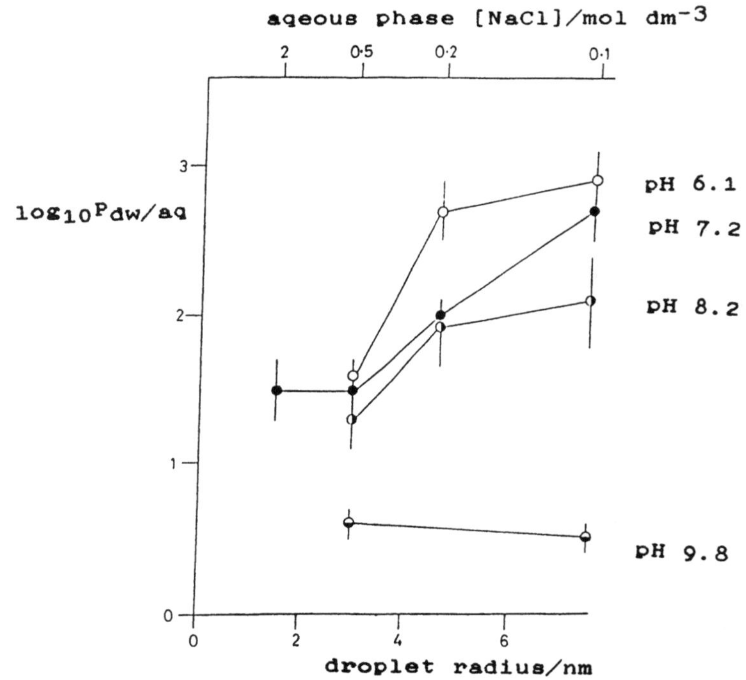

Fig. 3 *Variation of $P_{dw/aq}$ for chymotrypsin with droplet size (lower abscissa) or [NaCl] (upper abscissa). The numbers refer to the aqueous pH values*

monolayer coverage of this protein then virtually all the lysozyme is transferred to the oil phase. At lower AOT concentrations the lysozyme remains as a thick film at the interface separating the two phases. No protein is detected in the aqueous phase. Hence values of $P_{dw/aq}$ are indeterminate for this system. It appears that, for this protein, where the interactions are very strong, the microemulsion structure is significantly perturbed and the protein "creates" its own aggregate when sufficient AOT is present.

PARTITIONING AND ENZYME CATALYTIC ACTIVITY

The partition coefficient of a solute between two phases provides a measure of the free energy of transfer between the two phases. Hence equilibrium constants for a chemical reaction in two different phases are related through the partition coefficients of all species involved in the equilibrium. This principle has been applied to enzyme reactions in microemulsions [18].

In the example of the chymotrypsin catalysed hydrolysis of GPNA, the K_M is increased approximately one hundredfold in the water droplets as compared with bulk water. This effect could be the result of an effect of the microemulsion upon the enzyme E, the substrate GPNA or the enzyme/substrate complex ES. As described above, the K_M value in the dispersed water pseudo-phase K_M(dw) is related to the value in the corresponding aqueous phase K_M(aq) through the partition coefficients of E, GPNA and ES as shown in equation 3.

$$K_M(dw) = K_M(aq)P_{dw/aq}(E)P_{dw/aq}(GPNA)/P_{dw/aq}(ES) \qquad \text{eqn. 3}$$

Measurement of the terms in the RHS of eqn. 3 should allow the calculation of $K_M(dw)$ which can be compared with experimental values for this reaction [15]. The analysis is useful since it provides an answer to the question as to whether the K_M changes as a result of an effect upon E, GPNA or ES. The variation of the partition coefficient for GPNA with droplet size is shown in Fig. 4.

The partitioning of ES cannot be measured since the species is reactive. However, two limiting assumptions can be tested. Firstly, if the partitioning of ES is assumed to equal that of E, eqn. 4 is obtained.

$$K_M(dw) = K_M(aq)P_{dw/aq}(GPNA) \qquad \text{eqn. 4}$$

Alternatively, the partitioning of ES may be assumed to be the geometric mean of that of E and GPNA giving equation 5.

$$K_M(dw) = K_M(aq)\{P_{dw/aq}(E)P_{dw/aq}(GPNA)\}^{\frac{1}{2}} \qquad \text{eqn. 5}$$

Values of $K_M(dw)$ calculated using the measured values of $K_M(aq)$ at the corresponding salt concentrations, the partitioning data and the limiting assumptions (eqns. 4 and 5) are compared with the measured values in Fig. 5. It can be seen that the large difference between $K_M(dw)$ and $K_M(aq)$ is predicted from the partitioning data and arises mainly from an effect of the microemulsion upon the substrate and not the enzyme as reflected by the large partition coefficients for the substrate. A likely rationalisation of the data is that the substrate GPNA is located in the water droplet interfacial region where it is less available for reaction with the enzyme. If the substrate experienced a "water-like" environment in the microemulsion a partition coefficient of near unity would be expected.

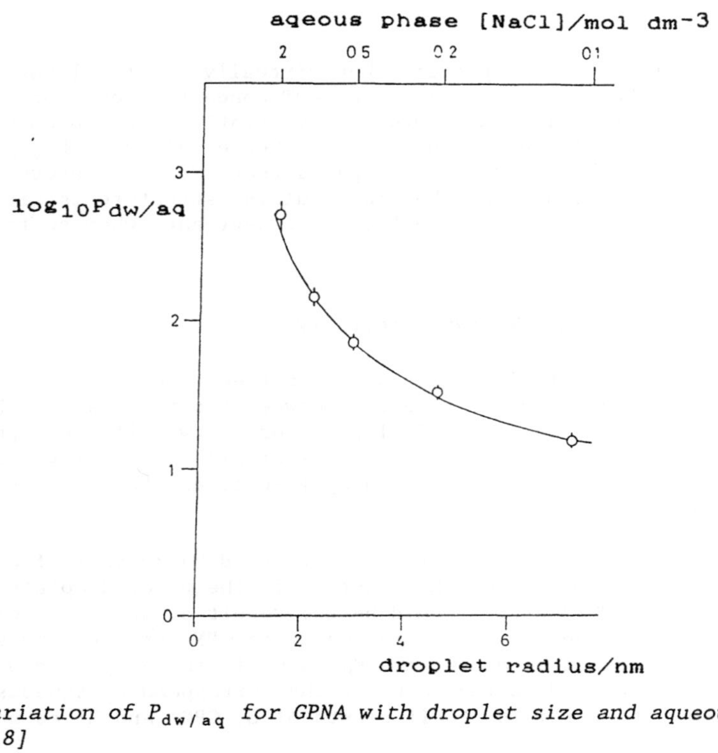

Fig. 4 *Variation of $P_{dw/aq}$ for GPNA with droplet size and aqueous [NaCl] [18]*

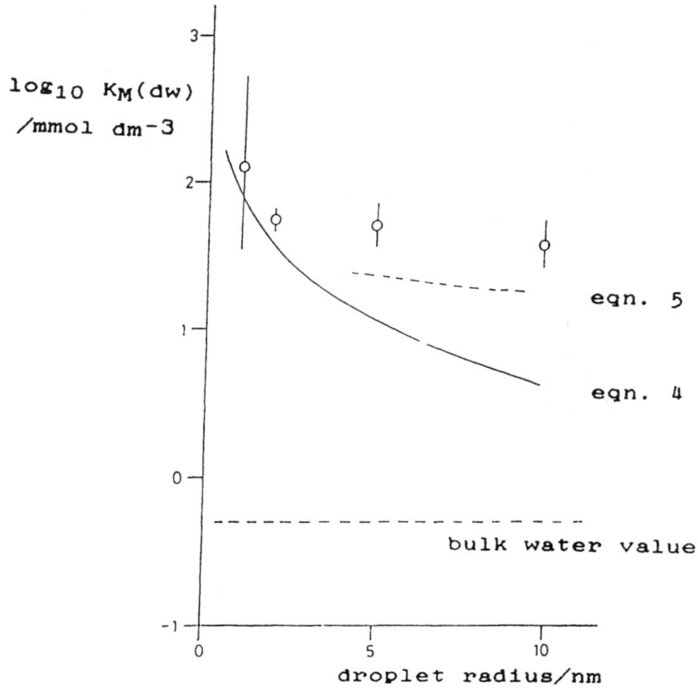

Fig. 5 *Comparison of $K_M(dw)$ values calculated according to eqns. 4 and 5 with experimental values. The lower dashed line is the value in bulk water at low ionic strength*

In conclusion, two-phase microemulsion systems have potential for the extraction of proteins. The partitioning can be modulated over three or four orders of magnitude through controlling pH, ionic strength and surfactant concentration. Partitioning data is also useful in understanding enzyme catalytic activity in microemulsions.

ACKNOWLEDGEMENTS

We thank the SERC for support of this work through provision of a studentship for D.P.

REFERENCES

1 P.D.I. Fletcher, B.H. Robinson, F. Bermejo-Barrera, D.G. Oakenfull, J.C. Dore and D.C. Steytler, in: "Microemulsions," I.D. Robb, ed., Plenum (1981)

2 M. Zulauf and H.F. Eicke, J. Phys. Chem. 83:480 (1979)

3 J.D. Nicholson and J.H.R. Clarke, in: "Surfactants in Soltuion," K. Mittal and B. Lindman, eds., Plenum, New York (1984)

4 C. Abos and P. Delord, J. Appl. Crystallogr., 12:502 (1979)

5 B.H. Robinson, C. Toprakcioglu, J.C. Dore and P. Chieux, J. Chem. Soc. Faraday Trans. I. (1984)

6 B.H. Robinson, C. Toprakcioglu, J.C. Dore, A.M. Howe and P. Chieux, J. Chem. Soc. Faraday Trans. I. (1984)

7 M. Kotlarchyk, S.H. Chen, J.S. Huang and M.W. Kim, Phys. Rev. A (1984)

8 E. Keh and B. Valeur, J. Colloid Interface Sci. (1981)
9 N.J. Bridge and P.D.I. Fletcher, J. Chem. Soc. Faraday Trans. I (1983)
10 R. Aveyard, B.P. Binks, S. Clark and J. Mead, J. Chem. Soc. Faraday Trans. I (1986)
11 R. Aveyard, B.P. Binks and J. Mead, J. Chem. Soc. Faraday Trans. I (1986)
12 P.D.I. Fletcher, J. Chem. Soc. Faraday Trans. I (1986)
13 P.L. Luisi, Angew. Chem. Int. Ed. Engl. 24:439 (1985)
14 K. Martinek, A.V. Levashov, N. Klyachko, Y.L. Khmelnitski and I.V. Berezin, Eur. J. Biochem. 155:453 (1986)
15 P.D.I. Fletcher, R.B. Freedman, J. Mead, C. Oldfield and B.H. Robinson, Colloids and Surfaces 10:193 (1984)
16 P.L. Luisi, F.J. Bonner, A. Pelligrini, P. Wiget and R. Wolf, Helv. Chim. Acta 62:740 (1979)
17 K.E. Goklen and T.A. Hatton, Separation Sci. and Technol. 22:831 (1987)
18 P.D.I. Fletcher and D. Parrott, submitted for publication in J. Chem. Soc. Faraday Trans. I

IMMISCIBLE LIQUID AFFINITY SUPPORTS

John W. Eveleigh

E.I. du Pont de Nemours & Co., Inc.
Medical Products Department
Glasgow Site, Building 100
Wilmington, Delaware 19898, USA

A new class of liquid affinity supports, based on the use of liquid perfluorocarbons, is reported. Macromolecular ligands are controllably substituted with perfluoroalkyl groups prior to addition to the liquid fluorocarbon and emulsification. Such substituted ligands act as efficient surfactants while retaining their biological activity and specificity. The resulting stable emulsions offer many advantages over conventional solid affinity supports, particularly in the area of large scale bioprocessing. High density and deformability of the fluorocarbon droplets allow processing of debris containing samples. In addition, packed emulsions can be easily pumped, thus offering the potential of continuous countercurrent affinity separations. Recovery or concentration of the ligand or its complex is possible by evaporation of the fluorocarbon support.

INTRODUCTION

Separations based upon the specific affinity of one molecule for another are well established, particularly in the fields of biochemistry and bioprocessing. The most common method of practically applying these specific interactions for separation purposes, involved immobilisation of one of the molecules, termed the ligand, on an insoluble support. The complementary molecules in an applied sample, often called the ligate, bind to the ligand on the support and can be recovered by subsequent dissociation.

Currently, the majority of affinity separations use solid supports contained in chromatographic columns. This approach is ideally suited to research and small-scale preparative applications, however, practical difficulties become apparent when attempts are made to scale-up to the process level. The soft gel supports, with their high porosity to ensure maximal ligand density, are structurally weak and easily compress at high flow rates. More rigid supports are prone to blockage by particulate matter invariably present in biological samples, thus necessitating prior intensive clarification of a process stream. Finally, the affinity chromatographic method is essentially a batch process; with the future emphasis on continuous culture methods, a compatible continuous separation and purification technology is highly desirable.

Liquid phase affinity partitioning, based upon the immiscibility of aqueous solution of dextran and poly(ethylene glycol), offers an alternative to the use of solid supports. Affinity ligands can be attached to one of the polymers, thus enhancing the partitioning of proteins between the two phases. One of the major problems of this approach in applying it to process scale separations is the small difference in densities of the two phases requiring lengthy settling times or the use of centrifugal devices.

The objective of this research was to investigate the feasibility of immobilising affinity ligands on the surface of water immiscible liquid droplets. Such emulsions have unique properties and, in some cases, advantages over their solid counterparts in the area of affinity separations. Their ease of transportation, for example, has the potential for the development of continuously operating countercurrent separation systems. Liquid perfluorocarbons were selected as supports because of their inherent high density (S.G. ~2.0) and their unique chemical and biological properties.

PREPARATION OF AFFINITY EMULSIONS USING THE 'DIRECT' METHOD

The simplest way of obtaining a dispersion of affinity droplets is to emulsify a liquid perfluorocarbon containing a soluble ligand-reactive reagent in an aqueous solution of the chosen ligand. Examples of the perfluorocarbons that we have used include perfluorodecalin, perfluorotributylamine and perfluoro-2-butyl-tetrahydrofuran. Provided they are soluble, the choice of reactive compound can be very wide, dependent upon the affinity ligand to be immobilised. In coating droplets with protein, reactive compounds found to be suitable, include perfluorooctanoyl chloride, (perfluorooctyl) acetyl chloride, and perfluroooctanoyl anhydride.

The concentration of the perfluoroalkylating reagent in the liquid perfluorocarbon is somewhat critical. Above about 20 μg per ml, multiple layers of ligand, particularly proteins, become deposited, distorting the spherical shape of the droplets and reducing the bioreactivity of the ligand. Other factors that dictate an empirical approach in this method, are the concurrent hydrolysis of the reagent, the degree of emulsification required for a desired size distribution of droplets and the long-term stability of the emulsion.

PREPARATION USING THE 'PARTITION' METHOD

A more controllable and efficient method of producing liquid affinity dispersions is offered by the finding that molecules containing a high proportion of fluorine atoms adsorb very strongly to fluorocarbon surfaces. To utilise this finding in the preparation of liquid affinity supports, the required ligand has to be modified with fluorine containing 'anchor' groups prior to emulsification. For low molecular weight ligands, such as biomimetic dyes, direct synthesis using fluoroalkyl or fluoroaryl intermediates is preferable. In the case of high molecular weight ligands, such as antibodies and other proteins, controlled conjugation with protein-reactive fluoroalkyl reagents has been found to be effective.

In preparing fluoroalkyl-substituted proteins, several amino group reactive compounds were used with equal success. Examples include perfluoroalkyl acid chlorides, anhydrides, imidazolides and isocyanates. Due to the poor solubility of these reagents in water, the reaction with protein is performed by adding a known amount dissolved in an organic solvents, such as tetrahydrofuran or acetonitrile, to a stirred buffered solution of protein. The final concentration of organic solvent should not exceed 20%, by volume, to avoid denaturation of the protein. The amount of fluoroalkylating reagent added is dependent upon its susceptibility to hydrolysis and the degree of substitution required. Imidazolides and isocyanates are preferred, as the concurrent hydrolysis is minimal, thus avoiding a subsequent clean-up step to remove by-products.

The degree of substitution, as measured by the loss of amino groups using trinitrobenzoic acid or fluorescamine assays, should be restricted to less than 20% to avoid excessive loss of ligand bioreactivty and aggregation. The number of fluoroalkyl chains on the ligand will affect the strength of the adsorption or partition to the fluorocarbon surface; again, about 15% amino group substitution appears more than adequate. From the aspect of availability and solubility, perfluoro-octyl based reagents appear to be optimal; the inclusion of a pair of methylene groups between the fluoroalkyl chain and the reactive group facilitates synthesis and coupling. Most fluoroalkyl-substituted proteins appear to be quite stable to long-term storage in sterile solution, and lyophilised storage is not recommended.

Liquid affinity dispersions are prepared by simply vortexing the fluorocarbon liquid with a buffered solution of the substituted protein. More homogeneous dispersions can be prepared by vibrating needles or by ultrasonically dispersing the components. Generally, speaking, the size of the droplets formed is dependent upon the applied energy and the total amount of perfluoroalkylated ligand available. The surface of the drop has to be completely coated with ligand or partial coalescence will occur. In practice, this means that a large excess of ligand will result in an emulsion containing many small droplets, whilst a deficiency results in re-coalescence to a few, much larger, drops. Stability of the emulsions can be improved by the addition of controlled amounts of inert proteins or a neutral fluorosurfactant, such as Zonyl^R FSN (perfluoroalkyl poly(ethylene glycol)), as co-surfactants.

CAPACITY OF LIQUID SUPPORTS

The amount of ligand that can be immobilised on a support is directly proportional to the available surface area. Most conventional solid affinity supports are based on a porous matrix - with the obvious objective of increasing the surface area and, hence, the ligand loading. What is not often appreciated, however, is that kinetic constraints, imposed by the porosity and steric hindrance of densely packed ligands, limit operational efficiency. The surface area of liquid supports is purely a function of the size of the droplets. Table 1 shows the calculated number of packed spherical drops, the surface areas and the expected loading of a monolayer of immunoglobulin, as a function of drop diameter:

Table 1 *Protein loading as a function of drop diameter*

Diameter (*micron*)	*Number* (*per ml*)	*Surface area* (cm^2 per ml)	*Ligand loading* (mg per ml)
50	1.5×10^7	8.8×10^2	0.44
10	1.9×10^9	4.4×10^3	2.22
5	1.1×10^{10}	8.8×10^3	4.44
1	1.4×10^{12}	4.4×10^4	22.20
.5	1.1×10^{13}	8.8×10^4	44.40

The ligand loading and, hence, the capacity for ligate of affinity emulsions prepared by the 'direct' method is extremely variable as previously mentioned. The 'partition' approach yields more reproducible ligand loadings provided that the energy of emsulsification remains fairly constant. Protein loads vary between 0.5 mg per ml resulting from conventional vortexing for 30 seconds up to 2.5 mg per ml after longer and more vigorous agitation - this variation obviously reflects the ranges of droplet diameters obtained. The capacity for antigen of an antibody support is usually around 300 μg per millilitre of a vortexed prepared emulsion. Similar capacities are obtained with ion-exchange fluorosurfactant coated droplets prepared in a similar manner, reflecting the influence of droplet surface area rather than reactivity of ligand.

STABILITY OF DISPERSIONS

The majority of perfluoroalkylated ligands act as effective surfactants, thus stabilising the emulsions as they are formed. If, however, there is insufficient ligand to fully coat the droplets, commiserate with the energy being supplied, then coalescence will occur until a more stable emulsion, consisting of larger drops, is reached. The ratio of ligand to liquid fluorocarbon, therefore, broadly determines the size and stability of the dispersions. To improve upon stability with a weakly surface active ligand, or to conserve ligand, a non-ionic fluorosurfactant can be added prior to or following emulsification. The presence of this surface active agent also serves to reduce nonspecific adsorption of unwanted proteins.

REMOVAL OF SUPPORT

The wide range of liquid fluorocarbons suitable for preparing affinity supports allows the choice of volatile compounds. After preparing the dispersion and reacting it with ligate, the ligand-ligate complex can be recovered by simply evaporating the support. We have demonstrated the utility of this strategy in an analytical application involving the extraction of DNA from clinical samples and subsequent assaying by established hybridisation techniques. Calf thymus histones were perfluoroalkylated and an emulsion with perfluorotetrahydrofuran was prepared. 50 μl aliquots of the emulsion were gently shaken for 30 minutes with samples containing dilutions of a Hae II digest of pBR322. After gentle washing, the liquid supports were transferred to Genescreen hybridisation membrane, retained in a Hybri-slot apparatus, in several consecutive additions. The liquid carrier was allowed to evaporate between additions. After air-drying, denaturation, neutralisation and prehybridisation, the membrane was hybridised overnight with an excess of labelled pBR322. Autoradiography of the washed membrane showed hybridisation of the pBR322 digest down to a concentration of 200 pg per ml in the original sample.

MOBILITY OF LIGAND

A significant difference between liquid and solid supports is the nature of ligand immobilisation. On solids, the ligand molecule is firmly anchored to the matrix, usually by one or more covalent bonds, and is sterically confined to its initial contact orientation. Ligands partitioned onto a liquid surface, however, are not fixed in space and are free to diffuse over the surface. They have a further degree of freedom in that rotational diffusion of the substituted ligand can also occur, albeit rather slowly. Two effects have been observed resulting from these characteristics; a higher avidity for multivalent ligates and increases in ligate capacity with time.

In the experiments involving interactions of DNA with histone-coated dispersions, it was observed that it was very difficult, if not impossible, to displace the DNA from the droplets. Displacement from a conventional solid histone support can be readily accomplished by increasing the salt concentration; this was not observed using liquid supports. It is suggested that the histones migrate to line up under the DNA to effectively create multiple attachments resulting in very high avidity binding. Whilst measuring the capacity of liquid supports for ligate capacity, it was noted that there appeared to be three reaction rates involved. An initial, relatively rapid, rate reflecting ligate mass transport; a second slower rate, presumably representing ligand redistribution, and a final, very slow, increase in capacity as a possible result of translational diffusion of ligand. More careful studies are planned to further demonstrate these effects.

CONTINUOUS SEPARATION POTENTIAL

The most significant difference between conventional solid supports and liquid dispersions, offering unique potentials in the area of bioprocessing, is the ability of the latter to be continuously transported. Whereas it is difficult to move packed solid supports in a continuous stream without considerable attrition, liquid emulsions can be readily pumped for long periods of time without degradation. This attribute allows the conceptual design of continuously operating separation systems as opposed to batch processing dictated by solid supports. Another significant advantage of the solid packed chromatographic column is its susceptibility to blockage by cellular and other debris in the process sample. This necessitates intensive sample clarification, by centrifugation or membrane filtration, prior to column processing. Liquid droplets are inherently deformable and, used in a fluidised bed configuration, will accept turbid process streams with no problems.

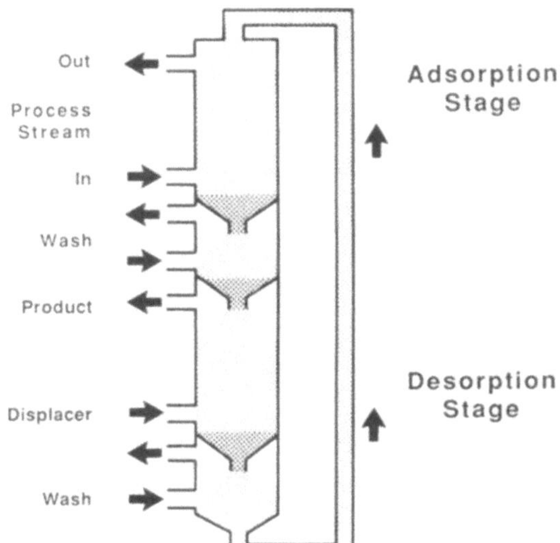

Fig. 1 *Continuous countercurrent extraction concept*

In designing truly continuous operating separation systems, one of the major engineering challenges is to prevent 'carry-over' in the mobile fluid phase with the mobile separation media. The two components have to be separated before commencing a subsequent stage. This has been done on a large scale by using gates or by adopting 'mix and settle' strategies, which, of course, is approaching intermittently operating batch processing. We have found that fluorocarbon dispersions, by virtue of their high density and deformability, settle at restrictions forming what are essentially one way fluidic valves. The fluorocarbon droplets can descend through the restriction, as sand flows through an hour glass, but the aqueous fluids on either side remain separated, even under slight pressure differentials. This finding considerably simplifies the design of a truly continuous separation system.

The most efficient means of continuously extracting a desired component from a process stream is by countercurrent separation. In this approach, the denser phase is introduced at the top of an upward flowing lighter stream, creating a descending fluidised bed. The dense phase is collected at the bottom, separated from the light phase, and transported to another stage. Affinity separations are ideally suited to this separation strategy by virtue of the lack of a need for high resolution, e.g. low theoretical plate requirements. Affinity separations basically require only four stages: adsorption of ligate, washing away excess or trapped sample, displacement of ligate and a final washing. All of these steps can be performed sequentially in a countercurrent mode with the liquid affinity support being passed through each stage before being re-circulated to the first stage. This concept is schematically shown in the Fig. 1.

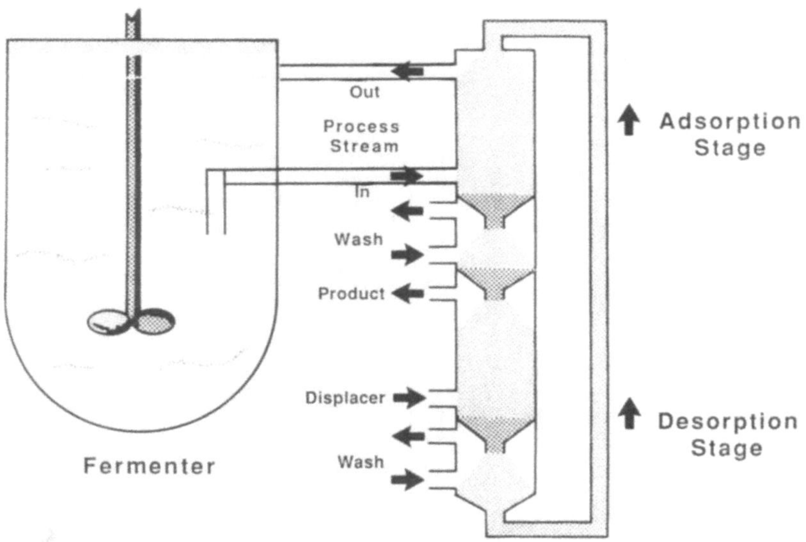

Fig. 2 *Fermenter respiration concurrent with continuous extraction*

The major requirement in designing such a system for use in a standard gravitational field is a reasonable sedimentation rate of the heavier affinity phase. Liquid fluorocarbons, with densities approaching 2, are ideally suited; a 50 micrometre drop, for example, has a Stoke's sedimentation of 7.5 cm per minute. Below 10 micrometres diameter, however, a centrifugal system is required to achieve reasonable throughputs. Other advantages offered by the use of fluorocarbons as liquid support matrices involve their biological and chemical properties. Fluorocarbons are biocompatible, as exemplified by their application as blood substitutes. Being chemically inert, they can be autoclaved, thus offering sterile affinity supports when prepared by the partition method using filter-sterilised perfluoroalkylated ligand solutions. Fluorocarbons have a high affinity for oxygen and carbon monoxide, offering the potential for fermenter respiration concurrent with continuous extraction of secreted product (Fig. 2).

A detailed analysis of the design criteria and operating requirements of continuous countercurrent affinity separation systems is far beyond the scope of this contribution -and it's author! As the bioprocessing industry, through advances in genetic engineering, moves towards continuous culture of secreted bioproducts, the need for associated separation facilities becomes increasingly apparent. Integration of the extraction system with the continuous culture unit will be an additional challenge for the future. In presenting our preliminary results obtained in preparing and using liquid fluorocarbon based affinity supports, it is hoped that such studies will be stimulated.

MAGNETICALLY ENHANCED AQUEOUS TWO-PHASE SEPARATION

P. Wikström, S. Flygare and P.-O. Larsson

Pure and Applied Biochemistry
University of Lund, Chemical Center
P.O. Box 124, S-221 00 Lund, Sweden

A new technique to increase the rate of phase separation in aqueous two-phase systems is described. The technique is based on the use of magnetically susceptible additives, ferrofluids or iron oxide particles. In a magnetic field these additives will induce a faster phase separation. When dextran-covered ferrofluid or iron oxide particles were mixed with an aqueous two-phase system (PEG and dextran), it was found that the magnetic additive was totally distributed to the dextran phase. After mixing of the two-phase system, it was possible to reduce the time for phase separation by a factor of 35 (ferrofluid) or 70 (iron oxide particles) by applying a magnetic field (0.3 Tesla) to the system. It was found that neither the activity, nor the partitioning of the enzymes tested, was influenced by the addition of magnetically susceptible material. When ferrofluid was used it was possible to keep the dextran-phase stationary in a column system, while the PEG-phase could be pumped through the column.

The ferrofluid concept

A ferrofluid is a magnetic liquid that will be magnetised in the presence of an external magnetic field. The particles of the ferrofluid are small enough to be kept from settling by the random thermal motion of the molecules in the surrounding liquid. To achieve a stable ferrofluid it is important to create small magnetically inducible particles around 100 Å in size. It is also necessary to stabilise the ferrofluid with a molecule that has an affinity for the carrier fluid as well as for the particle. In this way the coating will prevent the particles in the ferrofluid from agglomerating [1].

Preparation of dextran-stabilised ferrofluid

Ferrofluid stabilised with dextran was prepared as follows (a modification of a method used by Molday and MacKenzie [2]). Iron(II) chloride and iron(III) chloride (7.55 g $FeCl_3$ 6 H_2O and 3.20 g $FeCl_2$ 4 H_2O) were dissolved in distilled water (50 ml) and a 20% solution of Dextran T500 (50 ml) was added. The temperature of the mixture was kept constant at 60°C and aqueous ammonia (100 ml, 7.5%) was added in portions while stirring continuously. After the last addition the stirring was continued for another

15 minutes at 60°C. The suspension was then dialysed against distilled water to neutral pH, and centrifuged (15 min at 12,000 g) to remove larger aggregates. The iron content of the preparation was 7 mg/ml (determined by atomic absorption) and the dry weight contents was 55 mg/ml.

Preparation of two-phase systems containing ferrofluid

An aqueous two-phase system containing dextran-coated ferrofluid was prepared in the same way as an ordinary two-phase system. If larger amounts of ferrofluid were added to the system, the content of dextran in the ferrofluid had to be considered.

Partitioning of ferrofluid and iron oxide particles in aqueous two-phase systems

The dextran-coated ferrofluid described above was found to be totally partitioned to the dextran phase when added to a PEG/dextran two-phase system. We also found that uncoated iron oxide particles, which were considerably larger than the ferrofluid particles (approx. 1 μm), also partition to the dextran phase. In Fig. 1 test tubes are shown containing PEG/dextran two-phase systems to which have been added either iron oxide particles or different amounts of ferrofluid. The two phases in this figure are just beginning to separate, and in Fig. 2 the same systems are shown after complete phase separation. The ferrofluid is seen to be homogeneously and totally distributed to the dextran phase. It is also shown that it is possible to completely remove the larger iron oxide particles from the phase system, unlike the ferrofluid particles.

Both the ferrofluid and the larger iron oxide particles can be manipulated by an external magnetic field. In this way a faster separation of the phase system can be achieved. In Fig. 3 a well-mixed PEG/dextran system is shown, containing 100 mg/ml of ferrofluid. In Fig. 4 the same system is shown, but now phase separated by the use of a permanent magnet.

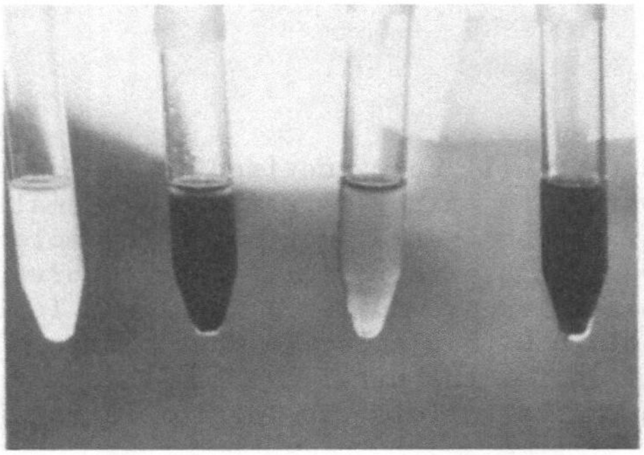

Fig. 1 *Test tubes containing PEG/dextran aqueous two-phase systems during separation. The tubes contain, from left to right: no additive, iron oxide particles (10 mg/ml), ferrofluid (15 mg/ml) and ferrofluid (100 mg/ml)*

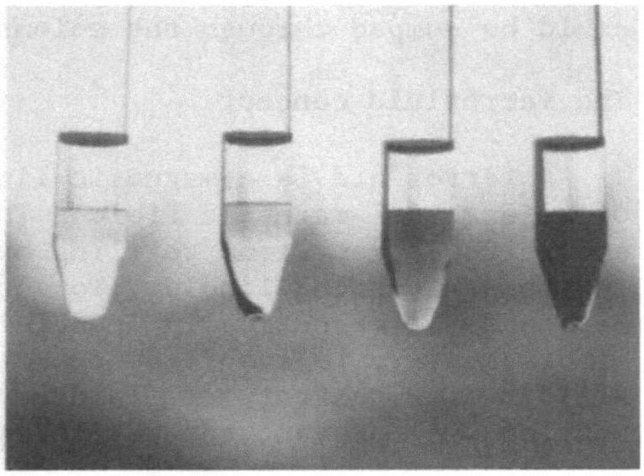

Fig. 2 *The same system as in Fig. 1 totally separated*

Enhanced phase separation in systems containing magnetically susceptible additives

The phase separation of well-mixed two-phase systems with and without added magnetic material was compared. The well-mixed two-phase systems (contained in test-tubes) were either allowed to separate under gravity or placed in a non-homogeneous magnetic field (obtained by a small laboratory-built electromagnet) and the time needed for separation (visual inspection) was recorded.

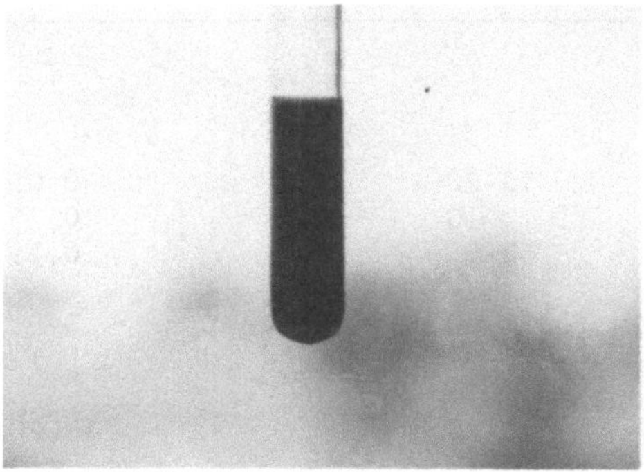

Fig. 3 *A well-mixed PEG/dextran aqueous two-phase system containing ferrofluid (100 mg/ml)*

Fig. 4 *The same system as in Fig. 3 separated by the use of a permanent magnet*

Table 1 shows that a very substantial reduction in the time needed for phase separation was generally obtained in the presence of magnetic additives. Addition of ferrofluid to a concentration of 100 mg/ml reduced the separation time by a factor of about 35. Similarly, the addition of iron oxide particles (1 μm particle size) reduced the separation time by a factor of about 70.

Table 1 also shows that the time needed for separation was, as expected, influenced by the amount of ferrofluid added. In Fig. 5 this is also shown to be the case for added iron oxide particles. It was also noticed that a small reduction in separation time occurred even in the absence of a magnetic field (open circles; Fig. 5). This is probably due to the increased density of the dextran phase caused by the addition of the particles. This effect was also seen in systems containing high concentrations of ferrofluid, although it was not as pronounced as in systems with iron oxide particles. This was to be expected since the density of the ferrofluid is not very different from that of the dextran phase.

Table 1 *Time required for phase separation of well-mixed aqueous two-phase systems containing ferrofluid or iron oxide particles*

	Volume ratio PEG:dextran (v:v)	Separation time (min)	
		Without magnetic field	With magnetic field
Ferrofluid (mg/ml)			
100 a/	9:1	15-20	0.05
	2:1	10	0.75
	1:1	3-5	0.75
15 b/	9:1	15-20	3-5
	1:1	3-5	1
0	9:1	15-20	15-20
	2:1	13	13
Iron oxide particles (mg/ml)			
10	9:1	15-20	0.25
	1:1	3-5	0.15

Phase systems containing 7% PEG 6000, 5% Dextran T500 and different amounts of ferrofluid or iron oxide particles were equilibrated and allowed to separate. Preparations with different volume ratios between top and bottom phases were made in the test tubes and thoroughly mixed. The electro-magnet used gave a non-homogenous magnetic field (maximum field strength of 1.2 Tesla) with a field strength of about 0.1 Tesla in the middle of the test tube.

a/ Fe content, 1 mg/ml

b/ Fe content, 0.15 mg/ml (Reproduced with permission from [4])

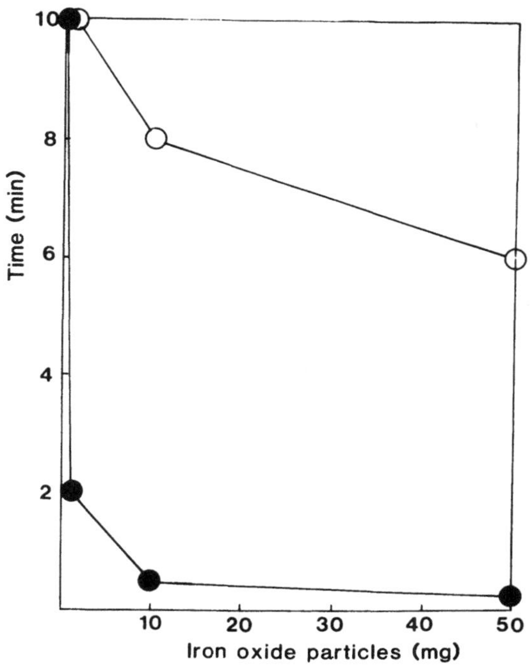

Fig. 5 *Time required for phase separation of an aqueous two-phase system containing iron oxide particles. Two millilitres of a two-phase system containing 0-50 mg iron oxide particles were thoroughly mixed on a Vortex mixer and the time required for phase separation was measured. ○ without a magnetic field, ● with a magnetic field (about 0.3 Tesla). Reproduced with permission from [4]*

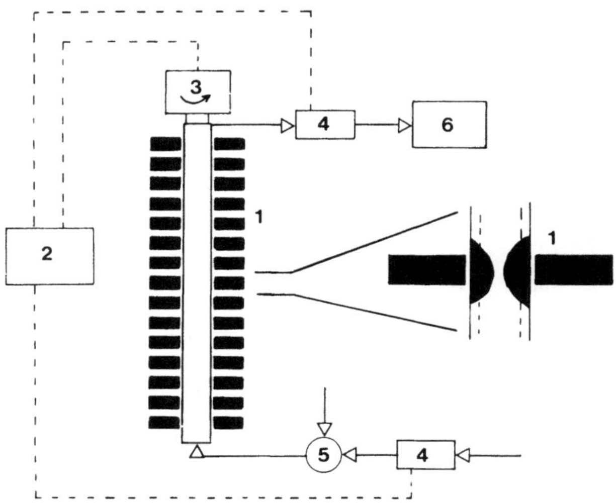

Fig. 6 *Apparatus for semi continuous separation: 1. column with permanent magnets and attached dextran-ferrofluid; 2. electronic switch; 3. stirrer; 4. pump; 5. injector; 6. fraction collector Reproduced with permission from [4]*

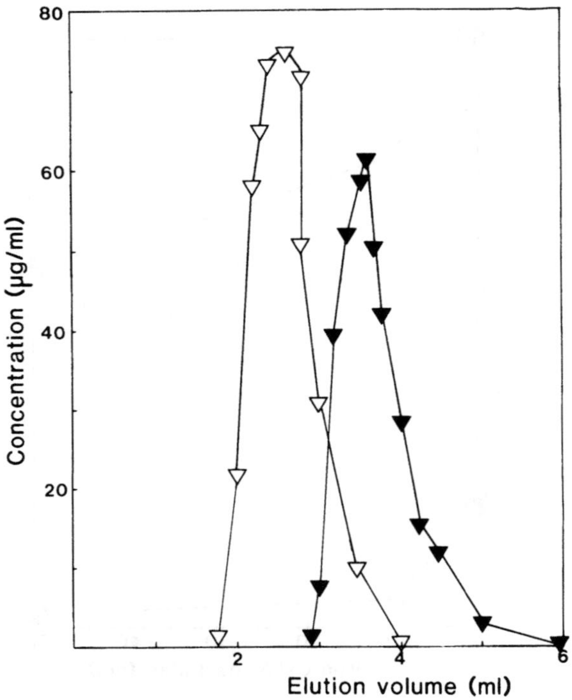

Fig. 7 *Separation of ß-galactosidase ▽ and lactate dehydrogenase ▼ . The apparatus shown in Fig. 6 was used. Time settings for stirring, pauses and pumping were 10, 5 and 15 seconds, respectively. The flow rate was 0.5 ml/min and stirring speed 400 rpm. Lactate dehydrogenase (92 µg) and ß-galactosidase (90 µg), dissolved in PEG phase (100 µl), were injected and 100 µl fractions were collected and assayed. Reproduced with permission from [4]*

Enzyme separation in a column containing a two-phase system with ferrofluid

The fact that the dextran phase containing ferrofluid is able to be entrapped in a magnetic field makes it possible to construct a simple two-phase column.

In Fig. 6 the apparatus is shown which we used for semi-continuous separation of enzymes (lactate dehydrogenase and ß-galactosidase). The column consisted of a glass tube, to which permanent magnets were fixed in pairs and inside the tube was a glass rod stirrer. The separation cycle is divided into three steps, and is controlled by an electronic switch. During the first step, the stirrer is mixing the two phases. Then there is a pause during which the dextran phase is drawn towards the magnets and the phases are thus separated. During the third step PEG-phase is pumped through the column while the dextran phase remains attached to the magnets. Then a new cycle starts.

In Fig. 7 a chromatogram is shown where lactate dehydrogenase and ß-galactosidase were separated. The plate number of the column was calculated (according to liquid chromatography theory [3] to be around 50-100, which means that each magnet-pair corresponds to more than one separation step

($N = 5.5(V_e/V_{\frac{1}{2}})^2$, N - plate number, V_e - elution volume, $V_{\frac{1}{2}}$ - volume corresponding to the half-width of the elution peak). The separation described in Fig. 7 was achieved in less than one hour and the time needed may be decreased further.

We also investigated whether or not the ferrofluid influenced the partitioning or the activity of the enzymes used. If the ferrofluid contained well stabilised particles, there was no significant difference between a phase system containing ferrofluid and a system without ferrofluid.

Partitioning of cell particles

Partitioning of chloroplasts and *Escherichia coli* was attempted in two-phase systems containing ferrofluid. The partitioning of the bacteria was not affected by the presence of ferrofluid, but the chloroplasts partitioned to the dextran phase instead of to the PEG-phase. A modified ferrofluid was thus made [4] and in a two-phase system, containing this modified ferrofluid, the chloroplast partitioning was not affected.

ACKNOWLEDGEMENT

This work was supported by a grant from the National Swedish Board for Technical Development.

REFERENCES

1 R.E. Rosenweig, Magnetic fluids, Scientific American 247(4):124 (1982)
2 R.S. Molday and D. MacKenzie, Immunospecific ferromagnetic iron-dextran reagents for the labeling and magnetic separation of cells, J. Immunol. Methods 52:353 (1982)
3 L.R. Snyder and J.J. Kirkland, in: "Introduction to Modern Liquid Chromatography," 2nd ed., p 277, Wiley, New York (1979)
4 P. Wikström, S. Flygare, A. Gröndalen and P.-O. Larsson, Magnetic aqueous two-phase separation: A new technique to increase rate of phase-separation, using dextran-ferrofluid or larger iron oxide particles, Analytical Biochemistry 331:339 (1987)

APPLICATION OF PARTITION TECHNOLOGY TO PARTICLE ELECTROPHORESIS

James M. Van Alstine [1,2], J. Milton Harris [3], Laurel J. Karr [1]
Stephan Bamberger [1], Helen C. Matsos [1] and Robert S. Snyder [1]

[1] Biophysics Branch ES76, Space Science Laboratory
NASA/Marshall Space Flight Center, Huntsville, AL 35812
[2] Universities Space Research Association
[3] Department of Chemistry, University of Alabama at
Huntsville, Huntsville, AL 35899, USA

Recent investigations have led to the development of polymer coatings to control electroosmosis which adversely affects electrophoretic separation processes on Earth and in space. Similar chemistry has been employed to create polymer derivatised antibodies for use in immunoaffinity electrophoresis of particles. The effects of polymer-ligand concentration on particle electrophoretic mobility and partition in aqueous polymer two-phase systems are compared.

INTRODUCTION

NASA-related research on aqueous polymer two-phase system partitioning presently encompasses three areas of effort [1]:

1 Theoretical and practical studies directed towards gaining a better understanding of phase partitioning in unit-gravity (g) and in low-g, Brooks et al [2,3].
2 Development of novel methods to control the localisation of demixed aqueous polymer phases in low-g.
3 Development of novel applications of "partition-related" technology and new polymer phase systems.

Research has progressed in all of these areas. For example, Sharp et al [4] and Karr et al [5] have developed Immunaffinity (IA) partition, discussed in more detail elsewhere in these Proceedings. In this technique poly(ethylene glycol) (PEG) derivatised antibodies (Ab's) are employed to specifically increase the partition of desired material into the PEG-rich phase in aqueous polymer two-phase systems. PEG-derivatised Protein A has been developed so that commercial Ab preparations can be utilised without prior derivatisation [6].

IA partition boasts much of the sophistication of Fluorescence Activated Cell Sorting with increased throughput, cell survival, and hardware cost-efficiency. To date immunoaffinity partition has been employed for particle partition but the technique should be readily applicable to the separation of macromolecules [7]. In this respect, IA partition is similar to hydrophobic affinity partition in which PEG-fatty acid esters or similar PEG-ligands are

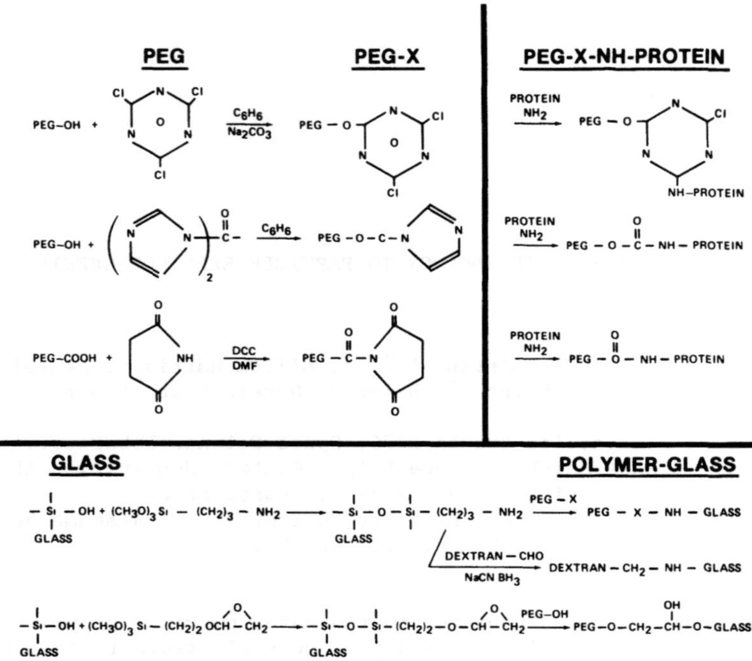

Fig. 1 *Polymer derivatisation of glass surfaces and proteins*

used to differentially partition cells and/or macromolecules [8-11]. Hydrophobic affinity partition features inexpensive ligands and readily reversible ligand-substance interaction, but it lacks the specificity of IA partition.

Harris et al [12,14] have developed covalent surface coatings based on the PEG and dextran polymers typically utilised to compound aqueous polymer two-phase systems (APTPS's). The chemistry involved, outlined in Fig. 1, is similar to that developed to produce polymer-ligands for affinity partition [15,16]. For instance, prior activation of PEG with cyanuric chloride facilitates its reaction with free amino containing substances, such as the lysine residues of proteins [4,5] or glass surfaces whose silanol groups have previously been derivatised with aminopropylsilane [14]. Dextran can be attached to "amino-glass" using sodium cyanoborohydride [15,16].

Polymer coatings have been shown to control which APTPS phase prefers to "wet" the container wall [13,17]. They can be used to control the disposition of demixed phases in microgravity; a prerequisite for Space Bioprocessing via partition. On Earth, wetting experiment results can be quantitated via contact angle measurements [17]. PEG coatings can also be evaluated electrophoretically. They have been shown to control (i.e. reduce or eliminate) the electroosmosis that adversely affects electrophoretic separation processes on Earth and in space [14]. In a similar manner, particles coated with PEG molecules, either covalently as in the case of standard particles, or via the use of PEG-ligands as in the case of cells, exhibit greatly reduced electrophoretic mobility.

This paper introduces two techniques currently under patent development; Covalent Polymer Coatings to Control Electroosmosis, and Affinity

Electrophoresis of Particles Using PEG-Ligands. Both techniques involve the application of partition-based research to electrophoresis, another separation technique influenced by gravity and liquid-surface interactions.

EXPERIMENTAL

Compounding and physicochemical characterisation of aqueous polymer two-phase systems has been described previously, as has cell partition in APTPS's [5,11,18]. Glutaraldehyde fixed erythrocyte partition and cell electrophoresis were carried out as described by Seaman [19] and Walter et al [20]. A Model I Cytopherometer (Rank Bros., Cambridge) was used in all electrophoretic mobility studies. Cells were suspended in isotonic phosphate buffered saline, pH 7.2, 25°C, and exposed to a field of 40 volts DC per cm. Electroosmosis measurements were performed in a similar manner using standard particles suspended in 7.5 mM NaCl buffer, pH 5.8, 890 micromhos/cm, and 2.0 mm diameter quartz capillaries (Vitro). The buffer and capillaries were chosen to produce electroosmosis at significant levels. Standard particles consisted of 2.2 micron diameter poly(styrene latex) spheres (NASA) or 2 micron glass spheres (Duke Scientific) [14]. Particle electrophoretic mobility is expressed as velocity (μm/sec) per unit field strength (V/cm).

Polymer coating chemistry and affinity ligand synthesis, purification, and analysis were performed as described previously [5,14-16]. PEG coated glass surfaces contained approximately 1.4 aminopropyl groups (AP) per 100A^2 of surface. Subsequently one PEG was bounded per 10 AP groups [14].

PEG-antibody preparations were prepared using polyclonal immunoglobulin G (IgG) anti-human erythrocytes [5]. Approximately half of the 90 lysine groups on each Ab molecule were modified via reaction with cyanuric chloride activated PEG 5000 monomethyl ether. Erythrocytes incubated with PEG-IgG in upper phase for 15 minutes at 37°C, were washed and then partitioned, or subjected to cytopherometry.

RESULTS AND DISCUSSION

Polymer Coatings to Control Surface Zeta Potential Expression

Electroosmosis (EO) is the movement of a conducting liquid, relative to a charged container surface in the presence of an electric field. It is a manifestation of the zeta potential of the liquid-wall interaction, and is analogous to the electrophoretic mobility which characterises the behaviour of particles under similar conditions [19]. The surface of the untreated glass or plastic chambers used in electrophoretic processes are usually charged and electroosmotic flow of liquid near the chamber walls accompanies the desired electrophoresis of particles or macromolecules. This flow can adversely affect electrophoretic separation process in a variety of ways [14,20].

EO is readily detected in small diameter cylindrical capillaries with closed electrode ends, containing standard particles suspended in dilute salt solution (Fig. 2). Fluid movement at the walls, coupled with a compensatory back flow down the centre of the capillary, causes a parabolic distribution of particle velocities across the diameter of the capillary. Only at the "stationary level", where the net fluid flow is zero, is the apparent mobility of a sample characteristic of its surface charge.

In the past a variety of coatings have been utilised to control EO. None have proven ideal [14]. Methylcellulose of high molecular weight (i.e. 110,000) works well, as indicated by the linear mobility profile in Fig. 2.

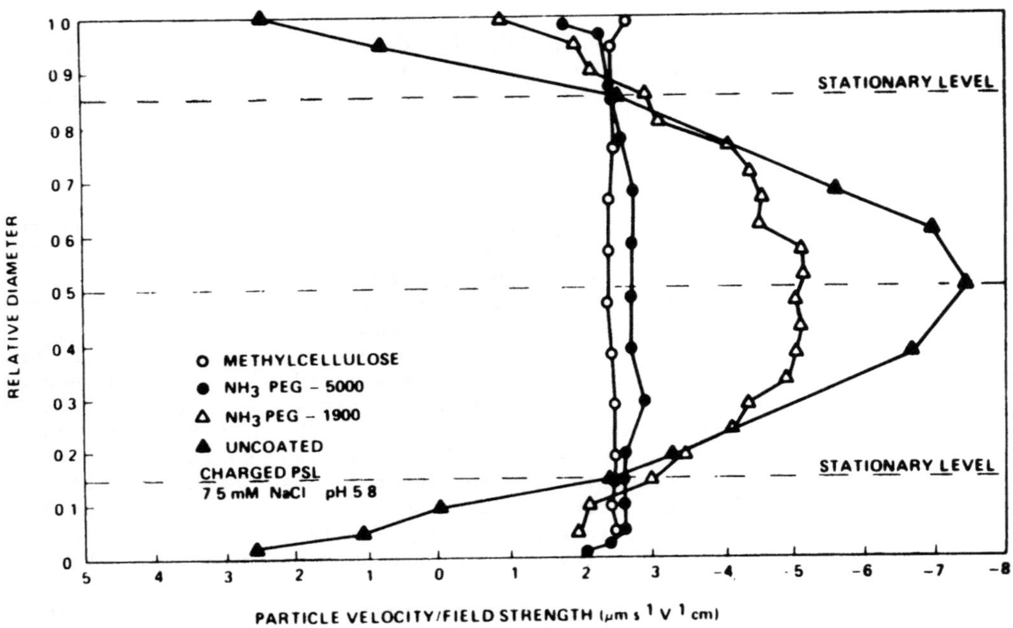

Fig. 2 *Degree of electroosmosis in uncoated and polymer coated capillaries*

Yet methylcellulose coatings often desorb during storage; EO becomes random and methylcellulose free in solution may adsorb onto particles altering their mobility. PEG coatings applied to aminopropylsilane activated glass have been shown to be very effective at controlling EO. Silane coating reduces EO in quartz capillaries by about 30% (not shown), whereas covalently coupling PEG 20,000 to the glass eliminates EO analogous to the initial effects of the methylcellulose coating in Fig. 2. PEG coatings appear to possess all the characteristics of ideal electrophoresis coatings including covalent, nonionic, monolayer, long life, ease of application, reduction of protein surface adsorbtion, pH stability. They can be utilised to control Eo or to produce standard particles of varied electrophoretic mobility [14]. We are currently testing a PEG coating that can be applied in one reaction step to glass or plastic surfaces.

The ability of neutral, high MW, PEG coatings to significantly reduce surface EO or particle electrophoretic mobility appears to be due to the hydrophilic nature of PEG. This property endows PEG-coated surfaces with a large resistance to local fluid flow through the coating network [21]. The "plane of shear" (i.e. region of appreciable fluid flow) is shifted to a region of near zero effective zeta potential [19].

Affinity Electrophoresis of Particles Using PEG-Ligands

The electrophoretic and chromatographic separation behaviour of macromolecules are altered by their PEG-derivatisation [4,5,15,16]. Similar alterations are expected of macromolecules and cells interacting with PEG-derivatised affinity ligands. Müller et al [22] demonstrated that PEG-derivatised DNA affinity ligands can be used to fractionate nucleic acids by gel electrophoresis, as well as via partition in APTP's. We have been studying the use of PEG-ligands for the affinity electrophoresis of particles.

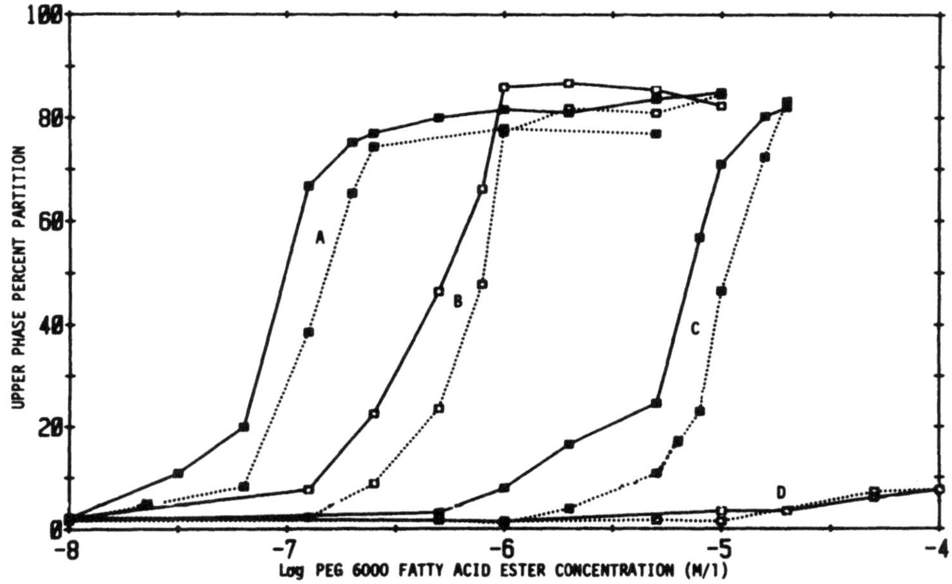

Fig. 3 *Partition of fresh (solid line) and glutaraldehyde fixed (dotted line) human erythrocytes in a (5,4) V system containing D T500, PEG 8000 and the following PEG 8000 esters: A - 18:0, B - 16:0, C - 14:0, D - 12:0*

Fig. 3 indicates the ability of PEG-fatty acid esters to increase cell partition in a manner sensitive to the cell surface as well as the hydrophobicity of the ester acyl tail (i.e. length). The effectiveness of hydrophobic affinity partition ligands is also related to acyl tail unsaturation and PEG molecular weight [8,10,11]. Fig. 4 demonstrates the ability of PEG-esters to coat erythrocytes and reduce their electrophoretic mobility in a concentration dependent manner (J.M. Van Alstine, Ph.D. Thesis, University of British Columbia, Vancouver, 1984). It should be noted that under these conditions, 0.3 represents the lower level of mobility which can be accurately detected.

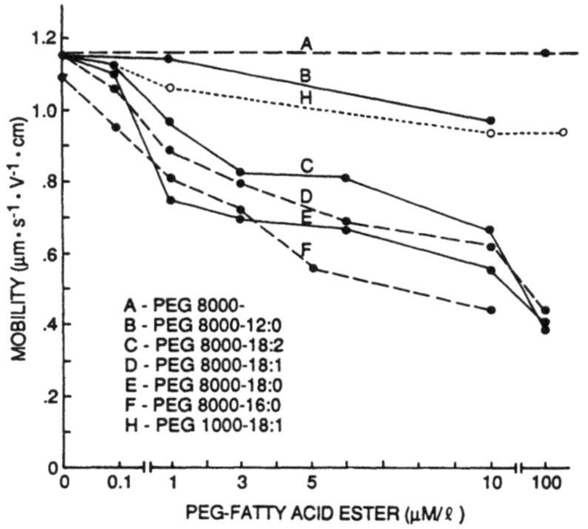

Fig. 4 *Effect of PEG-fatty acid esters on the electrophoretic mobility of human erythrocytes*

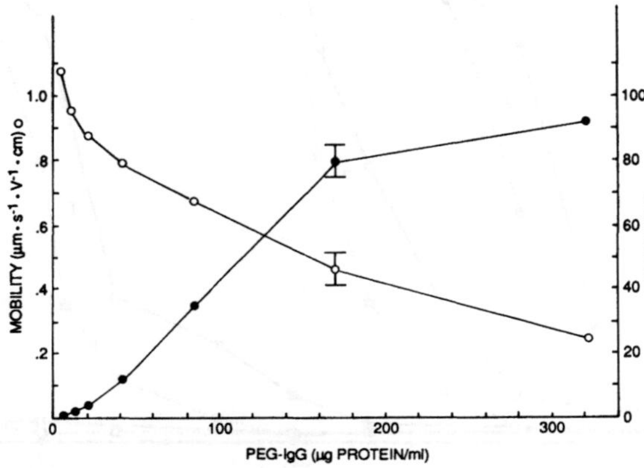

Fig. 5 *Effect of PEG-IgG (anti-human erythrocyte) on the electrophoretic mobility and partition in (5,4) V aqueous polymer two phase system of human erythrocytes*

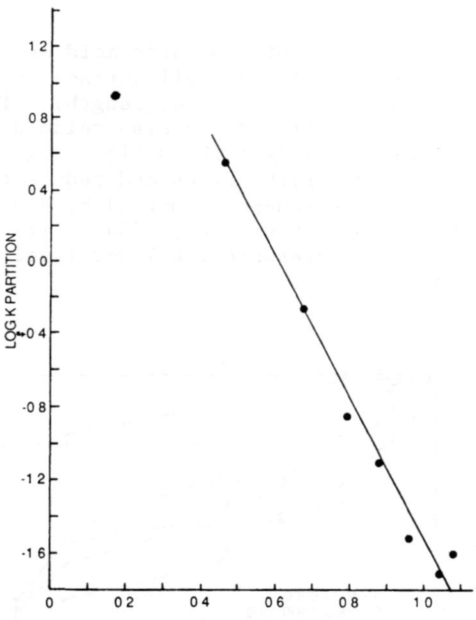

Fig. 6 *Comparison of the data presented in Fig. 5. If % partition = x;* $K = \frac{x}{100 - x}$

Immunoaffinity electrophoresis can be performed using PEG-modified antibodies. The concentration dependence, of both erythrocyte partition and electrophoresis, with regard to prior exposure of cells to PEG-modified anti-erythrocyte IgG, is shown in Fig. 5. PEG derivatisation does not appear to alter IgG specificity although it significantly reduces antibody induced cell aggregation [5]. Such aggregation is the principal reason why immuno-electrophoresis of particles has not been practically demonstrated since the idea was first proposed in the 1920's [19].

Partition is theoretically expected to vary exponentially with the surface concentration of adsorbed affinity ligand, while cell electrophoresis is expected to vary in a more direct manner [6,9,19]. The plot of log K (partition coefficient) verus mobility for various concentrations of PEG-affinity ligand is expected to yield a straight line of negative slope. Such dependence has been found for both PEG hydrophobic affinity ligands and PEG immunoaffinity ligands (Fig. 6).

ACKNOWLEDGEMENTS

This work was supported by the National Aeronautics and Space Administration (NASA). S.B. holds a fellowship of the National Research Council and J.V.A. is a Visiting Scientist of the Universities Space Research Association. The authors wish to thank Shelby Morris, Susan Burrer and Kate Bunker for editorial assistance.

REFERENCES

1 J.M. Van Alstine, L.J. Karr, J.M. Harris, R.S. Snyder, S. Bamberger, H.C. Matsos, P.A. Curreri, J. Boyce and D.E. Brooks, Phase partitioning in space and on Earth, in: "Immunobiology of Proteins and Peptides IV," M.Z. Atassi, ed., Plenum Press, New York

2 S. Bamberger, J.M. Van Alstine, J.M. Harris, J.K. Baird, R.S. Snyder, J. Boyce and D.E. Brooks, Demixing of aqueous polymer two-phase systems in low gravity, Sepn. Sci. and Technol. 23:17 (1988)

3 S. Bamberger, D.E. Brooks, J.M. Van Alstine, J.M. Harris and R.S. Snyder, Demixing kinetics of phase separated polymer solutions in microgravity, in: "Separations Using Aqueous Phase Systems: Applications in Cell Biology and Biotechnology," D. Fisher and I.A. Sutherland, eds., Plenum, New York (1989)

4 K.A. Sharp, M. Yalpani. S.J. Howard and D.E. Brooks, Synthesis and application of a poly(ethylene glycol) antibody affinity ligand for cell separation in aqueous polymer two-phase systems, Anal. Biochem. 154:110 (1986)

5 L.J. Karr, S.G. Shafer, J.M. Harris, J.M. Van Alstine and R.S. Snyder, Immuno-affinity partition of cells in aqueous polymer two-phase systems, J. Chromatog. 354:269 (1986)

6 L.J. Karr, J.M. Van Alstine, R.S. Snyder, S.G. Shafer and J.M. Harris, Cell separation by immunoaffinity partitioning with poly(ethylene glycol)-modified Protein A in aqueous polymer two-phase systems, J. Chromatog. 441:219 (1988)

7 M. Wilcheck, Affinity chromatography, in: "Methods in Enzymology: Enzyme Purification and Related Techniques," Vol. 104C, W.J. Jakoby, eds., Academic Press, New York (1984)

8 E. Eriksson, P.-Å. Albertsson and G. Johansson, Hydrophobic surface properties of erythrocytes studied by affinity partition in aqueous two-phase systems, Molec. Cell. Biochem. 10:123 (1976)

9 G. Johansson, Partition of proteins, in: "Partitioning in aqueous two phase systems," H. Walter, D.E. Brooks and D. Fisher, eds., Academic Press, New York (1985)

10 J.M. Van Alstine and D.E. Brooks, Cell membrane abnormality detected in erythrocytes from multiple sclerosis patients by partition in two-polymer aqueous-phase systems, Clin. Chem. 30:441 (1983)

11 J.M. Van Alstine, P. Sorensen, T.J. Webber, R. Greig, G. Poste and D.E. Brooks. Heterogeneity in the surface properties of B16 melanoma cells from sublines with differing metastatic potential detected via two-polymer aqueous phase partition, Exptl. Cell. Res. 164:366 (1986)

12 J.M. Harris, D.E. Brooks, J.F. Boyce, R.S. Snyder and J.M. Van Alstine, Hydrophillic polymer coatings for control of electroosmosis and wetting, in: "Polymer Surface Dynamics," J.D. Andrade, eds., Plenum Press, New York, (1988)

13 J.M. Van Alstine, J.M. Harris, R.S. Snyder, P.A. Curreri, S. Bamberger and D.E. Brooks, Separation of aqueous two-phase polymer systems in microgravity, in: Proceedings of the 5th European Symposium on Material Sciences under Microgravity, European Space Agency Publication SP-222 (1984)

14 B.J. Herren, S.G. Shafer, J.M. Van Alstine, J.M. Harris and R.S. Snyder, Control of electroosmosis in coated quartz capillaries, J. Colloid. Interf. Sci. 115:46 (1978)

15 J.M. Harris, E.C. Struck, M.G. Case, M.S. Paley, J.M. Van Alstine and D.E. Brooks, Synthesis and characterisation of poly(ethylene glycol) derivatives, J. Polymer Sci., Polymer Chem. Edn. 22:341 (1984)

16 J.M. Harris and M. Yalpani, Polymer-ligands used in affinity partitioning and their synthesis, in: "Partitioning in Aqueous Two-Phase Systems," H. Walter, D.E. Brooks and D. Fisher, eds., Academic Press, New York (1985)

17 J.F. Boyce, J.M. Harris, J.M. Van Alstine and D.E. Brooks, Dextran coated glass slides as a model for cell surface interactions with aqueous two phase polymer systems, in: "Separations Using Aqueous Phase Systems: Applications in Cell Biology and Biotechnology," D. Fisher and I.A. Sutherland, eds., Plenum, New York (1989)

18 S. Bamberger, D.E. Brooks, K.A. Sharp, J.M. Van Alstine and T.J. Webber, Preparation of phase systems and measurement of their phsicochemical properties, in: "Partitioning in Aqueous Two-Phase Systems," H. Walter, D.E. Brooks and D. Fisher, eds., Academic Press, New York (1985)

19 G.V.F. Seaman, The electrokinetic behaviour of red cells, in: "The Red Blood Cell, Vol. 1," D. MacN. Surgenor, ed., Academic Press, New York (1975)

20 H. Walter, E.J. Krob, D.E. Brooks and G.V.F. Seaman, Effects of acetaldehyde and glutaraldehyde fixation on the surface properties of red blood cells as determined by partition in aqueous phases, Exptl. Cell Res. 80:415 (1973)

21 K.A. Sharp and D.E. Brooks, Calculation of the electrophoretic mobility of a particle bearing bound polyelectrolyte using the nonlinear Poisson-Boltzmann equation, Biophys. J. 47:563 (1985)

22 W. Müller, I. Hattesohl, H.J. Schuetz and G. Meyer, PEG derivatives of base and sequence specific DNA ligands: DNA interaction and application for the base specific separation of DNA fragments by gel electrophoresis, Nucleic Acid Res. 9:95 (1981)

FORMATION OF BEADS IN AQUEOUS TWO-PHASE SYSTEMS

Torbjörn G.I. Ling and Bo Mattiasson

Department of Biotechnology, Chemical Center
University of Lund
P.O. Box 124, S-221 00 Lund, Sweden

INTRODUCTION

Spheres of agarose and other polymers are generally made in phase systems consisting of one aqueous phase and one organic phase. However, several attempts have been made to avoid the use of organic solvents, which is often hazardous to the biological function of enzymes and cells. Although the first aqueous two-phase system described in the literature was composed of starch and agar [1], such phase systems have not hitherto found any known practical use.

EXPERIMENTAL

Agarose beads have been produced in aqueous two-phase systems having a top phase consisting of a copolymer of PEG and PPG (Synperonic F-68, ICI Petrochemicals and Plastics Division) and a bottom phase consisting of hydroxyethylated agarose, gelling at low temperature (SeaPlaque, gelling temperature 40°C and SeaPrep, 20°C, from Marine Colloids).

Agarose 3% w/w is melted at an elevated temperature (70°C). The temperature is then decreased to 50°C and in case any cells are to be included, they are now mixed with the agarose solution. The agarose is then mixed with the Synperonic phase at constant temperature and constant stirring. After a short time (1-3 min) the temperature is rapidly changed to 20°C to make the agarose gel.

RESULTS

Beads have been made in 10 ml systems using a normal magnetic stirrer in a beaker and in a 1 l thermostatted reaction vessel. Beads with diameters from 20 to 200 μm could be made by controlling the stirrer speed.

DISCUSSION

This method has several advantages. One can void the use of organic solvents and, furthermore, aqueous two-phase systems have a suitable interfacial tension for production of beads with a size relevant for use both in chromatography and for entrapments of cells. It is therefore not necessary to add surfactants to obtain an emulsion. The well-established suitability of these two-phase systems for separation of biological material makes this method attractive for entrapment of labile components. The method is, of course, not restricted to agarose, but has also been employed for production of beads made of other gel forming polymers, for example, alginate and karragenan.

REFERENCES

1 M.W. Beijerinck, Zbl. Bakt. 2 (1896)

COMPARISON OF CELL AND MACROMOLECULAR PARTITION IN PEG 8000/DEXTRAN T500, AQUAPHASE PPT OR REPPAL PES 200 PHASE SYSTEMS

D.R. Skuse, R. Norris-Jones, D.E. Brooks [1] and M. Yalpani [2]

[1] Departments of Chemistry and Pathology
University of British Columbia
2211 Wesbrook Mall, Vancouver, B.C., Canada V6T 1W5
[2] Domtar Research Centre, P. O. Box 300
Senneville, Quebec, Canada H9X 3L7

Aquaphase PPT and Reppal PES 200 are starch-based polymers which are being marketed as less expensive alternatives to dextran in two-phase partitioning. We have investigated the phase forming and partitioning behaviour of these materials with PEG 8000 as the second polymer.

Phase diagrams for each polymer with PEG were determined using a combination of polarimetric and refractive index measurements. The partition behaviour of a representative cell (human erythrocyte), bacterium (*Salmonella entenditis*) and protein (rabbit IgG) in PEG/dextran, PEG/Reppal and

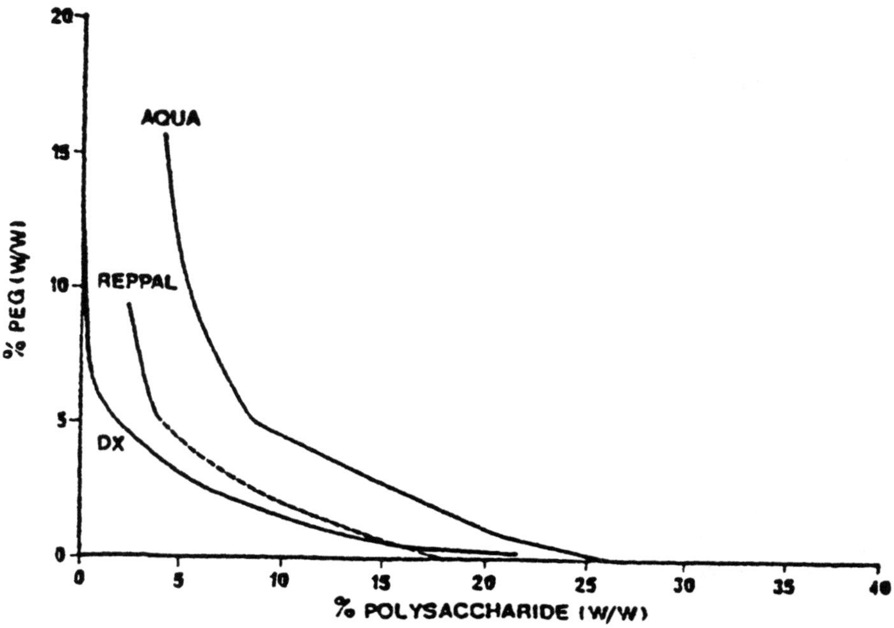

Fig. 1 *Phase diagrams for PEG/dextran, PEG/Reppal and PEG/Aquaphase*

PEG/Aquaphase phase systems of tie line length 15% wt/wt were determined (systems of equal tie-line length ought to have comparable interfacial tensions [1]. Partitions based on charge effects (using a phosphate buffer system), effects other than charge (using a chloride based buffer system) and affinity interactions (for erythrocytes only, using PEG-linoleate affinity ligand) were investigated. Phase viscosities were also measured.

Phase Diagrams are shown in Fig. 1. It was necessary to use higher concentrations of Aquaphase and Reppal than dextran. The diagram for Aquaphase does not agree with that supplied by the manufacturer.

Partition Experiments

It should be noted that a variation of 5% in the average value of partition values was experienced.

Charge Effects

Erythrocytes. Partition is markedly higher in Reppal (91.5% into upper phase) and Aquaphase (104%) than in dextran (6.8%) systems. However for Reppal this is also true in non-charged systems). This is probably an electrostatic effect which suggests a difference in the phosphate exclusion properties of Aquaphase (and perhaps Reppal).

Salmonella enteriditis. Partition changes to a much smaller extent: Reppal (103.0% into upper phase), Aquaphase (96.1%), dextran (87.7%). This implies a less highly charged surface.

IgG. Partition does not change with change in phase polymer: Reppal (54.4%), Aquaphase (46.1%), dextran (49.5%). This may be because the protein is near its isoelectric point.

Non-charge Effects

Erythrocytes. Partition is much higher for Reppal (74.4% into upper phase) than Aquaphase (19.5%) or dextran (3.9%). This suggests that either dextran or Aquaphase have an affinity for erythrocytes or Reppal is particularly incompatible with the red cell surface.

Salmonella enteriditis. Partition is high in all cases: Reppal (96.2%), Aquaphase (87.6%), dextran (81.9%). Evidently *Salmonella* either has an affinity for PEG or is incompatible with all of the polysaccharides.

Table 1

Phase System	Viscosity (cp) Upper Phase	Lower Phase
Dextran/PEG/NaCl	3.76	58.15
Aquaphase/PEG/NaCl	6.07	37.51
Reppal/PEG/NaCl	4.96	39.67
Dextran/PEG/PO_4	4.01	68.53
Aquaphase/PEG/PO_4	6.40	34.86
Reppal/PEG/PO_4	5.33	44.88

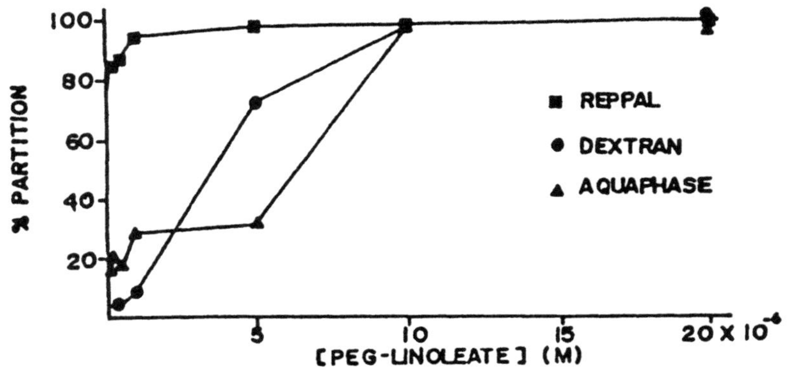

Fig. 2 *Affinity partitioning of erythrocytes in PEG/dextran, PEG/Reppal and PEG/Aquaphase*

IgG. Even partition between the phases: Reppal (54.5%), Aquaphase (50.9%), dextran (51.7%), suggests none of the polymers interact specifically with the protein.

Affinity Interactions

The partition of erythrocytes in all of the phase systems considered was altered strongly in favour of the upper phase by the inclusion of PEG-linoleate into the system (Fig. 2).

Phase Viscosities

Upper phase viscosities are higher for non-dextran containing systems. This probably reflects the higher concentrations of Reppal and Aquaphase present in the upper phase. This is related to their relatively lower molecular masses. Lower phase viscosities are considerably lower in non-dextran containing systems. This again reflects the relatively lower molecular mass.

CONCLUSIONS

Aquaphase and Reppal form phase systems with PEG 8000 which can be characterised using polarimetric and refractive index measurements. The upper phase viscosities are higher and the lower phase viscosities lower than in the corresponding dextran system.

The partitioning behaviour of human erythrocytes, *Salmonella entenditis* and rabbit IgG in these systems is generally comparable to the dextran system, except that erythrocytes partition more strongly into the upper phase in both instances. The effect of phosphate is particularly marked in the Aquaphase system.

It seems from these results that both Aquaphase PPT and Reppal PES200 can be useful substitutes for dextran in aqueous two-phase separations.

REFERENCES

1 S. Bamberger, G.F.V. Seaman, K.A. Sharp and D.E. Brooks, J. Colloid Interface Sci. (1984)

SOME RHEOLOGICAL PROPERTIES OF PERFLUOROCARBON EMULSIONS AND THEIR POTENTIAL USE IN COUNTERCURRENT CHROMATOGRAPHY DISTRIBUTION AND CHROMATOGRAPHY SEPARATION SYSTEMS

Ian A. Sutherland [1] and John Eveleigh [2]

[1] Engineering Department, National Institute for Medical Research, Mill Hill, London NW7 1AA, UK

[2] Department of Medical Products, E.I. du Pont de Nemours & Co Inc, Glasgow Site, Bldg 100 Wilmington, Delaware 19898, USA

INTRODUCTION

Aqueous two phase systems utilising polymers such as dextran and poly(ethylene) glycol have offered a unique method of purifying viable biological material for more than three decades. Perhaps the most inhibiting factors limiting their widespread use and application today is their cost, poor physical characteristics and variable reproducibility. New cheaper polymers are now coming on the market but their high viscosity, low interfacial tension and low density difference still result in long settling times which make high resolution countercurrent distribution laborious and continuous countercurrent extraction almost impossible.

Perfluorocarbon emulsions preserve the aqueous/aqueous nature and separation potential of the polymer phase systems [1], and yet offer physical properties and fluid handling characteristics more akin to aqueous/organic phase systems. Their development could lead to new and improved separation processes in the laboratory with potential of scale-up using liquid/liquid extraction techniques if commercially attractive material is isolated. The object of this preliminary study is to explore the principles of their potential use in countercurrent distribution and chromatography separation systems.

COUNTERCURRENT DISTRIBUTION

The settling behaviour of the perfluorocarbon emulsions (PE) was compared with a conventional aqueous phase system (PEG 6000/potassium phosphate 11% w/w) and an aqueous/organic one (BuOH/Acetic/Water 4:1:5). Each phase system, with equal volumes of upper and lower phase, was placed in an identical 25 ml stoppered measuring cylinder and mounted on a common framework. The proportion of upper phase settled was recorded as a percentage of upper phase volume (12.5 ml). The phases were mixed by twice inverting the framework containing the tubes and starting the stop watch. The proportion of upper phase settle increased linearly with time for both the non-PE systems, the rate of the aqueous/organic system being at least an order of magnitude greater than the aqueous/aqueous one. The PE system was non-linear, initially settling the fastest, but by 25 secs was the same as the aqueous/organic one. There then follows a period of quasi-equilibrium, when the perfluorocarbon emulsion had settled out, but not settled down. 75% of upper phase formation occurred within a minute while the remaining 25% took as long as 10-15 minutes.

The relative importance of the inter-droplet matrix volume (approximately 16%) would become less significant as the volume ratio (aqueous/emulsion phase) becomes larger and a wider container will help accelerate the bedding down of the emulsion. Nevertheless, the extent of fluidisation of the emulsion will have to be taken into account before assessing the optimum extraction procedure. For this configuration we estimate that between 3 to 6 extractions will be necessary, for long and short settling times respectively, to keep contamination from the supernatant to below 1%. It would appear from these results that a large number of short settling time extractions would be preferable to a small number of longer ones.

COUNTERCURRENT CHROMATOGRAPHY

The principle of countercurrent operation using PEP systems can be demonstrated by rotating them in a sealed glass coil (capacity 290 ml - 14 coils of 11.5 mm i.d.) mounted on a 50 mm diameter former. The head end of the coil is defined as the end to which a bubble or bead would travel by the Archimedean screw action. When the emulsion is initially positioned at the head end of the coil and the water phase at the tail, rotation of the coil will cause the emulsion to distribute towards the tail and the water phase towards the head until equilibrium occurs. If the opposite condition is set up with the emulsion phase initially at the tail and the water phase initially at the head, then rotation will cause the emulsion phase to move towards the head and the water phase towards the tail. In either case the excess of either phase will move to the tail. This behaviour is identical to that observed with an aqueous/organic phase system and looks extremely promising from the point of view of flow-through countercurrent chromatography. However, further study is required to assess the relative stability of the retained phase relative to rotational speed, mobile phase flow and sample throughput.

CONCLUSIONS

While both CCD and CCC appear feasible using PEP systems, the necessity for multiple extractions, even when partitioning completely favours the emulsion phase, suggests countercurrent processes as the most attractive option. If these studies can be extended to continuous flow-through countercurrent chromatography, then the perfluorocarbon phase systems could lead to new and improved separation processes in the laboratory, with considerable potential for scale-up.

REFERENCE

1 J. Eveleigh, Immiscible liquid affinity supports, in: "Separations Using Aqueous Phase Systems: Applications in Cell Biology and Biotechnology," D. Fisher and I.A. Sutherland, eds., Plenum, New York (1989)

INDEX

MIX
Papier aus verantwortungsvollen Quellen
Paper from responsible sources
FSC® C105338

If you have any concerns about our products,
you can contact us on
ProductSafety@springernature.com

In case Publisher is established outside the EU,
the EU authorized representative is:
Springer Nature Customer Service Center GmbH
Europaplatz 3, 69115 Heidelberg, Germany

Printed by Libri Plureos GmbH
in Hamburg, Germany